FUNGAL ENZYMES

FUNGAL ENZYMES

Editors

Maria de Lourdes T.M. Polizeli

Department of Biology
Faculty of Philosophy, Sciences and Letters of Ribeirão Preto
São Paulo University
Ribeirão Preto, SP
Brazil

Mahendra Rai

Department of Biotechnology
SGB Amravati University
Amravati, Maharashtra
India

and

Department of Chemical Biology
Institute of Chemistry
University of Campinas
Campinas, SP
Brazil

CRC Press
Taylor & Francis Group
Boca Raton London New York

CRC Press is an imprint of the
Taylor & Francis Group, an **informa** business
A SCIENCE PUBLISHERS BOOK

CRC Press
Taylor & Francis Group
6000 Broken Sound Parkway NW, Suite 300
Boca Raton, FL 33487-2742

Cover photographs reproduced by kind courtesy of Dr. Rosane Peralta and Dr. Maria de Lourdes T.M. Polizeli

International Standard Book Number: 978-1-4665-9454-8 (Hardback)

Library of Congress Cataloging-in-Publication Data

Fungal enzymes / editors: Maria de Lourdes T.M. Polizeli, Mahendra Rai.
p. cm.
Includes bibliographical references and index.
ISBN 978-1-4665-9454-8 (hardcover : alk. paper) 1. Fungi--Biotechnology. 2. Fungal enzymes. 3. Fungal enzymes--Utilization. I. Polizeli, Maria de Lourdes T. M., 1960- II. Rai, Mahendra.

TP248.27.F86F877 2013
579.5--dc23

2013001642

Visit the Taylor & Francis Web site at
http://www.taylorandfrancis.com

CRC Press Web site at
http://www.crcpress.com

Science Publishers Web site at
http://www.scipub.net

Foreword

The book *"Fungal enzymes: Progress and Prospects"*, edited by Profs. Maria T.M. Polizeli and Mahendra Rai includes the latest advances in fungal enzymes. It reminded me of famous old books on enzymes and also my own experience in this area of biotechnology. Way back in 1974 after my Ph.D., I found a fascinating book by S. Gutcho *"Microbial Enzyme Production"*, in which I read the wonderful world of bacterial, fungal and yeast enzymes. Later in 1980, during the visiting professorship, although not working directly with enzymes production, I found that microbial enzymes play a crucial role in biotransformation (A.H. Rose "Microbial enzymes and bioconversion", 1980). Another book which impressed me, was *"Fungal biology"* by H.J. Hudson, which deals with huge diversity of fungi and their extreme functional enzymes. Then, my research was focused on enzyme production and their role in bioremediation, new aspects merged in term of molecular genetics of filamentous fungi and their applications in the pharmaceutical, agricultural and enzyme industries (*"Advances in fungal biotechnology for industry, agriculture and medicine"*, by J.S. Tkacz and L. Lange, 2004). Recently, the trend in fungal research has been changed and this has been reflected in the book *"Fungi: Biology and Applications"* by Kavanagh, 2005, in which the author has described biotechnological applications of fungal enzymes, fungal proteomics, and fungal diseases of humans. It is clear to me that excellent books like these can influence our future research.

During the past four decades relevant books, as mentioned above, have been published on enzymes with valuable commercial applications, but the book edited by Profs. Polizeli and Rai represents fabulous contribution into considering new aspects of production and application of such fungal enzymes, as invertases, cellulases, hemicelullases, proteases, laccases, amylases, phytases, pectinases, mannanases, glucuronidases and lipases. The focus on application of biodegradation of vinasses in biodiesel, protein engineering and genetics show the great advances in this area.

As a matter of fact, this book exhibits a new vision of industrial enzymes. I am pretty sure that scientists and students will find this book as an excellent piece of work.

April 1, 2013

Nelson Durán
Professor, Institute of Chemistry
University of Campinas, SP, Brazil

Preface

Enzymes, "the biological catalysts" play a crucial role in life of all organisms. These catalysts, are responsible for accelerating the specific form of metabolic reactions. Taking advantage of this property, many enzymes are applied exogenously in the vast industrial field. These are primarily produced by microorganisms such as bacteria, yeast and fungi. The latter has saprophytic as well as a parasitic mode of nutrition. In both cases, the fungi secrete enzymes in order to obtain food from the host/ substrate for nutrition and also to digest the food outside the body, which is absorbed across the cell membranes. There are multiple applications of "fungal enzymes" in industries. The enzymes are used in agriculture, pulp and paper industry, waste management, improving quality of detergents, food processing, biofuels, brewing and wine production, pharmaceuticals, textiles and leather processing, etc.

The idea for this book came from the approval of the project "Biopropection of filamentous fungi producing holoenzymes with application in biorefinery" with financial support of Foundation for Advancement of Science in the State of São Paulo (FAPESP) and the National Council for Scientific and Technological Development (CNPq) (processes 2010-52322 - 3/563260/2010-6), in response to the Public Notice No. 47/2010 - Call 2 - Research on the Thematic Networks to expand knowledge on the biota, the functional role, use and conservation of Brazilian biodiversity. In this context, collection, isolation, taxonomic identification and study of different enzymes with applications in biorefineries, as cellulolytic, xylanolytic and ligninolytic systems was proposed.

Considering the diverse applications of fungal enzymes, the present book includes pectinolytic and amylolytic systems, invertases, cellulases and hemicellulases, pectinases, proteases, laccases, phytases, alpha-glucuronidases, mannanases, lipases and its many applications, including the transesterification and biodiesel production.

The book presents different approaches of diverse enzyme systems, for better understanding of the substrates involved in catalysis, and also different processes of enzyme production, purification processes, gene and metabolic regulations and engineering of enzyme.

We believe that the book is of multidisciplinary interest for researchers of fungal enzymes as well as educators and students.

The book would be very useful for mycologists, microbiologists, enzymologists, biotechnologists, pharmacologists and industrialists, who are interested in production and application of these enzymes.

We thankfully acknowledge the financial assistance provided by FAPESP and CNPq Brazil.

Dr. Maria de Lourdes T.M. Polizeli
Dr. Mahendra Rai

Contents

1

Biotechnological Advances in Fungal Invertases

Marina Kimiko Kadowaki,[1,a,*] *Rita de Cássia Garcia Simão,*[1,b] *Jose Luís da Conceição Silva,*[1,c] *Clarice Aoki Osaku*[1,d] and *Luis Henrique Souza Guimarães*[2]

Introduction

Invertase or β-D-fructofuranoside fructohydrolase (EC 3.2.1.26), is an enzyme that irreversibly hydrolyzes the β 1-2 linkage of disaccharide sucrose to produce glucose and fructose. The name "invertase" was given to this enzyme because the reaction product is a mixture of glucose and fructose, also known as "invert sugar." Invertase was one of the first carbohydrases to be reported over a century ago, and its enzymatic activity was described in 1828 in the fermentation of sucrose in yeast bread (Alberto et al. 2004).

[1]Centro de Ciências Médicas e Farmacêuticas, Universidade Estadual do Oeste do Paraná. Rua Universitária, 2069, CEP: 85819-110, Cascavel, Paraná, Brazil.
[a]E-mail: marina.kadowaki@unioeste.br
[b]E-mail: ritabioq@yahoo.com.br
[c]E-mail: jlcsilva@gmail.com
[d]E-mail: cla_osaku@yahoo.com.br
[2]Departamento de Biologia, Faculdade de Filosofia Ciências e Letras de Ribeirão Preto, Universidade de São Paulo. Av. Bandeirantes, 3900, CEP: 14040-901, Ribeirão Preto, São Paulo, Brazil.
E-mail: lhguimaraes@ffclrp.usp.br
*Corresponding author

This enzyme is also known as β-fructofuranosidase, sucrase, invertin or saccharase and belongs to the family glycoside hydrolase (GH) according to the sequence-based glycoside hydrolase classification (http://www.cazy.org). Additionally, it belongs to a group of proteins that have the common feature of two acidic residues: aspartate (Asp) and glutamate (Glu). Invertase showed two amino acid residues (aspartate and glutamate) strategically located in the region of the catalytic machinery. Aspartate, which is close to the N-terminus, acts as the nucleophile, and glutamate acts as the general acid/base catalyst; both are involved in the cleavage of glycosidic bonds (Reddy and Maley 1996, Linde et al. 2009, Polo et al. 2009). One of the methods used to understand the catalytic mechanism of enzymes is X-ray crystallography. However, there are very few reports on the three-dimensional structures from yeast invertase solved by X-ray crystallography. So far, only the crystal structures for β-fructofuranosidase from yeast *Schwanniomyces occidentalis* (Polo et al. 2009, Álvaro-Benito et al. 2010) and basidiomycetous yeast *Xanthophyllomyces dendrorhous* (Polo et al. 2010) have been reported. These studies have shown a common bimodular folding for the GH32 family, with an N-terminal five-fold β-propeller catalytic domain and a C-terminal β-sandwich domain. In addition, this β-sandwich domain was shown to be involved in dimerization of the invertase and was directly implicated in shaping its active site (Álvaro-Benito et al. 2010, Polo et al. 2010). Furthermore, an exoinulinase from the fungus *Aspergillus niger* (Nagem et al. 2004) and the structure of fructosyltransferase from *Aspergillus japonicus* have also been reported (Chuankhayan et al. 2010). In general, fungal invertases have been classified within the family 32 of the GH on the basis of their amino acid sequences (Henrissat 1991, Linde et al. 2009). This family includes enzymes that catalyze the release of β-fructose from the non reducing termini of various β-D-fructofuranoside substrates, such as invertases or β-fructofuranosidases, inulinases and levanases, which act on sucrose, raffinose, inulin and levans, respectively (Belcarz et al. 2002, Alberto et al. 2004, Polo et al. 2010). This is a group of enzymes, together with the GH68 (inulosucrase) family included in the GH-J clan.

Interestingly, some invertases produced by fungi or yeast have the versatility to not only hydrolyze sucrose but may also catalyze the synthesis of short-chain fructooligosaccharides (FOS) (Gutíerrez-Alonso et al. 2009). One to three fructosyl moieties are linked to the sucrose skeleton by different glycosidic bonds catalyzed by fructosyl transferases. Thus, the fructosyl transferase produces kestose (GF2), nystose (GF3) and fructofuranosyl nystose (GF4), and these FOS have great biotechnological importance as functional food ingredients due to their prebiotic properties (Nguyen et al. 2005).

Occurrence and Isoforms of Fungal Invertase

Several microorganisms can be used as sources of invertases with a potential for biotechnological application. Nevertheless, fungi are the best producers of industrial invertases because these organisms are able to secrete higher levels of the enzyme. There are many examples of fungi that have been widely studied as invertase producers, such as *Thermomyces lanuginosus* (Chaudhuri et al. 1999), *A. japonicus* (Chen and Liu 1996), *A. niger* (Nguyen et al. 2005, Goosen et al. 2007), *Aspergillus ochraceus* (Guimarães et al. 2007), *Fusarium oxysporum* (Wolska-Mitaszko et al. 2007), *Aspergillus oryzae* KB (Kurakake et al. 2010), *Aspergillus niveus* (Guimarães et al. 2009), *Aspergillus caespitosus* (Alegre et al. 2009), *Aspergillus phoenicis* (Rustiguel et al. 2010), *Aspergillus fumigatus* (Uma et al. 2010a) and *Paecilomyces variotii* (Giraldo et al. 2012).

Another relevant organism is the yeast *S. cerevisiae*, which is most extensively used for the industrial production of invertase. Commercially, invertases have also been produced on a large scale using yeast strains like *Aureobasidium* spp., *Rhodotorula glutinis* and *Saccharomyces carlsbergenesis*. Likewise, there are also other yeasts reported to be producers of invertase, such as *Schwanniomyces occidentalis* (Alvaro-Benito et al. 2007), *Pichia anomala* (Rodriguez et al. 1995), and *Candida utilis* (Chavez et al. 1997).

In fungi, different isoforms of the invertase can be found, they are normally monomeric, dimeric or multimeric. The molecular mass of the monomer ranges from approximately 41 to 172 kDa. An important feature of invertase is the carbohydrate content, which ranges from 14–53 percent of the total molecular mass. Usually, these enzymes are called intracellular and extracellular invertases, and each one exhibits individual biochemical properties and amounts of glycosylation. For example, isoforms of invertase from *A. niger* (Chen et al. 1996) and *C. utilis* (Belcarz et al. 2002) that are called high- and low-molecular weight, represent the slow-migrating form (S-form) and, fast-migrating form (F-form), respectively. In *Aureobasidium pullulans* DSM2404, five isoforms of β-fructofuranosidases are found (I, II, III, IV and V) when grown with sucrose (Yoshikawa et al. 2006). Wolska-Mithaszko et al. (2007) have reported the occurrence of three intracellular and two extracellular isoforms of invertase when cultured in the presence of fructose by *F. oxysporum*. Usually, in yeast such as *S. cerevisiae* it is common to find two isoforms. One external invertase form is highly N-glycosylated; its expression is regulated by glucose repression and is secreted into the periplasmic space. The internal invertase, or cytoplasmic form, is non-glycosylated, and its expression is constitutive; in other words, it is not inhibited by the products of sucrose hydrolysis. However, both enzymes are derived from the same gene (Sturm 1999). Many yeast isoforms have been described previously, such as in *Schizosaccharomyces pombe* (Moreno et al. 1990), *P. anomala* (Rodríguez et al. 1995), *C. utilis* (Belcarz et al. 2002) and *X. dendrorhous* (Linde et al. 2009). Recently, four external invertase isoforms (Einv1, Einv2, Einv3, and Einv4) have been isolated from *S. cerevisiae*, and they exhibited similar structural features but different thermodynamic stabilities (Andjelković and Lah 2011). Unfortunately, little is known about the specific function and the occurrence of these invertase isoforms, but they appear to control the entry of sucrose into different utilization pathways (Sturm 1999). In fungi, researchers have described the probable physiological advantage of having multiple isoenzymes. These different isoforms may be related to the development and infection process of fungi, or they may be required for the rapid changes in carbohydrate metabolism (Foster et al. 2003, Recorbet et al. 2003, Wolska-Mithaszko et al. 2007).

Fermentative Methods and Optimization for Invertase Production

In the last several years, studies on the invertase production by fungal or yeast have been carried out primarily using two methods: submerged fermentation (SmF) and solid-state fermentation (SSF). The SmF is the most cited by researchers for invertase production by fungi compared to SSF, despite the known advantages of solid-state fermentation, such as playing a natural environment and prevent bacterial contamination. Moreover, SSF requires less space, lower operating costs and offers simplicity of operation and high yield of the fermentation product (Sangeetha et al. 2005).

Some researchers have also adopted statistical experimental designs for the enhanced production of invertase, such as the central composite experimental design (CCD) and the response surface methodology (RSM) to optimize nutritional requirements such as the carbon source, incubation period, pH, salts, and additional nitrogen sources (Herwig et al. 2001, Ul-Haq et al. 2005), as the traditional one-dimensional optimization may be incapable of reaching the true optimum due to interactions among factors. Thus, statistical experimental design has proven to be a useful and powerful tool in developing optimum fermentation conditions. It was observed that the yields of invertase from *A. japonicus* were further maximized using an experimental design (Chen and Liu 1996). Likewise, a significant increase in invertase production was also obtained by hyperproducing *S. cerevisiae*, through the addition of urea to the culture medium (Ul-Haq et al. 2005). Driouch et al. (2010) achieved a three-fold increased fructofuranosidase activity after performing medium optimization using central composite design experiments and quadratic modeling methods with the recombinant strain *A. niger* SKAn1015. Recently, an optimal fermentative condition was also established using the Plackett Burman design for fructofuranosidases (inulinase and invertase) from *Rhizoctonia* spp. (Neagu et al. 2011).

Comparative studies of fermentation using SSF and SmF methods for invertase production have also been performed. Ashokkumar et al. (2001) compared the invertase production from *A. niger* under SmF and SSF using a fractional experimental design, and they concluded that the culture media should be optimized for each culture condition (either SmF or SSF). Romero-Gómez et al. (2000) cultured three strains of *A. niger* with high concentrations of sucrose and obtained higher titers of invertase under SSF. Similarly, there are also studies reporting high enzyme titers using SSF when high glucose concentrations were used (Aranda et al. 2006, Patil et al. 2012). However, the filamentous fungus *A. caespitosus* was able to produce both intracellular and extracellular invertases using either SmF or SSF with wheat bran as the carbon source (Alegre et al. 2009).

Carbon Sources for Invertase Production

The carbon sources used to produce invertase in fungi or yeast are sucrose as well as some non-traditional substrates, such as agro industrial residues (Table 1.1). A wide range of fungi produce invertase using sucrose as the only carbon source and as the inducer of this enzyme because the product of the sucrose hydrolysis is used for both fungal growth and enzyme production. Likewise, an extracellular invertase from *S. cerevisiae* was induced by sucrose or raffinose when grown in a medium containing these substrates. However, in *A. japonicus* (Hayashi et al. 1992), glucose was a better inducer than sucrose, while in *P. glabrum*, invertase synthesis increased when fructose was used as the sole carbon source (Rubio et al. 2003).

The commercial use of sucrose as a substrate is very expensive when used to produce invertase on a large scale. Thus, replacing it by a cheaper carbon source or a renewable substrate is more cost effective. Over the years, researchers have screened a variety of substrates for the production of invertase (Patil et al. 2011). These substrates include corn-steep liquor (Barlikova et al. 1991), industrial molasses media (Zech and Goerisch 1995), wheat bran and corn cob (Rajoka and Yasmeen 2005), food processing wastes (okra, orange pulp, apple pomace and red carrot residue) (Rashad and Nooman 2009), sugar cane bagasse (Guimarães et al. 2009), fruit peel waste (Uma et al. 2010a, Uma et al. 2010b), lemongrass (Madhan et al. 2010) and banana leaves (Patil et al. 2011). Thus, the optimization process for the production of invertase is very important for making it more economical and viable.

Table 1.1 Summary of the Culture Conditions and Purification Strategy used for Fungal Invertases.

Fungus	Carbon source	Culture condition (growth temp)	Purification procedure	Purification steps	Purification factor (fold)	Yield (%)	Reference
Aspergillus flavus	Sucrose	SmF (30°C)	Precip./DEAE – C	2	5.8	3.2	Uma et al. 2010a
Aspergillus fumigatus	Sucrose	SmF (30°C)	Precip./DEAE – C	2	4.73	4.91	Uma et al. 2010b
Aspergillus japonicus	Sucrose	SmF (28°C)	Precip./DEAE-C/F.C/A.C	4	39.2	24.9	Wang and Zhou 2006
Aspergillus niger IMI 303386	Sucrose	SmF (28°C)	Precip./DEAE – C/F.C.	3	49.8	41.8	Nguyen et al. 2005
Aspergillus niveus	Glucose	SmF (40°C)	DEAE – C/F.C.	2	8.1	9.8	Guimarães et al. 2009
Aspergillus ochraceus	Sugarcane bagasse	SmF (40°C)	DEAE – C/F.C.	2	7.1	24.0	Guimarães et al. 2007
Aspergillus oryzae	Guar gum	SmF (37°C)	Precip./TPP (2x)	3	12	54	Dhananjay and Mulimani 2008
Aureobasidium pullulans DSM 2404	Sucrose	nr	DEAE–C/F.C./HC	3	51.4	8.98	Yoshikawa et al. 2006
Cladosporium cladosporioides	Sucrose	SmF (30°C)	Precip./DEAE – C	2	6.35	3.96	Uma et al. 2012
Fusarium solani	Molasses	SmF (30°C)	Precip./DEAE–C/F.C.	3	3.11	9.33	Bhatti et al. 2006
Fusarium sp	Wheat bran	(30°C)	Precip./DEAE–C/F.C.	3	20.25	5.53	Shaheen et al. 2008
Paecylomices variotii	Soy bran	SmF (40°C)	DEAE – C/F.C.	2	10.5	33.3	Giraldo et al. 2012
Pycnoporus sanguineus	Sucrose	(30°C)	Precip./F.C./A.C	3	130	11	Quiroga et al. 1995

Pichia anomala	Glucose	SmF (28°C)	Precip./DEAE–C(2x)	3	51	46	Rodriguez et al. 1995
Candida utilis	nr	(30°C)	Precip./F.C./A.C.	3	97	34	Chávez et al. 1997
Rhodotorula glutinis	nr	(30°C)	Precip./F.C./DEAE–C	3	5	9.4	Rubio et al. 2002
Schwanniomyces occidentalis	Lactose	(29°C)	Filtration/DEAE-C(2x)	3	300	2	Alvaro-Benito et al. 2007
Rhodotorula dairenensis	Maltose	(30°C)	Filtration/DEAE–C	2	nr	nr	Gutiérrez-Alonso et al. 2009
Sacharomyces cerevisiae NRRL Y-12632	Food processing wastes	SmF (30°C)	DEAE-C/F.C.	2	29	4.7	Rashad and Nooman 2009

SmF - Submerged Fermentation; nr: not reported; Precip.: Precipitation Ammonium Sulphate / Acetone / Ethanol; DEAE - C: DEAE Chromatographic; HC-: Hydrophobic Chromatography; TPP: Three-phase partitioning (interfacial precipitate - TPP is a non-chromatographic); A.C.: Affinity Chromatographic; F.C.: Filtration Chromatographic.

Purification Strategy and Some Properties of Invertases

There are many published studies available in the literature on the purification and biochemical characteristics of fungi and yeast invertases. The data regarding invertase production as well as the purification procedures (purification steps, recovery field and factors) have been compiled from published articles and are summarized in Table 1.1.

Over the years, the purification strategy of fungal invertase has been through traditional purification methods, such as chromatographic methods (ion exchange chromatography and gel filtration chromatography), that are performed in two or three steps, including the concentration of salt (such as the precipitation of ammonium sulfate) or acetone/ethanol, and provide recovery yields ranging from 2 to 54 percent (Table 1.1).

Other methods of purification have also been adopted by researchers to isolate invertase, such as the use of polymeric carrier poly (p-chloromethyl styrene) (PCMS) beads for invertase purification from *S. cerevisiae* (Bahar and Tuncel 2004). Another technique employed by Dhananjay and Mulimani (2008) was the three-phase partitioning system under different conditions (ammonium sulfate, ratio of tert-butanol to crude extract, temperature and pH) to isolate invertase from *A. oryzae*. The enzyme was purified 12-fold with 54 percent activity recovery. Likewise, a Baker's yeast invertase was partitioned using similar techniques with a liquid–liquid partitioning with PEG/salt aqueous two-phase systems (ATPSs), giving a purification factor of 6.2-fold and an activity recovery of 217.7 percent (Karkas and Önal 2012).

The biochemical properties, such as pH and temperature, as well as the stability of the enzyme are important issues when determining the potential industrial application. An overall analysis of the biochemical characteristics of invertase is summarized in Table 1.2. The data show that the optimum pH for the activity of most fungal invertases is acidic, ranging from 4 to 6, with rare exceptions below pH 4, and these invertases generally have a broad range of optimum stability from pH 3 to 10.

Interestingly, upon analysis of the invertases produced by most fungi, they were found to display a high optimum temperature, between 50 to 60°C, except for *C. utilis* and *X. dendrorhous*, which exhibited higher temperatures of 70°C; however, the thermal stability for these fungal invertases at higher temperatures appears to be low (Table 1.2).

Invertase Applications

Fungal invertases have widespread applications in their use in biotechnology, especially in the production of invert sugars, which are sweeter than sucrose, due to a high degree of sweetness from fructose and also because they do

Table 1.2 Optimal pH, Temperature and Stability of the Fungal Invertases.

Fungal strain	pH_{Opt}	pH Stability	$Temp_{Opt}$ (°C)	Temperature Stability (time)	Reference
Aspergillus caespitosus					Alegre et al. 2009
I(extracellular- SSF)	4.0 – 6.0	nr	50	50 °C (1 h)	
II (intracellular- SmF)	5.0	nr	60	50 °C (1 h)	
III (extracellular-SmF)	5.0 – 6.0	nr	60	50 °C (1 h)	
Aspergillus flavus	6.0	nr	50	50°C (30 min)	Uma et al. 2010a
Aspergillus fumigatus	6.0	5.0 – 7.0	50	50°C (30 min)	Uma et al. 2010b
Aspergillus niger	5.5	5.5 (30 min)	50	20 – 55°C	Madhan et al. 2010
Aspergillus niger AS0023	4.4	4.0-11	55	nr	L'Hocine et al. 2000
Aspergillus niger IMI *303386*	5.5	4.0 – 8.0	50	55°C (5 h)	Nguyen et al. 2005
Aspergillus niveus	4.5	nr	60	60°C (53 min)	Guimarães et al. 2009
Aspergillus ochraceus	4.5	nr	60	60°C (60 min)	Guimarães et al. 2007
Aspergillus spp. M1	6.0	nr	60	nr	Patil et al. 2012
Aureobasidium pullulans DSM 2404	5.0	6.0 – 10.0	50	< 45°C	Yoshikawa et al. 2006
Chrysonilia sitophila PSSF84	5.0	4.5 – 5.5 (48 h)	50	40 – 50°C (48 h)	Patil et al. 2011
Cladosporium cladosporioides	6.0	nr	50	nr	Uma et al. 2012
Fusarium solani	2.6	nr	50	nr	Bhatti et al. 2006
Fusarium sp.	5	nr	50	nr	Shaheen et al. 2008
Paecilomyces variotii	4.0–4.5	nr	60	60°C (1 h)	Giraldo et al. 2012
Pycnoporus sanguineus	3.85	3.0 – 7.5	nr	0 – 50°C	Quiroga et al. 1995

Table 1.2 contd....

Table 1.2 contd.

Yeast strain	pH_{Opt}	pH Stability	Temp $_{Opt}$ (°C)	Temperature Stability (time)	Reference
Candida utilis	5.5	3.0 – 6.0	60 – 75	60°C (1 h)	Chávez et al. 1997
Kluyveromyces fragilis	4.5	nr	50°C	50 – 55°C	Workman and Day 1983
Leucosporidium antarcticum	4.5 – 4.7	nr	30	12°C (30 min)	Turkiewicz et al. 2005
Pichia anomala	4.5	4.5 – 6.5 (4 h)	38	37°C (15 min)	Rodriguez et al. 1995
Rhodotorula dairenensis	5.0	nr	55–60	40 – 60°C (24 h)	Gutiérrez-Alonso et al. 2009
Rhodotorula glutinis	4.5	2.6 – 5.5 (30 min)	60	60°C (30 min)	Rubio et al. 2002
Saccharomyces cerevisae	3.5 – 5.0	nr	60	60°C (35 min)	Andjelkovic et al. 2011
Saccharomyses cerevisiae NRRL Y-12632	6.0	5.0 – 7.0	50	< 50°C	Rashad and Nooman 2009
Schwanniomyces occidentalis	5.5	nr	45–55	nr	Álvaro-Benito et al. 2007
Xanthophyllomyces dendrorhous	5.0 – 6.5	nr	65-70	66-71°C	Linde et al. 2009

pH Opt: optimum pH for enzyme activity;
Temp Opt: optimum temperature for enzyme activity;
nr: note reported.

not crystallize. In addition, fructose is considered a safer sugar from a health point of view and may be used by diabetic individuals (Rustiguel et al. 2010). This enzyme is widely used in the processed food and confectionery industry to produce candies and soft-centered chocolates. Other uses of the enzyme include lactic acid production, fermentation of cane sugar molasses, calf-feed production, ethanol production and food for honeybees (Ul-Haq et al. 2005, Rashad and Nooman 2009). This enzyme is also used in the pharmaceutical industry as digestive aid tablets and powdered milk for infant food (Uma et al. 2012).

Fungal Invertases with Transfructosylating Activity

Many microorganisms such as yeasts and filamentous fungi are able to produce invertases with fructosyltransferase activity under high concentration of sucrose (generally ≥ 30% m/V). The enzymes with this interesting property catalyze the transference of the fructose residue to the sucrose acceptor molecule in order to produce fructooliogosaccharides (FOS). FOS are non caloric sugars with important functional properties involved in cholesterol reduction and growth of bifidobacteria, among others. The main FOS found and used by food industry have been studied, such as the 1-kestose (GF_2), that is constituted by one sucrose molecule added to only one fructose residue by glycosidic linkage, the nystose (GF_3) that is constituted by one sucrose molecule and two sucrose residues; and finally the 1-fructosylnistose (GF_4), constituted by one sucrose molecule added with three fructose residues. The 1-kestose and nystose are fructose oligomers that constitute the ^{1}F-FOS group with β(2-1) linkages, while other FOS, as 6-kestose constitute the ^{6}F-FOS group and the neokestose, and neonystose comprise the ^{6}G-FOS group. In these last two groups, the β(2-6) linkage is observed (Kilian et al. 2002). The β(2-1) and β(2-6) linkages characterize the inulin-like and levan-like fructose oligomers, respectively.

Considering that many microorganisms are able to produce enzyme with fructosyltransferase activity, a previous screening of the best producer strains is an important procedure. A simple plate methodology for this purpose was developed by Dominguez et al. (2006) using the principles of glucose oxidase (GOD)-peroxidase (POD) system employed for glucose determination as well as fructose dehydrogenase (FDH) for fructose detection.

Invertases with transfructosylating activity have been characterized from yeast as *Rhodotorula dairenenses* (Gutiérrez-Alonso et al. 2009). This homotetrameric enzyme has subunits of 172 kDa, and it was able to catalyze the production of different fructose oligomers with β(2-1) and β(2-6) linkages as for example 6-kestose, neokestose, 1-kestose and tetrasaccharides. Dimeric and multimeric structures have also been observed for other yeast strains like *X. dendrorhous* (Linde et al. 2009), *R. glutinis* (Rubio et al. 2002) and *Arxula adeninivorans* (Boer et al. 2004), among others. On the other hand, Rubio and colleagues (2002) reported the production of invertase by *R. glutinis* without transfructosylating activity. Other yeast strains have also been reported as fructosyltransferase producers like *S. cerevisiae* (Farine et al. 2001) and *S. occidentalis* (Álvaro-Benito et al. 2007).

Among the filamentous fungi *A. niger* is one of the most studied invertase producer with both activities, i.e., hydrolytic and fructosyltransferase. In addition, this fungus has been used as the main FOS producer with industrial application because of its improved prebiotic property (Kilian et al. 2002, Ghazi et al. 2007). Three genes (*SucA, SucB* and *Suc*C) were identified in this microorganism. The *SucB* gene was cloned, expressed and biochemically characterized (Goosen et al. 2007). The intracellular enzyme encoded by this gene was able to catalyze both hydrolytic and fructosyltransferase activity. The 1-kestose and nystose were obtained from sucrose and 1-kestose, respectively, using the transfructosylating reaction, although the enzyme has the ability to also hydrolyze 1-kestose, nystose and raffinose but not inulin and levan (Goosen et al. 2007). According to the biochemical analysis of the kinetic parameters, both hydrolytic (Km of 2.5 mM and Vmax of 5.5 μmol mg^{-1} min^{-1}) and transfructosylating (Km of 0.9 mM and Vmax of 1.2 μmol mg^{-1} min^{-1}) activities have similar patterns. Other fungi are able to produce multiples isoforms of β-fructofuranosidase as observed for *A. pullulans* (Yoshikawa et al. 2006), which produces five isoforms (I, II, III, IV and V). The FFase I, responsible for the transfructosylation, is predominantly expressed at the initial period in the sucrose medium and glucose and FOS is accumulated in the medium while the synthesis of the others (II–V) is increased in the absence of glucose (after been used by the microorganism). All isoforms were found in the medium containing fructose and FOS. However, in the glucose medium the FFase II–V is inhibited while the FFase I is not. So, the FFase I has an important role in the FOS production by *A. pullulans*. The FFase IV is involved in the hydrolysis of the FOS showing predominant hydrolytic activity. The function of FFase II, III and V is unknown since they have transfructosylation activity and are predominantly expressed at the FOS-degrading periods.

The production of FOS using high sucrose concentration as substrate can be conducted using isolated enzymes with fructosyltranseferase activity obtained from different steps. These steps consist of i) microbial cultivation, ii) obtaining the enzyme and iii) the fructosyltranseferase reaction. The β-fructosidase with fructosyltransferase activity from *A. aculeatus* was added in a 50 percent (m/V) sucrose solution in sodium acetate buffer pH 5.5. The FOS obtained after 10 hr of synthesis reaction were added to *Pichia pastoris* cells and incubated at 30°C for 24 hr. The quantification of FOS and other sugars were performed with an interval of 2 hr using high performance anion-exchange chromatography (Yang et al. 2008). The sucrose is converted to glucose and 1-kestose at the initial period of reaction and after a few hours nystose was found. After adding *P. pastoris* cells, the glucose content is exhausted and the fructosyltransferase repression by glucose disappears. According to this, the yield of FOS is increased (Yang et al. 2008). However, it was recently demonstrated that the one-step production of 1-kestose, 1-nystose and fructosylnystose by *A. phoenicis* biofilm on polyethylene as inert support in high sucrose medium (Aziani et al. 2012). This is an interesting way to produce FOS because the process cost can be reduced since two steps were eliminated.

Additionally, the use of fungal biofilms is an interesting alternative to produce FOS if compared to submerged and solid-state fermentations.

The FOS production by fungi has been optimized by a Plackett Burman design using different parameters as independent variables, for example pH, time of cultivation, substrate concentration, nitrogen and phosphate sources. This methodology allows not only the analysis of the influence of each variable separately, but also the interaction between the variables. The FOS production can be increased by the analysis of Response Surface Methodology as obtained for *A. aculeatus* (Nemukula et al. 2009). The enzyme produced by *A. aculeatus* is able to catalyze both hydrolytic and transfructosylating activity. The latter is predominant at high sucrose concentration. High levels of GF_3 and GF_4 were obtained using 400 mg mL^{-1} sucrose, temperature of 60°C and pH 5.6. The GF_2 was also obtained from the GF_4 hydrolysis (Nemukula et al. 2009). In addition, the production of FOS has been studied using substrate and enzyme engineering (Beine et al. 2008). In this situation, an engineering levansucrase (from *Bacillus subtilis*) and sucrose analogs were applied. It is an interesting approach that can be also applied for fungal fructosyltransferases. Imobilized fungal cells on lignocellulosic materials have been used for FOS production as described by Mussato et al. (2009) using *A. japonicus*. Under this condition, the cells are more stable maintaining their biological functions and allowing the increase of cell productivity and consequently maximizing the process efficiency.

Immobilized Invertase

The study of the properties of immobilized invertase is of great interest for the application of the enzyme, especially in the food industry (Mirzarakhmetova et al. 2006). Enzyme immobilization offers several advantages, such as the improvement of the enzymatic properties, including stability at different temperatures and pH, as well as easy recovery of the products and enzyme reuse. Remarkably, invertase has also been a model to study enzyme immobilization. Hence, a variety of support materials as well as methods for invertase immobilization have been reported. Usually, yeast invertases have been used for immobilization in their free form or in yeast cells with enzymatic activity. The methods used for the immobilization of fungal invertases include adsorption, covalent binding, cross-linking and entrapment (Uzun et al. 2011).

Nelson and Griffin (1916) reported that invertase was the first enzyme to be immobilized, and it was bound on charcoal and aluminum hydroxide. Subsequently, many fungal invertases have been immobilized onto a variety of media, including DEAE- and CM-Sephadex and Con A agarose (Woodward and Wiseman 1978), gelatin (Parascandola et al. 1982), liquid-core alginate capsules (Chang et al. 1996), methacrylamide-based polymeric beads (Chiang et al. 1997), chitosan beads (Cheng et al. 2005) and porous silica matrix (David et al. 2006).

Among the immobilized enzymes, *S. cerevisiae* invertase is most commonly used in industry, primarily for the commercial production of invert syrup (Basha and Palanivelu 2000), but there have been other studies on the immobilization of fungal and yeast invertases in organisms such as *C. utilis* (Dickensheets et al. 1977), *Aspergillus athecius* (Madyastha et al. 1987), *Sclerotium rolfsii* (Kotwal and Shankar 1997), *A. niger* ATCC 20611 and *A. japonicus* TIT-KJ1 (Chiang et al. 1997), *A. japonicus* (Cheng et al. 2005), *Cladosporium cladosporioides* (de Almeida et al. 2005) and *Kluyveromyces marxianus* (Karandikar et al. 2006).

Recently, Raj and co-workers (2011) used nanogels as supports for invertase immobilization. A hydrogel composed of methacrylic acid (MAAc), N-vinyl pyrrolidone (N-VP) and ethyleneglycol dimethacrylate (EGDMA) was converted to a nanogel and used as the support for the covalent immobilization of invertase from *S. cerevisiae*. The biochemical characterization of both free and immobilized invertase showed that the V_{max}, k_m, and E_a values for the immobilized invertase were comparable to those of the free invertase. In addition, the immobilized invertase enhanced the pH and thermal stability and was efficiently reused in up to eight cycles.

There are several studies on the use of immobilized invertase to obtain invert sugar at the laboratory scale, but there are still many difficulties that must be overcome for industrial application, one of which is the high cost of the enzymes. Furthermore, another limiting factor is the competition with immobilized glucose isomerase, which costs less than immobilized invertase. Thus, efforts will be necessary to reduce the cost of the production of immobilized invertase to allow the successful application in industry (Kotwal and Shankar 2009).

The Regulation of Invertase

The regulation of invertase synthesis has been studied in different strains of fungi and is important for the commercial production of enzymes.

Invertase can be induced by different carbohydrates in certain filamentous fungi (Vainstein and Peberdy 1991, Rubio and Navarro 2006). Although fungi and some yeast can flourish in different carbon sources, glucose and fructose are the best carbon source for their growth. However, when the medium contains glucose the biosynthesis of enzymes catalyzing the degradation of other carbon sources will be strongly reduced or halted, in a phenomenon known as glucose repression. The comprehension of the different mechanisms responsible for glucose repression in fungi is extremely important and has many applications for industrial fermentation (Chi and Gao 1999).

The production of high levels of an extracellular invertase in *Aspergillus nidulans* depends on the addition of sucrose or raffinose to the medium, indicating that synthesis is dependent on induction by the substrate. In the presence of other sugars, the enzyme is produced only at low constitutive levels. The repressive effect of other sugars (e.g., 1 percent glucose and xylose) on invertase production has also been observed (Vainstein and Peberdy 1991).

To determine the mechanism that controls the synthesis of invertase in *A. niger* (Fig. 1.1), the inducive or repressive effects of different sugars have been studied. In the presence of sugars such as sucrose, turanose, raffinose and inulin, the highest levels of specific production of invertase were observed (Rubio and Navarro 2006). These sugars have a β-linkage and a fructose located at the end of the molecule as a common property in their structure.

The proposed mechanism (Fig. 1.1) of induction for the synthesis of invertase in *A. niger* identifies the means by which a sucrose molecule interacts with the receptor in the cell membrane. The contact generates a chemical signal inside the cell that is amplified by cAMP in the cell nucleus, thus commencing the induction of invertase synthesis at the transcriptional level. In fact, in the absence of an inducer, cAMP enhances the expression of invertase, increasing the total enzyme production. The synthesized mRNA then carries the information to be translated to the ribosomes, and the enzyme is later secreted into the periplasmic space, which borders the cell wall (Rubio and Navarro 2006).

In *A. niger*, glucose and fructose are repressors of invertase synthesis and act at the translational level (Fig. 1.1). This repression is reversible and can be avoided with modified sucrose molecules (Rubio and Navarro 2006). On the contrary, in yeast, glucose represses the expression of invertase at the transcriptional level (Chu and Maley 1980). In fact, the mechanism of induction for the synthesis of invertase in *A. niger* is different from that of other filamentous fungi (Sestak and Farkas 1993, Chen 1996a, Romero-Gómez et al. 2000, Rubio et al. 2003) and yeast (Costaglioli et al. 1997, Rubio and Navarro 2002).

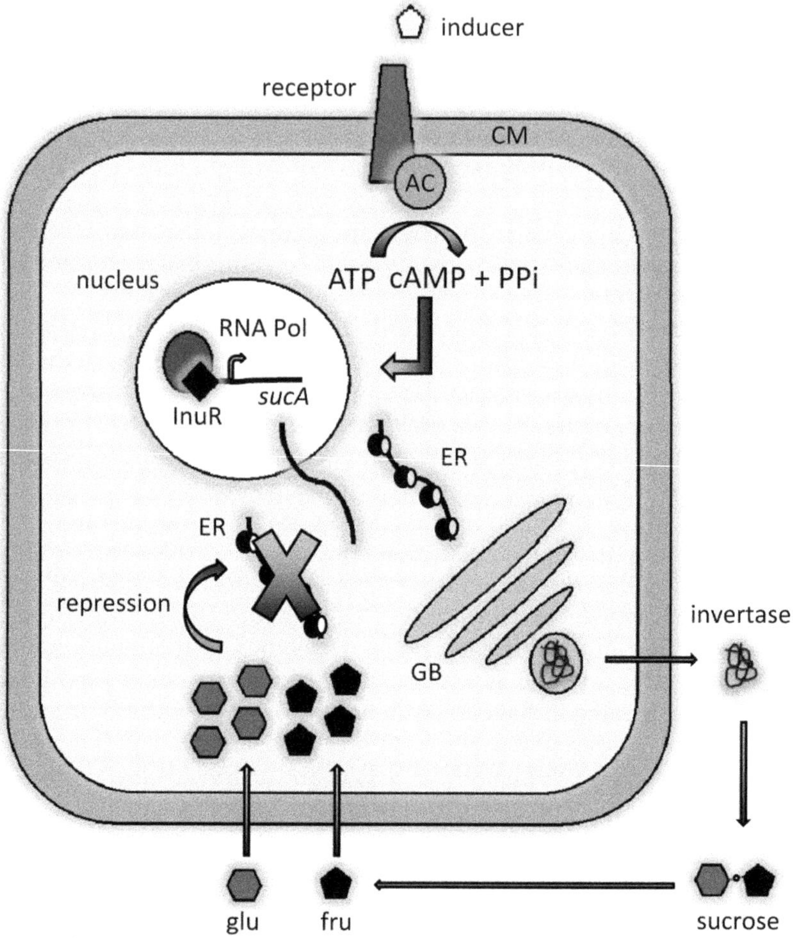

Figure 1.1 Mechanism of regulation of invertase expression in *Aspergillus niger*. CM: cell membrane; AC: adenylatecyclase; ATP: adenosine triphosphate; RNA Pol: RNA Polymerase; InuR: transcriptional regulator of the inulolytic genes; *suc*A: extracellular invertase gene; ER: endoplasmic reticulum; GB: Golgi body; glu: glucose; fru: fructose.

As a soil fungus, *A. niger* hosts a wide variety of enzymes capable of degrading plant-derived polysaccharides. This fungus has a protein with invertase activity, which is encoded by the *sucA* gene, and two additional intracellular invertase-like proteins encoded by the *sucB* and *sucC* genes (Yuan et al. 2006). The SucB protein has transglycosylating activity (Goosen et al. 2007), that is expressed at low levels on all the carbon sources tested and its expression is slightly upregulated in the presence of inulin and sucrose (Yuan et al. 2006).

In the *A. niger* genome, the *sucB* gene is located next to a putative sugar transporter gene (An15g00310) and the *inuR* gene, which encodes a transcriptional regulator required for the induced expression of inulinolytic genes (Yuan et al. 2008). Besides sharing the same promoter region, *sucB* and the putative sugar transporter gene are not coregulated, indicating that the common regulatory region does not act bidirectionally.

The function of SucB in relation to inulin or sucrose metabolism is unknown. The mutant strain, constructed by deletion of the *sucB* gene, did not result in a phenotype related to growth on inulin or sucrose (Goosen et al. 2007). The low levels of expression of the *sucB* gene suggest that SucB could act as a sensing enzyme to "note" the presence of a carbon source by generating an inducer molecule that can further activate the transcription of inulinolytic genes, including *sucB* (Yuan et al. 2008). The presence of an additional intracellular invertase encoded by the *sucC* gene could be a possible explanation for the lack of phenotype in the *sucB*-deletion mutant. The depletion of InuR causes *A. niger* cells to show a severe reduction in growth on both sucrose and inulin medium. In addition, the expression of the sucrolytic gene *sucA* as well as the putative sugar transporter gene An15g00310, depends on the presence of InuR (Fig. 1.1) (Yuan et al. 2008).

The synthesis of invertase by *S. cerevisiae* and *Neurospora crassa* is primarily regulated by catabolic repression by hexoses. Despite the repressive effects of different sugars (e.g., glucose and xylose) in filamentous fungi such as *A. nidulans* (Rubio and Navarro 2006), in another species within the same genus (*A. japonicus*), glucose is a better inducer than sucrose (Hayashi et al. 1992).

S. cerevisiae invertase is encoded by a family of *suc* genes with several closely related members. In fact, there are six polymeric *suc* genes coding for invertases, namely *suc*1 to *suc*5 and *suc*7. The *suc*2 gene is located at the end of chromosome IX, and the remaining *suc* genes are located in the highly mobile telomere regions of different chromosomes: *suc*1 in chromosome VII, *suc*3 in chromosome II, *suc*4 in chromosome XIII, *suc*5 in chromosome IV and *suc*7 in chromosome VIII (Carlson and Botstein 1983). However, recent comparative studies have indicated that the natural strains of *S. cerevisiae* may possess a pool of new genes not found in the standard laboratory genetic lines studied in the framework of the international genome sequencing project (Naumov and Naumova 2011).

As mentioned above, three new *suc* genes were identified in natural strains of *S. cerevisiae*: *suc*8 located on the chromosome X (Naumov and Naumova 2010a), *suc*9 located on the chromosome XIV and *suc*10 on the chromosome XVI/XIII doublet (Naumov and Naumova 2010b). Usually a strain of *S. cerevisiae* may have no, one or several *suc* genes. However, one functional gene is sufficient to ensure sucrose fermentation. In fact, the sequenced genome of the yeast does not reflect the genetic diversity found in natural populations, thus limiting evolutionary studies and selection (Korshunova et al. 2005, Naumov and Naumova 2011).

The expression of *S. cerevisiae suc* genes is exclusively regulated by glucose repression. The most characterized *suc*2 gene, for instance, presents a greater than 100-fold difference between the repressed and derepressed levels of its mRNA (Carlson and Botstein 1983). The genes involved in glucose repression are not specific for invertase regulation and have been identified via mutant strains of yeast that are unable to produce invertase or have a constitutive expression. These mutant strains allow the identification of a pathway for general glucose repression, which has several key elements: hexokinase PII, which appears to play a role in sensing glucose levels; the SNF1-protein kinase complex (*snf*, sucrose non-fermenting), required for derepression of many glucose-repressible genes; protein phosphatase PP2A, which is also involved in the derepression of the *suc*2 gene; and MIG1/MIG2, which are zinc-finger transcriptional repressors that bind in the upstream regions of *suc*2 and other repressible genes (Rolland et al. 2002, Chi et al. 2004, Castermans et al. 2012).

The majority of isolated *snf* mutants are defective for invertase derepression under low glucose conditions. Regarding SNF proteins, two classes can be distinguished: 1-SNF1 and 4, which are involved in glucose signaling (Jiang and Carlson 1997), and 2-SNF2, 5, 6 and 11, which are part of the SWI-SNF chromatin-remodeling complex (Vignali et al. 2000). In addition, SNF7 and SNF8/VPS22 belong to the collection of class-E VPS (vacuolar protein sorting) proteins.

The general co-repressor proteins SSN6 and TUP1 are required for the repressive effect of MIG1. In fact, MIG1 is responsible for recruiting both SSN6 and TUP1 (Rolland et al. 2002), which repress several diverse gene families and their family-specific transcriptional inducer genes (Treitel and Carlson 1995, Rolland et al. 2002). In the presence of high levels of glucose, MIG1 rapidly moves into the nucleus where it binds to the promoters of glucose-repressive genes. On the contrary, when the levels of glucose are low or when the cells are deprived of glucose, MIG1 is rapidly transferred back to the cytoplasm (Fig. 1.2) (De vit et al. 1997, Gancedo 1998).

The SNF1 kinase activity is inhibited by glucose and stimulates when glucose is limited (Wilson et al. 1996). In the absence of glucose, it is most likely that SNF1 phosphorylates and induces the translocation of MIG1 to the cytoplasm, such that the glucose repression on the *suc*2 gene is relieved, leading to the initiation of transcription (Fig. 1.2) (Treitel and Carlson 1995, De vit et al. 1997, Ostling and Ronne 1998). There is evidence that when MIG1 is in the phosphorylated state, the protein HXK2 can regulate its nucleo-cytoplasmic distribution (Ahuatzi et al. 2007), and similar to MIG1,

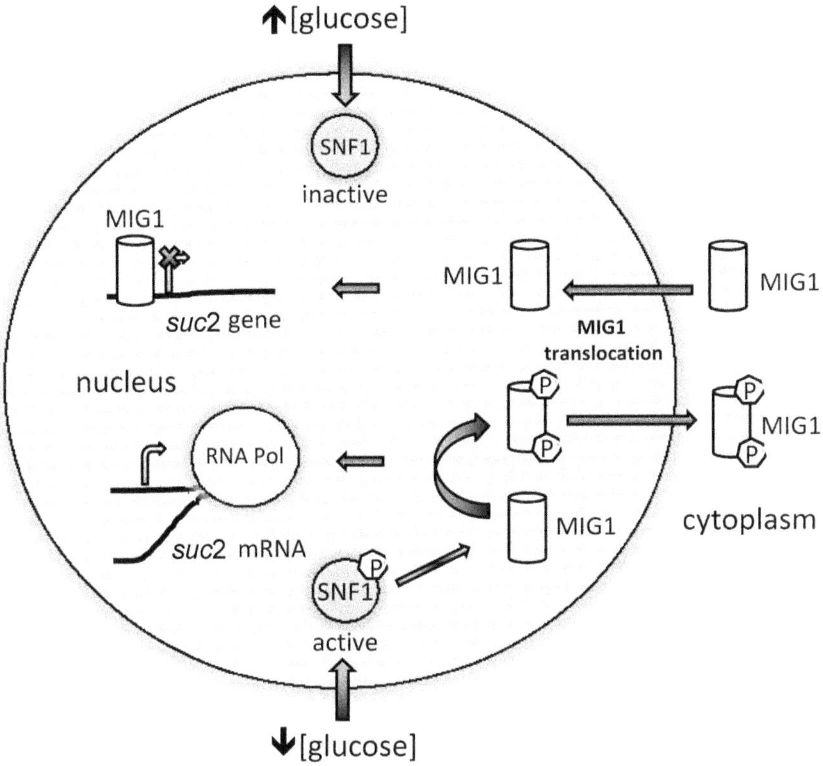

Figure 1.2 The mode of action of SNF1 and MIG1 in yeast. RNA Pol: RNA Polymerase; *suc*2: invertase gene.

MIG2 can also bind to the promoter of *suc*2 and contribute to its repression by glucose (Chi et al. 2004).

SNF1 is also regulated and during its activation the protein is phosphorylated on Thr210 by three upstream protein kinases: ELM2, TOS3 and SAK1 (Hong et al. 2003, Sutherland et al. 2003, Nath et al. 2003). However, the glucose-sensing mechanism that is involved in the activation of SNF1 during the glucose repression process remains unclear (Castermans et al. 2012). It is hypothesized that SNF1 is deactivated through dephosphorylation by the PP1 protein phosphatase Glc7 and its regulatory subunit Reg1 (Tu and Carlson 1994, Tu and Carlson 1995).

The biding of MIG1 to the regulatory region of a variety genes repressed by glucose requires a GC-box with the consensus sequence (G/C)(C/T) GGGG and an AT-rich region on the 5'side of the GC-box. It is believed that finger 1 of MIG1 recognizes a G(G/A)G triplet, finger 2 recognizes a (G/C)(C/T)G triplet and the AT region is required to stabilize the interaction (Chi et al. 2004).

SNF7 is involved in the regulation of *suc*2 transcription by regulating the level of the transcription factor NRG1 instead of MIG1/MIG2. SNF7 affects the NRG1 levels through activation of the transcription factor RIM101 (Weiss et al. 2008).

Glucose represses not only the expression of extracellular invertase but also the secretion of this protein (Chi et al. 2004). In *Saccharomyces* sp., inositol supplementations can derepress invertase secretion and cell growth. In *S. pombe,* similar to *Saccharomyces* sp., the invertase secretion and cell growth can be derepressed when the yeast cells are grown in higher concentrations of inositol. This derepression may be related to a high content of phosphatidylinositol (PI) in the cells and less glucose in the medium (Chi et al. 2004).

During carbon catabolic repression in yeast, the main effect on the expression of *suc*2 occurs at the transcriptional level through the binding of regulatory repressor proteins to the promoter region (Rolland et al. 2002, Chi et al. 2004). It is believed that in *S. pombe,* the high PI content may induce the phosphorylation of MIG1 via a PI-type signaling pathway, and its translocation to the cytoplasm (Tanaka et al. 1998, Chi et al. 2004). Thus, in the presence of high concentrations of sucrose, the transcriptional activators can bind to the promoter region of the inv^+ gene (encoding invertase) and lead to an increase in mRNA encoding for secreted invertase. When the PI content decreases, MIG1 is dephosphorylated and returns to the nucleus to repress the transcription of the inv^+ gene (Chi et al. 2004).

The protein phosphatases PP2A and PP1 are major regulators of a variety of cellular processes in yeast, and both enzymes are direct targets of glucose sensing. The activation of both PP2A and PP1 depend on the glucose-mediated activation of the cAMP-PKA (protein kinase cAMP-dependent) pathway (Castermans et al. 2012).

S. cerevisiae has two PP2A catalytic subunits, PPH21 and PPH22, and three PP2Ac-like catalytic subunits, PPH3, PPH1 and SIT4 (Sneddon et al. 1990, Stark 1996). PP2A plays a role in a variety of cellular processes in yeast, including metabolic regulation, cell-cycle progression and DNA metabolism (Stark 1996, Zabrocki et al. 2002).

PP2A activity is regulated by phosphatase 2A phosphatase activator (PTPA) (Van Hoof et al. 2005), for which yeast has two orthologs: RRD1/YPAI and RRD2/YPA2. PP1, on the other hand, it has a catalytic subunit encoded by the *inv* gene, but its substrate specificity is regulated by a number of regulatory subunits, including SIP5 and REG1, which are involved in glucose repression (Frederick and Tatchell 1996, Huang et al. 1996, Sanz et al. 2000); REG2 and SDS22, which affect growth and cell-cycle progression (MacKelvie et al. 1995, Frederick and Tatchell 1996, Huang et al. 1996); and SHP1, GAC1, GIP2 and PIG1, which appear to affect glycogen accumulation (Stuart et al. 1994, Zhang et al. 1995, Cheng et al. 1997, Williams-Hart et al. 2002).

The interaction of REG1 and GLC7 is responsible for the involvement of PP1 in the glucose repression pathway, while deletion of SHP1 causes a strong derepression of the invertase gene *suc*2. On deletion of the PP2A subunits PPH21 and PPH22, RRD1 and RRD2 enhance the derepression level of *suc*2, indicating that PP2A disrupts *suc*2 derepression. The regulatory subunit RTS1 interacts with PPH21 and GLC7 in a manner that is dependent on REG1. Interestingly, the effect of REG1 is consistent with its function as a subunit of both PP2A and PP1, affecting the derepression and repression of *suc*2, respectively (Castermans et al. 2012).

The comprehensive knowledge of various regulatory mechanisms of the invertase genes in the different fungi presented here has contributed to a lot of information and led to further exploration of the enzyme in a variety of biotechnological processes that have been highlighted throughout this chapter.

Heterologous Expression of Invertase

Many molecular biology tools have been applied in the cloning of invertase to study the biochemical characteristics of the enzyme as well as to improve the strains used in the production of invertases. The characterization of heterologous invertase has also been reported. Heterologous gene expression has been performed using the *inv*1 gene (from *P. anomala*) in *S. cerevisiae* and resulted in two different forms of hypoglycosylated invertases. The invertase was efficiently secreted by *S. cerevisiae,* and the enzyme content reached levels comparable to those of the *inv*1 gene. However, these heterologous invertases displayed different subcellular locations and molecular mass. One invertase was found in the culture medium, and the other in the periplasm, such that they were referred to as extracellular (81.5 kDa) and intracellular (78.3 kDa) invertases, respectively. The recombinant invertases (22–26 percent carbohydrate content) have shown a lower degree of glycosylation than that exhibited by the native *P. anomala* invertase, which has 30 percent carbohydrate content (Pérez et al. 2001). The availability of the complete genome sequence of *A. niger* (Pel et al. 2007) enabled the identification of two novel putative intracellular invertases (SucB and SucC) (Yuan et al. 2006). Subsequently, Goosen and co-workers (2007) have heterologously expressed the *A. niger suc*B gene in *E. coli* as well as in an invertase-negative strain of *S. cerevisiae,* and its biochemical characteristics have been analyzed. The invertase (SucB) was able to hydrolyze substrates such as sucrose, kestose, and nystose. In addition, SucB displayed an apparent Km of 2.0 mM for sucrose, which was lower than that reported for the extracellular *A. niger* invertase Suc1, with a Km of 30 mM. However, SucB showed a lower optimal temperature and pH when compared to Suc1.

Challenges in the Post-genomic Era for Fungi Producing Invertase

In the past decade, the full-genome sequencing of several fungi has been completed, and recently, attention has focused on filamentous fungi, which have had a significant impact in areas such as biotechnology and medicine. *Aspergillus* is the preferred genus that is studied at the genomic level, highlighted primarily by the number of genomes sequenced (http://www.ncbi.nlm.nih.gov/genome/browse/). Nevertheless, with the evolution of next-generation nucleic acid sequencing and the availability of powerful proteomics technologies, the goal now is to understand the expression, function and regulation of the entire set of proteins encoded by fungal genomes (Bhadauria et al. 2007, Braaksma et al. 2010, Costa et al. 2010). In recent years, there have been several proteomics advances related to filamentous fungi due to the availability of multiple fungal genome sequences (Doyle 2011, de Oliveira and de Graaff 2011). However, there are few studies on fungal invertases using the tool of proteomics. Recently, it was reported that the cultivation of *Candida utilis* in a xylose-based medium generated a secreted protein pattern (secretome) containing the invertase Inv1 as the major extracellular protein (Buerth et al. 2011). By analyzing the genomic data, Parrent et al. (2009) determined the number of GH32 genes in the fungi kingdom and traced the evolutionary history of this gene family through PCR-based and bioinformatics analyses. Ceccaroli et al. (2011) showed, for the first time, with the application of genomic and transcriptomic tools, the presence of an acid invertase in the symbiotic fungi *Tuber melanosporum*. However, the post-genomic era is still in the early stages for filamentous fungi, and advanced methodologies will greatly improve the understanding of the metabolic pathways in fungal producers of invertase.

Conclusions and Future Perspectives

Despite the biotechnological advances achieved in the last decade, which have primarily focused on the production, biochemical characterization and regulation of fungal invertase, many important questions remain to be investigated and answered, such as why there are multiple forms of invertase synthesized by fungi and yeast, each with individual properties and located in different cellular compartments. Are these isoforms synthesized at different times of cell growth? And what is the mechanism for the regulation and expression of these isoforms? Additionally, why are there no invertase crystals from filamentous fungus using X-ray crystallography? Another question is why do some invertases have fructosyltransferase activity? Is there an evolutionary relationship for these enzymes with this dual catalytic function? And is there a mechanism for regulation of the invertase and fructosyltransferase activities? Therefore, there are many areas of fungal invertases that need to be investigated, and to answer these questions, studies must be conducted with the help of advanced techniques combined with multiple disciplines of physiology, chemistry, biochemistry, genetics and molecular biology.

References

Ahuatzi, D., A. Riera, R. Pelaez, P. Herrero and F. Moreno. 2007. Hxk2 regulates the phosphorylation state of Mig1 and threrefore its nucleocytoplasmic distribution. J. Biol. Chem. 282: 4485–4493.

Alberto, F., C. Bignon, G. Sulzenbacher, B. Henrissat and M. Czjzek. 2004. The three-dimensional structure of invertase (β-fructosidase) from *Thermotoga maritima* reveals a bimodular arrangement and an evolutionary relationship between retaining and inverting glycosidases. J. Biol. Chem. 279: 18903–18910.

Alegre, A.C.P., M.L.T.M. Polizeli, H.F. Terenzi, J.A. Jorge and L.H.S. Guimarães. 2009. Production of thermostable invertases by *Aspergillus caespitosus* under submerged or solid state fermentation using agroindustrial residues as carbon source. Braz. J. Microbiol. 40: 612–622.

Álvaro-Benito, M., M. Abreu, L. Fernández-Arrojo, F.J. Plou, J. Jiménez-Barbero, A. Ballesteros, J. Polaina and M. Fernández-Lobato. 2007. Characterization of a β-fructofuranosidase from *Schwanniomyces occidentalis* with transfructosylating activity yielding the prebiotic 6-kestose. J. Biotechnol. 132: 75–81.

Álvaro-Benito, M., A. Polo, B. González, M. Fernández-Lobato and J. Sanz-Aparicio. 2010. Structural and kinetic analysis of *Schwanniomyces occidentalis* invertase reveals a new oligomerization pattern and the role of its supplementary domain in substrate binding. J. Biol. Chem. 285: 13930–13941.

Andjelković, U. and J. Lah. 2011. Thermodynamics and structural features of the yeast *Saccharomyces cerevisiae* external invertase isoforms in guanidinium-chloride solutions. J. Agric. Food Chem. 59 (2): 727–732.

Aranda, C., A. Robledo, O. Loera, J.C. Contreras-Esquivel, R. Rodríguez and C.N. Aguilar. 2006. Fungal invertase expression in solid-state fermentation. Food Technol. Biotechnol. 44: 229–233.

Ashokkumar, B., N. Kayalvizhi and P. Gunasekaran. 2001. Optimization of media for β-fructofuranosidase production by *Aspergillus niger* in submerged and solid state fermentation. Process Biochem. 37: 331–338.

Aziani, G., H.F. Terenzi, J.A. Jorge and L.H.S. Guimarães. 2012. Production of fructooligosaccharides by *Aspergillus phoenicis* biofilm on polyethylene as inert support. Food Technol. Biotechnol. 50: 40–45.

Bahar, T. and A. Tuncel. 2004. Concanavalin A carrying reactive beads for yeast invertase purification. React. Funct. Polym. 61: 203–210.

Barlikova, A., J. Svore and S. Miertus. 1991. Hybrid biosensor for determinations of sucrose. Anal. Chim. Acta. 247: 83–87.

Basha, S.Y. and P. Palanivelu. 2000. A novel method for immobilization of invertase from the thermophilic fungus, *Thermomyces lanuginosus*. World J. Microb. Biot. 16: 151–154.
Bhatti, H., N.M. Asgher, A. Abbas, R. Nawaz and M.A. Sheikh. 2006. Studies on kinetics and thermostability of a novel acid invertase from *Fusarium solani*. J. Agric. Food Chem. 54, 4617–4623.
Beine, R., R. Moraru, M. Nimtz, S. Na'amnieh, A. Pawlowski, K. Buchholz and J. Seibel. 2008. Synthesis of novel fructooliosaccharides by substrate and enzyme engineering. J. Biotechnol. 138: 33–41.
Belcarz, A., G. Ginalska, J. Lobarzewski and C. Penel. 2002. The novel non-glycosylated invertase from *Candida utilis* (the properties and the conditions of production and purification). Biochim. Biophys. Acta. 1594: 40–53.
Bhadauria, V., W.S. Zhao, L.X. Wang, Y. Zhang, J.H. Liu, J, Yang, L.A. Kong and Y.L. Peng. 2007. Advances in fungal proteomics. Microbiol. Res. 162: 193–200.
Boer, E., T. Wartmann, B. Luther, R. Manteuffel, R. Bode, G. Gellissen and G. Kunze. 2004. Characterization of the *AINV* gene and the encoded invertase from the dimorphic yeast *Arxula adeninivorans*. Anton. Leeuw. 86: 121–134.
Braaksma, M., E.S. Martens-Uzunova, P.J. Punt and P.J. Schaap. 2010. An inventory of the *Aspergillus niger* secretome by combining in silico predictions with shotgun proteomics data. BMC Genomics 11: 584.
Buerth, C., C.J., Heilmann, F.M. Klis, C.G. de Koster, J.F. Ernst and D. Tielker. 2011. Growth-dependent secretome of *Candida utilis*. Microbiology 157: 2493–2503.
Carlson, M. and D. Botstein. 1983. Organization of the SUC gene family in *Saccharomyces*. Mol. Cell. Biol. 3: 351–359.
Castermans, D., I. Somers, J. Kriel, W. Louwet, S. Wera, M. Versele, V. Janssens and J.M. Thevelein. 2012. Glucose-induced posttranslational activation of protein phosphatases PP2A and PP1 in yeast. Cell Res. 2012: 1–20.
Ceccaroli, P., M. Buffalini, R. Saltarelli, E. Barbieri, E. Polidori, S. Ottonello, A. Kohler, E. Tisserant, F. Martin and V. Stocchi. 2011. Genomic profiling of carbohydrate metabolism in the ectomycorrhizal fungus *Tuber melanosporum*. New Phytologist 189: 751–764.
Chang, H.N., G.H. Seong, I.K. Yoo, J.K. Parks and J.H. Seo. 1996. Microencapsulation of recombinant *Saccharomyces cerevisiae* cells with invertase activity in liquid-core alginate capsules. Biotechnol. Bioeng. 51: 157–162.
Chaudhuri, A., G. Bharadwaj and R. Maheshwari. 1999. An unusal pattern of invertase activity development in the thermophilic fungus *Thermomyces lanuginosus*. FEMS Microbiol. Lett. 177: 39–45.
Chávez, F.P., L. Rodriguez, J. Díaz, J.M. Delgado and J.A. Cremata. 1997. Purification and characterization of an invertase from *Candida utilis*: comparison with natural and recombinant yeast invertases. J. Biotechnol. 53: 67–74.
Chen, J., J. Saxton, F.W. Hemming and J.F. Peberdy. 1996. Purification and partial characterization of the high and low molecular weight form (S- and F-form) of invertase secreted by *Aspergillus nidulans*. Biochim. Biophys. Acta. 1296: 207–218.
Chen, W. 1996. β-Fructofuranosidase production by *Aspergillus japonicus* in shaking batch cultures as effected by initial sucrose concentration. Biotechnol. Lett. 18: 69–72.
Chen, W. and C. Liu. 1996. Production of β-fructofuranosidase by *Aspergillus japonicus*. Enzyme Microb. Tech. 18: 153–160.
Cheng, C., D. Huang and P.J. Roach. 1997. Yeast PIG genes: PIG1 encodes a putative type 1 phosphatase subunit that interacts with the yeast glycogen synthase Gsy2p. Yeast 13: 1–8.
Cheng, T.C., K.J. Duan and D.C. Sheu. 2005. Immobilization of β-fructofuranosidase from Aspergillus japonicus on chitosan using tris(hydroxymethyl) phosphine or glutaraldehyde as a coupling agent. Biotechnol. Lett. 27: 335–8.
Chi, Z. and J. Gao. 1999. Recent advances in the mechanisms of ethanol tolerance in yeasts. Microbiology 26: 373–376.

Chi, Z-M, J.-F. Li, X.-H. Wang and S.-M. Yao. 2004. Inositol and phosphatidylinositol mediated glucose derepression, gene expression and invertase secretion in yeasts. Acta Bioch. Bioph. Sin. 36: 443–449.

Chiang, C.J., W.C. Lee, D.C. Sheu and K.J. Duan. 1997. Immobilization of β-fructofuranosidases from Aspergillus on methacrylamide-based polymeric beads for production of fructooligosaccharides. Biotechnol. Prog. 13: 577–82.

Chu, F. and F. Maley. 1980. The effect of glucose on the synthesis and glycosylation of the polypeptide moiety of yeast external invertase. J. Biol. Chem. 255: 6392–6397.

Chuankhayan, P., C.Y. Hsieh, Y.C. Huang, Y.Y. Hsieh, H.H. Guan, Y.C. Hsieh, Y.C. Tien, C.D. Chen, C.M. Chiang and C.J. Chen. 2010. Crystal structures of *Aspergillus japonicus* fructosyltransferase complex with donor/acceptor substrates reveal complete subsites in the active site for catalysis. J. Biol. Chem. 285: 23251–23264.

Costa, V., C. Angelini, I. De Feis and A. Ciccodicola. 2010. Uncovering the complexity of transcriptomes with RNA-Seq. J. Biomed. Biotechnol. 2010: 853–916.

Costaglioli, P., E. Meilhoc, I. Janatova, R. Klein and J. Masson. 1997. Secretion of invertase from *Schwanniomyces occidentalis*. Biotechnol. Lett. 19: 623–627.

David, A.E., N.S. Wang, V.C. Yang and A.J. Yang. 2006. Chemically surface modified gel (CSMG): An excellent enzyme-immobilization matrix for industrial processes. J. Biotechnol. 125: 395–407.

de Almeida, L.S., A.M. de Araújo, C.A.M. Costa, M.A.G.A. de Abreu, A.C.S. Lima and M.L.A.P.F. Palha. 2005. Sucrose hydrolysis catalysed by auto-immobilized invertase into intact cells of *Cladosporium cladosporiodes.* Electron. J. Biotechnol. 8(1): 54: 58.

de Oliveira, J.M. and L.H. de Graaff. 2011. Proteomics of industrial fungi: trends and insights for biotechnology. Appl. Microbiol. Biotechnol. 89: 225–237.

De vit, M.J., J.A. Waddle and M. Johnston. 1997. Regulated nuclear translocation of the Mig1 glucose repressor. Mol. Biol. Cell. 8: 1603–1618.

Dhananjay, S.K. and V.H. Mulimani. 2008. Purification of α-galactosidase and invertase by three-phase partitioning from crude extract of *Aspergillus oryzae*. Biotechnol. Lett. 30: 1565–1569.

Dickensheets, P.A., L.F. Chen and G.T. Tsao. 1977. Characteristics of yeast invertase immobilized on porous cellulose beads. Biotechnol. Bioeng. 19: 365–375.

Dominguez, A., I.M. Santos, J.A. Teixeira and N. Lima. 2006. New and simple plate test for screening relative transfructosylation activity of fungi. Rev. Iberoam. Micol. 23: 189–191.

Doyle, S. 2011. Fungal proteomics: from identification to function. FEMS Microbiol. Lett. 321: 1–9.

Driouch, H., A. Roth, P. Dersch and C. Wittmann. 2010. Optimized bioprocess for production of fructofuranosidase by recombinant *Aspergillus niger*. Appl. Microbiol. Biotechnol. 87: 2011–2024.

Farine, S., C. Versluis, P.J. Bonnici, A. Heck, C. L'homme, A. Puigserver and A. Biagini. 2001. Application of high performance anion exchange chromatography to study invertase-catalyzed hydrolysis of sucrose and formation of intermediate fructan products. Appl. Microbiol. Biotechnol. 55: 55–60.

Foster, A.J., J.M. Jenkinson and N.J. Talbot. 2003. Trehalose synthesis and metabolism are required at different stages of plant infection by *Magnaporthe grisea*. EMBO J. 22: 225–235.

Frederick, D.L. and K. Tatchell. 1996. The REG2 gene of *Saccharomycescerevisiae* encodes a type 1 protein phosphatase-binding protein that functions with Reg1p and the Snf1 protein kinase to regulate growth. Mol. Cell. Biol. 16: 2922–2931.

Gancedo, J.M. 1998. Yeast carbon catabolite repression. Microbiol. Mol. Biol. Rev. 62: 334–361.

Ghazi, I., L. Fernández-Arrojo, H. Garcia-Arellano, F.J. Plou and A. Ballesteros. 2007. Purification and kinetic characterization of a fructosyltransferase from *Aspergillus aculeatus*. J. Biotechnol. 128: 204–211.

Giraldo, M.A., T.M. Silva, F. Salvato, H.F. Terenzi, J.A. Jorge, and L.H. Guimarães. 2012. Thermostable invertases from *Paecilomyces variotii* produced under submerged and solid-state fermentation using agroindustrial residues. World J. Microb. Biot. 28: 463–472.

Goosen, C., X. Yuan, J.M. van Munster, A.F.J. Ram, M.J.E.C. van der Maarel and L. Dijkhuizen. 2007. Molecular and biochemical characterization of a novel intracellular invertase from *Aspergillus niger* with transfructosylating activity. Eukaryot. Cell. 6: 674–681.

Guimarães, L.H.S., H.F. Terenzi, M.L.T.M. Polizeli and J.A. Jorge. 2007. Production and characterization of a thermostable extracellular β-D-fructofuranosidase produced by *Aspergillus ochraceus* with agroindustrial residues as carbon sources. Enzyme Microb. Tech. 42: 52–57.

Guimarães, L.H.S., A.F. Somera, H.F. Terenzi, M.L.T. M. Polizeli and J.A. Jorge. 2009. Production of β-fructofuranosidases by *Aspergillus niveus* using agroindustrial residues as carbon sources: Characterization of an intracellular enzyme accumulated in the presence of glucose. Process Biochem. 44: 237–241.

Gutiérrez-Alonso, P., L. Fernández-Arrojo, F.J. Plou and M. Fernández-Lobato. 2009. Biochemical characterization of a β-fructofuranosidase from *Rhodotorula dairenensis* with transfructosylating activity. FEMS Yeast Res. 9: 768–773.

Hayashi, S., K. Matsuzaki, Y. Takasaki, H. Ueno and K. Imada. 1992. Production of β-fructofuranosidase by *Aspergillus japonicus*. World J. Microbiol. Biotechnol. 8: 155–159.

Henrissat, B. 1991. A classification of glycosyl hydrolases based on amino-acid-sequence similarities. Biochem. J. 280: 309–316.

Herwig, C., C. Doerries, I. Marison and U. Stockar. 2001. Quantitative analysis of the regulation scheme of invertase expression in *Saccharomyces cerevisiae*. Biotechnol. Bioeng. 76.

Hong, S.P., F.C. Leiper, A. Woods, D. Carling and M. Carlson. 2003. Activation of yeast SNF1 and mammalian AMP-activated protein kinase by upstream kinases. P. Natl. Acad Sci. USA 100: 8839–8843.

Huang, D., K.T. Chun, M.G. Goebl and P.J. Roach. 1996. Genetic interactions between REG1/HEX2 and GLC7, the gene encoding the protein phosphatase type 1 catalytic subunit in *Saccharomyces cerevisiae*. Genetics 143: 119–127.

Jiang, R. and M. Carlson. 1997. The Snf1 protein kinase and its activating subunit, Snf4, interact with distinct domains of the Sip1/Sip2/Gal83 component in the kinase complex. Mol. Cell. Biol. 17: 2099–2106.

Karandikar, S., A. Prabhune, S.A. Kalele, S.W. Gosavi and S.K. Kulkarni. 2006. Immobilization of thermotolerant *Kluyveromyces marxianus* on silica aerogel for continuous production of invert syrup. Res. J. Biotech. 1: 16–9.

Karkas, T. and S. Önal. 2012. Characteristics of invertase partitioned in poly(ethylene glycol)/magnesium sulfate aqueous two-phase system. Biochem. Eng. J. 60: 142–150.

Kilian, S., S. Kritzinger, C. Rycroft, G. Gibson and J. du Preez. 2002. The effects of the novel bifidogenic trisaccharide, neokestose, on the human colonia microbiota. World J. Microbiol. Biotechnol. 18: 637–644.

Korshunova, I.V., E.S. Naumova and G.I. Naumov. 2005. Comparative molecular genetic analysis of β- Fructosidases of yeasts *Saccharomyces*. Mol. Biol. 39: 366–371.

Kotwal, S.M. and V. Shankar. 1997. Preparation and properties of *Sclerotium rolfsii* invertase immobilized on Indion 48-R. Appl. Biochem. Biotechnol. 62: 151–8.

Kotwal, S.M. and V. Shankar. 2009. Immobilized invertase. Biotechnol. Adv. 27: 311–322.

Kurakake, M., R. Masumoto, K. Maguma, A. Kamata, E. Saito, N. Ukita and T. Komaki. 2010. Production of fructooligosaccharides by β-fructofuranosidases from *Aspergillus oryzae* KB. J. Agric. Food Chem. 58: 488–492.

L'Hocine, L., Z. Wang, B. Jiang and S. Xu. 2000. Purification and partial characterization of fructosyltransferase and invertase from *Aspergillus niger* AS0023. J. Biotechnol. 81: 73–84.

Linde, D., I. Macias, L. Fernández-Arrojo, F.J. Plou, A. Jiménez and M. Fernández-Lobato. 2009. Molecular and biochemical characterization of a β-Fructofuranosidase from *Xanthophyllomyces dendrorhous*. Appl. Environ. Microbiol. 75: 1065–1073.

MacKelvie, S.H., P.D. Andrews and M.J. Stark. 1995. The *Saccharomyces cerevisiae* gene SDS22 encodes a potential regulator of the mitotic function of yeast type 1 protein phosphatase. Mol. Cell. Biol. 15: 3777–3785.

Madhan, S.S.R., R. Sathyavani and B. Niket. 2010. Production and partial purification of invertase using *Cympopogan caecius* leaf powder as substrate. Indian J. Microbiol. 50: 318–324.

Madyastha, K.M., A.R. Ganguli, V.G. Kubair, N. Kowser and D. Vidya. 1987. Extracellular invertase from *Aspergillus athecius*: isolation and immobilization. Biotechnol. Lett. 9: 555–60.

Mirzarakhmetova, D.T., M.M. Rakhimov, S.Kh. Abdurazakova and Z.R. Akhmedova. 2006. Enzymatic conversion of invertase substrates in a water-organic medium. Appl. Biochem. Micro. 42: 150–155.

Moreno, S., Y. Sanchez and L. Rodriguez. 1990. Purification and characterization of the invertase from *Schizosaccharomyces pombe*. Biochem. J. 267: 697–702.

Mussato, S.I., C.N. Aguilar, L.R. Rodrigues and J.A. Teixeira. 2009. Fructooligosaccharides and β-fructofuranosidase production by *Aspergillus japonicus* immobilized on lignocellulosic materials. J. Mol. Cat. B: Enzymatic 59: 76–81.

Nagem, R.A.P., A.L. Rojas, A.M. Golubev, O.S. Korneeva, E.V. Eneyskaya, A.A. Kulminskaya, K.N. Neustroev and I. Polikarpov. 2004. Crystal structure of exo-inulinase from *Aspergillus awamori*: The enzyme fold and structural determinants of substrate recognition. J. Mol. Biol. 344: 471–480.

Nath, N., R.R. MacCartney and M.C. Shimidt. 2003. Yeast Pak1 kinase associates with and activates SNF1. Mol. Cell. Biol. 23: 3909–3917.

Naumov, G.I. and E.S. Naumova. 2010a. Comparative genetics of yeasts: A novel β-fructosidase gene SUC8 in *Saccharomyces cerevisiae*. Russ. J. Genet. 46: 323–330.

Naumov, G.I. and E.S. Naumova. 2010b. Polygenic control for fermentation of β-fructosides in the yeast *Saccharomyces cerevisiae*: new genes SUC9 and SUC10. Microbiology 79: 160–166.

Naumov, G.I. and E.S. Naumova. 2011. Comparative genetics of yeast *Saccharomyces cerevisiae*. chromosomal translocations carrying the SUC2 marker. Russ. J. Genet. 47: 147–152.

Neagu Bonciu, C., O. Constantin and G. Bahrim. 2011. Screening of biotechnological parameters for fructofuranosidases production by a newly isolated fungal strain using plackett-burman design. Not. Bot. Hort. Agrobot. 39: 271–275.

Nelson, J.M. and E.G. Griffin. 1916. Adsorption of invertase. J. Am. Chem. Soc. 38: 1109–15.

Nemukula, A., T. Mutanda, B.S. Wilhelmi and C.G. Whiteley. 2009. Response surface methodology: synthesis of short chain fructooligosaccharides with fructosyltransferase from *Aspergillus aculeatus*. Biores. Technol. 100: 2040–2045.

Nguyen, Q.D., J.M. Rezessy-Szabó, M.K. Bhat and A. Hoschke. 2005. Purification and some properties of b-fructofuranosidase from *Aspergillus niger* IMI303386. Process Biochem. 40: 2461–2466.

Ostling, J. and H. Ronne. 1998. Negative control of the Mig1p repressor by Snf1p-dependent phosphorylation in the absence of glucose. Eur. J. Biochem. 252: 162–168.

Parascandola, P., S. Salvadore and V. Scardi. 1982. Tuff as a convenient material for supporting immobilized invertase-active whole cells of *Saccharomyces cerevisiae*. J. Ferment. Technol. 60: 477–80.

Parrent, J.L., T.Y. James, R. Vasaitis and A.F.S. Taylor. 2009. Friend or foe? Evolutionary history of glycoside hydrolase family 32 genes encoding for sucrolytic activity in fungi and its implications for plant-fungal symbioses. BMC Evol. Biol. 9: 148.

Patil, M., R. Bhamre and U. Patil. 2012. Invertase production from *Aspergillus* spp. M1 isolated from honeycomb. Int. J. Appl. Biores. 4: 1–5.

Patil, P.R., G.S.N. Reddy and M.B. Sulochana. 2011. Production, optimization and characterization of β-fructofuranosidase by *Chrysonilia sitophila* PSSF84—A novel source. Indian J. Biotechnol. 10: 56–64.

Pel, H.J., J.H. de Winde, D.B. Archer, P.S. Dyer, G. Hofmann, P.J. Schaap, G. Turner, R.P. de Vries, R. Albang, K. Albermann, M.R. Andersen, J.D. Bendtsen, J.A.E. Benen, M. van den Berg, S. Breestraat, M.X. Caddick, R. Contreras, M. Cornell, P.M. Coutinho, E.G.J. Danchin, A.J.M. Debets, P. Dekker, P.W.M. van Dijck, A. van Dijk, L. Dijkhuizen, A.J.M. Driessen, C. d'Enfert, S. Geysens, C. Goosen, G.S.P. Groot, P.W.J. de Groot, T. Guillemette, B. Henrissat, M. Herweijer, J.P.T.W. van den Hombergh, C.A.M. J.J. van den Hondel, R.T.J.M. van der Heijden, R.M. van der Kaaij, F.M. Klis, H.J. Kools, C.P. Kubicek, P.A. van Kuyk, J. Lauber, X. Lu, M.J.E.C. van der Maarel, R. Meulenberg, H. Menke, A.M. Mortimer, J. Nielsen, S.G. Oliver, M. Olsthoorn, K. Pal, N.N.M.E. van Peij, A.F.J. Ram, U. Rinas, J.A. Roubos, C.M.J. Sagt, M. Schmoll, J. Sun, D. Ussery, J. Varga, W. Vervecken, P.J.I. van de Vondervoort and H. Wedler. 2007. Genome sequencing and analysis of the versatile cell factory *Aspergillus niger* CBS 513.88. Nat. Biotechnol. 25: 221–231.

Pérez, J. A., J. Rodríguez, T. Ruiz and L. Rodríguez. 2001. Expression of *Pichia anomala* INV1 gene in *Saccharomyces cerevisiae* results in two different active forms of hypoglycosylated invertase. Arch. Microbiol. 175: 189–197.

Polo, A., M. Álvaro-Benito, M. Fernández-Lobato and J. Sanz-Aparicioa. 2009. Crystallization and preliminary X-ray diffraction analysis of the fructofuranosidase from *Schwanniomyces occidentalis*. Acta Crystallogr. 65: 1162–1165.

Polo, A., D. Linde, M. Estévez, M. Fernández-Lobato and J. Sanz-Aparicio. 2010. Crystallization and preliminary X-ray diffraction analysis of the fructofuranosidase from *Xanthophyllomyces dendrorhous*. Acta Crystallogr. 66: 1441–1444.

Quiroga, E.N., M.A. Vattuone and A.R. Sampietro. 1995. Purification and characterization of the invertase from *Pycnoporus Sanguines*. Biochim. Biophys. Acta. 1251: 75–80.

Raj, L., G.S. Chauhan, W. Azmi, J.H. Ahn and J. Manuel. 2011. Kinetics study of invertase covalently linked to a new functional nanogel. Bioresource Technol. 102: 2177–2184.

Rajoka, M.I. and A. Yasmeen. 2005. Improved productivity of β-fructofuranosidase by a derepressed mutant of *Aspergillus niger* from conventional and non-conventional substrates. World J. Microbiol. Biotechnol. 21: 471–478.

Rashad, M.M. and M.U. Nooman. 2009. Production, purification and characterization of extracellular *Saccharomyses Cerevisiae* NRRL Y-12632 by solid-fermentation of red carrot residue. Aust. J. Basic Appl. Sci. 3(3): 1910–1919.

Recorbet, G., C. Steinberg, C. Olivain, V. Edel, S. Trouvelot, E. Dumas-Gaudot, S. Gianinazzi and C. Alabouvette. 2003. Wanted: pathogenesis-related marker molecules for *Fusarium oxysporum*. New Phytol. 159: 73–92.

Reddy, A. and F. Maley. 1996. Studies on identifying the catalytic role of glu-204 in the active site of yeast invertase. J. Biol. Chem. 271: 13953–13958.

Rodriguez, J., J.A. Perez, T. Ruiz and L. Rodriguez. 1995. Characterization of the invertase from *Pichia anomala*. Biochem. J. 306: 235–239.

Rolland, F., J. Winderickx and J.M. Thevelein. 2002. Glucose-sensing and -signaling mechanisms in yeast. FEMS Yeast Res. 2: 183–201.

Romero-Gómez, S., C. Augur and G. Viniegra-González. 2000. Invertase production by *Aspergillus niger* in submerged and solid-state fermentation. Biotechnol. Lett. 22: 1255–1258.

Rubio, M.C., R. Runco and A. Navarro. 2002. Invertase from a strain of *Rhodotorula glutinis*. Phytochemistry 61: 605–609.

Rubio, M.C. and A.R. Navarro. 2006. Regulation of invertase in *Aspergillus niger*. Enzyme Microb. Tech. 39: 601–606.

Rubio, M.C., M.C. Maldonado, P.Y. Aznar and A. Navarro. 2003. Production y caracterizacion de una invertasa extracelular de *Penicillium glabrum*. La Alimen Latinoamer 247: 40–45.

Rustiguel, C.B., H.F. Terenzi, J.A. Jorge and L.H.S. Guimarães. 2010. A novel silver-activated extracellular β-D-fructofuranosidase from *Aspergillus phoenicis*. J. Mol. Catal. B-Enzym. 67: 10–15.

Sangeetha, P.T., M.N. Ramesh and S.G. Prapulla. 2005. Recent trends in the microbial production, analysis and application of Fructooligosaccharides. Trends Food Sci. Tech. 16: 442–457.

Sanz, P., K. Ludin and M. Carlson. 2000. Sip5 interacts with both the Reg1/Glc7 protein phosphatase and the Snf1 protein kinase of *Saccharomyces cerevisiae*. Genetics 154: 99–107.

Sestak, S. and V. Farkas. 1993. Metabolic regulation of endoglucanase synthesis in *Trichoderma reesei:* participation of ciclic AMP and glucose 6- phosphate. Can. J. Microbiol. 39: 342–347.

Shaheen, I., H.N. Bhatti and T. Ashraf. 2008. Production, purification and thermal characterisation of invertase from a newly isolated *Fusarium* sp. under solid-state fermentation. Int. J. Food Sci. Tech. 43: 1152–1158.

Sneddon, A.A., P.T. Cohen and M.J. Stark. 1990. *Saccharomyces cerevisiae* protein phosphatase 2A performs an essential cellular function and is encoded by two genes. EMBO J 9: 4339–4346.

Stark, M.J. 1996. Yeast pretein serine/threonine phosphatases: multiple roles and diverse regulation. Yeast 12: 1647–1675.

Stuart, J.S., D.L. Frederick, C.M. Varnera and K. Tatchell. 1994. The mutant type 1 protein phosphatase encoded by glc7-1 from *Saccharomyces cerevisiae* fails to interact productively with the GAC1- encoded regulatory subunit. Mol. Cell. Biol. 14: 896–905.

Sturm, A. 1999. Invertases. Primary structures, functions, and roles in plant development and sucrose partitioning. Plant Physiol. 121: 1–7.

Sutherland, C.M., S.A. Hawley, R.R. MacCartney, A. Leech, M.J. Stark, M.C. Schimidt and D.G. Hardie. 2003. Elm1p is one of three upstream kinases for the *Saccharomyces cerevisiae* SNF1complex. Curr. Biol. 13: 1299–1305.

Tanaka, N., N. Ohuchi, Y. Mukai, Y. Osaka, Y. Ohtani, M. Tabuchi, M.S. Bhuiyan, H. Fukui, S. Harashima and K. Takegawa. 1998. Isolation and characterization of an invertase and its repressor genes from *Schizosaccharomyces pombe.* Biochem. Biophys. Res. Commun. 245: 246–253.

Treitel, M.A. and M. Carlson. 1995. Repression by SSN6-TUP1 is directed by MIG1, a repressor activator protein. P. Natl. Acad. Sci. USA 92: 3132–3136.

Tu, J. and M. Carlson. 1994. The GLC7 type 1 protein phosphatase is required for glucose repression in *Saccharomyces cerevisiae*. Mol. Cell Biol. 14: 6789–6796.

Tu, J. and M. Carlson. 1995. REG1 binds to protein phosphatase type 1 and regulates glucose repression in *Saccharomyces cerevisiae*. EMBO Journal 14: 5939–5946.

Turkiewicz, M., M. Pazgier, S.P. Donachie and H. Kalinowska. 2005. Invertase and α-glucosidase production by the endemic Antarctic marine yeast *Leucosporidium antarcticum*. Pol. Polar Res. 26(2): 125–136.

Ul-Haq, I., M.A. Baig and S. Ali. 2005. Effect of cultivation conditions on invertase production by hyperproducing *Saccharomyces cerevisiae* isolates. World J. Microb. Biot. 21: 487–492.

Uma, C., D. Gomathi, C. Muthulakshmi and V.K. Gopalakrishnan. 2010a. Optimization and characterization of invertase by *Aspergillus fumigatus* using fruit peel waste as substrate. Res. J. Pharm. Biol. Chem. Sc. 1: 93–101.

Uma, C., D. Gomathi, C. Muthulakshmi and V.K. Gopalakrishnan. 2010b. Production, purification and characterization of invertase by *Aspergillus flavus* using fruit peel waste as substrate. Adv. Biol. Res. 4: 31–36.

Uma, C., D. Gomathi, G. Ravikumar, M. Kalaiselvi and V.K. Gopalakrishnan. 2012. Production and properties of Invertase from a *Cladosporium cladosporioides* in SmF using pomegranate peel waste as substrate. Asian Pac. J. Trop. Biomed. 1–7.

Uzun, K., E. Çevik and M. Senel. 2011. Invertase immobilization on a metal chelated triazole-functionalized Eupergit® C K. Am. J. Chem. 1(1): 16–21.

Vainstein, M.H. and J.F. Peberdy. 1991 Regulation of invertase in *Aspergillus nidulans*: effect of different carbon sources. J. Gen. Microbiol. 137: 315–321.
Van Hoof, C., E. Martens, S. Login, J. Jordens, I. Stevens, V. Janssens and J. Goris. 2005. Specific interactions of PP2A and PP2A-like phosphatases with the yeast PTPA homologues, Ypa1 and Ypa2. Biochem. J. 386: 93–102.
Vignali, M., A.H. Hassan, K.E. Neely and J.L. Workman. 2000. ATP-dependent chromatin-remodeling complexes. Mol. Cell. Biol. 20: 1899–1910.
Wang, L. and H. Zhou. 2006. Isolation and identification of a novel *Aspergillus japonicus* JN19 producing β-fructofuranosidase and characterization of the enzyme. J. Food Biochem. 30: 641–658.
Weiss, P., S. Huppert and R. Kölling. 2008. ESCRT-III Snf7 mediates high level expression of the SUC2 gene via the RIM101 pathway. Eukaryot. Cell 7: 1888–1894.
Williams-Hart, T., X. Wu and K. Tatchell. 2002. Protein phosphatase type 1 regulates ion homeostasis in *Saccharomyces cerevisiae*. Genetics 160: 1423–1437.
Wilson, W.A., S.A. Hawley and D.G. Hardie. 1996. Glucose repression/derepression in budding yeast: SNF1 protein kinase is activated by phosphorylation under derepressing conditions, and this correlates with a high AMP: ATP ratio. Curr. Biol. 6: 1426–1434.
Wolska-Mitaszko, B., J. Jaroszuk-Scise and K. Pszeniczna. 2007. Isoforms of trehalase and invertase of *Fusarium oxysporum*. Mycol. Res. 456–465.
Woodward, J. and A. Wiseman. 1978. The involvement of salt links in the stabilization of baker's yeast invertase. Evidence from immobilization and chemical modification studies. Biochim. Biophys. Acta 527: 8–16.
Workman, W.E and D.F. Day. 1983. Purification and properties of the β-fructofuranosidase from *Kluyveromyces fragile* 160: 16–20.
Yang, Y.-L., J.-H. Wang, D. Teng and F. Zhang. 2008. Preparation of high-purity fructo-oligosaccharides by *Aspergillus japonicus* β-fructofuranosidase and successive cultivation with yeast. J. Agric. Food Chem. 56: 2805–2809.
Yoshikawa, J., S. Amachi, H. Shinoyama and T. Fujii. 2006. Multiple β- fructofuranosidases by *Aureobasidium pullulans* DSM2404 and their roles in fructooligosaccharide production. FEMS Microbiol. Lett. 265: 159–163.
Yuan, X.-L., C. Goosen, H. Kools, M.J.E.C van der Maarel, C.A.M.J.J. van der Hondel, L. Dijkhuizen and A.F.J. Ram. 2006. Database mining and transcriptional analysis of genes encoding inulin-modifying enzymes of *Aspergillus niger*. Microbiology 152: 3061–3073.
Yuan, X.-L., J.A. Roubos, C.A.M.J.J. van der Hondel and A.F.J. Ram. 2008. Identification of InuR, a new Zn(II)2Cys6 transcriptional activator involved in the regulation of inulinolytic genes in *Aspergillus niger*. Mol. Genet. Genomics 279: 11–26.
Zabrocki, P., C. Van Hoof, J. Goris, J.M. Thevelein, J. Winderickx and S. Wera. 2002. Protein phosphatase 2A on track for nutrient-induced signaling in yeast. Mol. Microbiol. 43: 835–842.
Zech, M. and H. Goerish. 1995. Invertase from *Saccharamyces cerevisiae;* reversible inactivation by components of industrial molasses media. Enzyme Microb. Tech. 17: 41–46.
Zhang, S., S. Guha and F.C. Volkert. 1995. The *Saccharomyces* SHP1 gene, which encodes a regulator of phosphoprotein phosphatase 1 with differential effects on glycogen metabolism, meiotic differentiation, and mitotic cell cycle progression. Mol. Cell. Biol. 15: 2037–2050.

2

Application of Lignocelulosic Residues in the Production of Cellulase and Hemicellulases from Fungi

Michele Michelin,[1,*] *Maria de Lourdes T.M. Polizeli,*[2] *Denise S. Ruzene,*[3] *Daniel P. Silva*[4] and *José A. Teixeira*[1]

Introduction

Lignocellulosic residues are abundant, renewable and inexpensive energy sources, and the increasing expansion of the agro-industrial sector has led to the accumulation of a large quantity of these residues from agricultural (e.g., sugarcane bagasse, corn stover, rice straw, wheat straw, barley straw and olive stones), forestry (e.g., sawdust, thinnings, and mill waste), herbaceous (e.g., switchgrass, coastal Bermuda grass and alfalfa hay), hardwood (aspen,

[1]IBB Institute for Biotechnology and Bioengineering, Center of Biological Engineering, University of Minho, Campus Gualtar, 4710-057 Braga, Portugal.
[2]Department of Biology, Faculty of Philosophy, Sciences and Letters of Ribeirão Preto, University of São Paulo, 14040-901 Ribeirão Preto, SP, Brazil.
[3]Institute of Technology and Research, University Tiradentes, Campus Farolândia, 49032-490 Aracaju, SE, Brazil.
[4]Federal University of Sergipe, Nucleus of Production Engineering NPR/UFS, São Cristovão-SE, 49100-000, Brazil.
*Corresponding author: mimichelin.bio@gmail.com

poplar), softwood (pine, spruce), municipal solid wastes (e.g., waste paper) and various industrial wastes all over the world (Sánchez 2009). It is worth noting that crop residues encompass all agricultural wastes such as straw, husk, shell, peel, stem, stalk, leaves, seed and others.

More than 200 billion tons of lignocellulosic biomass are estimated to be produced annually worldwide. In relation to residues from 27 food crops, the world annual production is estimated at about 4 billion tons, from which 75 percent are cereal lignocellulosic residues that include wheat residues, rice straw and husk, barley residue, corn stalks and leaves, millet and sorghum stalks. Sugarcane, after cereals, provides the highest amount of residue with two major wastes-leaves and stalk, and bagasse. Significant amounts are also obtained from cotton crops, in the form of stalks and husks. Although in smaller amounts residues supplied by others crops such as sunflower, oil palm, coconut, banana, vines, groundnut and coffee must also be considered (Philippoussis 2009).

The largest producers of agricultural residues are the countries with a significant agro-industrial activity and large areas of cultivated or cultivable land. China and India will become large producers as a consequence of their expanding agricultural activity.

Cereal straw is one of the most abundant residues generated in the world, with an annual production of 2.9 billion tons, and China alone (East Asia) generates 700 million tons per year, including mainly by-products of rice, corn and wheat. The amount of residues produced in India (South Asia) is 600 million tons, including 480 tons of crop residues (rice, wheat, millet, sorghum, pulses, oilseed crops, corn stalks and cobs, cotton stalks, sugarcane trash etc.) and 120 tons of processing-based residues (mainly groundnut shells, rice husk, sugarcane bagasse, cotton waste, coconut shell and coir pith) (Philippoussis 2009).

In the USA, the United State Department of Agriculture (USDA 2005) estimates a production of 1.3 billion tons of biomass per year, including agricultural (998 million tons) and forest resources (368 million tons). Europe produces 120 million tons/year, with wheat straw as the major residue and smaller amounts of residues from barley, maize, sunflower, rapeseed, cotton, olive trees and vines (Philippoussis 2009). In Brazil, the amount of lignocellulosic wastes generated annually is approximately 350 million tons (Pereira Jr. et al. 2008).

Africa and Australia characterized by their desert areas are therefore, less favorable to agricultural production. The Australian production of agro-industrial biomass reaches 100 million tons/year and in Africa it is estimated to be about 40 million tons per year of lignocellulosic biomass (Philippoussis 2009).

As the chemical composition of these residues is based on sugars and other compounds of interest, significant efforts have been made to convert these enormous amounts of agricultural and agro-industrial lignocellulosic wastes into value added products, such as biofuels, chemicals, animal feed, enzymes and others. It is clear that, besides the environmental pollution caused by their accumulation in the environment, the non-use of these materials constitutes a loss of potentially valuable materials.

In the mid to long-term scenarios, 2020–2050, one of the most promising and important applications of lignocellulosic biomass, especially from sugarcane bagasse, is ethanol production (Lin and Tanaka 2006, Hamelinck and Faaij 2006). These days, one of the main advantages of using lignocellulosic biomass is their easy availability and the fact that its application is not directly related to food production. Therefore, bioethanol can be produced without the need of using vast extensions of fertile cultivable land for cropping cane or corn exclusively dedicated to bioenergy production.

It is well documented that cellulosic ethanol offers important environmental benefits and sustainability. Nevertheless, the main limiting factor for its production is the high degree of complexity inherent to the processing of these lignocellulosic biomass, and its long-term economic feasibility will depend on major technological innovations centered on feedstocks, effective pretreatment processes and micro-organisms, integration of processes, scale economies, fermentation of hexoses and pentoses, and mainly production of effective and low cost enzymes, such as cellulases and hemicellulases.

In order to solve this problem, it is necessary to obtain low cost and high efficiency cellulase and xylanase enzymes for the hydrolysis of cellulose and hemicellulose in biomass. Once these technological limitations are overcome, lignocellulosic biomass will be the main feedstock for ethanol production.

Composition of Lignocellulosic Residues

Lignocellulose is a renewable organic material and is the major structural component of all plants. Its main component is cellulose that represents about 40–50 percent of the cell wall material of plants, followed by hemicellulose (25–30 percent) and lignin (15–20 percent) that are closely associated with each other constituting the cellular complex of the vegetal biomass (Menon and Rao 2012). Cellulose and hemicellulose are macromolecules constructed from different sugars, whereas lignin is an aromatic compound synthesized from phenylpropanoid precursors. In addition, small amounts of other materials such as pectin, protein, extractives (soluble nonstructural materials, such as nonstructural sugars, nitrogenous material, chlorophyll, and waxes) and ash can be found in lignocellulosic residues, in different degrees according to the residue considered (Jorgensen et al. 2007). Table 2.1 presents the composition of some lignocellulosic residues.

Cellulose, the main constituent of plant cell walls and the most abundant organic molecule on Earth, is a high molecular weight linear homopolymer composed of repeted D-glucose units linked by β-1,4 glycosidic bonds forming the dimer cellobiose. The long-chain cellulose polymers are linked together by hydrogen bonds and van der Waals forces, which cause the cellulose to be packed into microfibrils. Cellulose microfibrils present a crystalline region (around 2/3rds of the total cellulose) and a small amount of less ordered, noncrystalline structure referred to as amorphous region (Arantes and Saddler 2010). In the latter conformation, cellulose is more soluble and susceptible to enzymatic hydrolysis. However, the molecules of individual microfibrils in crystalline cellulose are packed so tightly that not only enzymes but even small molecules like water cannot enter the complex framework (Menon and Rao 2012). The high molecular weight and ordered tertiary structure make natural cellulose insoluble in water.

Hemicellulose, the second most abundant component of lignocellulosic biomass, is a linear and branched heterogeneous polymer, with a lower molecular weight than cellulose, typically made up of five different sugars

Table 2.1 Composition of some Lignocellulosic Residues.

Lignocellulosic Residues	Cellulose (%)	Hemicellulose (%)	Lignin (%)	Reference
Sugarcane bagasse	44.7	36.9	11	Song and Wei 2010
Wheat straw	31.5	25.2	15.9	Nabarlatz et al. 2007
Corn cobs	35.4	35.2	14.8	Michelin et al. 2012
Rice straw	36.2	19	9.9	Nigam et al. 2009
Rice husks	36.7	20	21.3	Parajó et al. 2004 Garrote et al. 2007
Barley straw	30.8	25.3	16.7	Nabarlatz et al. 2007
Barley husks	21.4	36.6	19.2	Parajó et al. 2004
Oat straw	39.4	27.1	17.5	Nigam et al. 2009
Rye straw	37.6	30.5	19	Nigam et al. 2009
Almond shells	26.8	32.5	27.4	Nabarlatz et al. 2007
Olive stones	24	27.8	31.3	Nabarlatz et al. 2007
Cotton stalks	58.5	14.4	21.5	Nigam et al. 2009
Soya stalks	34.5	24.8	19.8	Nigam et al. 2009
Sunflower stalks	42.1	29.7	13.4	Nigam et al. 2009

—D-xylose and L-arabinose as pentoses, and D-mannose, D-glucose and D-galactose as hexoses—as well as other components such as acetic, glucuronic and ferulic acids. The backbone of the chains of hemicelluloses can be a homopolymer or a heteropolymer with short branches linked by β-(1,4)-glycosidic bonds and occasionally β-(1,3)-glycosidic bonds (Kuhad et al. 1997, Kumar et al. 2009). The classification of these hemicellulose fractions depends on the types of sugar moieties present, e.g., xylans, mannans, glucans, glucuronoxylans, arabinoxylans, glucomannans, galactomannans, galactoglucomannans, β-glucans and xyloglucans. The composition of hemicelluloses is very variable in nature and depends on the plant source. The main differences between cellulose and hemicellulose are that hemicellulose has branches with short lateral chains consisting of different sugar units and has an amorphous structure (Fengel and Wegener 1989), which makes it easier to hydrolyze than cellulose. Hemicellulose has been proposed to act as a bonding agent between lignin and cellulose. Its covalent linkage to lignin and its non-covalent interaction with cellulose may be important in maintaining the integrity of the cellulose *in situ* and help to protect the fibers against degradation by cellulases (Uffen 1997).

Among hemicelluloses, xylan is the most abundant on the Earth surface, representing generally up to 30–35 percent of the total dry weight (Joseleau et al. 1992). Xylans are heteropolymers consisting principally of D-xylose as its monomeric unit and traces of L-arabinose. Most xylans occur as heteropolysaccharides, containing different substituent groups in the backbone chain such as acetyl, arabinosyl and glucuronysyl residues.

Lignin is an amorphous compound, non-water soluble and optically inactive that is formed from phenylpropane units joined in a large three-dimensional structure by non-hydrolyzable linkages. Three phenyl propionic alcohols exist as precursors of lignin: coniferyl alcohol (guaiacyl propanol), coumaryl alcohol (p-hydroxyphenyl propanol) and sinapyl alcohol (syringyl propanol). Lignin is synthesized by coupling of these components and their ratio varies among different plants, wood tissues and cell wall layers. This heterogeneous structure is linked by C–C and aryl-ether linkages, with aryl-glycerol β-aryl ether being the predominant structures. Lignin is closely linked to both hemicellulose and cellulose, and its role is to provide rigidity and cohesion to the cell wall (structural support), to confer impermeability and to form a physic–chemical barrier against microbial attack (Fengel and Wegener 1989, Kumar et al. 2009) and oxidative stress. Due to its molecular configuration, lignins are extremely resistant to chemical and enzymatic degradation. Not surprisingly, it is the most recalcitrant component of lignocellulosic material to degradation (Himmel et al. 2007, Sánchez 2009).

In general, the amount of carbohydrate polymers and lignin vary from species to species, and even from tree to tree (Fengel and Wegener 1989). In addition, the ratio between the various constituents in a single plant varies with age, stage of growth and other conditions (Perez et al. 2002).

Pretreatments

Glucose, as cellulose, is present in large quantities in agro-industrial residues. However, the cellulose in the plant cell wall is not readily available to enzymatic hydrolysis (cellulases) mainly due to the low accessibility of microcrystalline cellulose fibers and the presence of lignin (mainly) and hemicellulose on the surface of cellulose, which prevents cellulases to act efficiently. Thus, pretreatment of lignocellulosic residues before hydrolysis is a prerequisite and it can be performed by different methods, which can be classified into physical, physicochemical, chemical and/or biological (Dashtban et al. 2009).

The pretreatment methods consist of a series of operations that when applied to the lignocellulosic materials are capable of breaking the bonds that join the macrostructures (Fig. 2.1). This process involves the delignification of the feedstock in order to make cellulose more accessible during hydrolysis. It removes the lignin and hemicellulose components from cellulose, resulting in the reduction of the crystallinity of cellulose, increases the pore size of fibers of the lignocellulosic materials as well as makes cellulose more accessible to the cellulolytic enzymes, which in

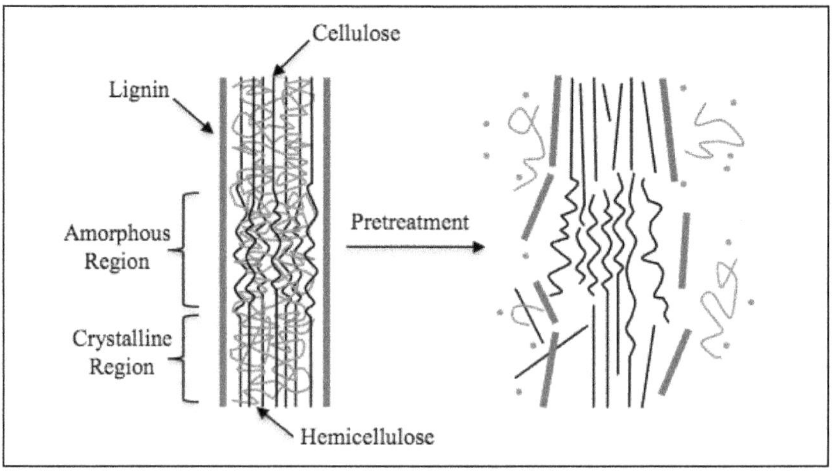

Figure 2.1 Scheme of pretreatment action on lignocellulosic residues (Adapted from Mosier et al. 2005).

Color image of this figure appears in the color plate section at the end of the book.

return reduces enzyme requirements and thus the process cost, e.g., for the production of ethanol or any other bioproduct (as enzyme production). Typical requirements of pretreatment include:

1. to improve the release of sugars or the capacity to subsequently form sugars by enzymatic hydrolysis;
2. to avoid/minimize the degradation of sugars;
3. to avoid/minimize the formation of compounds which are inhibitory to the subsequent hydrolysis and fermentation processes;
4. to develop a cost-effective technology.

Pretreatment of the lignocellulosic biomass is necessary because hydrolysis of non-pretreated materials is slow and results in low product yield. This topic has been an active field of research for several decades, and a wide variety of pretreatment approaches (and combinations thereof) have been investigated and reported in the scientific literature (Dashtban et al. 2009). Some promising technologies that have been developed for the cost-effective pretreatment of lignocellulosic biomass for biological conversion to fuels and chemicals will be described.

Physical Pretreatments

Most of the lignocellulosic biomass requires some mechanical processing for size reduction, such as milling or grinding, irradiation (using gamma rays, ultrasound, electron beam, microwave radiations) and extrusion to improve the enzymatic hydrolysis or biodegradation of lignocellulosic materials. These processes do not alter biomass composition (Menon and Rao 2012).

Mechanical Comminution

Comminution of lignocellulosic materials can be done through a combination of chipping, grinding and/or milling. This procedure reduces the particle size and cellulose cristallinity in order to increase the specific surface area of cellulosic biomass by breaking it down into smaller particles, and reducing the degree of polymerization, thus making the substrate more favorable to subsequent enzymatic hydrolysis. The size of the materials is usually 10–30 mm after chipping and 0.2–2.0 mm after milling or grinding (Sun and Cheng 2002, Kumar et al. 2009, Alvira et al. 2010).

Different milling processes, such as ball milling, hammer milling, compression milling, two-roll milling, colloid milling and vibro energy milling, agitation bead milling, pan milling (Taherzadeh and Karimi 2008, Zheng et al. 2009) can be used to improve the enzymatic hydrolysis of lignocelullosic materials. The power requirement of the mechanical pretreatment in lignocellulosic materials depends on the final particle size and of the biomass characteristics. However, currently, it is rarely used as a unique pretreatment method, because of the high energy requirements of milling and the continuous rise of energy prices (Hendriks and Zeeman 2009). Furthermore, mechanical comminution is much less effective than chemical pretreatments since it does not result in lignin removal, which has been proved to significantly restrict accessibility of cellulose and inhibit cellulases activity. In most cases, the required energy consumption is higher than the theoretical energy content available in the biomass.

Physicochemical Pretreatments

Physicochemical pretreatments combine both chemical and physical processes and consist in the submission of the material to high temperatures, catalyzed or not by chemical agents. It is characterized by the severe conditions of temperature used (Menon and Rao 2012). The most important processes of this group include: steam explosion, liquid hot water and ammonia fiber explosion.

Steam Explosion

Steam explosion is the most widely employed method for the pretreatment of lignocellulosic biomass. In this process, biomass is treated at high temperature (160–260°C) and with pressurized steam for a short period of time (from seconds to a few minutes), and then the system is rapidly depressurized, which makes the materials undergo an explosive decompression that acts on lignocellulose structure (Sun and Cheng 2002, Menon and Rao 2012).

The high temperature causes hemicellulose hydrolysis and solubilization. The hemicellulose is thought to be hydrolyzed by acetic acid (generated from hydrolysis of acetyl groups associated with the hemicellulose) and other acids released during this process (Kumar et al. 2009). Lignin is, to some extent, removed from the material during the pretreatment and is redistributed on the fiber surfaces as a result of melting and depolymerization/repolymerization reactions (Li et al. 2007). Depending on the severity of the pretreatment, some degradation of the cellulose to glucose can occur (Jorgensen et al. 2007).

Thus, steam provides an effective way to rapidly heat the material to the target temperature without excessive dilution of the resulting sugars. The rapid descompression reduces the temperature and stops the reaction at the end of the pretreatment. The rapid thermal expansion used to finish the reaction opens up the structure of the biomass; however the enhancement of the digestibility of the cellulose in the pretreated solid is only slightly correlated with this physical effect (Jorgensen et al. 2007, Kumar et al. 2009).

Steam explosion can be referred as "uncatalyzed steam explosion", when there is no addition of any chemical, and "catalyzed steam splosion", when some catalyst is added to the process.

Factors that affect the effectiveness of steam-explosion pretreatment are particle size, temperature, residence time and moisture content (Duff and Murray 1996). The particle size is the main factor contributing to the effectiveness of the process and it has been shown that relatively large particle sizes are able to yield maximum sugar concentrations.

Optimal hemicellulose hydrolysis and solubilization can be achieved by either high temperature and short residence time (270°C, 1 min) or lower temperature and longer residence time (190°C, 10 min) (Duff and Murray 1996).

Addition of an acid catalyst in the steam explosion process has been done in diluted quantities in order to improve hemicellulose hydrolysis during the pretreatment and cellulose digestibility in the further process. Sulfuric acid (H_2SO_4) is a strong catalyst that greatly improves the hemicellulose removal, but also easily yields inhibitory substances. Thus, limitations of steam explosion pretreatment include destruction of a portion of the xylan fraction, incomplete disruption of the lignin-carbohydrate matrix and generation of toxic compounds that may be inhibitory to micro-organisms growth used in downstream processes (Sun and Cheng 2002, Oliva et al. 2003). In relation to the use of acids, the main drawbacks are related to equipment requirements and higher formation of degradation compounds (Mosier et al. 2005, Alvira et al. 2010).

Liquid Hot Water

Liquid hot water (LHW) is another hydrothermal treatment that uses compressed liquid hot water to promote disintegration and separation of the lignocellulosic structure. This process is able to hydrolyze hemicellulose in minutes with high yield of hemicellulosic sugars, low inhibitors formation, and no significant lignin solubilization (Carvalheiro et al. 2008). It is similar to steam explosion, but in this case, rapid decompression does not occur. Besides, this method does not employ any catalyst or chemicals. Pressure is utilized to maintain water in the liquid state at elevated temperatures (160–240°C) and occurs in the range of time from a few minutes up to an hour (Yu et al. 2010, Brodeur et al. 2011).

The goal of the LHW is to completely solubilize the hemicellulose, and separate it from the rest of the solid material to make the cellulose more susceptible to enzymatic attack and to avoid or reduce the formation of inhibitors. With this pretreatment, two fractions are obtained: a solubilized fraction rich in hemicellulose derived sugars and one solid cellulose-enriched fraction. These fractions can be separated from each other by filtration (Pérez et al. 2008, Alvira et al. 2010, Brodeur et al. 2011). Lignin is partially broken and solubilized during hot water pretreatment but complete delignification is not possible using hot water alone, because of the recondensation of soluble components originating from lignin. The cellulose becomes more accessible by the pemeation and hydration of cell walls that cause swelling and disruption of the matrix. The physico-chemical alteration provoked by treatments on lignin and cellulose facilitates further separation and processing of these fractions.

LHW pretreatment has been reported to have the potential to enhance cellulose digestibility, sugar extraction, and pentose recovery, with the advantage of producing prehydrolyzates containing little or no inhibitor of sugar fermentation. LHW has been shown to remove up to 80 percent of the hemicellulose and to enhance the enzymatic digestibility of pretreated biomass, such as corn fiber (Allen et al. 2001), sugarcane bagasse (Laser et al. 2002, Mosier et al. 2005) and wheat straw (Pérez et al. 2008). The two-step pretreatment has been studied to optimize hemicellulosic sugars recovery and to enhance enzymatic hydrolysis yields.

LHW pretreatments are generally attractive from a cost-savings standpoint, since no catalyst is required and expensive reactor systems are not necessary due to the low corrosive nature of this pretreatment technique. Another advantage is that the concentration of the solubilized hemicellulose and lignin products is lower compared to steam pretreatment (Bobleter 1994), probably due to higher water input in LHW pretreatment, and subsequent reduction of the concentration of degradation products. This process also reduces the need for neutralization of liquid streams and conditioning chemicals since acid is not added, making it environment-friendly (Negro et al. 2003, Mosier et al. 2005, Kim et al. 2009). In comparison to steam explosion, higher pentosan recovery and lower formation of inhibitory products are obtained. Additionally, biomass size reduction is not needed because the particles are broken apart during pretreatment; however, the amount of water applied in the process and energetic requirement are higher.

The yield of solubilized (monomeric) xylan is generally higher for LHW pretreatment, though this result diminishes when the solid concentration increases, because (monomeric) xylan is then further degraded by hydrolytic reactions to, for example, xylose and furfural (Laser et al. 2002). LHW is also denominated as "autohydrolysis" process, in other words, during LHW pretreatment, the cleavage of *O*-acetyl and uronic acid substitutions from hemicellulose produces acetic acid and other organic acids which help catalyze the hydrolysis of polysaccharide such as hemicellulose into soluble oligosaccharides first, and then monomeric sugars. Under acidic conditions, these monomeric sugars are subsequently partially degraded to aldehydes such as furfural and HMF (5-hydroxymethyl-2-furaldehyde) that are inhibitors to fermenting micro-organisms. Additionally, hot water has an unusually high dielectric constant and enables the dissolution of almost all hemicellulose and a certain amount of lignin, depending on the temperature. Therefore, hot water, itself, plays a role like an acid in the hydrolysis of hemicellulose to release sugars and acids (Antal 1996).

To avoid the formation of inhibitors, the pH should be kept between 4 and 7 during the pretreatment because at this pH hemicellulosic sugars are retained in oligomeric form and monomers formation is minimized. Therefore the formation of degradation products is also lower (Mosier et al. 2005).

Temperature also plays an important role in the properties of the liquid fraction of this pretreatment since the amount of inhibitor formed is seen to mainly rise with an increase in temperature. For example, 180ºC and 30 min have been the most cost-effective pretreatment condition for rice straw conversion to glucose (Yu 2010). It has been seen that lower process temperatures during the LHW treatment will increase the solubilization of the lignin allowing for higher enzymatic hydrolysis of the solid fraction. Like many other pretreatments, the severity of the process will depend primarily on the type of lignocellulosic material that is being used and will have to be tailored as such (Brodeur et al. 2011).

Ammonia Fiber Explosion

Ammonia Fiber Explosion (AFEX) is another type of physicochemical pretreatment in which lignocellulosic materials are treated with liquid anhydrous ammonia under high pressure and moderate temperatures (60 to 100ºC) for a period of time, and then the pressure is rapidly released (steam explosion) (Alvira et al. 2010, Brodeur et al. 2011, Menon and Rao 2012). This rapid expansion of the ammonia gas causes swelling of the lignocellulosic biomass creating a disruption of the lignin-carbohydrate linkages and partial decrystallization of cellulose structure (Carvalheiro et al. 2008).

AFEX pretreatment does not significantly remove hemicellulose compared to acid pretreatment and acid-catalyzed steam explosion (Sun and Cheng 2002, Zheng et al. 2009). During the pretreatment only a small amount of the solid material is solubilized. The hemicellulose is degraded to oligomeric sugars and deacetylated (Gollapalli et al. 2002), which is most likely the reason that the hemicellulose is not soluble. This system does not directly liberate any sugars, but allows a higher accessible surface area of the polymers (hemicellulose and cellulose) to be attacked enzymatically.

Both cellulases and hemicellulases will be required in hydrolysis process due to the considerable remaining hemicellulose in the pretreated material (Galbe and Zacchi 2007).

However, while some other pretreatments such as steam explosion and LHW produce a slurry that can be separated in solid and liquid fractions, AFEX produces only a pretreated solid material (Alvira et al. 2010). Besides, the temperatures are significantly less than those of the steam explosion and LHW processes, meaning less energy input and less overall costs associated with the process.

Strategies to optimize the conditions in the AFEX pretreatment have been reported for different materials, being observed as more effective on agricultural residues and herbaceous crops with limited effectiveness on woody biomass and other feedstocks with high lignin content (Alvira et al. 2010).

Important advantages of AFEX include the production of negligible amounts of inhibitors for the downstream biological processes, due to the moderate conditions, making water wash not necessary and no requirement for particle size reduction (Sun and Cheng 2002, Zheng et al. 2009). A small number of disadvantages can be seen in the form of costs, since ammonia is expensive and requires recycling after the pretreatment to reduce the cost and protect the environment (Holtzapple et al. 1992).

Chemical Pretreatment

Chemical treatment is generally used to remove lignin from agro-industrial residues. This pretreatment, by alkali or acid hydrolysis, has been extensively used in the pulp and paper industries for delignification of cellulosic materials to produce high quality paper products (Dashtban et al. 2009). It is the most studied pretreatment method, but this process tends to be expensive and, therefore, is not used for bioconversion purposes. The need for corrosion-resistant equipments is the main disadvantage of chemical pretreatments for lignin removal (Nigam et al. 2009). The commonly used chemical pretreatments include mainly acid and alkali hydrolysis, organosolvents, ozonolysis and ionic liquids.

Acid Hydrolysis

Acid pretreatments normally aim for high yields of sugars from lignocellulosic biomass. This process involves the use of concentrated or diluted acids (depending on the purpose) to break the rigid structure of the lignocellulosic material (Brodeur et al. 2011, Menon and Rao 2012). Besides, different acids have been evaluated including sulfuric acid (Sun and Cheng

2005, Martin et al. 2007), hydrochloric acid (Kurakabe et al. 2005, Wang et al. 2010), peracetic acid (Zhao et al. 2007, 2008), nitric acid (Himmel et al. 1997) and phosphoric acid (Zhang et al. 2007).

In this process, hemicellulose (converted to sugars) and lignin are solubilized, making the cellulose more accessible for the action of the hydrolytic enzymes. Thus, an advantage of acid pretreatment is that this subsequent enzymatic hydrolysis is sometimes not required since the acid itself hydrolyzes the biomass to desired sugars. However, washing and/or detoxification steps are required to remove the acid before the fermentation process. Due to the corrosive nature and toxicity of most acids, the process also requires more expensive (resistant to corrosion) reactors. Another drawback of acid hydrolysis is the formation of fermentation inhibitors (sugar degradation products) like furfural and hydroxymethyl furfural (HMF) that reduce the effectiveness of the fermentation and downstream processes (Brodeur et al. 2011). Besides, the neutralization of the pH value is necessary for the subsequents steps of the process (Sun and Cheng 2002, Kumar et al. 2009).

Among all the acid pretreatment methods, dilute sulphuric acid (H_2SO_4) is the most widely used since it is inexpensive and effective (Zheng et al. 2009). This pretreatment can be performed at high temperature (e.g., 180°C) during a short period of time, or at lower temperature (e.g., 120°C) for longer retention time (30–90 min). In general, the use of high temperatures and short reaction times results in higher soluble xylose recovery yields and higher enzymatic cellulose digestibility (Balat et al. 2008).

Although dilute-acid pretreatment can significantly improve cellulose hydrolysis, its cost is generally higher than those of physicochemical processes, such as steam explosion or AFEX. However, the operating conditions to be used during acid pretreatment must be adapted in accordance with the objective to be reached and the composition of the biomass (residue). Dilute H_2SO_4 has traditionally been used to manufacture furfural from cellulosic materials by hydrolyzing the hemicellulose to simple sugars, such as xylose, that is further broken down to form furfural (Balat et al. 2008, Kumar et al. 2009, Brodeur et al. 2011).

Alkaline Hydrolysis

Alkaline pretreatment is basically a delignification process, its effect depending on the lignin content of the materials (McMillan 1994). The action mechanism of alkaline hydrolysis is believed to be saponification of intermolecular ester bonds crosslinking xylan hemicelluloses and other components, for example, lignin and other hemicellulose. The porosity of the lignocellulosic materials increases with the removal of the crosslinks (Sun and Cheng 2002, Zheng et al. 2009), increasing cellulose digestibility.

Alkaline hydrolysis is more effective for lignin solubilization, exhibiting minor cellulose and hemicellulose solubilization than acid or hydrothermal processes; in other words, lignin can be removed without much effect on other components (McMillan 1994, Carvalheiro et al. 2008).

The effectiveness of alkaline pretreatment depends on the substrate and process conditions. In general, alkaline pretreatment is more effective on hardwood, herbaceous crops, and agricultural residues with low lignin content than on softwood with high lignin content (Zheng et al. 2009).

The conditions of temperature and pressure in alkaline pretreatment are less severe than those applied in other pretreatment technologies (Mosier et al. 2005, Kumar et al. 2009, Brodeur et al. 2009, Menon and Rao 2012). It can be carried out at ambient conditions with longer pretreatment times required (hours or days rather than minutes or seconds when higher temperatures are used) (Balat et al. 2008, Brodeur et al. 2009). Neutralization of the pH value, and/or removal of the lignin and inhibitors (salts, phenolic acids, furfural and aldehydes) are required before enzymatic hydrolysis (Menon and Rao 2012).

Bases as sodium hydroxide (Carrillo et al. 2005, Silverstein et al. 2007), calcium hydroxide (Chang et al. 2001, Kim and Holtzapple 2005), potassium hydroxide (Chang and Holtzapple 2000), and ammonium hydroxide (Prior and Day 2008) can be used for the pretreatment of lignocellulosic materials, of which sodium hydroxide has been the one most studied (Kumar et al. 2009, Brodeur et al. 2011).

Dilute NaOH treatment causes swelling of lignocellulosic biomass, increase of the internal surface area, decrease of the degree of polymerization and cristallinity, separation of structural linkages between lignin and carbohydrates, and disruption of the lignin structure (Sun and Cheng 2002, Taherzadeh and Karimi 2008, Balat et al. 2008, Kumar et al. 2009) increasing the accessibility of enzymes to cellulose and hemicellulose (Brodeur et al. 2011). Another alkali that has been widely used for the pretreatment of biomass is calcium hydroxide (also denominated Lime pretreatment), mainly used to pretreatment of corn stover, switchgrass, bagasse, wheat, and rice straw (Sun and Cheng 2002, Liang et al. 2010, Park et al. 2010).

Compared with acid processes, alkaline hydrolysis cause less sugar degradation and many of the caustic salts can be recovered and/or regenerated (Kumar et al. 2009). However, conversion of some alkali into irrecoverable salts and/or the incorporation of salts into the biomass during the pretreatment reactions can occur (Balat et al. 2008, Zheng et al. 2009). A number of studies have combined alkaline pretreatment with other pretreatment methods, such as the steam explosion and ammonia fiber explosion (Brodeur et al. 2011).

Organosolvents

The organosolv pretreatment is a promising method that has attracted much attention for utilization in lignocellulosic residues. It can be an important contribution for the decrease of the environmental impact caused by conventional delignification processes (Ruzene et al. 2007). In this process, an organic or aqueous organic solvent mixture with inorganic acid catalysts (usually diluted HCl or H_2SO_4 solutions) is used to break the internal lignin and hemicellulose bonds (Sun and Cheng 2002, Kumar et al. 2009). Thus, the cellulose is partially hydrolyzed into smaller fragments that still remain insoluble in the liquor, while the hemicellulose is hydrolyzed mostly into soluble components, such as oligosaccharides, monosaccharides and acetic acid. Acetic acid lowers the liquor pH, stimulating acid-catalyzed hydrolysis of the other components. Some of the pentose sugars are subsequently dehydrated under the operating conditions to form furfural. Under the conditions employed in the process, lignin is mainly hydrolyzed into lower-molecular-weight fragments that dissolve in the aqueous ethanol (solvent) liquor (Kumar et al. 2009). Therefore, organosolv is a delignification process, with simultaneous hemicellulose solubilization. The presence of a catalyst can increase the solubilization of hemicelluloses allowing for a high recovey of xylose (Sun and Cheng 2002, Kumar et al. 2009, Alvira et al. 2010).

The usual operation temperature of the organosolv process is in the range of 150–200°C. However, at high temperatures (above 185°C), it is possible to obtain a satisfactory delignification without acid addition (Sun and Cheng 2002, Zheng et al. 2009, Alvira et al. 2010).

Numerous organic solvents are commonly utilized, including methanol, ethanol, acetone, ethylene glycol, and tetrahydrofurfuryl alcohol in order to solubilize lignin and provide treated cellulose suitable for enzymatic hydrolysis (Alvira et al. 2010).

The use of organsolvents for lignin removal is an attractive process because solvents can be recovered and recycled to reduce the operation costs (Nigam et al. 2009). Removal of solvents from the system is necessary because the solvents may be inhibitory to enzymatic hydrolysis and fermentation.

The main advantage of organosolv over other chemical pretreatments is the recovery of relatively pure lignin as a by-product. However, the process is still too complex and expensive to attract the attention of the industry. Organic solvents are expensive and their use requires high-pressure equipment due to their high volatility. For economic reasons, the use of low-molecular-weight alcohols such as ethanol and methanol has been favored (Zheng et al. 2009, Alvira et al. 2010).

Biological and Enzymatic Pretreatment

Biological pretreatment involves the use of wood degrading micro-organisms, including white-, brown-, soft-rot fungi to degrade lignin and hemicellulose in waste materials, while leaving the cellulose intact or with a small degradation. Considered the most effective basidiomycetes for biological pretreatment of lignocellulosic materials, the white-rot fungi have been used on different lignocellulosic biomass showing high delignification efficiency, which occurs through the action of the lignin degrading enzymes (ligninolytic biodegradation) secreted by the fungi: lignin peroxidase (LiP), manganese peroxidase (MnP) and laccase (Sánchez 2009).

White-rot fungi include *Phanerochaete chrysosporium*, *Trametes versicolor*, *Trametes hirsute*, *Ceriporia lacerata*, *Cyathus stercolerus*, *Ceriporiopsis subvermispora*, *Pycnoporus cinnarbarinus* and *Pleurotus ostreaus* (Alvira et al. 2010). Brown- and soft-rot fungi mainly attack cellulose while imparting minor modifications to lignin (Zheng et al. 2009, Menon and Rao 2012).

Some basidiomycetes such as *Phlebia radiata*, *Phlebia floridensis* and *Daedalea flavida*, selectively degrade lignin in wheat straw and are good choices for delignification of lignocellulosic residues (Arora and Chander 2002).

Although biological pretreatments offer some advantages such as low-capital cost (no expensive equipment, as other physical and thermochemical process), low energy requirements, no chemical requirement (safe and environment friendly) and mild environmental conditions (Sun and Chen 2002, Kumar et al. 2009, Alvira et al. 2010), important disadvantages also exist such as the low rates of hydrolysis and long pretreatment times required compared to other technologies (Sun and Cheng 2002). A period of two to five weeks may be required for sufficient delignification. The direct application of ligninolytic enzymes has also been investigated in order to reduce the treatment period, but the direct use of enzymes for delignification is expensive (Nigam et al. 2009).

Cellulase and Hemicellulase System

Cellulases and hemicellulases are usually a complex of several hydrolytic enzymes with diverse specificity and modes of action, that act synergistically to breakdown cellulose and hemicelluloses and produce oligosaccharides and sugar. This complex is necessary because of the heterogeneity and complex chemical nature of plant material.

Cellulolytic Complex

1. Endoglucanase (EG, endo-1,4-β-D-glucanohydrolase; EC 3.2.1.4) are also referred as carboxymethylcellulases (CMCase), because of the artificial substrate used to measure the enzyme activity. This enzyme is responsible for starting cellulose breakdown by randomly attacking the interior of the amorphous regions of the cellulose fiber, releasing oligosaccharides of varying degrees of polymerization (DP) and consequently, terminals with free reducing and non-reducing ends, which makes it more accessible for cellobiohydrolases action (Dashtban et al. 2009, Sánchez et al. 2009). The EG is the cellulolytic enzyme responsible for the rapid solubilization of cellulosic polymer (DP reduction) due to their fragmentation in oligossacharides. This has been shown by the effect of the enzyme on carboxymethylcellulose and amorphous cellulose (Saha 2000).
2. Cellobiohydrolase (CBH, 1,4-β-D-glucan cellobiohydrolase; EC 3.2.1.91) often referred as exoglucanase, degrades preferentially β-1,4-glycosidic bonds from the non-reducing end of the cellulose chain, releasing cellobiose as the main product (Dashtban et al. 2009, Sánchez et al. 2009). Although IUBMB (International Union of Biochemistry and Molecular Biology) defines the CBH as a catalyst of the hydrolysis of only the non-reducing terminal of the cellulosic fiber and oligosaccharides with DP> 3 in cellobiose, there are reports of the enzyme to attack the reducing end (Lynd et al. 2002). CBH I and II from *T. reesei* act on reducing and non-reducing ends of cellulose, respectively (Zhang and Lynd 2004). Microcrystalline cellulose (Avicel) has been used as a substrate to measure its activity. These enzymes often suffer inhibition by its hydrolysis product (cellobiose).
3. β-glucosidase (BG, β-D-glucoside glucohydolase or cellobiase; EC 3.2.1.21) hydrolyzes cellobiose and cellodextrins in glucose, and are competitively inhibited by glucose (Dashtban et al. 2009).

Figure 2.2 represents the action mode of the cellulolytic complex.

Xylanolytic Complex

1. Endo-1,4-β-xylanase (1,4-β-D-xylan xylanohydrolase; EC 3.2.1.8) cleaves the glycosidic bonds in the xylan backbone, causing a reduction in its degree of polymerization. Xylan is not hydrolyzed randomly, the bonds to be hydrolyzed depend on the nature of the substrate molecule, i.e. on the chain length, the degree of branching and the presence of substituents. Initially, the main hydrolysis products are

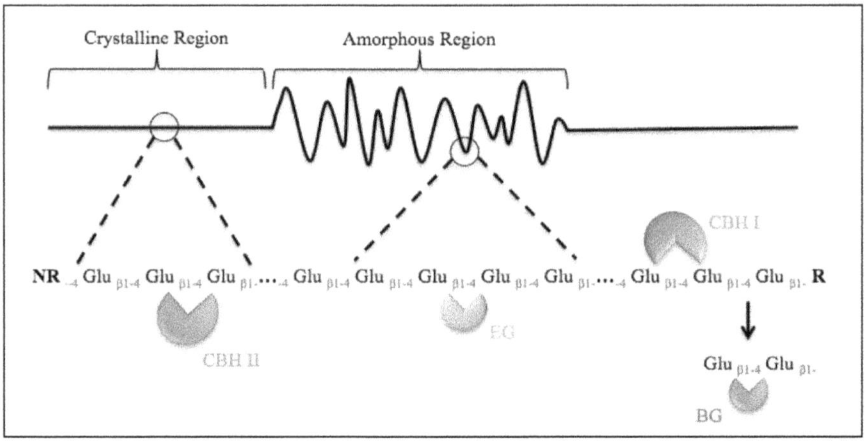

Figure 2.2 Schematic representation of cellulose structure and the action points of cellulolytic enzymes. CBH I: cellobiohydrolase I; CBH II: cellobiohydrolase II; EG: endoglucanase; BG: β-glucosidase; NR: non-reducing end; R: reducing end; Glu: D-glucose.

Color image of this figure appears in the color plate section at the end of the book.

xylooligomers but, at a later stage, smaller molecules may be produced (Polizeli et al. 2005). These enzymes have been classified by Wong et al. (1988) in non-debranching enzymes, which do not hydrolyze at the 1,3-α-L-arabinofuranosyl branch-points of arabinoxylans, and thus do not liberate arabinose, and debranching enzymes, which do hydrolyze these side-branches, liberating arabinose.

2. Exo-1,4-β-D-xylosidase (1,4-β-D-xylan xylohydrolase; EC 3.2.1.37) hydrolyzes xylobiose and short chain xylooligosaccharides generated by the action of endoxylanases, releasing D-xylose residues from the non-reducing end. Usually β-xylosidases do not hydrolyze xylan; their best substrate is xylobiose and their affinity for xylooligosaccharides decreases with the increasing degree of polymerization. *p*-Nitrophenyl-β-D-xylopyranoside is usually used as a substrate for the detection of β-xylosidase activity. As regards the location of β-xylosidases, they appear to be mainly cell-associated, though many extracellular β-xylosidases have also been reported (Polizeli et al. 2005, Javier Pastor et al. 2007).
3. α-L-Arabinofuranosidases (EC 3.2.1.55) hydrolyze the terminal arabinose residues from the side chains of xylan and other arabinose-containing polysaccharides (Saha 2000). There are two types of α-L-arabinofuranosidases with distinct modes of action: exo-α-L-arabinofuranosidase (EC 3.2.1.55) which degrades *p*-nitrophenyl-α-L-arabinofuranosides and branched arabinans, and endo-1,5-α-L-arabinase (EC 3.2.1.99) which only hydrolyzes linear arabinans (Polizeli et al. 2005). Most arabinases investigated so far are the exo type.

4. α-D-glucuronidases (EC 3.2.1.-) hydrolyzes the α-1,2 linkages between the 4-*O*-methylglucuronic/D-glucuronic acid and xylose residues in glucuronoxylan. The hydrolysis of the stable α-(1,2)-linkage is the bottleneck in the enzymatic hydrolysis of xylan. Similar to lignin carbohydrate linkage, 4-*O*-methylglucuronic acid linkage forms a barrier in wood degradation (Subramaniyan and Prema 2002). It has also been noted that acetyl groups close to the glucuronosyl substituents can partially hinder the α-glucuronidase activity (Polizeli et al. 2005). Despite their role in the biodegradation of xylan, some show activity only on short glucuronoxylan, while others can release glucuronic acid from polymeric xylan (Javier Pastor et al. 2007).
5. Acetyl xylan esterase (EC 3.1.1.6) removes the *O*-acetyl groups from acetyled xylan. This enzyme plays an important role in the hydrolysis of xylan, since the acetyl groups can interfere in the action of enzymes that cleave the xylan backbone, and so their removal facilitates the action of xylanases (Polizeli et al. 2005).
6. Ferulic acid esterase (EC 3.1.1.-) and *p*-coumaric acid esterase (EC 3.1.1.-) cleave ester bonds between arabinose side chains and ferulic acid or *p*-coumaric acid residues, respectively (Polizeli et al. 2005, Javier Pastor et al. 2007). Cleavage of feruloyl and *p*-coumaroyl groups from the xylan is also helpful in the removal of lignin. They may contribute to lignin solubilization by cleaving the ester linkages between lignin and hemicelluloses (Subramaniyan and Prema 2002).

Figure 2.3 represents the action mode of xylanolytic complex.

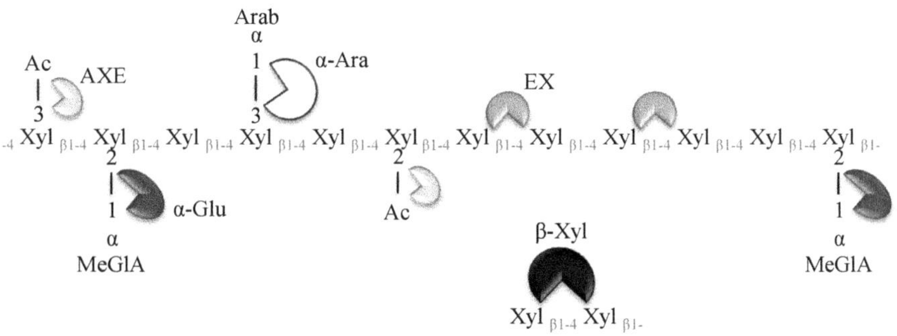

Figure 2.3 Schematic representation of xylan structure and the action points of xylanolytic enzymes. Ac: acetyl group; Arab: L-arabinofuranoside; MeG1A: 4-*O*-methyl-glucuronic acid; Xyl: D-xylose; AXE: acetyl xylan esterase; α-Glu: α-glucuronidase; α-Ara: α-arabinofuranosidase; EX: endoxylanases; β-Xyl: β-xylosidase. (Adapted from Biely 1985).

Color image of this figure appears in the color plate section at the end of the book.

Fungi Producing Cellulases and Xylanases

Several micro-organisms including fungi and bacteria have been reported to degrade cellulose and hemicellulose, especially xylans, by synthesizing enzymes of the cellulolytic and xylanolytic complex. Among microbial sources, filamentous fungi are especially interesting since they secrete these enzymes into the medium, with levels superior to those found in yeasts and bacteria (Polizeli et al. 2005). These micro-organisms can degrade lignocellulosic residues better than other microbes as they closely resemble their natural habitat.

Filamentous fungi are well known as a cost effective resource for industrial cellulases and xylanases. However, in relation to cellulase production, relatively few fungi secrete large amounts of proteins capable of solubilizing crystalline cellulose extensively (Bhat and Bhat 1997). One of the most extensively studied mesophilic fungi is *Trichoderma reesei,* which is capable of hydrolyzing native cellulose. Inspite of being an excellent producer of extracellular cellulases, it may not be the most effective cellulase system for use in biomass conversion processes that require the complete hydrolysis of the feedstock. *Trichoderma* and other cellulase-producing microbes produce very little amounts of β-glucosidase compared to other cellulolytic enzymes. Fungi such as *Humicola, Aspergillus* and *Penicillium* are also actively involved in cellulase production and have the ability to hydrolyze native cellulose (Singhania 2009).

Although *T. reesei, Aspergillus, Penicillium,* and *Humicola* can hydrolyze native cellulose, the reaction may be very slow due to the recalcitrant nature of biomass. Since cellulose very rarely occurs in a pure state in nature, sometimes it is necessary to remove lignin from cellulose with a suitable pretreatment method to make cellulose accessible to the action of the enzymes produced by the micro-organisms.

Agro-industrial residues such as wheat straw, corncobs and sugarcane bagasse (pretreated or not) have been widely included in the nutrient media composition as the main strategy in microbial cellulase and xylanase biosynthesis (Table 2.2), due to their high cellulose and xylan contents, wide availability and low cost.

In recent years, a lot of effort has been put into the isolation of thermophilic micro-organisms, such as *Sporotrichum thermophile, Thermoascus aurantiacus, Chaetomium thermophile* and *Humicola insolens*. They are of particular interest because of their ability to produce thermostable enzymes that are generally stable under a variety of severe conditions including highly acidic and alkaline pH values as well as elevated temperatures. These micro-organisms are also capable of fermenting a wide range of substrates with minimal risk of contamination by other pathogens. Because of these

Table 2.2 Fungi-producing cellulases and xylanases and the substrates used for their production.

Fungal strain	Enzymes	Substrate	References
Trichoderma reesei QM9414	Cellulases (FPase, avicelase, CMCase)	rice bran	Rocky-Salimi and Hamidi-Esfahani 2010
Trichoderma sp.	Cellulase (CMCase), xylanase	corn fiber pulp	Vlaev et al. 1997
Trichoderma spp., *Phanerochaete chrysosporium*	Cellulases (FPase, CMCase)	rice straw	Khan et al. 2007
Aspergillus niger	Xylanases	rice straw	Park et al. 2002
Aspergillus niger B03	Xylanase	corn cobs, wheat bran	Dobrev et al. 2007
A. niger KK2	Cellulases (CMCase, FPase, β-glucosidase), xylanases (endoxylanase, β-xilosidase)	rice straw	Kang et al. 2004
T. reesei RUT C30, *A. niger* MTCC 7956	Cellulases (FPase, β-glucosidase)	wheat bran	Sukumaran et al. 2009
Penicillium echinulatum	Cellulases (CMCase, FPase, β-glucosidase), xylanase	sugarcane bagasse treated with white-rot fungus or not	Camassola and Dillon 2009
Penicillium pinophilum	Cellulases (CMCase, FPase, β-glucosidase)	steam exploded sugarcane bagasse	Singh et al. 2009
P. janthinellum NCIM 1171, *T. viride* NCIM 1051	Cellulases (CMCase, FPase, β-glucosidase), xylanase	sugarcane bagasse pretreated with chemicals	Adsul et al. 2004

FPase: cellulase that hydrolyzes Filter Paper (used to quantify total cellulase); CMCase: cellulase that hydrolyzes Carboxymethyl Cellulose

advantages, thermophilic cellulolytic micro-organisms have attracted considerable research interest in recent years (Bhat and Bhat 1997).

In relation to xylanases, these are synthesized by mesophiles and thermophiles. Among the mesophilic fungi, the genera *Aspergillus* and *Trichoderma* are pre-eminent in xylanase production. Noted thermophilic fungi include *Chaetomium thermophile*, *Humicola insolens*, *Humicola lanuginosa*, *Humicola grisea*, *Melanocarpus albomyces*, *Paecylomyces variotii*, *Talaromyces byssochlamydoides*, *Talaromyces emersonii*, *Thermomyces lanuginosus* and *Thermoascus aurantiacus* (Polizeli et al. 2005).

Therefore, the hydrolysis of these lignocellulosic materials by cellulose- and hemicellulose-hydrolyzing microbial strains has received an increasing interest in research. The incorporation of cheap carbon sources into cultivation media for the production of lignocellulolytic enzymes should help decrease the production costs of enzymatic complexes.

Industrial Applications

Microbial cellulases and hemicellulases have a wide range of potential applications in several sectors of biotechnology such as food and beer, animal feed, textile, fuels, and pulp and paper industries, due to their enormous potential to modify and transform the lignocellulosic biomass. Currently, xylanases and cellulases, together with pectinases, account for 20 percent of the world enzyme market.

Food Processing Industry

Cellulases and xylanases have a wide range of potential applications in food biotechnology. In fruit and vegetable juices, these enzymes have an important application on the extraction and clarification process, as part of the complex called macerating enzymes (cellulases, xylanases, and pectinases) (Bhat 2000, Kuhad et al. 2011). The main advantages of the macerating enzymes involve the increase of juice yield and improvement of the extraction of valuable fruit components; reduction of the processing time and of the viscosity of fruit juice prior to concentration; hydrolysis of substances that may cause cloudiness in the concentrate; and increase of the filtration rate and stability of the final product. The use of macerating enzymes increases both—yield and process performance, without additional capital investment (Bhat 2000).

These macerating enzymes are also used in the production and preservation of fruit nectars and purees, especially from fruits which are easily perishable. They not only improve the stability and texture of nectars and purees, but also decrease their viscosity rapidly (Grassin and Fauquembergue 1996). Furthermore, infusion of pectinases and β-glucosidases has also shown to alter the texture, flavor and other sensory properties such as aroma and volatile characteristics of fruits and vegetables (Kuhad et al. 2011). Enzyme mixtures containing pectinases, cellulases and hemicellulases are also used for improved extraction of olive oil, their main advantages are an increased oil yield, obtention of oil with high levels of anti-oxidants and vitamin E and reduction of the induction of rancidity (Galante et al. 1998a).

Xylanases, like other hemicellulases may be employed in industrial baking, together with amylases and proteases. They break down the hemicelluloses (arabinoxylan) in wheat-flour, facilitating the redistribution of water in dough and favoring dough handling, bread volume, texture and stability. Besides, the addition of xylanases during dough processing increases the concentration of arabino xylo-oligosaccharides in bread, with beneficiary effects on human health (Bhat 2000, Polizeli et al. 2005).

Wine and Brewery Industry

Microbial β-glucanases and related polysaccharidases play an important role in fermentation processes for the production of alcoholic beverages including beers and wines and are known to improve the quality, yield and overall production efficiency. In wine production, enzymes such as glucanases, hemicellulases and pectinases play an important role by improving color extraction and skin maceration, facilitating wort clarification and filtration rate, and finally improving wine quality and stability (Galante et al. 1998a). β-glucosidase has also attracted considerable attention in the wine industry because of its ability to improve the aroma of wines by modifying naturally present glycosylated precursors (Bhat 2000). Beer brewing is based on the action of enzymes activated during malting and fermentation. Malting of barley depends of the action of several endogenous enzymes which act synergistically during the seed germination process to produce high quality malt. Nevertheless, many breweries end up using unmalted or poor quality barley, due to seasonal variations, different cultivars or poor harvest, which contains low levels of endogenous β-glucanase activity. The problem associated with the use of this barley and other cereals in combination with malt is the presence of 6–10 percent non-starch polysaccharide (NSP), mainly a soluble β-glucan. This forms gels during the brewing process and leads to poor filtration of the wort, slow run-off times, low extract yields and/or the development of haze in the final product. To overcome these problems, microbial β-glucanases are added either during mashing or primary fermentation to hydrolyze β-glucan (decrease in β-glucan content) and reduce the viscosity of the wort, improving its filterability (Bhat 2000, Kuhad et al. 2011). Arabinoxylans are also present in cereals like barley, resulting in a highly viscous aqueous solution and problems in brewing, as decreased rate of filtration or haze formation in beer. A better solution for this problem could be the application of xylanases for pretreating the arabinoxylan containing substrates (Subramaniyan and Prema 2002).

Animal Feed Industry

The animal feed industry is an important agribusiness sector with an annual production superior to 600 million tonnes of feed corresponding to more than 50 billion US$ (Bhat 2000, Polizeli et al. 2005). The major share of the feed produced is taken by poultry, pigs and ruminants (up to 90 percent), while pet foods and fish farming account for 10 percent. Cellulases and hemicellulases have received much attention in the animal feed industry because of their potential to improve feed value and performance of animals (Bhat 2000, Kuhad et al. 2011).

Pretreatment of agricultural silage and grain feed by cellulases or xylanases can improve its nutritional value (mainly low-quality feedstuffs). The enzymes can also eliminate antinutritional factors (ANF) present in the feed grains, degrade certain feed constituents in order to improve its nutritional properties, and provide supplementary digestive enzymes (e.g., proteases, amylases and glucanases). Dietary fiber consists of non-starch polysaccharides (NSP) such as arabinoxylans, cellulose, and many other plant components that can act as ANF for several animals such as swine and poultry. In this case, the cellulases effectively hydrolyze the cellulose anti-nutritional factor in the feed materials, improving the animal health and performance (Kuhad et al. 2011).

β-Glucanases and xylanases have been successfully used in the feed of monogastric animals to hydrolyze NSP such as barley β-glucans and arabinoxylans. They reduce intestinal viscosity of fiber rye- and barley-based feeds in poultry and pigs, causing weight gain in broiler chickens, egg laying hens and piglets by degrading NSP and improving digestion and absorption of feed components (Bhat 2000, Kuhad et al. 2011).

The forage diet of ruminants, which contains cellulose, hemicellulose, pectin and lignin, is more complex than the cereal-based diet of poultry and pigs. Enzyme preparations containing high levels of cellulase, hemicellulase and pectinase have been used to improve the digestibility and nutritive value of forages (Bhat 2000). Some studies have shown substantial improvements in feed digestibility and weight gain of cattle, by addition of commercial cellulase and xylanase (Beauchemin et al. 1995). Nevertheless, others reported no significant results (Lewis et al. 1996). Thus, the success in applying these enzymes on improving the fiber digestion in the ruminant may be limited.

Textile and Laundry Industry

Most of the natural materials used in fabric manufacturing contain cellulosic fibers, such as cotton, linen, ramie, viscose and lyocell. Cellulases have achieved worldwide success in textile and laundry because of their ability to modify cellulosic textile fibers in a controlled and desired manner, so as to improve the quality and appearance of fabrics. Bio-stoning and bio-polishing are the best-known current textile applications of cellulases on cotton and other cellulosic fabrics (Bhat 2000, Karmakar and Ray 2011).

Traditional washing of jeans (stone-washing) to produce a faded effect involves treatment (abrasion) with pumice stone for partial removal of the dye, which results in a faded, worn and old look. However, this technique caused several problems and the inclusion of microbial cellulases provided a perfect alternative for stone-washing, later known as "bio-stoning". During the bio-stoning process, cellulases act on the cotton fabric and break off the small fiber ends on the yarn surface, thereby loosening the dye (indigo), which is easily removed by mechanical abrasion in the wash cycle. The main advantages in the replacement of pumice stones by a cellulase-based treatment include less damage of fibers (better quality garments), increased productivity of the machines, less work-intensive and safer working conditions and safe environment (no pumice powder production). However, a major drawback in bio-stoning is the strong tendency of the released dye to redeposit on the garments (back-staining), masking the blue/white contrast of the finished product. This problem can be resolved by the application of both cellulase and some protease (e.g., subtilisin) (Galante et al. 1998b, Bhat 2000). The protease is believed to prevent the cellulase from binding the dye back to the surface of the denim, not affecting the abraded look caused by the action of the cellulase.

Bio-polishing is currently a key step in the textile industry for producing high quality garments, and cellulases are applied in this process to reduce the tendency for "fuzz" (short fibers protruding from the surface) and "pill" formation (fluffy/loosened fuzz attached to the surface) (Sreenath et al. 1996) and increase the commercial value of cellulosic fabrics. During this process, the cellulases act on small fiber ends that protrude from the fabric surface, removing these fibers and polishing the fabrics. Therefore, their main advantages include the removal of short fibers and surface fuzziness, creation of a smooth and glossy appearance, improvement in color brightness and uniformity, high hydrophilicity and moisture absorbance and being an environmentally friendly process (Bhat 2000). Cellulase preparations rich in endoglucanases are best suited for bio-polishing enhancing the fabric's look, feel and color without needing any chemical coating of fibers (Galante et al. 1998b).

Use of cellulases along with protease and lipase in detergents is a recent innovation of this sector (Kuhad et al. 2011). Cellulases are added to laundry detergents to enhance the detergent performance and improve the appearance and color brightness, besides improving the feel on the hand. The degradation of microfibrils by cellulase also softens the garment and removes dirt particles trapped within the microfibril network (Bhat 2000). These days, liquid laundry detergent containing anionic or nonionic surfactant, citric acid or a water-soluble salt, protease, cellulase, and a mixture of propanediol and boric acid or its derivative have been used to improve the stability of cellulases (Kuhad et al. 2011).

Pulp and Paper Industry

Cellulases and hemicellulases have been used in the pulp and paper industry for different purposes. Commercial enzyme preparations contain various enzyme activities, some important for a given process, while others may be harmful for a specific application. Therefore, enzyme mixtures or purified enzymes should be well characterized with respect to their substrate specificity and mode of action before being used for a particular application in the pulp and paper industry.

In this sector, the main purpose is the application of xylanases in the bleaching of Kraft pulp. These enzymes are gaining importance as alternatives to chlorine-containing chemicals that are toxic and pollutants to the environment (Viikari et al. 1994, Bajpai 1999, Ruzene and Gonçalves 2003).

The Kraft process involves the cooking of wood chips, in a solution of $Na_2S/NaOH$ under 8 kgf/cm^2 and at 165°C, accelerating the degradation and solubilization of lignin with recovery of the cellulose fibers. It can be assumed that 90–95 percent of the hemicellulose and lignin are dissolved and partially degraded during this process, resulting in a brown pulp which is mainly due to the presence of residual lignin and lignin derivatives (Polizeli et al. 2005). The residual lignin is difficult to remove due to its covalent binding to the hemicellulose and perhaps to cellulose fibers. However, to obtain pulp of very high brightness and brightness stability, all the lignin must be removed. For this, the pulp is submitted to a bleaching process usually using chlorine (Subramaniyan and Prema 2002).

Hydrolysis of hemicellulose in pulps by hemicellulases (mainly xylanases) can modify pulp-refining properties, increasing the extractability of lignin from the Kraft pulps and reducing the chlorine required in subsequent bleaching. Although the exact mechanism of action of xylanase in bio-bleaching is not fully understood, it has been proposed that the xylanase either hydrolyzes the xylan precipitated on the surface of the fiber after Kraft cooking, apparently rendering the struture of the fiber more permeable and allowing a more efficient extraction of lignin by bleaching chemicals used in later treatment stages (Viikari et al. 1994, Bhat 2000, Polizeli et al. 2005, Javier Pastor et al. 2007) or removes the xylan from the lignin-carbohydrate complexes (Buchert et al. 1992).

In the Kraft process, polluting reagents are used in large amounts. The use of chlorine results in organochlorines by-products from the degradation products of lignin, which are highly toxic, mutagenic and bioaccumulate requiring treatment of the effluents from the pulp and paper industries. In response to environmental regulations, paper industries are currently changing practices to minimize the use of chlorine compounds in bleaching processes, especially in Western Europe and North America (Beg et al. 2001, Polizeli et al. 2005).

Xylanase enzymes can reduce the requirement of chlorine-based chemicals by up to 30 percent and a reduction of 15–20 percent in the generation of organochlorines in the effluents. The utilization of xylanases could lead to the replacement of 5–7 kg of chlorine dioxide per ton of Kraft pulp and an average fall of 2–4 units in the kappa number, a measure of the lignin content in the cellulose pulp (Polizeli et al. 2005).

The main enzyme needed to enhance the delignification of Kraft pulp is endoxylanase, but enrichment with other enzymes, such as mannanase has been shown to improve the effect of enzymatic treatment of Kraft pulp (Beg et al. 2001). Mannanase facilitates bleaching, eliminating residual lignin and increasing paper brightness, though the effect of mannanases is usually less pronounced than that of xylanases (Bhat 2000, Javier Pastor 2007). The xylanases from *A. terricola* and *A. ochraceus* have been reported to act on defibrillation of the cellulose pulp, resulting in 14.3 and 36.4 percent delignification, respectivelly, and brightness 2.4–3.4 percent ISO higher than the control (Michelin et al. 2010).

Bioethanol Industry

Bioethanol production involves several steps starting from the selection of proper feedstock, its pretreatment, cellulase production, hydrolysis of feedstock using cellulases and finally fermentation of hydrolysate to obtain ethanol. Depending on the type and the conditions employed during the pretreatment, a portion of hemicelluloses can remain in the pretreated material and harm enzymatic hydrolysis of cellulose. In this case, efficient hydrolysis of the pretreated material requires, in addition to cellulase, the use of hemicellulase (Jorgensen et al. 2007).

Enzymatic saccharification of lignocellulosic materials by cellulases and hemicellulases for biofuel production is perhaps the most efficient method of biomass hydrolysis and currently the most investigated application. There is an increased interest in commercializing technologies for ethanol production from inexpensive biomass (Singhania 2009, Kuhad et al. 2011). According to the Brazilian scenario, sugarcane bagasse could be the most probable feedstock for long-term application.

However, the process of converting low value biomass to ethanol via fermentation depends on the development of economically viable cellulolytic enzymes to achieve effective depolymerization of the cellulosic content of the biomass. An enzyme cocktail can be prepared using cellulases from different sources to achieve maximum efficiency. Besides, potent cellulase genes from different filamentous fungi can be isolated, cloned and expressed in the host organism to get a better combination or synergism. Other powerful tools that can facilitate the development of more efficient cellulases include expression cassettes, site directed mutagenesis and antisense technology, which have been successfully employed in designing cellulases (Singhania 2009).

Strategies for recycling and reusage of the enzymes may also be used to reduce enzymatic hydrolysis costs (Sun and Cheng 2002, Mosier et al. 2005). However, there are several reports where the nonspecific and irreversible adsorption of cellulase to lignin has been observed, influencing enzyme recovery; although some studies report the use of lignin resistant cellulase, these recycling techniques have only been tested at laboratory scale (Kuhad et al. 2011).

The main advantage in the use of cellulases in relation to chemical methods, such as acid or alkaline hydrolysis, is that enzyme hydrolysis is usually conducted at mild conditions (pH 4–6 and temperature 45–50°C) and does not have corrosion issues (Gupta et al. 2011). Besides, chemical methods are inefficient and generate toxic by-products and effluents.

Other Applications

Cellulases in association with pectinases can be used in carotenoid extraction. These enzymes disrupt the cell wall of orange peel, sweet potato and carrot, and release carotenoids in the chloroplasts and in cell fluids (Kuhad et al. 2011). In addition, cellulases or mixture of glucanases (Bhat and Bhat 1997) and some xylanases (Wong et al. 1988) can be used for the production of plant and fungal protoplasts, to produce hybrid strains with desirable properties. Products of the hydrolysis of xylan, such as xylose residues, can be converted into solvents, furfural and xylitol, an artificial low-calorie sweetener with a sweetening power comparable to sucrose (Parajó et al. 1998) and with non-cariogenic properties (Polizeli et al. 2005, Javier Pastor 2007).

Perspectives and Future Applications

Lignocellulosic residues have become a substrate with great biotechnological value. They present potential application in biofuels, animal feed, enzymes and chemicals, but exploitation of this renewable resource requires either physical/chemical or biological pretreatment, due to its recalcitrant nature. Pretreatment breaks the lignocellulosic structure and leaves the cellulose and hemicellulose more available to be hydrolyzed by enzymes. However, appropriate pretreatment processes of biomass must be developed so that toxic compounds that reduce the hydrolysis efficiency and subsequent fermentation are not produced.

The use of the current commercial preparations of cellulases and xylanases for bioconversion of lignocellulosic biomass is economically not feasible. The high production cost of the enzymes and low yields are the main limiting factors for industrial application. Therefore, reduction in the production costs and improvement in the enzymatic efficiency are major goals for future research to industrial applications.

Recent research has focused on the improvement of the fermentation processes for microbial enzymes production, as well as in the development of hyper-producing microbial strains, in order to reduce the costs and make the process commercially viable. Among the various substrates, lignocellulosic biomass has been considered as a promising raw material because of its abundance, low cost and wide availability. Besides, an enzyme complex prepared from the same lignocellulosic material has proved to perform better.

Currently research is being directed towards the discovery of enzymes that are more robust with respect to temperature and pH. Techniques used include protein engineering by identification of active site residues through chemical modification, X-ray crystallography and site-directed mutagenesis.

One mix of mesophilic and thermophilic enzymes in a multienzymatic cocktail would ensure enzymatic hydrolysis of lignocellulosic substrates in a wide temperature and pH range. This would imply versatility in the hydrolysis processes, and consequently more efficient and lower costs production process, because the majority of the available commercial cocktails act mainly at 50–55°C and pH around 6.0. Mixed cultures have also been used with improved production and high efficiency enzyme complexes.

Investigation on microbial strains able to utilize inexpensive substrate has also been done. However, although cellulase and xylanase enzymes costs have decreased due to improvements in expression vectors and on-site production, there is still a necessity of engineering a new generation of enzymatic cocktails that would further reduce the cost of these enzymes.

References

Adsul, M.G., J.E. Ghule, R. Singh, H. Shaikh, K.B. Bastawde, D.V. Gokhale and A.J. Varma. 2004. Polysaccharides from bagasse: applications in cellulase and xylanase production. Carbohydr. Polymers 57: 67–72.

Allen, S.G., D. Schulman, J. Lichwa, M.J. Antal, M. Laser and L.R. Lynd. 2001. A comparison between hot liquid water and steam fractionation of corn fiber. Ind. Eng. Chem. Res. 40: 2934–2941.

Alvira, P., E. Tomás-Pejó, M. Ballesteros and M.J. Negro. 2010. Pretreatment technologies for an efficient bioethanol production process based on enzymatic hydrolysis: A review. Bioresour. Technol. 101: 4851–4861.

Antal, M.J. Water: A traditional solvent pregnant with new applications. pp. 24–32. *In*: H.J. White [ed.]. 1996. Proceedings of the 12nd International Conference on the Properties of Water and Steam. Begell House, New York.

Arantes, V. and J.N. Saddler. 2010. Acess to cellulose limits the efficiency of enzymatic hydrolysis: the role of amorphogenesis. Biotechnol. Biofuels. 3–4: 1–11.

Arora, D.S. and M.K.G.P. Chander. 2002. Involvement of lignin peroxidase, manganese peroxidase and laccase in degradation and selective ligninolysis of wheat straw. Int. Bioterior. Biodegrad. 50: 115–120.

Bajpai, P. 1999. Application of enzymes in the pulp and paper industry. Biotechnol. Prog. 15: 147–157.
Balat, M., H. Balat and C. Oz. 2008. Progress in bioethanol processing. Prog. Energy Comb. Sci. 34: 551–573.
Beauchemin, K.A., L.M. Rode and V.J.H. Sewalt. 1995. Fibrolytic enzymes increase fiber digestibility and growth rate of steers fed dry forages. Can. J. Animal Sci. 75: 641–644.
Beg, Q.K., M. Kapoor, L. Mahajan and G.S. Hoondal. 2001. Microbial xylanases and their industrial applications: a review. Appl. Microbiol. Biotechnol. 56: 326–338.
Bhat, M.K. 2000. Cellulases and related enzymes in biotechnology. Biotechnol. Adv. 18: 355–383.
Bhat, M.K. and S. Bhat. 1997. Cellulose degrading enzymes and their potential industrial applications. Biotechnol. Adv. 15(3-4): 583–620.
Biely, P. 1985. Microbial xylanolytic systems. Trends Biotechnol. 3: 286–290.
Bobleter, O. 1994. Hydrothermal degradation of polymers derived from plants. Prog. Polym. Sci. 19: 797–841.
Brodeur, G., E. Yau, K. Badal, J. Collier, K.B. Ramachandran and S. Ramakrishnan. 2011. Chemical and Physicochemical Pretreatment of Lignocellulosic Biomass: A Review. Enzyme Res. 2011: 1–17.
Buchert, J., M. Ranua, A. Kantelinen and L. Viikari. 1992. The role of two *Trichoderma reesei* xylanases in the bleaching of pine kraft pulp. Appl. Microbiol. Biotechnol. 37: 825–829.
Camassola, M. and A.J.P. Dillon. 2009. Biological pretreatment of sugar cane bagasse for the production of cellulases and xylanases by *Penicillium echinulatum*. Ind. Crops Prod. 2(9): 642–647.
Carrillo, F., M.J. Lis, X. Colom, M. López-Mesas and J. Valldeperas. 2005. Effect of alkali pretreatment on cellulase hydrolysis of wheat straw: Kinetic study. Process Biochem. 40: 3360–3364.
Carvalheiro, F., L.C. Duarte and F.M. Girio. 2008. Hemicellulose biorefineries: a review on biomass pretreatments. J. Scientific Ind. Res. 67: 849–864.
Chang, V.S. and M.T. Holtzapple. 2000. Fundamental factors affecting biomass enzymatic reactivity. Appl. Biochem. Biotechnol. 84–86(1-9): 5–37.
Chang, V.S., M. Nagwani, C.H. Kim and M.T. Holtzapple. 2001. Oxidative lime pretreatment of high-lignin biomass—poplar wood and newspaper. Appl. Biochem. Biotechnol. 94: 1–28.
Dashtban, M., H. Schraft and W. Qin. 2009. Fungal Bioconversion of Lignocellulosic Residues: Opportunities & Perspectives. Int. J. Biol. Sci. 5(6): 578–595.
Dobrev, G.T., I.G. Pishtiyski, V.S. Stanchev and R. Mircheva. 2007. Optimization of nutrient medium containing agricultural wastes for xylanase production by *Aspergillus niger* B03 using optimal composite experimental design. Bioresour. Technol. 98: 2671–2678.
Duff, S.J.B. and W.D. Murray. 1996. Bioconversion of forest products industry waste cellulosics to fuel ethanol: A review. Bioresour. Technol. 55: 1–33.
Fengel, D. and G. Wegener. 1989. Wood: Chemistry, ultrastructure, reactions. Walter de Gruyter, New York.
Galante, Y.M., A. De Conti and R. Monteverdi. Application of *Trichoderma* enzymes in food and feed industries. pp. 327–342. *In*: G.F. Harman and C.P. Kubicek [eds.]. 1998a. *Trichoderma & Gliocladium*—Enzymes, biological control and commercial applications, vol. 2. Taylor & Francis, London.
Galante, Y.M., A. De Conti, R. Monteverdi. Application of *Trichoderma* enzymes in textile industry. pp. 311–326. *In*: G.F. Harman and C.P. Kubicek [eds.]. 1998b. *Trichoderma & Gliocladium*—Enzymes, biological control and commercial applications, vol. 2. Taylor & Francis, London.
Galbe, M. and G. Zacchi. 2007. Pretreatment of lignocellulosic materials for efficient bioethanol production. Adv. Biochem. Eng. Biotechnol. 108: 41–65.
Garrote, G., E. Falqué, H. Domínguez and J.C. Parajó. 2007. Autohydrolysis of agricultural residues: Study of reaction byproducts. Bioresour. Technol. 98: 1951–1957.

Gollapalli, L.E., B.E. Dale and D.M. Rivers. 2002. Predicting digestibility of ammonia fiber explosion (AFEX)-treated rice straw. Appl. Biochem. Biotechnol. 98: 23–35.
Grassin, C. and P. Fauquembergue. Fruit juices. pp. 226–264. *In*: T. Godfrey and S. West [eds.]. 1996. Industrial Enzymology. 2nd edn. MacMillan Press, London, UK.
Gupta, R., Y.P. Khasa and R.C. Kuhad. 2011. Evaluation of pretreatment methods in improving the enzymatic saccharification of cellulosic materials. Carbohydr. Polymers 84: 1103–1109.
Hamelinck, C. and A.P.C. Faaij. 2006. Outlook for advanced biofuels. Energy Policy 34(17): 3268–3283.
Hendriks, A.T.W.M. and G. Zeeman. 2009. Pretreatments to enhance the digestibility of lignocellulosic biomass. Bioresour. Technol. 100: 10–18.
Himmel, M.E., W.S. Adney, J.O. Baker, R. Elander, J.D. McMillan, R.A. Nieves, J.J. Sheehan, S.R. Thomas, T.B. Vinzant and M. Zhang. 1997. Advanced bioethanol production technologies: a perspective. Fuels Chem. Biomass 666: 2–45.
Himmel, M.E., S.Y. Ding, D.K. Johnson, W.S. Adney, M.R. Nimlos, J.W. Brady and T.D. Foust. 2007. Biomass recalcitrance: engineering plants and enzymes for biofuels production. Science 315: 804–807.
Holtzapple, M.T., J.E. Lundeen, R. Sturgis, J.E. Lewis and B.E. Dale. 1992. Pretreatment of lignocellulosic municipal solid-waste by ammonia fiber explosion (AFEX). Appl. Biochem. Biotechnol. 34(5): 5–21.
Javier Pastor, F.I., O. Gallardo, J. Sanz-Aparicio and P. Díaz. Xylanases: Molecular Properties and Applications. pp. 65–82. *In*: J. Polaina and A.P. MacCabe [eds.]. 2007. Industrial Enzymes. Springer, Germany.
Jorgensen, H., J.B. Kristensen and C. Felby. 2007. Enzymatic conversion of lignocellulose into fermentable sugars: Challenges and opportunities. Biofuels, Bioprod. Bioref. 1: 119–134.
Joseleau, J.P., J. Comtat and K. Ruel. Chemical structure of xylans and their interactions in the plant cell walls. pp. 1–15. *In*: J. Visser, G. Beldman, M.A.K. vanSomeren, A.G.J. Voragen [eds.].1992. Xylans and xylanases. Elsevier, Amsterdam.
Kang, S.W., Y.S. Park, J.S. Lee, S.I. Hong and S.W. Kim. 2004. Production of cellulases and hemicellulases by Aspergillus niger KK2 from lignocellulosic biomass. Bioresour. Technol. 91: 153–156.
Karmakar, M. and R.R. Ray. 2011. Current trends in research and application of microbial cellulases. Res. J. Microbiol. 6(1): 41–53.
Khan, M.H., S. Ali, A. Fakhru'l-Razi and Z. Alam. 2007. Use of fungi for the bioconversion of rice straw into cellulase enzyme. J. Environ. Sci. Health B 42(4): 381–386.
Kim, S. and M.T. Holtzapple. 2005. Lime pretreatment and enzymatic hydrolysis of corn stover. Bioresour. Technol. 96: 1994–2006.
Kim, Y., N.S. Mosier and M.R. Ladisch. 2009. Enzymatic digestion of liquid hot water pretreated hybrid poplar. Biotechnol. Progr. 25: 340–348.
Kuhad, R.C., A. Singh, and K.E. Eriksson. 1997. Microorganisms and enzymes involved in the degradation of plant fiber cell walls. Adv. Biochem. Eng. Biotechnol. 57: 45–125.
Kuhad, R.C., R. Gupta and A. Singh. 2011. Microbial Cellulases and Their Industrial Applications. Enzyme Res. 2011: 1–10.
Kumar, P., D.M. Barrett, M.J. Delwiche and P. Stroeve. 2009. Methods for Pretreatment of Lignocellulosic Biomass for Efficient Hydrolysis and Biofuel Production. Ind. Eng. Chem. Res. 48(8): 3713–3729.
Kurakake, M., K. Ouchi, W. Kisaka and T. Komaki. 2005. Production of L-arabinose and xylose from corn hull and bagasse. J. Appl. Glycosci. 52: 281–285.
Laser, M., D. Schulman, S.G. Allen, J. Lichwa, M.J. Antal Jr. and L.R. Lynd. 2002. A comparison of liquid hot water and steam pretreatments of sugar cane bagasse for conversion to ethanol. Bioresour. Technol. 81: 33–44.

Lewis, G.E., C.W. Hunt, W.K. Sanchez, R. Treacher, G.T. Pritchard and P. Feng. 1996. Effect of direct-fed fibrolytic enzymes on the digestive characteristics of a forage-based diet fed to beef steers. J. Animal Sci. 74: 3020–3028.

Li, J., G. Henriksson and G. Gellerstedt. 2007. Lignin depolymerization/repolymerization and its critical role for delignification of aspen wood by steam explosion. Bioresour. Technol. 98: 3061–3068.

Lin, Y. and S. Tanaka. 2006. Ethanol fermentation from biomass resources: current state and prospects. Appl. Microbiol. Biotechnol. 69: 627–642.

Liang, Y., T. Siddaramu, J. Yesuf and N. Sarkany. 2010. Fermentable sugar release from Jatropha seed cakes following lime pretreatment and enzymatic hydrolysis. Bioresour. Technol. 101: 6417–6424.

Lynd, L.R., P.J. Weimer, W.H. van Zyl and I.S. Pretorius. 2002. Microbial cellulose utilization: fundamentals and biotechnology. Microbiol. Mol. Biol. Rev. 66: 506–577.

Martin, C., B. Alriksson, A. Sjode, N-O. Nilvebrant and L.J. Jonsson. 2007. Dilute sulfuric acid pretreatment of agricultural and agro-industrial residues for ethanol production. Appl. Biochem. Biotechnol. 137: 339–352.

McMillan, J.D. Pretreatment of lignocellulosic biomass. pp. 292–324. *In*: M.E. Himmel, J.O. Baker and R.P. Overend [eds.]. 1994. Enzymatic Conversion of Biomass for Fuels Production. American Chemical Society, Washington, DC.

Menon, V. and M. Rao. 2012. Trends in bioconversion of lignocellulose: Biofuels, platform chemicals & biorefinery concept. Prog. Energy Comb. Sci. pp. 1–29 (in press).

Michelin, M., S.C. Peixoto-Nogueira, J.H.A. Betini, T.M. Silva, J.A. Jorge, H.F. Terenzi and M.L.T.M. Polizeli. 2010. Production and properties of xylanases from *Aspergillus terricola* and their use on cellulose pulp bleaching. Bioprocess Biosyst. Eng. 33: 813–821.

Michelin, M., M.L.T.M. Polizeli, D.S. Ruzene, D.P. Silva, H.A. Ruiz, A.A. Vicente, J.A. Jorge, H.F. Terenzi and J.A. Teixeira. 2012. Production of xylanase and β-xylosidase from autohydrolysis liquor of corncob using two fungal strains. Bioprocess Biosyst. Eng. (in press).

Mosier, N., C. Wyman, B. Dale, R. Elander, Y.Y. Lee, M. Holtzapple and M. Ladisch. 2005. Features of promising technologies for pretreatment of lignocellulosic biomass. Bioresour. Technol. 96: 673–686.

Nabarlatz, D., A. Ebringerová and D. Montané. 2007. Autohydrolysis of agricultural by-products for the production of xylo-oligosaccharides. Carbohydr. Polymers 69: 20–28.

Negro, M.J., P. Manzanares, I. Ballesteros, J.M. Oliva, A. Cabanãs and M. Ballesteros. 2003. Hydrothermal pretreatment conditions to enhance ethanol production from poplar biomass. Appl. Biochem. Biotechnol. 105: 87–100.

Nigam, P.S., N. Gupta and A. Anthwal. Pre-treatment of Agro-Industrial Residues. pp. 13–33. *In*: P.S. Nigam and A. Pandey [eds.]. 2009. Biotechnology for Agro-Industrial Residues Utilisation—Utilisation of Agro-Residues. Springer, Germany.

Oliva, J.M., F. Sáez, I. Ballesteros, A. Gónzalez, M.J. Negro, P. Manzanares and M. Ballesteros. 2003. Effect of lignocellulosic degradation compounds from steam explosion pretreatment on ethanol fermentation by thermotolerant yeast *Kluyveromyces marxianus.* Appl. Microbiol. Biotechnol. 105: 141–154.

Parajó, J.C., H. Domíngues and J.M. Domíngues. 1998. Biotechnological production of xylitol, part 1: Interest of xylitol and fundamentals of its biosynthesis. Bioresour. Technol. 65: 191–201.

Parajó, J.C., G. Garrote, J. M. Cruz and H. Dominguez. 2004. Production of xylooligosaccharides by autohydrolysis of lignocellulosic materials. Trends Food Sci. Technol. 15: 115–120.

Park, J.Y., R. Shiroma, M.I. Al-Haq, Y. Zhang, M. Ike, Y. Arai-Sanoh, A. Ida, M. Kondo and K. Tokuyasu. 2010. A novel lime pretreatment for subsequent bioethanol production from rice straw—calcium capturing by carbonation (CaCCO) process. Bioresour. Technol. 101: 6805–6811.

Park, Y.S., S.W. Kang, J.S. Lee, S.I. Hong and S.W. Kim. 2002. Xylanase production in solid state fermentation by Aspergillus niger mutant using statistical experimental designs. Appl. Microbiol. Biotechnol. 58: 761–766.

Pereira Jr., N., M.A.P.G. Couto, L.M.M. and Santa Anna. 2008. Biomass of Lignocelulosic Composition for fuel ethanol production within the context of biorefinery. Rio de Janeiro: Escola de Química/UFRJ, Series on Biotechnology 2: 47.

Perez, J., J.M. Dorado, T.D. Rubia and J. Martinez. 2002. Biodegradation and biological treatment of cellulose, hemicellulose and lignin: An overview. Int. Microbiol. 5: 53–63.

Pérez, J.A., I. Ballesteros, M. Ballesteros, F. Sáez, M.J. Negro and P. Manzanares. 2008. Optimizing Liquid Hot Water pretreatment conditions to enhance sugar recovery from wheat straw for fuel-ethanol production. *Fuel* 87(17-18): 3640–3647.

Philippoussis, A.N. Production of Mushrooms Using Agro-Industrial Residues as Substrates. pp. 163–196. *In*: P.S. Nigam and A. Pandey [eds.]. 2009. Biotechnology for Agro-Industrial Residues Utilisation—Utilisation of Agro-Residues. Springer, Germany.

Polizeli, M.L.T.M., A.C.S. Rizzatti, R. Monti, H.F. Terenzi, J.A. Jorge and D.S Amorim. 2005. Xylanases from fungi: properties and industrial applications. Appl. Microbiol. Biotechnol. 67: 577–591.

Prior, B.A. and D.F. Day. 2008. Hydrolysis of ammonia-pretreated sugar cane bagasse with cellulase, beta-glucosidase, and hemicellulase preparations. Appl. Biochem. Biotechnol. 146: 151–164.

Rocky-Salimi, K. and Z. Hamidi-Esfahan. 2010. Evaluation of the effect of particle size, aeration rate and harvest time on the production of cellulase by *Trichoderma reesei* QM9414 using response surface methodology. Food Bioprod. Processing 88: 61–66.

Ruzene, D.S. and A.R. Gonçalves. 2003. Effect of Dose of Xylanase on Bleachability of Sugarcane Bagasse Ethanol/Water Pulps. Appl. Biochem. Biotechnol. 105–108: 769–774.

Ruzene, D.S., A.R. Gonçalves, J.A. Teixeira and M.T. Pessoa de Amorim. 2007. Carboxymethylcellulose obtained by ethanol/water organosolv process under acid conditions. Appl. Biochem. Biotechnol. 136–140: 573–582.

Saha, B.C. 2000. Alpha-L-arabinofuranosidases: biochemistry, molecular biology and application in biotechnology. Biotechnol. Adv. 18: 403–423.

Sánchez, C. 2009. Lignocellulosic residues: Biodegradation and bioconversion by fungi. Biotechnol. Adv. 27: 185–194.

Singh, R., A.J. Varma, R.S. Laxman and M. Rao. 2009. Hydrolysis of cellulose derived from steam exploded bagasse by Penicillium cellulases: Comparison with commercial cellulase. Bioresour. Technol. 100: 6679–6681.

Silverstein, R.A., Y. Chen, R.R. Sharma-Shivappa, M.D. Boyette and J. Osborne. 2007. A comparision of chemical pretreatment methods for improving saccharification of cotton stalks. Bioresour. Technol. 98: 3000–3011.

Singhania, R.R. Cellulolytic Enzymes. pp. 371–381. *In*: P.S. Nigam and A. Pandey [eds.]. 2009. Biotechnology for Agro-Industrial Residues Utilisation—Utilisation of Agro-Residues. Springer, Germany.

Song, J.-M. and D.-Z. Wei. 2010. Production and characterization of celluloses and xylanases of *Cellulosimicrobium cellulans* grown in pretreated and extracted bagasse and minimal nutrient medium M9. Biomass and Bioenergy 34: 1930–1934.

Sreenath, H.K., A.B. Shah, V.W. Yang, M.M. Gharia and T.W. Jeffries. 1996. Enzymatic polishing of jute/cotton blended fabrics. J. Ferment. Bioeng. 81(1): 18–20.

Subramaniyan, S. and P. Prema. 2002. Biotechnology of Microbial Xylanases: Enzymology, Molecular Biology and Application. Crit. Rev. Biotechnol. 22(1): 33–46.

Sukumaran, R.K., R.R. Singhania, G.M. Mathew and A. Pandey. 2009. Cellulase production using biomass feed stock and its application in lignocellulose saccharification for bio-ethanol production. Renewable Energy 34: 421–424.

Sun, Y. and J. Cheng. 2002. Hydrolysis of lignocellulosic materials for ethanol production: a review. Bioresour. Technol. 83: 1–11.

Sun, Y and J.J. Cheng. 2005. Dilute acid pretreatment of rye straw and bermudagrass for ethanol production. Bioresour. Technol. 96: 1599–1606.

Taherzadeh, M.J. and K. Karimi. 2008. Pretreatment of lignocellulosic wastes to improve ethanol and biogas production: a review. Int. J. Mol. Sci. 9: 1621–1651.

Uffen, R.L. 1997. Xylan degradation: a glimpse at microbial diversity. J. Ind. Microbiol. Biotechnol. 19: 1–6.

USDA-US DOE. 2005. Biomass as feedstock for a bioenergy and bioproducts industry: The technical feasibility of a billion-ton annyal supply. http: //feedstockreview.ornl.gov/ pdf/ billion ton vision.pdf. Accessed 4 May 2012.

Viikari, L., A. Kantelinen, J. Sandquist and M. Linko. 1994. Xylanases in bleaching: from an idea to the industry. FEMS Microbiol. Rev. 13: 338–350.

Vlaev, S.D., G. Djejeva, V. Raykovska and K. Schugerl. 1997. Cellulase production by *Trichoderma sp.* grown on corn fibre substrate. Process Biochem. 32(7): 561–565.

Wang, H., J. Wang, Z. Fang, X. Wang and H. Bu. 2010. Enhanced bio-hydrogen production by anaerobic fermentation of apple pomace with enzyme hydrolysis. Int. J. Hydrogen Energy 35: 8303–8309.

Wong, K.K.Y., L.U.L. Tan and J.N. Saddler. 1988. Multiplicity of β-1,4-xylanase in microorganisms: functions and applications. Microbiol. Rev. 52: 305–317.

Yu, G., S. Yano, H. Inoue, S. Inoue, T. Endo and S. Sawayama. 2010. Pretreatment of rice straw by a hot-compressed water process for enzymatic hydrolysis. Appl. Biochem. Biotechnol. 160: 539–551.

Zhang, Y.H. and L.R. Lynd. 2004. Toward an aggregated understanding of enzymatic hydrolysis of cellulose: noncomplexed cellulase systems. Biotechnol. Bioeng. 88: 797–824.

Zhang, Y.-H.P., S.-Y. Ding, J.R. Mielenz, J.-B. Cui, R.T. Elander, M. Laser, M.E. Himmel, J.R. McMillan and L.R. Lynd. 2007. Fractionating recalcitrant lignocellulose at modest reaction conditions. Biotechnol. Bioeng. 97: 214–223.

Zhao, X., L. Wang and D. Liu. 2007. Effect of several factors on peracetic acid pretreatment of sugarcane for enzymatic hydrolysis. J. Chem. Technol. Biotechnol. 82: 1115–1121.

Zhao, X.B., L. Wang and D. Liu. 2008. Peracetic acid pretreatment of sugarcane bagasse for enzymatic hydrolysis: a continued work. J. Chem. Technol. Biotechnol. 83: 950–956.

Zheng, Y., Z. Pan and R. Zhang. 2009. Overview of biomass pretreatment for cellulosic ethanol production. Int. J. Agric. & Biol. Eng. 2(3): 51–68.

3

Biodegradation of Vinasse: Fungal Lignolytic Enzymes and their Application in the Bioethanol Industry

Luiz Fernando R. Ferreira,[1,*] *Ana Maria Q. Lopez,*[2] *Regina T.R. Monteiro,*[1] *Denise S. Ruzene*[3] and *Daniel P. Silva*[4]

Introduction

Alcohol distilleries are extensively growing due to widespread industrial applications of alcohol in pharmaceuticals, perfumery, food and alternative fuel, among others (Mohana et al. 2009). One of the main problems faced by the sugar-alcohol industries is the large volume of residues generated,

[1]Center for Nuclear Energy in Agriculture, University of São Paulo, Piracicaba—SP, P.O. Box 96 134000-970, Brazil.
[2]Federal University of Alagoas, Institute of Chemistry and Biotechnology, Laboratory of Biochemistry of Parasitism and Environmental Microbiology, Maceio—AL, 57072-970, Brazil.
[3]Institute of Technology and Research, University Tiradentes, Campus Farolândia, Aracaju—SE, 49032-490, Brazil.
[4]Federal University of Sergipe, Nucleus of Production Engineering NPR/UFS, São Cristovão-SE, 49100-000, Brazil.
*Corresponding author: romanholobio@gmail.com

especially vinasse, which leads to a potential pollutant to water bodies when inadequately managed (Machado 1998, Wadt 2008, Ferreira 2009). An average of eight to 18 liters of this effluent is generated for every liter of alcohol produced. Its composition varies according to the raw material and equipment utilized in the production process (Kumar et al. 1998, Naik et al. 2008).

Because of its composition, vinasse has been applied in the agricultural sector as a fertilizing agent, mainly in sugarcane crops. It consists of 94 to 97 percent water, a high amount of organic matter, potassium, phosphorous, sulfate, calcium and magnesium. It is characterized by a low pH (4.5 to 4.8), with a typical odor, and a dark-brownish color due to the presence of melanoidins (Prada et al. 1998, Wadt 2008, Ferreira 2009). Its color components disrupt the flow of sun rays in surface waters, reducing the photosynthetic activity and damaging aquatic life. Besides this, the high chemical and biochemical demand of oxygen (CDO and BDO) load results in eutrophication of contaminated waterways, and such discharge can also increase the temperature of the receiving waters, reducing their amount of dissolved oxygen content. In addition, acidic effluent decreases the pH by dissolving some metals in flowing water, and disposal on land is equally hazardous by reducing soil alkalinity and manganese availability, which in part inhibits seed germination and destroys crops and vegetation.

In Brazil, methodologies have been developed for utilizing the stillage nutrients and land application on sugarcane fields. However, treatment options are being considered for decolorization, odor control, and nutrient removal prior to acceptable discharge into water bodies (Miranda et al. 1996, Benito et al. 1997, Benito et al. 1999, Kahraman and Yeşilada 2003). Vinasse quantities applied to soil may not surpass the ion retention capacity, in other words the used dosage depends on the characteristics of the soil.

Waste treatment methods target the removal of unwanted compounds in wastewater for a safe discharge into the environment. This can be achieved by using physical, chemical and biological methods either alone or in combination. Physical treatment methods such as screening, sedimentation and skimming, remove floating solid residues. Chemical treatments such as precipitation, pH adjustment, coagulation etc., work by removing toxic materials and colloidal impurities. Coloring compounds are more difficult to treat due to their common synthetic origin and complex aromatic molecular structures. Such structures resist fading on exposure to water, light or oxidizing agents, and this renders molecules less amenable to biodegradation (Naik et al. 2008). Physical or chemical methods of decolorization are invariably cost intensive and are not easily accepted by industries.

In recent years, the importance of biological wastewater treatment systems has attracted the attention of companies worldwide and also helped in developing efficient low cost waste treatment systems. Increased attention has been directed towards utilization of microbial activity for the mineralization and decolorization of vinasse (Pant and Adholeya 2009a).

Since the increased alcohol production in Brazil, there was a need to find alternatives in the reutilization and/or treatment for the vinasse generated, thus leading to many new technologies (Mane et al. 2006, Silva et al. 2007, Ferreira 2009, Mohana et al. 2009).

Until very recently, the literature concerning biodegradation of recalcitrant biomolecules such as lignocellulose, and bioremediation of organic chemical wastes (xenobiotics) dealt almost exclusively with bacteria. It is now becoming apparent that fungi, mainly ascomycetes and basidiomycetes, also play an important role in degrading organic materials in the ecosystem, and that they have potential for remediating contaminated soils and waters (Barr and Aust 1994, Kumar et al. 1998, Balan and Monteiro 2001, Coulibaly et al. 2003). They are able to grow under environmental stress conditions which limit bacterial growth and also have a different mode of growth. Fungi growth is induced by chemotaxis toward the source of organic carbon, through stretching and branching of their mycelial hyphae. This allows the colonization of large areas, and surface contact with the contaminant is optimized, increasing its bioavailability and, consequently, its biodegradation (Dupont et al. 1997).

These fungi produce several types of enzymes. The peroxidases containing heme group, such as lignin peroxidase (LiP, EC 1.11.1.14) and manganese peroxidase (MnP, EC 1.11.1.13), flavin oxidases such as glucose oxidases (EC 1.1.3.4), pyranose-2-oxidase (EC 1.1.3.10), methanol oxidase (EC 1.1.3.13) and aryl alcohol oxidase (AAO, EC 1.1.3.7). These enzymes reduce dissolved oxygen to hydrogen peroxide with the simultaneous oxidation of hydroxyl groups of substrates specific to the carbonyl. Copper containing oxidases such as laccases (EC 1.10.3.2) reduce dissolved oxygen to water and utilize the oxidation process of non-phenolic and phenolic substrates, forming radicals cationic quinone or phenoxy radicals, in addition to the glyoxal oxidase, and galactose oxidase (EC 1.1.3.9). Further, galactose oxidase reduces oxygen to hydrogen peroxide, with oxidation of a specific alcohol or aldehyde and cellobiose dehydrogenase (EC 1.1.99.18), are flavohemeproteins reducing phenoxy radicals and radicals cationic ion quinones or iron, and simultaneously oxidizing cellobiose to cellobiolactone (Shan and Nerud 2002).

However, the mechanism for degradation of lignin and recalcitrant organic pollutants is not the same for all the fungi because of the composition and particularities found in such residue. Galliano et al. (1991) studied the degradation of lignin (solubilization) by *Rigidoporus lignosus*, an organism that does not produce LiP, but synthesizes laccase and MnP. When these two enzymes were purified and their properties studied *in vitro*, none were able to solubilize radiolabeled lignin. However, when both were added simultaneously to the reaction mixture, the solubilization of lignin was extensive, MnP and laccase acted synergistically. Moreover, glucose oxidase increased solubilization of lignin by preventing re-polymerization caused by radicals of MnP and laccase. Also, nano-technologies offer great opportunities in production of enzymes-related nanolignocellulosic materials and composites such as cellulose-quoted sensors, cellulose nano-crystals and kraft lignin nano-fibers (Rojas 2011). Thus, it is important to study the ligninolytic enzyme system of fungi, especially basidiomycetes used in bioremediation processes.

Lignocellulolytic Enzymes-Producing Fungi

Lignocellulose is a renewable organic material and is the major structural component of all plants, consisting of three major components: cellulose, hemicellulose and lignin. In addition, small amounts of other materials such as ash, proteins and pectin can be found in lignocellulosic residues in different degrees based on the source (Sanchez 2009). A better understanding of the lignocellulolytic mechanism of fungi contributed to the generation of "clean" environmental technologies for industrial application (Hammel 1996, Chagas and Durrant 2001, Moreira Neto 2006).

Briefly, ascomycetous fungi produce spores within microscopic cells called asci. The asci vary in shape from cylindric to spherical forms. Usually each ascus holds eight spores—however there are species with just one spore per ascus and others with over a 100 spores per ascus. In basidiomycetes the spores develop on projections that grow out from microscopic cells called basidia, rather than being enveloped within cells. In most cases the basidia are elongated and club-like, with some variation in shape. Usually each basidium has four projections and four spores—some species may have just one projection and spore per basidium and others up to eight. In most basidiomycetes the basidia have no dividing walls (or septa), but in a small number of genera the basidia are septated. The projections from the basidia are called sterigmata (singular: sterigma) (Hibbett and Binder 2007).

Lignocellulolytic enzymes-producing fungi are widespread, and include species from ascomycetes (e.g., *T. reesei*), basidiomycetes as white-rot fungi (e.g., *P. chrysosporium*) and brown-rot fungi (e.g., *Fomitopsis palustris*) and, finally, a few anaerobic species (e.g., *Orpinomyces sp.*) which degrade cellulose in gastrointestinal tracts of ruminant animals (Yoon et al. 2007, Ljungdahl 2008). Table 3.1 summarizes a few fungi producing different lignocellulolytic enzymes.

Biomass degradation by these fungi is performed by a complex mixture of cellulases (Bayer et al. 1998), hemicellulases (Ljungdahl 2008, Sanchez

Table 3.1 Examples of fungi producing different lignocellulolytic enzymes. (CMCase: Carboxymethylcellulases (endoglucanase), CBH: Cellobiohydrolases, BGL: β-glucosidases, CDH: Cellobiose dehydrogenase, MnP: Manganese peroxidises, LiP: Lignin peroxidises).

Fungus	Group	Fungal strain	Enzymes
Aerobic fungi (Extracellular ligno-cellulolytic enzymes)	Ascomycetes	*Trichoderma reesei*	Cellulases (CMCase, CBH, BGL), Hemicellulases (xylanases)
		T. harzianum	Cellulases (CMCase, CBH), Hemicellulases (β-1,3-glucanases)
		Aspergillus niger	Cellulases, Hemicellulase (xylanases)
		Pestalotiopsis sp.	Cellulases (CMCase, CBH), Laccases
		Podospora anserina	Laccases
		Neurospora crassa	Laccases
	Basidiomycetes	*Phanerochaeta chrysosporium*	Cellulases (CMCase, CBH, BGL), CDH, LiP, MnP, Hemicellulases (xylanases)
		Pleurotus ostreatus	Laccases
		Fusarium palustris	Cellulases (CMCase, CBH, BGL)
		Trametes versicolor	Laccases
		Rhizoctonia solani	Laccases
Anaerobic rumen fungi (Cell-wall associated ligno-cellulolytic enzymes, "cellulosome") (Chytridiomycetes)	*Anaeromyces*	*Anaeromyces mucronatus 543*	Cellulase (CMCase), Hemicellulases (xylanases)
	Caecomyces	*Caecomyces communis*	Cellulases, Hemicellulases (xylanases, β-D-xylosidases)
	Cyllamyces	*Cyllamyces aberensis*	Cellulases, Hemicellulases (xylanases)
	Neocallimastix	*Neocallimastix frontalis*	Cellulases, Hemicellulases (xylanases, β-galactosidases)
	Orpinomyces	*Orpinomyces sp.*	Cellulase (CMCase, CBH, β-glucosidase), Hemicellulases (xylanases, mannanases)
	Piromyces	*Piromyces sp.*	Cellulases (CMCase, CBH, β-glucosidase) Hemicellulases (xylanase, mannanases)

2009) and ligninases (Wood and Garcia-Campayo 1990, Kerem et al. 1992, Tuor et al. 1995, Weng et al. 2008). However, despite their proficient ability to degrade lignin, white-rot fungi, for instance, are unable to utilize lignin as the sole carbon or energy source, and depolymerization provides access to cellulose and hemicellulose encrusted into the lignin matrix, and then to saccharides which can be used as carbon and energy source.

Fungal Extracellular Cellulases, Hemicellulases and "Cellulosomes"

Cellulases and most hemicellulases belong to a group of enzymes known as glycoside hydrolases (GH). Currently more than 2500 GH have been identified and classified into 115 families (Cantarel et al. 2009), and one enzyme family may contain members from bacteria, fungi and plants with several different activities and substrate specifications. However, fungal cellulases (hydrolysis of β-1,4-glycosidic bonds) have been mostly found within a few GH families including 5, 6, 7, 8, 9, 12, 44, 45, 48, 61 and 74 (Cantarel et al. 2009, Sandgren et al. 2005).

Cellulases use two different catalytic mechanisms: retaining and inverting mechanisms. All GH 12 cellulases, for example, hydrolyze glycosidic bonds by the retaining mechanism, whereas family 6 cellulases use the inverting mechanism (Bayer et al. 1998, Sandgren et al. 2005). In both mechanisms, two catalytic carboxylate residues are involved and they catalyze the reaction by acid-base catalysis. Many different fungal species have the ability to degrade cellulose by producing extracellular fungal cellulose-degrading enzymes including endo-cleaving (endoglucanases) and exo-cleaving (cellobiohydrolases). Endoglucanases can hydrolyze glycosidic bonds internally in cellulose chains whereas cellobiohydrolases act preferentially on chain ends. The products of the enzymatic reaction are mostly a disaccharide known as cellobiose and, to a lesser extent, cello-oligosaccharides, which is then further hydrolyzed by the third group of enzymes called β-glucosidases (Kumar et al. 2008). Cellulases mostly have a small independently folded carbohydrate binding module (CBM) which is connected to the catalytic domain by a flexible linker. The CBMs are responsible for binding the enzyme to the crystalline cellulose and thus enhancing the enzyme activity (Bayer et al. 1998). Currently many CBMs have been identified and classified into 54 families, however only 20 families (1, 13, 14, 18, 19, 20, 21, 24, 29, 32, 35, 38, 39, 40, 42, 43, 47, 48, 50 and 52) have been found in fungi.

On the other hand, several different enzymes are needed to hydrolyze hemicelluloses, due to their heterogeneity (Saha 2003). Xylan is the largest component of hemicellulose contributing over 70 percent of its structure. Xylanases are able to hydrolyze β-1,4 linkages in xylan and produce oligomers which can be further hydrolyzed into xylose by β-xylosidase. Not surprisingly, additional enzymes such as β-mannanases, arabinofuranosidases or α-L-arabinanases are needed depending on the hemicellulose composition which can be mannan-based or arabinofuranosyl-containing (Shallom and Shoham 2003). Similar to cellulases, hemicellulases are usually modular proteins and have other functional modules, such as CBM, in addition to their catalytic domains. Also similar to cellulases, most hemicellulases are glycoside hydrolases (GHs), although some hemicellulases belong to carbohydrate esterases (CEs), which hydrolyze ester linkages of acetate or ferulic acid side groups (Bourne and Henrissat 2001, Shallom and Shoham 2003). Hemicellulases belong to 20 different GH families (1, 2, 3, 4, 5, 8, 10, 11, 26, 27, 36, 39, 43, 51, 52, 53, 54, 57, 62 and 67) and all of them except for 4 (families 4, 8, 52 and 57) have been found in fungi. All but 1 famiy (7) of the 7 different CE families (1, 2, 3, 4, 5, 6 and 7) reported for hemicellulases have been found in fungi (Cantarel et al. 2009). Similar to cellulases, aerobic fungi such as *Trichoderma* and *Aspergillus* secrete a wide variety of hemicellulases in high concentrations (8 and 12 hemicellulases, respectively) and work in a synergistic manner (Shallom and Shoham 2003).

Anaerobic gut fungi represent a special group of microorganisms inhabiting the gastro-intestinal tract of ruminants and most non-ruminant herbivores, along with some anaerobic bacteria (mainly Clostridia) which also produce a range of cellulolytic and hemicellulolytic enzymes in a multienzyme complex known as cellulosome (Lamed et al. 1983). The first report of an anaerobic gut fungi able to break down ingested lignocellulosic residues was in 1975 (Orpin 1975) and since then 6 genera and 18 species have been identified, some of which are shown in Table 3.1. However, the cellulosome from anaerobic fungi *Neocallimastix frontalis* was first described in 1992 (Wilson and Wood 1992). *N. frontalis* and *Piromyces* sp. cellulosomes include at least six or 10 polypeptides, respectively (cellulosome-type complex), against the 20 or more different cellulolytic/hemicellulolytic enzymes of the anaerobic bacterial cellulosome (Wilson and Wood 1992, Ali et al. 1995, Bayer et al. 2004). In addition to catalytic hydrolytic subunits, all proteins of fungal cellulosomes have noncatalytic subunits known as "fungal dockerin domains" (FDD), which allow binding to cohesin modules of the scaffolding proteins. These FDDs (at least 50 have been already identified) show different amino acid sequences than those found in bacterial dockerins (Nagy et al. 2007, Ljungdahl 2008). However, beyond cellulosomes, some free enzymes have also been identified in anaerobic gut fungi. In *Piromyces* sp. PC2, a cellulosome-producing anaerobic fungus, 17 lignocellulolytic enzyme encoding genes have been isolated (10 cellulases, one β-glucosidase, five hemicellulases and one enzyme facilitating protein folding), but FDD has been reported only for 11 genes, which indicates that these cellulases are cellulosome-associated. Moreover, CBM has been identified only in three of those 17 genes, including two cellulases and one hemicellulase (Ljungdahl 2008). The major product of cellulose digestion by fungal cellulosomes is glucose which eliminate the costly addition of β-glucosidase, whereas in the case of bacterial cellulosomes, cellobiose is the major product (Dijkerman et al. 1997). Despite many advantages of cellulosomes such as synergistic activity between the components and efficient hydrolytic activity on both cellulose and hemicellulose, fungal cellulosomes are much less well characterized compared to bacterial cellulosomes.

Fungal Ligninases

Lignin, the most abundant renewable aromatic polymer on Earth, is composed of non-phenolic (80–90 percent) and phenolic structures (Wang et al. 2008). It has been shown that fungi degrade lignin by secreting enzymes collectively termed "ligninases" or *Fungal Oxidative Lignin enzymes* (FOLe). The sequence and structure were classified (Table 3.2) and integrated in a database (FOLy). Two categories of potentially FOLe were defined according to their direct or indirect action on lignin degradation: Lignin Oxidases (LOs) and Lignin Degrading Auxiliary enzymes (LDAs), respectively (Levasseur et al. 2008).

The LOs include two ligninolytic families: i) phenol oxidase (laccase) and ii) peroxidases [lignin peroxidase (LiP) and manganese peroxidase (MnP)] (Martinez et al. 2005), besides an auxiliary lignin-oxidase family of enzymes. White-rot basidiomycetes such as *Trametes* (*Coriolus*) *versicolor* (Moreto et al. 2003, Wang et al. 2008) and *P. chrysosporium* (Moredo et al. 2003) have been found to be the most efficient lignin-degrading microorganisms studied.

a) Laccases (E.C. 1.10.3.2)

Laccases are widely distributed in higher plants and fungi and have also been found in insects and bacteria (Yaver et al. 1996, Enguita et al. 2003, Claus 2004), and are an interesting group of multi copper enzymes which have received much attention from researchers in last decades due to their ability to oxidize both phenolic and non-phenolic lignin related compounds, as well as highly recalcitrant environmental pollutants (Crecchio et al. 1995, Claus 2004, Mensah et al. 2012). According to Szklarz et al. (1989), these enzymes are extracellular, linked to the cell membrane, or intracellular, as it has been described for Lenzites trabea. Usually the enzyme originates in the cytoplasm, but many examples of its secretion have been described in the literature, although little attention has been given to the subcellular localization of this enzyme and its mechanism of secretion (Mayer and Staples 2002). The production of laccase is affected by many factors of fungal growth, such as the composition of the culture medium (the carbon and

Table 3.2 Description of families in FOLy (Fungal Oxidative Lignin Enzymes database). *Numbers of full-length total entries; **3D Structures (Levasseur et al. 2008).

Family	Known activities	E.C. numbers	Entry number	3D**	Reactional mechanism
Lignin oxidase					
LO1	Laccase	1.10.3.2	205*	7	4-benzendiol + H_2O_2 = 4-benzo-semiquinone + $2H_2O$
LO2	Chloroperoxidase	1.10.1.10	86	8	2 RH + 2 Cl^- + H_2O_2 = 2 RCl + 2 H_2O
	Manganese peroxidase	1.11.1.13			Catalytically dependent on H_2O_2 and Mn^{2+} ions
	Ligninperoxidase	1.11.1.14			Oxidations in alkyl side chains, C–C cleavage in side chain of lignin, cleavage of aromatic ring, oxidation of benzyl alcohol
	Versatile peroxidase	1.11.1.16			This ligninolytic peroxidase combines the substrate specificity characteristics of the two other ligninolytic peroxidases
LO3	Cellobiose dehydrogenase	1.1.99.18	15	1	Cellobiose + acceptor = cellobiono-1, 5-lactone + reduced acceptor. Presence of two redoxin domains, one containing a flavin and the other a protoheme group
Auxiliary enzymes					
LDA1	Aryl-alcohol oxidase	1.1.3.7	7	0	Aromatic primary alcohol + O_2 = aromatic aldehyde + H_2O_2
LDA2	Vanillyl-alcohol oxidase	1.1.3.38	1	1	Vanillyl alcohol + O_2 = vanillin + H_2O_2
LDA3	Glyoxal oxidase	1.1.3.-.	15	0	Oxidation of a number of simple aldehydes and α-hydroxy carbonyl compounds with concomitant production of H_2O_2
LDA4	Pyranose oxidase	1.1.3.10	11	2	D-glucose + O_2 = 2-dehydro-D-glucose + H_2O_2
LDA5	Galactose oxidase	1.1.3.9	6	1	D-galactose + O_2 = D-galacto-hexodialdose + H_2O_2
LDA6	Glucose oxidase	1.1.3.4	24	2	β-D-glucose + O_2 = D-glucono-1,5-lactone + H_2O_2
LDA7	Benzoquinone reductase	1.6.5.6	9	0	NADPH + H^+ + *p*-benzoquinone = $NADP^+$ + hydroquinone

nitrogen), pH, temperature, aeration, etc. (Kahraman and Gurdal 2002). In 2006, a novel polyphenol oxidase, with laccase like activity, was found from a metagenome expression library from bovine rumen microflora (Beloqui et al. 2006). These biocatalysts are useful in several biotechnological processes such as the detoxification of industrial effluents, mostly from the paper and pulp, textile and petrochemical industries, polymer synthesis, stabilization of wine and beverage, bioremediation of contaminated soils and cocoa residues, cattle nutrition (Mensah et al. 2012), manufacture of anti-cancer drugs and cosmetics by nanobiotechnology (Schneider et al. 1999, Durán et al. 2002, Freire et al. 2002, Mayer and Staples 2002, Moreira Neto 2006).

Laccase is a multicopper blue oxidase that couples the four electron reduction of oxygen with the oxidation of a broad range of organic substrates, including phenols, polyphenols, anilines, and even certain inorganic compounds by a "one-electron" transfer mechanism with a concomitant reduction of O_2 to H_2O (Malmström et al. 1975, Langh 1997, Ferraz 2004). These phenol-oxidases have a molecular mass varying from 45 to 100 kDa (Moreira Neto 2006), with different physico-chemical characteristics according to their isoform type (Dittmer et al. 1997, Brown et al. 2002, Mougin et al. 2003, Shleev et al. 2004), with a low redox potential, allowing direct oxidation of phenolic lignin units. The radicals formed act in the non-catalytic reactions as the radical-radical coupling, disproportion, deprotonation and nucleophilic attack by water. These reactions lead to polymerization, alkyl-aril cleavages, and Cα oxidations and demethylizations. In the presence of mediators (artificial substrates or fungal metabolites), laccases are able to oxidize high redox potential substrates (Hoff et al. 1985, Bollag et al. 1988, Bourbonnais and Paice 1990, Boyle et al. 1992, Bourbonnais et al. 1996, Eggert et al. 1996, Schliephake and Lonergan 1996, Call and Mücke 1997, Amitai et al. 1998, Majcherczyk et al. 1999, Shüitzendüel and Majancherczk 1999, D'Acunzo et al. 2002, Bollag et al. 2003, Mougin et al. 2003, Wesenberg et al. 2003), and have other physiological roles including fungal development, melanin synthesis, detoxification, and human and plant pathogenesis (Thurston 1994, Gianfreda et al. 1999, Tsai et al. 1999, Mayer and Staples 2002, Gómez and Nosanchuk 2003). Spectrophotometry can be used to detect laccases by using syringaldazine substrates, 2,6-dimethoxyphenol, dimethyl-p-phenylenediamine and 2,2'-azino-bis(3-ethylbenzothiazoline -6-sulphonic acid) (ABTS). Activity can also be monitored with an O_2 sensor, as the oxidation is paired with the reduction of O_2 to H_2O.

Laccases participate in other ligninolytic events of great importance, which were first assigned to other enzymes of the ligninolytic system. These events include the oxidation of non-phenolic units of lignin, the generation of H_2O_2, the formation of hydroxyl radicals and the production of Mn^{3+} from Mn^{2+}. This involvement of laccase in the generation of reduced oxygen species, e.g., radicals of anion superoxide ($O2^{\bullet -}$), H_2O_2, and •OH, have been demonstrated by oxidation of hydroquinones derived from lignin. In addition, the oxidation of Mn^{2+} can occur directly and also by oxidation of the hydroquinones as a consequence of generation of $O2^{\bullet-}$ (Höfer and Schlosser 1999, Gonzales et al. 2002, Saparrat et al. 2002, Schlosser and Höfer 2002).

Different laccase-producing fungi secrete several isoforms due to the presence of distinct aromatic compounds such as toluidine, vanillic acid, p-hydrobenzoic acid and aniline (Fähraeus 1962). *P. ostreatus* produces several of them, including POXA1b, POXA1w, POXA2 and POXC—the more abundant isoenzyme under all the studied growth conditions (Palmieri 1997).

POXA1w exhibits peculiar differences as to their metal content. This enzyme has two atoms of zinc, and one of iron and other of copper per molecule (Palmieri 1997). While the addition of $CuSO_4$ in a cultivar of *P. ostreatus* caused a large increase in total activity of laccase and production of the isoenzyme POXA1b, the production of POXA1w was not affected. The main structure of the laccase POXA1b is similar to the one of POXA1w, however, POXA1b contains four atoms of Cu per molecule, can be detected under UV light, and both are more stable than POXA2 and POXC (Palmieri 1997, Tanaka et al. 1986).

b) Lignin (E.C. 1.11.1.14), Manganese (E.C. 1.11.1.13) and Chloro (E.C. 1.11.1.10) Peroxidases

LiP and MnP were first described as true ligninases because of their high redox potential (Gold et al. 2000). LiP degrades non-phenolic lignin units (Kersten et al. 1990, Wesenberg et al. 2003), whereas MnP generates Mn^{3+}, which acts as a diffusible oxidizer on both phenolic and non-phenolic lignin units via lipid peroxidation reactions (Jensen et al. 1996). Versatile peroxidase represents a third type of peroxidase, combining catalytic properties of both LiP and MnP enzymes (Heinfling et al. 1998). More specifically, new peroxidases were reported in the FOLy database even when their roles in the lignin degradation were not biochemically shown (i.e., *Coprineus cinereus* peroxidase, CiP). Interestingly, LiP is able to oxidize the non-phenolic part of lignin, but it was not detected in many lignin-degrading fungi. In addition, it has been widely accepted that oxidative ligninolytic enzymes (Table 3.2) are not able to penetrate the cell walls due to their size. It is possible that prior to the enzymatic attack, low-molecular weight diffusible reactive oxidative compounds have to initiate changes to the lignin structure (Srebotnik et al. 1998, Tanaka et al. 1999).

The MnP is an extracellular glycosylated enzyme with a molecular weight ranging from 38 to 62.5 kDa, but most of the purified isoforms have a molecular mass close to 45 kDa (Hofrichter 2000), with a heme prosthetic group. This enzyme is dependent on hydrogen peroxide and Mn^{2+} ion, and α-keto acids such as lactate are responsible for stabilizing its activity (Kuhad et al. 1997, Ferraz 2004). The production of MnP or peroxidase-dependent manganese is apparently limited to certain Basidiomycete fungi, and until now, there was no evidence of bacteria, yeast and no basidiomycete mycorrhizal capable of producing this enzyme. The ability to synthesize MnP is distributed among different taxonomic groups of basidiomycetes. Colonizing species of wood belonging to following families Meruliaceae, Coriolaceae and Polyporaceae, as well as the litter decomposing basidiomycetes families Strophariaceae (family of *Psilocybe castanella*) and Tricholomataceae all express this enzyme. Up to 11 isoforms of MnP (Langh et al. 1997, Moreira Neto 2006) have been described for *Ceriporiopsis subvermispora*. These isoforms differ mainly in the isoelectric points (pI), which were in the acid range (pH 3–4), although isoforms in the neutral range and less acid were found in some fungi.

MnP, which generates Mn^{3+}— a powerful oxidizing agent, plays a key role in the first stage of the degradation of lignin. Mn^{3+} oxidizes certain non-phenolic aromatic structures of lignin, resulting in the formation of free radicals (such as peroxide or other organic peroxide) which reduces Fe^{3+}, forming a complex iron-peroxide. The subsequent breach of the bond O-O of hydrogen peroxide requires the transfer of two electrons from the heme group, which results in the formation of a complex radical Fe^{4+}–oxo-porphyrin (MnP-I). On breaking the link between the oxygen atoms, a water molecule is released. Reducing the complex, yields MnP-II (Fe^{4+} non-porphyrin-oxo radical). An ion Mn^{2+} acts as an electron donor for this intermediate complex, and is oxidized to Mn^{3+}. The reduction in MnP-II occurs in a similar manner and the other Mn^{3+} forms a Mn^{2+}, leading to a generation of the original form of the enzyme, releasing a second molecule of water. While the MnP-I acts as LiP, it can be reduced by other electron donors with Mn^{2+} ion. The MnP-II is somewhat reduced for other substrates and requires only Mn^{2+} to complete the catalytic cycle (Hofrichter 2002). The Mn^{3+} formed, is stabilized with organic acids such as oxalic acid, acting as a reducing agent oxy-diffuse, with low molecular weight organic which attacks nonspecifically molecules by subtracting an electron (Gold et al. 2000). Due to unspecific Mn^{3+}, the MnP system is efficient for oxidation of various organic pollutants (Moreira Neto 2006). Lipid peroxidation has also been suggested to participate in the process of degradation of lignin by MnP (Tien and Kirk 1984, Szklarz et al. 1989, Novikova 2002).

Therefore, during the oxidation of lignin, the Mn^{3+} is chelated and stabilized by organic acids, thus acting as a mediator of low molecular mass, far from the active site of MnP. The chelating Mn^{3+}-system generates radical reactive intermediates (peroxy radicals) from unsaturated fatty acids such as linoleic acid and other derivatives. The MnP-lipid system is strong enough to degrade Cα-Cβ bonds of non-phenolic structures of lignin, by hydrogen abstraction from the aromatic ring (Higuchi 1990, Collins et al. 1998, Hofrichter 2002, Hernández-Luna et al. 2008). Furthermore, MNP oxidize recalcitrant xenobiotics such as textile dyes (Heinfling et al. 1998) and nitroamine-toluenes (Vanacken et al. 1999).

There is a positive association of MnP with other extracellular oxidoreductases, for instance, MnP and laccase, MnP and LiP, and MnP and cellulases (cellobiose dehydrogenase) that resulted in increased lignin depolymerization. Future studies using 14C-lignins, 14C-polysaccharides, as well as different concentrations of Mn, *in vivo*, may help to better understand the mineralization process of lignin and organic pollutants by ligninolytic basidiomycetes (Hofrichter 2002, Moreira Neto 2006).

P. chrysosporium is a fungus widely studied for its ability to produce LiP and MnP. H_2O_2 in the presence of MnP is able to degrade lignin and other various phenolic substrates, however, LiP is able to degrade not only lignin, but also various non-phenolic aromatic compounds (Grgic and Perdih 2003). The veratryl alcohol (3,4 dimethoxybenzyl), a secondary metabolite, synthesized and accumulated in the extracellular environment, is present in lignin biodegradation process and induces the lignolytic system of *P. chrysosporium* by increasing the level of H_2O_2 production and the level of extracellular lignin peroxidase, oxidized itself (Lee and Moon 2003).

Chloroperoxidases (CPOs) are potential chlorinators of liginin and may thus account for some of the high-molecular-weight organochlorine residues. Therefore, CPOs may also contribute to the slow breakdown of lignin in plant cell-wall degradation (Ortiz-Bermúdez et al. 2003).

Oxidative Lignocellulose-degradation Mechanisms in Higher Fungi

Many white and brown-rot fungi have been shown to produce H_2O_2 which enters the Fenton reaction and results in release of free hydroxyl radicals (•OH) (Guillen et al. 1992, Suzuki et al. 2006). The free radicals attack polysaccharides as well as lignin in plant cell walls in a nonspecific manner, providing cleavages which make it easier for the lignocellulolytic enzymes to penetrate the cell (Wang et al. 2006). Three different pathways have been found for the generation of free radicals and cellulose-degrading fungi including basidiomycetes (mostly white-rot fungi) and ascomycetes growing on cellulose. The enzyme is able to oxidize cellobiose, higher cellodextrins and other disaccharides or oligosaccharides with β-1,4 linkages. In addition, CDH with (in ascomycetes) or without CBM (in basidiomycetes) have been identified and are able to bind to cellulose through hydrophobic interactions (Renganathan et al. 1990). It has been shown in some fungi that under cellulolytic conditions CDH production increases, helping cellulases and hemicellulases (Henriksson et al. 2000a,b, Baminger et al. 2001). It is now widely accepted that CDH is able to degrade and modify all three major components of the lignocellulosic residues (cellulose, hemicelluloses and lignin) by producing free hydroxyl radicals in a Fenton-type reaction (Baldrin and Valaskova 2008).

Cellobiose dehydrogenase (CDH) plays a role in carbohydrate metabolism, and some studies also suggest a role in lignin degradation (Henriksson et al. 1995, Kersten and Cullen 2007). This hemoprotein can generate hydroxyl radicals by Fenton-type reactions, thereby oxidizing lignin (Henriksson et al. 2000a, b). It is an extracellular monomeric protein and some glycosylation has been identified in a number of wood- and cellulose-degrading fungi including basidiomycetes (mostly white-rot fungi) and ascomycetes growing on cellulose. CDH is able to oxidize cellobiose, higher cellodextrins and other disaccharides or oligosaccharides with β-1,4 linkages. CDH with (in ascomycetes) or without CBM (in basidiomycetes) have been identified and are able to bind to cellulose through hydrophobic interactions (Renganathan et al. 1990). Under cellulolytic conditions CDH production increases (Henriksson et al. 2000a,b, Baminger et al. 2001), being able to degrade and modify all three major components of the lignocellulosic residues (cellulose, hemicelluloses and lignin) by producing free hydroxyl radicals in a Fenton-type reaction (Baldrin and Valaskova 2008).

It has been shown that white and brown-rot fungi produce low molecular weight chelators which are able to penetrate into the cell wall. For example, *Gloeophyllum trabeum* produces a low molecular weight peptide (known as short fiber generating factor, SFGF) which can degrade cellulose into short fibers by an oxidative reaction (Yang et al. 2004, Wang et al. 2006). It has also been reported that some of these low molecular weight compounds are quinones which have to be converted to hydroquinones by some fungal enzymes and then through a Fenton reaction free hydroxyl radicals will be produced (Wang et al. 2008).

Different glycopeptides with different molecular weight (ranging from 1.5 to 12 kDa) have been found in many brown-rot fungi such as *G. trabeum* (Enoki et al. 1999) and white-rot fungi such as *P. chrysosporium* (Tanaka et al. 1999, Tanaka et al. 2007). Similar to the other mechanisms, glycopeptides are able to catalyze redox reactions and thus produce free hydroxyl radicals.

Most enzymes classified as LDAs (auxiliary lignolytic enzymes) are H_2O_2 producers. Among H_2O_2-generating enzymes, aryl alcohol oxidase (LDA1, E.C. 1.1.3.7) and glyoxal oxidase (LDA3, E.C. 1.2.3.5) are thought to be the main enzymes responsible for the production of H_2O_2 (Kersten and Cullen 2007).

Another group of important ligninolytic enzyme complex includes the superoxide dismutases (SOD, EC 1.15.1.1). These enzymes, as well as glyoxal oxidases, generally act in a coordinated manner, working with the ligninolytic enzymes, cellulases and hemicellulases. However, they never attack the wood components alone. The superoxide dismutases are enzymes of universal distribution, containing iron (FeSOD), manganese (MnSOD) or copper and zinc (CuZnSOD) in its active site. They are the key enzymes in the fungal cell protection against oxidative stress caused by reactive oxygen species (ROS) such as hydrogen peroxide (H_2O_2), superoxide anion radicals ($O2^{\bullet-}$) and hydroxyl radicals (•OH). The SODs catalyze the reaction of superoxide radical anion to oxygen and hydrogen peroxide. These superoxide anion radicals commonly appear during the cyclical reduction of quinones and also during the action of laccase, MnP and LiP by ligninolytic fungi. ROS act as oxidizing or reducing agents, reacting with other radicals produced by ligninolytic enzymes, contributing to critical processes in the processing of lignin, such as breakage of the aromatic ring or demethoxilation (Leonowicz et al. 1999, Gonzales et al. 2002).

Melanoidins and Xenobiotic Compounds Degradation

Two strategies have been proposed for use of ligninolytic enzymes in order to degrade the recalcitrant compounds: (1) direct transformation of pollutants by ligninolytic active cultures of basidiomycetes and (2) the use of enzymes extracted from the culture medium. However, choosing the best strategy will depend on the objectives and the environmental conditions used during the bioremediation process (Novotny et al. 2001, Trupkin et al. 2003, Novotný et al. 2004).

The melanoidins present in the vinasse are high molecular weight heterogeneous polymers formed when low molecular weight carbohydrates or phenolic compounds (such as tannic and humic acids) combine with amino acids from proteins (through the Maillard reaction) at high temperatures and low water activity. They may act by decreasing oxygen concentration, intercepting singlet oxygen and preventing first-chain initiation by scavenging initial radicals such as hydroxyl radicals. Also by binding metal ions, such as iron, decomposing primary products to non-radical products, and chain-breaking to prevent continued hydrogen abstraction from substrates, synergism and chelating nutrient components (Rufian-Henares and Morales 2007). Consequently, melanoidins are frequently toxic including several microorganisms used in bio-treatments of effluents and are highly recalcitrant, persisting in the soil and holding antioxidant properties (Migo et al. 1993, Pérez et al. 2006, Chandra et al. 2008, Naik et al. 2008, Mohana et al. 2009).

The role of enzymes other than laccase or peroxidases in melanoidin decolorization by strains of *Coriolus* (*Trametes*) has been very well described during the last 30 yr. Aoshima et al. (1985) claimed the induction of intracellular enzymes by *Coriolus versicolor* Ps4a in the presence of melanoidin. Two-thirds of enzymes may be carbohydrates dependent (such as sorbose oxidase) and one-third is independent of these molecules. The decolorization of melanoidins may be caused by active oxygen produced by reaction with these carbohydrate oxidases.

There is also some knowledge about the role of fungal ligninolytic enzymes, such as those from white-rot fungi, in decolorization of effluents such as vinasse. This is due to the fact that the white-rot fungi hold a non-specific extracellular enzymatic system such as isoforms of extracellular oxidases including laccases, MnP and LiP, capable of "breaking" a large quantity of different chemical bonds (Rodríguez et al. 1999, Eggen 2000) present in their natural lignocellulosic substrate and in xenobiotic compounds such as dyes (Wesenberg et al. 2003, Mohana et al. 2009).

A culture filtrate of *T. versicolor* showed a mineralizing activity on melanoidins, and its purification yielded a 47 KDa extracellular protein Mn^{2+}-dependent that requires oxygen and is restricted by inhibitors of heme proteins. However, oxidation does not occur through the formation of H_2O_2 (Mansur et al. 1997). Miyata et al. (1998) verified that synthetic melanoidins are decolorized by the participation of manganese dependent peroxidase (MnP) and manganese-independent peroxidase (MnIP) of *T. hirsutus* pellets, and the extracellular H_2O_2 produced by glucose oxidase along with the partial participation of laccase. Increased expression of laccase genes in *Trametes* sp. I-62 (lcc1 and lcc2) and basidiomycete fungus NIOCC #2a upon exposure to molasses wastewater (MWW), accompanied by enhanced color removal, suggested the involvement of laccase in the melanoidins metabolism (D'Souza et al. 2006, González et al. 2008, Asgher et al. 2008).

Decolorization of sugar refinery effluent by *P. chrysosporium* matched with the time of detection of MnP in the medium (Guimarães et al. 1996). Color reduction by this fungus in the presence of VA (veratryl alcohol) is lower than when the compound is not present, thus confirming the major role of MnP. According with Vahabzadeh et al. (2004), the ability of *P. chrysosporium* to remove color is correlated not only to the activity MnP, but also with the activity of LiP.

The vinasse decoloration by *P. sajor-caju* CCB020 can be attributed to the degradation of smaller molecular weight fractions of the melanoidins (Kumar et al. 1998, Chandra et al. 2008, Ferreira et al. 2010, Ferreira et al. 2011). Aguiar et al. 2010 and Ferreira et al. (2010, 2011) observed the participation of the laccase enzyme at decolorization and in the reduction of turbidity (99 percent) and COD (≅90%) and BOD (≅75%) of the vinasse from the sugarcane processing plants, by the edible fungus *P. sajor-caju* CCB020, with a consequent participation of MnP in the decolorization of the effluent. The vinasse trigged the process of oxidative stress in the fungus, as observed through the elevation of the enzyme activities (superoxide dismutase, catalase and glutathione reductase) associated to mechanisms of detoxifying reactive oxygen species. When compared to the growth in synthetic medium, *P. sajor-caju* in submerged vinasse cultivation tests demonstrated an elevation of the biomass production (1.06 g 100 mL^{-1}) and in enzyme activities such as laccase (400 to 450 U L^{-1}) MnP (60 a 100 U L^{-1}), respectively reached at the 9th and 12th day of cultivation, indicating the participation of these enzymes in the depolymerization of phenolic compounds and melanoidins pigments. With the same treatments, reduction of 99.17 percent in color, and 99.73 percent in turbidity were observed. After the treatment with the fungus, a reduction in the toxicity was evident through the exposition of *Pseudokirchneriella subcapitata, Daphnia magna, Daphnia similis* and *Hydra attenuata*. As a result, the system *P. sajor-caju*/ vinasse can be utilized not only as a bioprocess for color removal and degradation of complex molecules for generating water for reuse, but also for biomass, laccase and MnP enzymes production. Similar results were seen by Pant and Adholeya (2007b) using *Pleurotus florida* EM1303. As mentioned before, fungi have been recognized by their superior ability to produce a large variety of extracellular proteins, organic acids and other metabolites and also to adapt to severe environmental constraints (Coulibaly et al. 2003).

Several reports have indicated that some fungi in particular have a potential for decolorizing vinasse (Kumar et al. 1998). The most studied fungi possessing the ability to degrade and decolorize distillery effluents are *Aspergillus* spp., *Aspergillus fumigatus* G-2-6, *Aspergillus niger, A. niveus,* and *A. fumigatus* UB260; with an average of 69–75 percent decolorization along with 70–90 percent COD reduction (Ohmomo et al. 1987, Miranda et al. 1996, Jimnez et al. 2003, Angayarkanni et al. 2003, Shayegan et al. 2004, Mohammad et al. 2006). The treatment of distillery residues with ascomycete fungi, such as *Penicillium* spp., *Penicillium decumbens* and *Penicillium lignorum*, reduced by 50 percent the color and the DBO of the effluent, and also reduced by 70% in phenolic compounds quantity (Jimnez et al. 2003). Sirianuntapiboon et al. (1995) reported an absorption mechanism for decolorization of melanoidins by *Rhizoctonia* spp. D-90. The pigments were accumulated in the cytoplasm and around the cell membrane as a melanoidin complex, which was then gradually decolorized by intracellular enzymes. Pant and Adholeya (2007b) isolated and identified three fungi by molecular methods. These include *Penicillium pinophilum* TERI DB1, *Alternaria gaisen* TERI DB6 and *Pleurotus florida* EM 1303. These fungi were capable of producing lignolytic enzymes and also to decolorize the effluent from sugarcane distilleries in 50, 47 and 86 percent, respectively.

There are promising developments in the application of solid-state fermentation (SSF) in enzymatic bioprocessing of lignocellulosic biomass. Due to their ability to produce ligninolytic extracellular enzymes, white rot fungi are the main depolymerizers of lignin constituents of wood, and therefore they are potential assets to biopulping processes (Singh et al. 2011).

Selective SSF degradation of pulp by tropical basidiomycetes (*Pycnoporus coccineus* and *Coriolus versicolor*) produces ligninolytic enzymes (including MnP) with activities as high as 275 U mL^{-1} in liquid cultures under stationary conditions. Biodegraded *Acacia mangium* wood chips show up to 10 percent lignin loss (Husaini et al. 2011), while strains of *Phellinus* sp., *Daedalea* sp., *T. versicolor* and *P. coccineus* show up to 30 percent delignification after 60 d of incubation (Liew et al. 2011).

Ligninolytic enzymes are very useful in biodegradation of synthetic dyes including biobleaching of colored effluents from pulp and paper and textile industries, de-inking of recycled paper and decolorization of wastewater. Tropical fungal strains (*Trichaptum* sp., *Datronia* sp. and *Trametes* sp.) from decomposed woods can grow on PDA agar containing synthetic lignin medium, and are efficient (upto 55 percent) decolorizers of pulp and paper effluents (Apiwatanapiwat et al. 2006). Laccases and peroxidases from white-rot fungi such as *Trametes* sp., *Pleurotus* sp., *Polystictus* sp. and *P. chrysosporiun* are also able to degrade a wide range of xenobiotic compounds including agricultural chemicals (pesticides and herbicides) such as Dieldrin, Simazine, DDT, Picloram, Triazines, Trifluralin, Diuron, Diazinon MCPA, Parathion and 2,4-D (Magan et al. 2010). Spent lignocellulosic substrates after *Pleurotus pulmonarius* cultivation are useful for the treatment of wastewater containing broad spectrum, non-systemic organochloride fungicide called Chlorothalonil (Tetrachloroisophthalonitrile). *P. pulmonarius* enzymes could remove upto 100 percent of this fungicide after 45 min of reaction (Juarez et al. 2011). Mineralization of 2-nitrophenol (2-NP), 3-nitrophenol (3-NP), 4-nitrophenol (4-NP) by white rot fungi *Bjerkandera adusta* and *Lentinus squarrosulus* in liquid cultures has been investigated by Tripathi et al. (2011). Both strains are able to completely remove 2-NP and 3-NP but fail to mineralize 4-NP. The main enzymes involved are aryl alcohol oxidase from *B. adusta* and MnP from *L. squarrosulus*. The ability of ligninolytic enzymes from *P. ostreatus* waste to degrade 1, 1, 1-trichloro-2, 2-bis (4-chlorophenyl) ethane (DDT) has been investigated by Purnomo et al. (2010). Upto 80 percent degradation and 8 percent mineralization of DDT-contaminated soils was achieved after 5 d of incubation. In addition, trans-chlordane has been metabolized with wood-rotting fungi (*Phlebia lindtneri, P. brevispora* and *P. aurea*). Over 50 percent of transchlordane was metabolized after 42 d of incubation (Xiao et al. 2011). Laccase from a tropical white-rot fungus (*Ganoderma lucidum*) subjected to 16 types of polycyclic aromatic hydrocarbons (PAHs) shows high degradability potential in the absence of a redox mediator (Punnapayak et al. 2009). Soil fungi including *Trichocladium canadense, Fusarium oxysporum, Aspergillus* sp., *Verticillium* sp. and *Achremonium* sp. are producers of ligninolytic enzymes and are able to degrade a wide range of recalcitrant organic pollutants such as polycyclic aromatic hydrocarbons (PAH), under low oxygen conditions (Silva et al.

2009). *P. pulmonarius* can degrade Pentachlorophenol (PCP) used as wood preservative and as a pesticide, in submerged cultures (Souza et al. 2011).

Conclusion and Perspectives

Lignocellulolytic fungi, especially ascomycetes and basidiomycetes, have attracted a great deal of interest as biomass degraders for large-scale applications due to their ability to produce large amounts of extracellular enzymes. The contribution of both ligninolytic and cellulotytic enzymes may be required for complete biodegradation of lignocellulosic biomass and similar recalcitrant compounds. Also, mediators such as 2, 5-xylidine and copper ions can be used as inducers of lignocellulolytic enzymes.

Many successful strategies have been made to improve fungal degradation of lignocellulose and recalcitrant phenols aiming at bioremediation of effluents. Vinasse is a pollutant residue resulting from sugarcane processing for generating alcohol, and is rich in melanoidins and phenolic compounds that can be utilized by lignolytic fungi for remediation. *P. sajor-caju* CCB 020, and edible fungus with high nutritional value, can be utilized to biodegrade and decolorize sugarcane vinasse without dilution. The vinasse increases the biomass and triggers the process of oxidative stress in this fungus. This fact has been observed through the elevation of superoxide dismutase, catalase and glutathione reductase. This means that these enzymes are associated with the mechanisms of detoxifying reactive oxygen species. The activities of laccase and MnP also increase, indicating the participation of these enzymes in the depolymerization of phenolic compounds and melanoidins pigments. In the same treatment is possible to see reduction of 75 to 99.73 percent in BOD, COD, color and turbidity of the effluent. After treatment with the fungus, a reduction in the toxicity was evident through the exposition of *Pseudokirchneriella subcapitata*, *Daphnia magna*, *Daphnia similis* and *Hydra attenuata*. Therefore, fungi such as, *P. sajor-caju* along with its lignolytic enzymes, as well as other fungi, are yet to be identified and studied for their potential in the bioprocess for color and toxicity removal of vinasse.

References

Aguiar, M.M., L.F.R. Ferreira and R.T.R. Monteiro. 2010. Use of vinasse and sugarcane bagasse for the production of enzymes by lignocellulolytic fungi. Braz. Arch. Biol. Technol. 53(5): 1245–1254.

Ali, B.R., L. Zhou, F.M. Graves, R.B. Freedman, G.W. Black, H.J. Gilbert and G.P. Hazelwood. 1995. Cellulases and hemicellulases of the anaerobic fungus *Piromyces* constitute a multiprotein cellulose-binding complex and are encoded by multigene families. FEMS Microbiol Lett. 125: 15–21.

Amitai, G., R. Adani, G. Sod-Moriah, I. Rabinovitz, A. Vincze and H. Leader. 1998. Oxidative biodegradation of phosphorothiolates by fungal laccase. FEBS Letters 438: 195–200.

Angayarkanni, J., M. Palaniswamy and K. Swaminathan. 2003. Biotreatment of distillery effluent using *Aspergillus niveus*. Bull. Environ. Contam. Toxicol. 70: 268–277.
Aoshima, I., Y. Tozawa, S. Ohmono and K. Ueda. 1985. Production of decolorizing activity for molasses pigment by *Coriolus versicolor* Ps4a. Agric. Biol. Chem. 49: 2041–2045.
Asgher, M., H.N. Bhatti, M. Ashraf and R.L. Legge. 2008. Recent developments in biodegradation of industrial pollutants by white rot fungi and their enzyme system. Biodegradation 19: 771–783.
Apiwatanapiwat, W., P. Siriacha and P. Vaithanomsat. 2006. Screening of fungi for decolorization of wastewater from pulp and paper industry. Kasetsart J. Nat. Sci. 40: 215–221.
Balan, D.S.L. and R.T.R. Monteiro. 2001. Decolorization of textile indigo dye by ligninolytic fungi. J. of Biotech. 89: 141–145.
Baldrian, P. and V. Valaskova. 2008. Degradation of cellulose by basidiomycetous fungi. FEMS Microbiol. Rev. 32: 501–521.
Baminger, U., S.S. Subramaniam, V. Renganathan and D. Haltrich. 2001. Purification and characterization of cellobiose dehydrogenase from the plant pathogen *Sclerotium* (*Athelia*) *rolfsii*. Appl. Environ. Microbiol. 67: 1766–1774.
Barr, D.P. and S.D. Aust. 1994. Mechanisms white rot fungi use to degrade pollutants. Environ. Sci. and Techn. 28: 78–87.
Bayer, E.A., H. Chanzy, R. Lamed and Y. Shoham. 1998. Cellulose, cellulases and cellulosomes. Curr. Opin. Struct. Biol. 8: 548–557.
Bayer, E.A., J.P. Belaich, Y. Shoham and R. Lamed. 2004. The cellulosomes: multienzyme machines for degradation of plant cell wall polysaccharides. Annu. Rev. Microbiol. 58: 521–554.
Beloqui, A., M. Pita and J. Polaina. 2006. Novel Polyphenol Oxidase Mined from Metagenome Expression Library of Bovine Rumen: Biochemical, Properties, Structural Analysis and Phylogenetic Relationship. J. Biol. Chem. 281: 22933–22942.
Benito, G.G., M.P. Miranda and D.R. Santos. 1997. Decolorization of wastewater from an alcoholic fermentation process with *Trametes versicolor*. Biores. Technol. 61: 33–37.
Benito, G.G., M.P. Miranda, M.T.G. Cubero and A.M.U. Alonso. 1999. Decolorization of molasses effluents by coagulation-flocculation process. Zuckerindustrie 5: 406–410.
Bollag, J.M., K.L. Shuttleworth and D.H. Anderson. 1988. Laccase-mediated detoxification of phenolic compounds. Appl. and Environ. Microb. 54: 3086–3091.
Bollag, J.M., H.L. Chu, M.A. Rao and L. Gianfreda. 2003. Enzymatic Oxidative Transformation of Chlorophenol Mixtures. J. of Environ. Qual. 32: 63–69.
Bourbonnais, R. and M.G. Paice. 1988. Veratryl alcohol oxidases from the lignin-degrading basidiomycete *Pleurotus sajor-caju*. Biochemical J. 255: 445–450.
Bourbonnais, R. and M.G. Paice. 1990. Oxidation of non-phenolic substrates: An expanded role for laccase in lignin biodegradation. FEBS Letters 267: 99–102.
Bourbonnais, R., M.G. Paice, B. Freiermuth, E. Bodie and S. Bornemann. 1996. Reactivities of various mediators and laccases with kraft pulp and lignin model compounds. Appl. and Environ. Microb. 63: 4627–4632.
Bourne, Y. and B. Henrissat. 2001. Glycoside hydrolases and glycosyltransferases: families and functional modules. Curr. Opin. Struct. Biol. 11: 593–600.
Boyle, C.D., B.R. Kropp and I.D. Reid. 1992. Solubilization and mineralization of lignin by white rot fungi. Appl. and Environ. Microb. 58: 3217–3224.
Brown, M.A., Z. Zhao and A.G. Mauk. 2002. Expression and characterization of a recombinant multi-copper oxidase: laccase IV from *Trametes versicolor*. Inorg. Chim. Acta 331: 232–238.
Bumpus, J.A. White-Rot Fungi and Their Potential Use in Soil Bioremediation Process. pp. 65–100. *In:* J.M. Bollag and G. Statzkim [eds.]. 1993. Soil Biochemistry. Marcel Dekker Inc., New York.
Bumpus, J.A. Microbial degradation of azo dyes. pp. 157–177. *In:* V.P. Singh [ed.]. 1995. Biotransformations: microbial degradation of health risk compounds. Elsevier Science, New York.

Call, H.P. and I. Mucke. 1997. History, overview and applications of mediated lignolytic systems, especially laccase-mediator-systems (Lignozym®-process). J. Biotech. 53: 163–202.

Cantarel, B.L., P.M. Coutinho, C. Rancurel, T. Bernard, V. Lombard and B. Henrissat. 2009. The Carbohydrate-Active Enzymes database (CAZy): an expert resource for Glycogenomics. Nucleic Acids Res. 37: D233–238.

Chagas, E.P. and L.R. Durrant. 2001. Decolorization of azo dyes by *Phanerochaete chrysosporium* and *Pleurotus sajorcaju*. Enzyme and Microbial Techn. 29: 473–477.

Chandra, R. and R. Rustgi. 1998. Degradable polymers. Progress in Polymer Science 23: 1273–1335.

Chandra, R., R.N. Bharagava and V. Rai. 2008. Melanoidins as major colourant in sugarcane molasses based distillery effluent and its degradation. Biores. Technol. 99: 4648–4660.

Claus, H. 2004. Laccases: structure, reactions, distribution. Micron. 35: 93–96.

Clemente, A.R. 2002. Degradação de hidrocarbonetos aromáticos policíclicos por fungos ligninolíticos e sua aplicação na biorremediação de solos contaminados. Tese (Doutorado em Ciências dos Alimentos)—Faculdade de Engenharia de Alimentos, Universidade Estadual de Campinas, Campinas.

Cohen, R., L. Persky and Y. Hadar. 2002. Biotechnological applications and potential of wood-degrading mushrooms of the genus *Pleurotus* Appl. Microb. and Biotech. 58: 582–594.

Collins, P., A.D.W Dobson and J.A. Field. 1998. Reduction of the 2,2′-Azinobis(3-Ethylbenzthiazoline-6-Sulfonate) Cation Radical by Physiological Organic Acids in the Absence and Presence of Manganese. Appl. and Environ. Microb. 64: 2026–2031.

Coulibaly, L., G. Germain and A. Spiros. 2003. Utilization of fungi for biotreatment of raw wastewaters. Afr. J. Biotechnol. 2: 620–630.

Crecchio, C., P. Ruggiero and M.D.R. Pizzigallo. 1995. Polyphenoloxidases inmobilized in organic gels: properties and applications in the detoxification of aromatic compounds. Biotech. and Bioengin. 48: 585–591.

D'Acunzo, F., C. Galli and B. Masci. 2002. Oxidation of phenols by laccase and laccase-mediator systems Solubility and steric issues. Eur. J. Bioch. 269: 5330–5335.

D'Souza, D.T., R. Tiwari, A.K. Sah and C. Raghukumar. 2006. Enhanced production of laccase by a marine fungus during treatment of colored effluents and synthetic dyes. Enzyme Microb. Technol. 38: 504–511.

Dellamatrice, P.M., R.T.R. Monteiro, H.M. Kamida, N.L. Nogueira, M.L. Rossi and C. Blaise. 2005. Decolourization of municipal effluent and sludge by *Pleurotus sajor-caju* and *Pleurotus ostreatus*. World J. Microb. and Biotech. 21: 1363–1369.

Dijkerman, R., D.C.P. Bhansing, H.J.M. Op den Camp, C. van der Drift and G.D. Vogels. 1997. Degradation of structural polysaccharides by the plant cell-wall degrading enzyme system from anaerobic fungi: An application study. Enz. Micr. Tech. 21: 130–136.

Dittmer, J.K., N.J. Patel, S.W. Dhawale and S.S. Dhawale. 1997. Production of multiple laccase isoforms by *Phanerochaete crysosporium* grown under nutrient sufficiency. FEMS Microb. Letters 149: 65–70.

Dupont, R.R., C.J. Bruell, M.C. Marley, D.C. Downey, R.D. Norris, S.G. Hulling and B. Pivets. 1997. Bioremediation. Annapolis: American Academy of Environ. Engineers and USEPA.

Durán, N. and E. Esposito. Biodegradação de lignina e tratamento de efluentes por fungos lignolíticos. *In:* I.S. Melo, J.L. Azevedo. 1997. Microbiologia ambiental, Jaguariúna: CNPM/EMBRAPA 269–292.

Durán, N., M.A. Rosa, A. D'Annibale, L and Gianfreda. 2002. Applications of Laccases and Tyrosinases (Phenoloxidases) Immobilized on Different Supports: A Review. Enzyme and Microb. Techn. 31: 907–931.

Eggen, T. 2000. Bioremediation of recalcitrant aromatic organic pollutants with white rot fungi. Dissertation, Agricultural University of Norway. Norway.

Eggert, C., U. Temp, J.F.D. Dean and K.E.L. Eriksson. 1996. A fungal metabolite mediates degradation of non-phenolic lignin structures and synthetic lignin by laccase. FEBS Letters 391: 144–148.

Enguita, F.J., L.O. Martins, A.O. Henriques and M.A. Carrondo. 2003. Crystal Structure of a Bacterial Endospore Coat Component. A Laccase with enhanced thermostability properties. J. Biol. Chem. 278: 19416–19425.

Enoki, A., T. Hirano and H. Tanaka. 1992. Extracellular substance from the brown rot basidiomycete *Gloeophyllum trabeum* that produces and reduces hydrogen peroxide. Mater Organism. 27: 247–261.

Eriksson, K.E. 1990. Biotech. in the pulp and paper industry. Wood Sci. and Techn. 24: 79–101.

Fähraeus, G. 1962. Aromatic compounds as growth substances for laccase producing rot fungi. Physiol. Plant. 15: 572–579.

Ferraz, A.L. Fungos decompositores de materiais lignocelulolíticos. *In:* E. Esposito and J.L. Azevedo (eds.). 2004. Fungos: uma introdução à biologia, bioquímica e biotecnológica. Educs. Caxias do Sul.

Ferreira, L.F.R. 2009. Fungi biodegradation of vinasse from industrial sugarcane processing. Thesis, University of São Paulo—ESALQ. São Paulo.

Ferreira, L.F., M. Aguiar, G. Pompeu, T.G. Messias and R.S. Monteiro. 2010. Selection of vinasse degrading microorganisms. World J. Microbiol. Biotechnol. 26: 1613–1621.

Ferreira, L.F.R., M.M. Aguiar, T.G. Messias, G.B. Pompeu, A.M.Q. Lopez, D.P. Silva and R.T.R. Monteiro. 2011. Evaluation of sugar-cane vinasse treated with *Pleurotus sajor-caju* utilizing aquatic organisms as toxicological indicators. Ecotoxicol. Environ. Saf. 74: 132–137.

Fragoeiro, S. and N. Magan. 2005. Enzymatic activity, osmotic stress and degradation of pesticide mixtures in soil extract liquid broth inoculated with *Phanerochaete chrysosporium* and *Trametes versicolor*. Environ. Microb. 7: 348–355.

Freire, R.S., N. Duran, J. Wang and L.T. Kubota. 2002. Laccase-based screen printed electrode for amperometric detection of phenolic compounds. Anal. Lett. 35: 29–38.

Fujian, X., C. Hongzhang and L. Zuohu. 2001. Solid-state production of lignin peroxidade (LiP) and manganese peroxidade (MnP) by *Phanerochaete chrysosporium* using steam-exploded straw as substrate. Bioresource Techn. 80: 149–151.

Galliano, H., G. Gas, J.L. Sevis and A.M. Boudet. 1991. Lignin degradation by *Rigidoporus lignosus* involves synergistic action of two oxidizing enzymes: Mn peroxidase and laccase. Enzyme Microb. Technol., 13: 478–482.

Gianfreda, L., F. Xu and J.M. Bollag. 1999. Laccases: A useful group of oxidoreductive enzymes. Bioremediation J. 3: 1–26.

Gold, M.H. and M. Alic. 1993. Molecular Biology of the Lignin-Degrading Basidiomycete *Phanerochaete chrysosporium*. Microbiol. Rev. 57: 605–622.

Gold, M.H., H.L. Youngs and M.D. Gelpke. 2000. Manganese peroxidase. Metal Ions in Biological Systems. 37: 559–586.

Gómez, B.L. and J.D. Nosanchuk. 2003. Melanin and fungi. Curr. Opin. Infect. Dis. 16: 91–96.

Gonzales, L., J.R. Hernández, F. Perestelo, A. Carnicero and M.A. Falcón. 2002. Relationship between mineralization of synthetic lignins and the generation of hydroxyl radicals by lacase and a low molecular weight substance produced by *Petriellidium fusoideum*. Enzyme and Microb. Techn. 30: 474–481.

González, T., M.C. Terrón, S. Yague, H. Junca, J.M. Carbajo, E.J. Zapico, R. Silva, A. Arana-Cuenca, A. Téllez and A.E. González. 2008. Melanoidin-containing wastewaters induce selective laccase gene expression in the white rot fungus *Trametes* sp. I-62. Res. Microbiol. 159: 103–109.

Grgic, I. and A. Perdih. 2003. Stimulation of ligninolytic enzyme production in *Phanerochaete chrysosporium* by polyxyalkanes. J. Appl. Microb. 94: 360–368.

Guillen, F., A.T. Martinez and M.J. Martinez. 1992. Substrate specificity and properties of the aryl-alcohol oxidase from the ligninolytic fungus *Pleurotus eryngii*. Eur. J. Biochem. 209: 603–611.

Guimarães, C., L.S.M. Bento and M. Mota. Assessment of the decolourization ability of *Phanerochaeta chrysosporium* towards different sugar colourants. pp. 259–262. *In:* E. Srebotnik and K. Messner [eds.]. 1996. Proceedings of the Sixty Int. Conf. on Biotech. in the Pulp and Paper Industry: Recent Advances in Appl. and Fundamental Research. Facultas-Universitätswerlag, Vienna, Austria.

Hammel, K.E. 1996. Extracellular free radical biochemistry of ligninolytic fungi. New J. Chem. 20: 195–198.

Hatakka, A. 1994. Lignin-modifying enzymes from selected white-rot fungi: production and role in lignin degradation. FEMS Microb. Reviews 13: 125–135.

Heinfling, A., M.J. Martínez, A.T. Martínez, M. Bergbauer and U. Szewzyk. 1998. Transformation of industrial dyes by manganese peroxidases from *Bjerkandera adusta* and *Pleurotus eryngii* in a manganese independent reaction. Appl. and Environ. Microb. 64: 2788–2793.

Henriksson, G., P. Ander, B. Pettersson and G. Pettersson. 1995. Cellobiose dehydrogenase (Cellobiose oxidase) as a wood degrading enzyme studies on cellulose, xylan and synthetic lignin. Appl. Microbiol. Biotechnol. 42: 790–796.

Henriksson, G., G. Johansson and G. Pettersson. 2000a. A critical review of cellobiose dehydrogenases. J. Biotechnol. 78: 93–113.

Henriksson, G., L. Zhang, J. Li, P. Ljungquist, T. Reitberger, G. Pettersson and G. Johansson. 2000b. Is cellobiose dehydrogenase from *Phanerochaete chrysosporium* a lignin degrading enzyme? Biochim. Biophys. Acta 1480: 83–91.

Hernández-Luna, C.E., G. Gutiérrez-Soto and S.M. Salcedo-Martínez. 2008. Screening for decolorizing basidiomycetes in México: Screening and selection of ligninolytic basidiomycetes with decolorizing ability in Northeast México. World J. Microb. and Biotech. 24: 465–473.

Hibbett, D.S. and M. Binder. 2007. A higher-level phylogenetic classification of the Fungi. Mycol. Res. 111: 509–547.

Higuchi, T. 1990. Lignin biochemistry: biosynthesis and biodegradation. Wood Science and Techn. 24: 23–63.

Hofer, C. and D. Schlosser. 1999. Novel enzymatic oxidation of Mn^{2+} to Mn^{3+} catalyzed by a fungal laccase. FEBS Letters 451: 186–190.

Hoff, T., S.Y. Liu and J.M. Bollag. 1985. Transformation of Halogen-, Alkyl-, and Alkol-Substituted Anilines by a Laccase of *Trametes versicolor*. Appl. and Environ. Microb. 49: 1040–1045.

Hofrichter, M. 2000. Mangan-Peroxidasen: enzymatische verbrennung von lignin. BioSpektrum 6: 198–199.

Hofrichter, M. 2002. Review: lignin conversion by manganese peroxidase (MnP). Enzyme and Microbial Techn. 30: 454–466.

Hofrichter, M., T. Vares, K. Scheibner, S. Galkin, J. Sipila and A. Hatakka. 1998. Mineralization and solubilization of synthetic lignin (DHP) by manganese peroxidases from *Nematoloma frowardii* and *Phlebia radiata*. J. Biotech. 67: 217–228.

Husaini, A., F.A. Fisol, L.C. Yun, M.H. Hussain, S. Muid and H.A. Roslan. 2011. Lignocellulolytic enzymes produced by tropical white-rot fungi during biopulping of *Acacia mangium* wood chips. J. Biochem Tech. 3: 245–250.

Jensen, K.A.J., W.B. Bao , S. Kawai, E. Srebotnik and K.E. Hammel. 1996. Manganese-dependent cleavage of nonphenolic lignin structures by Ceriporiopsis subvermispora in theabsence of lignin peroxidase. Appl. Environ. Microbiol. 62: 3679–3686.

Jimnez, A.M., R. Borja and A. Martin. 2003. Aerobic–anaerobic biodegradation of beet molasses alcoholic fermentation wastewater. Proc. Biochem. 38: 1275–1284.

Jönsson, L.J., E. Palmqvist, N.O. Nilvebrant and B. Hahn-Hägerdal. 1998. Detoxification of wood hydrolysates with laccase and peroxidase from the white-rot fungus *Trametes versicolor*. Appl. Microb. and Biotech. 49: 691–697.

Jørgensen, H., A. Mørkeberg, K. Krogh and L. Olsson. 2005. Production of cellulases and hemicellulases by three *Penicillium* species: effect of substrate and evaluation of cellulose adsorption by capillary electrophoresis. Enzyme and Microbial Techn. 36: 42–48.

Juarez, R.A., L.L.G. Dorry, R. Bello-Mendoza and J.E. Sanchez. 2011. Use of spent substrate after *Pleurotus pulmonarius* cultivation for the treatment of chlorothalonil containing wastewater. J. Environ. Manag. 82: 948–952.

Kahraman, S.S. and I.H. Gurdal. 2002. Effect of synthetic and natural culture media on laccase production by white rot fungi. Bioresource Techn. 82: 215–217.

Kahraman, S.S. and O. Yeşilada. 2003. Decolorization and bioremediation of molasses wastewater by white-rot fungi in a semi-solid-state condition. Folia Microbiol. 48: 525–528.

Kamida, H.M., L.R. Durrant, R.T.R. Monteiro and E.D. Armas. 2005. Biodegradação de efluentes têxteis por *Pleurotus sajo-caju*. Quím. Nova 28: 629–632.

Kerem, Z., D. Friesem and Y. Hadar. 1992. Lignocellulose degradation during solid-state fermentation: *Pleurotus ostreatus* versus *Phanerochaete chrysosporium*. Appl. and Environ. Microb. 58: 1121–1127.

Kersten, P.J. and D. Cullen. 2007. Extracellular oxidative systems of the lignin-degrading Basidiomycete *Phanerochaete chrysosporium*. Fungal Genet. Biol. 44: 77–87.

Kersten, P.J., B. Kalyanaraman, K.E. Hammel, B. Reinhammar and T.K. Kirk. 1990. Comparison of lignin peroxidase, horseradish peroxidase and laccase in the oxidation of methoxybenzenes. Bioch. J. 268: 475–480.

Kirk, T.K. and R.L. Farrel. 1987. Enzymatic "combustion": The microbial degradation of lignin. Annual Rev. Microb. 41: 465–505.

Kirk, T.K. and D. Cullen. Enzimology and molecular genetics of wood degradation by white-rot fungi. pp. 273–308. *In:* T.K. Kirk (ed.). 1998. Environmentally friendly technologies for the pulp and paper industry. John Wiley and Sons, New York.

Kirk, T.K., S. Croan and M. Tien. 1986. Production of multiple ligninases by *Phanerochaete chrysosporium*: effect of selected growth conditions and use of a mutant strain. Enzyme and Microbial Techn. 8: 27–32.

Klyosov, A.A. 1990. Trends in biochemistry and enzymology of cellulose degradation, Biochemistry 29: 10577–10585.

Kuhad, R.C., A. Singh and K.E.L. Eriksson. 1997. Microorganisms and enzymes involved in the degradation of plant fiber cell walls. Adv. Bioch. Engine./Biotech. 57: 45–126.

Kumar, R., S. Singh and O.V. Singh. 2008. Bioconversion of lignocellulosic biomass: biochemical and molecular perspectives. J. Ind. Microbiol. Biotechnol. 35: 377–391.

Kumar, V., L. Wati, P. Nigam, I.M. Banat, B.S. Yadav, D. Singh and R. Marchant. 1998. Decolorization and biodegradation of anaerobically digested sugarcane molasses spent wash effluent from biomethanation plants by white-rot fungi. Proc. Biochem. 33: 83–88.

Kuwahara, M., J.K. Glenn, M.A. Morgan and M.H. Gold. 1984. Separation and characterization of two extracellular H_2O_2 dependent oxidases from lignolytic cultures of *Phanerochaete chrysosporium*. FEBS Microb. Letters 169: 247–250.

Lamed, R., E. Setter and E.A. Bayer. 1983. Characterization of a cellulose-binding, cellulase-containing complex in *Clostridium thermocellum*. J. Bacteriol. 156: 828–836.

Langh, E., G. Eller and F. Zadrazil. 1997. Lignocellulose decomposition and production of ligninolytic enzymes during interaction of white rot fungi with soil microorganisms. Microb. Ecol. 34: 1–10.

Lee, K. and S.H. Moon. 2003. Electroenzymatic oxidation of veratryl alcohol by lignin peroxidase. J. Biotech. 102: 261–268.

Leonowicz, A., A. Matuszewska, J. Luterek, D. Ziegenhagen, M. Wojtaś-Wasilewska, N.S. Cho, M. Hofrichter and J. Rogalski. 1999. Biodegradation of lignin by white-rot fungi. Fungal Gen. and Biol. 27: 175–185.

Levasseur, A., F. Piumi, P.M. Coutinho, C. Rancurel, M. Asther, M. Delattre, B. Henrissat, P. Pontarotti, M. Asther and E. Record. 2008. FOLy: An integrated database for the classification and functional annotation of fungal oxidoreductases potentially involved in the degradation of lignin and related aromatic compounds. Fungal Gen. and Biol. 45: 638–645.

Liew, C.Y., A. Husaini, H. Hussain, S. Muid, K.C. Liew and H.A. Roslan. 2011. Lignin biodegradation and lignin enzyme studies during biopulping of *Acacia mangium* wood chips by tropical white rot fungi. World J. Microbiol. Biotechnol. 27:1457–1458.

Ljungdahl, L.G. 2008. The cellulase/hemicellulase system of the anaerobic fungus *Orpinomyces* PC-2 and aspects of its applied use. Ann. N Y Acad Sci. 1125: 308–321.

Lobarzenwski, J. 1990. The characteristics and functions of the peroxidases from *Trametes versicolor* in lignin biotransformation. J. Biotech. 13: 11–117.

Lobos, S., J. Larraín, L. Salas, D. Cullen and R. Vicuna. 1994. Isozymes of manganese dependent peroxidase and laccase produced by the lignin-degrading basidiomycete *Ceriporiopsis subvermispora*. Microb. 140: 1691–1698.

Machado, K.M.G. 1998. Biodegradação de Pentaclorofenol por fungos basidiomicetos lignocelulolíticos em solos contaminados com resíduos industriais. Thesis, Universidade Estadual Paulista—UNESP, Rio Claro, São Paulo, Brazil. 172 p.

Machado, K.M.G., D.R. Matheus and V.L.R. Bononi. 2005a. Biodegradation of pentachlorophenol by tropical basidiomycetes in soils contaminated with industrial residues. World J. Microb. and Biotech. 21: 297–310.

Machado, K.M.G., D.R. Matheus and V.L.R. Bononi. 2005b. Ligninolytic enzymes production and Remazol Brilliant Blue R decolorization by tropical Brazilian basidiomycetes. Braz. J. Microb. 36: 246–252.

Magan, N., S. Fragoeiro and C. Bastos. 2010. Environ. factors and bioremediation of xenobiotics using white-rot fungi. Microb. 38: 238–248.

Majcherczyk, A., C. Johannes and A. Huttermann. 1999. Oxidation of aromatic alcohols by laccase from *Trametes versicolor* mediated by the 2,2′-azino-bis-(3-ethylbenzothiazoline-6-sulphonic acid) cation radical and dication. Appl. Microb. and Biotech. 51: 267–276.

Malmström, B.G., L.E. Andéasson and B. Reinhammer. 1975. The enzymes. Academic Press, New York.

Mane, J.D., S. Modi, S. Nagawade, S.P. Phadnis and V.M. Bhandari. 2006. Treatment of spentwash using chemically modifed bagasse and colour removal studies. Biores. Technol. 97: 1752–1755.

Mansur, M., T. Suarez, J. Fernandez-Larrea, M.A. Brizuela and A.E. Gonzalez. 1997. Identification of a laccase gene family in the new lignin-degrading basidiomycete CECT 20197. Appl. Environ, Microbiol. 63: 2637–2646.

Martinez, A.T., M. Speranza, F.J. Ruiz-Duenas, P. Ferreira, S. Camarero, F. Guillen, M.J. Martinez, A. Gutierrez and J.C. del Rio. 2005. Biodegradation of lignocellulosics: microbial, chemical, and enzymatic aspects of the fungal attack of lignin. Int. Microbiol. 8: 195–204.

Matheus, D.R., V.L.R. Bononi and K.M.G. Machado. New basidiomycetes on bioremediation of organochlorine contaminated soil. pp. 99–106. *In:* V.S.M. Glenn, J.S.K. Ong and A. Leeson [eds.]. 2001. Bioremediation of Energetics, Phenolics, and Polycyclic Aromatic Hydrocarbons. Battelle Press, San Diego.

Mayer, A.M. and R.C. Staples. 2002. Laccase: new functions for an old enzyme. Phytochem. 60: 551–565.

Mensah, C.A., N.A. Adamafio, K. Amaning-Kwarteng and F.K. Rodrigues. 2012. Reduced Tannin Content of Laccase-treated Cocoa (Theobroma cacao) Pod Husk. Int. J. Biol. Chem. 6: 31–36.

Migo, V.P., M. Matsumura, E.J. del Rosario and H. Kataoka. 1993. Decolourization of molasses waste water using an inorganic flocculent. J. Ferment. Bioeng. 75: 438–442.

Miranda, P.M., G.G. Benito, N.S. Cristobal and C.H. Nieto. 1996. Color elimination from molasses wastewater by *Aspergillus níger*. Bioresour. Technol. 57: 229– 235.

Miyata, N., K. Iwahori and M. Fujita. 1998. Manganese-independent and -dependent decolorization of melanoidin by extracellular hydrogen peroxide and peroxidases from *Coriolus hirsutus* pellets. J. Ferment. Bioeng. 85: 550–553.

Mohammad, P., H. Azarmidokht, M. Fatollah and B. Mahboubeh. 2006. Application of response surface methodology for optimization of important parameters in decolorizing treated distillery wastewater using *Aspergillus fumigatus* UB2.60. Int. Biodet. Biodegrad. 57: 195–199.

Mohana, S., B.K. Acharya and D. Madamwar. 2009. Distillery spent wash: treatment technologies and potential applications. J. Hazard. Mater. 163: 12–25.

Moredo, N., M. Lorenzo, A. Domínguez, D. Moldes, C. Cameselle and A. Sanroman. 2003. Enhanced ligninolytic enzyme production and degrading capability of *Phanerochaete chrysosporium* and *Trametes versicolor*. World J. Microb. Biotechnol. 19: 665–669.

Moreira Neto, S.L. 2006. Enzimas ligninolíticas produzidas por *Psilocybe castanella* CCB444 em solo contaminado com hexaclorobenzeno. Dissertação (Mestrado em Biodiversidade Vegetal e Meio Ambiente)—Instituto de Botânica da Secretaria do Meio Ambiente, São Paulo.

Mougin, C., C. Jolivalt, P. Briozzo and C. Madzak. 2003. Fungal laccases: from structure-activity studies to Environ. applications. Environ. Chemistry Letters 1: 145–148.

Nagy, T., R.B. Tunnicliffe, L.D. Higgins, C. Walters, H.J. Gilbert and M.P. Williamson. 2007. Characterization of a double dockerin from the cellulosome of the anaerobic fungus *Piromyces equi*. J Mol Biol. 373: 612–622.

Naik, N.M., K.S. Jagadeesh and A.R. Alagawadi. 2008. Microbial decolorization of spentwash: a review. Indian J. Microbiol. 48: 41–48.

Novikova, L.N., A.S. Medvedeva, I.V. Volchatova and S.A. Bogatyreva. 2002. Changes in macromolecular characteristics and biological activity of hydrolytic lignin in the course of composting. Appl. Bioch. and Microb. 38: 181–185.

Novotný, C., P. Erbanova, V. Šašek, A. Kubatova, T. Cajthaml, E. Lang, J. Krahl and F. Zadražil. 1999. Extracellular oxidative enzyme production and PAH removal in soil by exploratory mycelium of white rot fungi. Biodegradation 10: 159–168.

Novotný, C., B. Rawal, M. Bhatt, M. Patel, V. Šašek and H.P. Molitoris. 2001. Capacity of *Irpex lacteus* and *Pleurotus ostreatus* for decolorization of chemically different dyes. J. Biotech. 89: 113–122.

Novotný, C., K. Svobodová, P. Erbanová, T. Cajthaml, A. Kasinath, E. Lang and V. Šašek. 2004. Ligninolytic fungi in bioremediation: extracellular enzyme production and degradation rate. Soil Biol. and Bioch. 36: 1545–1551.

Ohmomo, S., Y. Kaneko, S. Sirianuntapiboon, P. Somchal, P. Atthasampunna and I. Nakamura. 1987. Decolorization of molasses wastewater by a thermophilic strain, *Aspergillus fumigatus* G-2-6. Agric. Biol. Chem. 51: 3339–3346.

Orpin, C.G. 1975. Studies on the rumen flagellate *Neocallimastix frontalis*. J. Gen. Microbiol. 91: 249–262.

Ortiz-Bermúdez, P., E. Srebotnik and K.E. Hammel. 2003. Chlorination and cleavage of lignin structures by fungal chloroperoxidases Appl. Environ. Microbiol. 69: 5015–5018.

Palmieri, G., P. Giardina, C. Bianco, A. Scaloni, A. Capasso and G. Sannia. 1997. A novel white laccase from *Pleurotus ostreatus*. J. Biol. Chem. 272: 31301–31307.

Pant, D. and A. Adholeya. 2007a. Biological approaches for treatment of distillery wastewater: a review. Biores. Technol. 98: 2321–2334.

Pant, D. and A. Adholeya. 2007b. Identification, ligninolytic enzyme activity and decolorization potential of two fungi isolated from a distillery effluent contaminated site. Water Air Soil Pollut. 183: 165–176.

Pant, D. and A. Adholeya. 2007c. Enhanced production of ligninolytic enzymes and decolorization of molasses distillery wastewater by fungi under solid state fermentation. Biodegradation 18: 647–659.

Pant, D. and A. Adholeya. 2009. Concentration of fungal ligninolytic enzymes by ultrafiltration and their use in distillery effluent decolorization. World J. Microbiol. Biotechnol. 25: 1793–1800.

Prada, S.M., M. Guekezian and M.E.V. Suarez-Iha. 1998. Metodologia analítica para a determinação de sulfato em vinhoto. Quím. Nova 21: 249–252.

Pérez, S.R., R.C. Savón, M.S. Diaz and A. Kourouma. 2006. Selección de cepas de Pleurotus *ostreatus* para la decoloración de efluentes industriales. Rev. Mex. Micol. 23: 9–15.
Punnapayak, H., S. Prasongsuk, K. Messner, K. Danmek and A. Lotrakul. 2009. Polycyclic aromatic hydrocarbons (PAHs) degradation by laccase from a tropical white-rot frungus *Ganoderma lucidum*. Afr. J. Biotechnol. 8: 5897–59000.
Purnomo, A.S., T. Mori, I. Kamei, T. Nishii and R. Kondo. 2010. Application of mushroom waste medium from *Pleurotus ostreatus* for bioremediation of DDT-containing contaminated soil. Int. Biodet. and Biodegr. 64: 397–402.
Reddy, C.A. 1993. An overview of the recent advances on the physiology and molecular biology of lignin peroxidases of *Phanerochaete chrysosporium*. J. Biotech. 30: 91–107.
Renganathan, V., S.N. Usha and F. Lindenburg. 1990. Cellobiose oxidizing enzymes from the lignocellulose-degrading basidiomycete *Phanerochaete chrysosporium*—interaction with microcrystalline cellulose. Appl. Microbiol. Biotechnol. 32: 609–613.
Rodríguez, E., M. Pickard and R. Vazquez-Duhalt. 1999. Industrial dye decolorization by laccases from ligninolytic fungi. Curr. Microbiol. 38: 27–31.
Rojas, O. 2011. NanoBiotech. in lignocellulosic applications. North Carolina State University. Forest Biomaterials Science and Engineering. Available at http://www4.ncsu.edu/-ojrojas Accessed May 1, 2012.
Rufián-Henares, J.A. and F.J. Morales. 2007. Functional properties of melanoidins: *In vitro* antioxidant, antimicrobial and antihypertensive activities. Food Res. Int. 40: 995–1002.
Sanchez, C. 2009. Lignocellulosic residues: biodegradation and bioconversion by fungi. Biotechnol. Adv. 27: 185–194.
Saha, B.C. 2003. Hemicellulose bioconversion. J. Ind. Microbiol. Biotechnol. 30: 279–291.
Sandgren, M., J. Stahlberg and C. Mitchinson. 2005. Structural and biochemical studies of GH family 12 cellulases: improved thermal stability, and ligand complexes. Prog. Biophys. Mol Biol. 89: 246–291.
Saparrat, M., F. Guillén, A.M. Arambarri, A.T. Martinez and M.J. Martinez. 2002. Induction, Isolation, and Characterization of Two Laccases from the White Rot Basidiomycete *Coriolopsis rigida*. Appl. and Environ. Microb. 68: 1534–1540.
Sayadi, S. and R. Ellouz. 1995. Roles of lignin peroxidase and manganese peroxidase from *Phanerochaete crysosporium* in the decolorization of olive mill wastewaters. Appl. and Environ. Microb. 61: 1098–1103.
Schliephake, K. and G.T. Lonergan. 1996. Laccase variation during dye decolourisation in a 200 L packed-bed bioreactor. Biotech. Letters 18: 881–886.
Schlosser, D. and C. Höfer. 2002. Laccase-Catalyzed Oxidation of Mn^{2+} in the Presence of Natural Mn^{3+} Chelators as a Novel Source of Extracellular H_2O_2 Production and Its Impact on Manganese Peroxidase. Appl. and Environ. Microb. 68: 3514–3521.
Schneider, P., M.B. Caspersen, K. Mondorf, T. Halkier, L.K. Skov and P.R. Ostergaard. 1999. Characterization of a *Coprinus cinereus* laccase. Enzyme and Microbial Techn. 25: 502–508.
Shallom, D. and Y. Shoham. 2003. Microbial hemicellulases. Curr. Opin. Microbiol. 6: 219–228.
Shan, V. and F. Nerud. 2002. Lignin degrading system of white-rot fungi and its exploitation for dye decolorization. Canadian J. of Microb. 48: 857–870.
Shayegan, J., M. Pazouki and A. Afshari. 2004. Continuous decolorization of anaerobically digested distillery wastewater. Proc. Biochem. 40: 1323–1329.
Shleev, S.V., O.V. Morozova, O.V. Nikitina, E.S. Gorshina, T.V. Rusinova, V.A. Serezhenkov, D.S. Burbaev, I.G. Gazaryan and A.I. Yaropolov. 2004. Comparison of physico-chemical characteristics of four laccases from different basidiomycetes. Biochimie. 86: 693–703.
Shüitzendüel, A. and A. Majancherczk. 1999. A degradation of fluorene, phenenthrene, fluoranthene, and pyrene lacks conection to the production of extracellular enzymes by *Pleurotus ostreatus* and *Bjerkandera adusta*. Intern. Biodeterioration 43: 93–100.

Silva, I.S., M. Grossman and L.R. Durrant. 2009. Degradation of polycyclic aromatic hydrocarbons under microaerobic and very low oxygen conditions by soil fungi. Int. Biodet. and Biodegr. 63: 224–229.

Silva, V.L.M.M., W.C. Gomes and O.L.S. Alsina. 2007. Utilização do bagaço de cana-de-açúcar como biomassa adsorvente na adsorção de poluentes orgânicos. Rev. Eletr. de Mat. e Proc. 2: 27–32.

Singh, C.S., V.K. Singh, S.P. Tiwari and R. Sharma. 2011. Ligninolytic activity and lignocellulosic degradation profiles by natural isolates of *Pleurotus flabellatus* during mushroom cultivation on rice straw. Asian J. Exp. Biol. Sci. 2: 158–161.

Sirianuntapiboon, S., P. Sihanonth, P. Somchai, P. Atthasampunna and S. Hayashida. 1995. An adsorption mechanism for melanoidin decolorization by *Rhizoctonia* sp. Biosci. Biotechnol. Biochem. 59: 1185–1189.

Souza, D.F., S.C. Costa, A.S. Dacome, C.G. MSouza, A. Bracht and R.M. Peralta. 2011. Pentachlorophenol removal by *Pleurotus pulmonarius* in submerged cultures. Braz. Arch. Biol. Technol. 54(2): 357–362.

Srebotnik, E., K. Messner and R. Foisner. 1988. Penetrability of White Rot-Degraded Pine Wood by the Lignin Peroxidase of *Phanerochaete chrysosporium*. Appl. Environ. Microbiol. 54: 2608–2614.

Sugiura, M., H. Hirai and T. Nishida. 2003. Purification and characterization of a novel lignin peroxidase from white-rot fungus *Phanerochate sordida* YK-624. FEMS Microb. Letters 224: 285–290.

Szklarz, G., R.K. Antibus, R.L. Sinsabaugh and A.E. Linkins. 1989. Production of phenoloxidases and peroxidases by wood-rotting fungi. Mycol. 81: 234–240.

Suzuki, M.R., C.G. Hunt, C.J. Houtman, Z.D. Dalebroux and K.E. Hammel. 2006. Fungal hydroquinones contribute to brown rot of wood. Environ. Microbiol. 8: 2214–2223.

Tanaka, H., S. Itakura and A. Enoki. 1999. Hydroxyl radical generation by an extracellular low-molecular-weight substance and phenol oxidase activity during wood degradation by the white-rot basidiomycete *Phanerochaete chrysosporium*. Holzforschung. 53: 21–28.

Tanaka, H., G. Yoshida, Y. Baba, K. Matsumura, H. Wasada, J. Murata, M. Agawa, S. Itakura and A. Enoki. 2007. Characterization of a hydroxyl-radical-producing glycoprotein and its presumptive genes from the white-rot basidiomycete *Phanerochaete chrysosporium*. J. Biotechnol. 128: 500–511.

Tanaka, M., H. Nakamura, M. Taniguchi, T. Morita, R. Matsuno and T. Kamikubo. 1986. Elucidation of adsorption processes of cellulases during hydrolysis of crystalline cellulose. Appl. Microb. and Biotech. 23: 263–268.

Thurston, C.F. 1994. The structure and function of fungal laccases. Microb. 140: 19–26.

Tien, M. and T.K. Kirk. 1984. Lignin-degrading enzyme from *Phanerochaete chrysosporium*: purification, characterization and catalytic properties of a unique H_2O_2 requiring oxygenase. Proceedings of the National Academy of Science 81: 2280–2284.

Tripathi, A., R.C. Upadhyay and S. Singh. 2011. Mineralization of mononitrophenols by *Bjerkandera adusta* and *Lentinus squarrosulus* and their extracellular ligninolytic enzymes. J. Basic. Microbiol. 51: 1–15.

Trupkin, S., L. Levin, F. Forchiassin and A. Viale. 2003. Optimization of a culture medium for ligninolytic enzyme production and synthetic dye decolorization using response surface methodology. J. Ind. Microb. and Biotech. 30: 682–690.

Tsai, H.F., M.H. Wheeler, Y.C. Chang and K.J. Kwon-Chung. 1999. A developmentally regulated gene cluster involved in conidial pigment biosynthesis in *Aspergillus fumigatus*. J. Bacteriol. 181: 6469–6477.

Tuomela, M., M. Vikman, A. Hatakka and M. Itavaara. 2000. Biodegradation of lignin in a compost environment: a review. Bioresource Techn. 72: 169–183.

Tuor, U., K. Winterhalter and A. Fiechter. 1995. Enzymes of white-rot-fungi involved in lignin degradation and ecological determinants for wood decay. J. Biotech. 41: 1–17.

Vahabzadeh, F., M. Mehranian and A.R. Saatari. 2004. Color removal ability of *Phanerochaete chrysosporium* in relation to lignin peroxidase and manganese peroxidase produced in molasses wastewater. World. J. Microbiol. Biotechnol. 20: 859–864.

Vanacken, B., L.M. Godefroid, C.M. Peres, H. Naveau and S.N. Agathos. 1999. Mineralization of 14C- ring labeled 4-hydroxylamino-2,6-dinitrotoluene by manganese-dependent peroxidase of the white-rot basidiomycete *Phlebia radiata*. J. Biotech. 68: 159–169.

Vares, T., M. Kalsi and A. Hatakka. 1995. Lignin peroxidase, manganese peroxidase, and others ligninolytic enzymes produced by *Phlebia radiata* during solid-state fermentation of wheat straw. Appl. and Environ. Microb. 61: 3515–3520.

Wadt, L.C. 2008. Cultivo de *Pleurotus* spp. Em vinhaça visando à produção de biomassa e exopolissacarídeos. Dissertation, University of São Paulo—CENA.São Paulo.

Wang, L., W. Yan, J. Chen, F. Huang and P. Gao. 2008. Function of the iron-binding chelator produced by *Coriolus versicolor* in lignin biodegradation. Sci. China C. Life Sci. 51: 214–221.

Wang, W., F. Huang, X. Mei Lu and P. Ji Gao. 2006. Lignin degradation by a novel peptide, Gt factor, from brown rot fungus *Gloeophyllum trabeum*. Biotechnol. J. 1:447–453.

Wesenberg, D., I. Kyriakides and S.N. Agathos. 2003. White-rot fungi and their enzymes for the treatment of industrial dye effluents. Biotechnol. Adv. 22: 161–187.

Wessels, J.G.H. 1994. Development regulation of fungal cell wall formation. Annual Review of Phytopathology 32: 413–437.

Weng, J.K., X. Li, N.D. Bonawitz and C. Chapple. 2008. Emerging strategies of lignin engineering and degradation for cellulosic biofuel production. Curr. Opin. Biotechnol. 19: 166–172.

Wesenberg, D., I. Kyriakides and S.N. Agathos. 2003. White-rot fungi and their enzymes for the treatment of industrial dye effluents. Biotechnol. Adv. 22: 161–187.

Wilson, C.A. and T.M. Wood. 1992. The anaerobic fungus *Neocallimastix frontalis*: isolation and properties of a cellulosome-type enzyme fraction with the capacity to solubilize hydrogen-bond-ordered cellulose. Appl. Microbiol. Biotechnol. 37: 125–129.

Wood, T.M. and V. Garcia-Campayo. 1990. Enzymology of cellulose degradation. Biodegradation 1: 147–167.

Xiao, P., T. Mori and R. Kondo. 2011. Biotransformation of the organochloride pesticide trans-chlordane by food rot fungi. New Biotech. 29: 107–115.

Yang, W., J. Liu, W. Wang, Y. Zhang and P. Gao. 2004. Function of a low molecular peptide generated by cellulolytic fungi for the degradation of native cellulose. Biotechnol. Lett. 26: 1799–1802.

Yaver, D.S., F. Xu, E.J. Golightly, K.M. Brown, S.H. Brown, M.W. Rey, P. Schneider, T. Halkier, K. Mondorf and H. Dalboge. 1996. Purification, Characterization, Molecular Cloning, and Expression of Two Laccase Genes from the White Rot Basidiomycete *Trametes villosa*. Appl. and Environ. Microb. 62: 834–841.

Yoon, J.J., C.J. Cha, Y.S. Kim, D.W. Son and Y.K. Kim. 2007. The brown-rot basidiomycete *Fomitopsis palustris* has the endo-glucanases capable of degrading microcrystalline cellulose. J. Microbiol. Biotechnol. 17: 800–805.

4

Proteolytic Enzymes: Biochemical Properties, Production and Biotechnological Application

Fabiana F. Zanoelo,[1,a,*] *Giovana C. Giannesi*[1,b] and *Hamilton Cabral*[2]

Introduction

Proteases are enzymes that have attracted the interest of many researchers, since antiquity. Studies of these enzymes began in the 16th century and were empirically already used in the preparation of cheese and derivatives and in the preparation of leather. Currently, these enzymes are used in many industrial sectors, especially in textiles, pharmaceuticals, food and detergents (Gupta et al. 2002, Salleh et al. 2006). The peptidases, peptide hydrolases

[1]Biological Sciences and Health Center, Federal University of Mato Grosso do Sul. Brazil.
[a]E-mail: fabiana.zanoelo@ufms.br
[b]E-mail: giannesigiovana@hotmail.com
[2]Department of Pharmaceutical Science, School of Pharmaceutical Science of Ribeirão Preto, University of São Paulo, Brazil.
E-mail: hamilton@fcfrp.usp.br
*Corresponding author

or just proteases as they are known, are hydrolytic enzymes which cleave peptide bonds in other proteins and peptide fragments (Barrett 1994). Proteases play an important role in all physiological processes, ranging from the general downturn of the protein as a nutrient to the regulation of programmed cell death. They are produced in all living organisms, actively participating in metabolic pathways and complex cell signaling pathways (Turk 2006). In general, the extracellular proteases catalyze the hydrolysis of various proteins into smaller molecules that are subsequently absorbed by the cells, while intracellular proteases play an important role in regulating the metabolism (Rao et al. 1998).

Proteases extracted from plants and animals are not able to meet industrial demand and therefore microbial proteases have been given greater importance. Microbial proteases have almost all the desired characteristics for biotechnological applications, and are caused mainly by the biochemical diversity and ease of genetic manipulation of microorganisms as well as their rapid growth. Additionally many microorganisms secrete proteases to the external environment in order to degrade proteins whose hydrolysis products serve as a source of nitrogen and carbon for cell multiplication. Due to rapid growth, filamentous fungi are capable of producing proteases by fermentation processes using various substrates. This chapter will discuss fungal proteases, emphasizing the biochemical characteristics, production and biotechnological applications.

Biochemical Characteristics

Peptidases (EC 3.4) are enzymes that are able to hydrolyze peptide bonds in proteins and peptides. They represent approximately 2 percent of the total protein present in all types of organisms (Polgár 2005). The MEROPS database (http://merops.sanger.ac.uk) is a rich source of information on peptidases.

These days, the terms peptidase, protease, proteinase and proteolytic enzymes are used interchangeably. This results in a redundancy of unnecessary terminology (Barrett and Rawlings 2007).

The term peptidase is favored the most as it is an obvious contraction of "hydrolysis of peptide bond", and also provides the root of the whole range of terminology used, including exopeptidase, endopeptidase, aminopeptidase and carboxypeptidase. The term peptidase is recommended by the Nomenclature Committee of IUBMB (http://www.chem.qmul.ac.uk/iubmb/enzyme/EC34/) (Barrett and Rawlings 2007).

Specificity

Schechter and Berger (1967) developed the nomenclature of the peptidases of the catalytic site, dividing it into a series of sub sites, each subsite accommodates a single and consecutive amino acid residue of the substrate.

The amino acid residues of the substrate are numbered from the point of cleavage of the peptide bond P_1---P'_1 position. The portion P_n-...-P_3-P_2-P_1 is the amino terminal direction, since the portion P'1-P'2-P'3-...P'_n is toward the carboxy-terminal substrate, while the corresponding sub-sites in peptidase are S_n-...-S_3-S_2-S_1-S'_1-S'_2-S'_3-...S'_n (Schechter and Berger 1967).

The determination of specificity is the first step when characterization of new peptidases is desired. Obtained results demonstrate the capability of the tested enzyme to discern between different substrates (Fretweel et al. 2008).

The specificity of an enzyme for a substrate is extremely important in biological systems. Substrate selection depends on several factors, including its location within the cell and regulation of enzyme expression. The amino acid sequence of the substrate is the determining factor for many endopeptidases as to cleave the substrate or not. This information can be used for the construction of small peptide substrates (Wildeboer et al. 2009).

Peptide substrates with intra-molecular fluorescence suppression include the most versatile class of synthetic substrates for peptidases currently available. This class of substrate is useful for a wide application in enzymology, due to its simplicity, speed and accuracy in the determination of reaction constants (Alves et al. 2007).

These substrates may be considered as a non radioactive process in which energy is transferred from a donor to an acceptor when they have an overlap of emission/absorption spectrum with an appropriate orientation and distance (generally between 10 and 100Å) (Zhang 2004).

According Hatanaka et al. (2009) studies on preferences of different peptidase substrates have been investigated using combinatorial libraries of peptide substrates with intra-molecular fluorescence suppressed.

Classification of Peptidases

The peptidases are classified based on three main criteria: type of catalyzed reaction, chemical nature of the catalytic site and the relationship of the evolution of the primary structure (Rao et al. 1998).

Regarding the type of catalyzed reaction, the peptidases are endopeptidases and exopeptidases. Endopeptidases cleave peptide bonds for which amino acid residues are internal to the polypeptide chain. Exopeptidases cleave peptide bonds at the ends which are amino (aminopeptidase) and carboxy-terminal (carboxypeptidase) (Garcia-Carreño and Del Toro 1997).

The peptidases can be grouped into six catalyst types based on the nature of the nucleophile of the reaction. Rawlings and Barrett in (1993) recognized four catalytic types, which are serine, cysteine, aspartic and metallo-peptidases, a concept which recaptures the ideas of Hartley (1960) and Rawlings et al. (2011).

In serine, threonine and cysteine peptidases, the nucleophile is the oxygen of the hydroxyl of the side chain of serine of the active site, the sulfur of the side chain of cysteine of the active site and the hydroxyl oxygen belonging to the side chain of threonine of the active site. In the case of metallo, aspartic and glutamic peptidases, the nucleophile is a molecule of water in aspartic peptidase which is activated by the two aspartic of the active site and metalloproteinase peptidase by one or two metal ions (usually zinc, but also cobalt, manganese, nickel, copper and iron) (Rawlings et al. 2011).

In 1995 the fifth type of peptidase was discovered with the resolution of the structure of the proteasome. Three of the 14 sub-units of peptidases showed threonine at the N-terminal region, which works as a nucleophile, and thus sets the fifth class as threonine peptidase (Rawlings et al. 2011).

The sixth peptidase catalytic type or class was identified in 2004 when certain endopeptidases from fungus now known as eqilisinas were discovered, called glutamic peptidase (Rawlings et al. 2011).

The active site of serine peptidases that is constituted by the catalytic triad consists of the following amino acids, Ser/His/Asp, however modifications can occur because of the position of the primary sequence, although the serine residue is found in all variations. These atypical serine peptidases use a new catalytic triad such as Ser/His/Glu, Ser/His/His or Ser/Glu/Asp and also have a catalytic dyad composed of Ser/Lys, Ser/His, or the mere presence of the catalytic residue Ser (Polgár 1989, Ekici et al. 2008).

The active site of cysteine peptidases consists of the catalytic triad composed of the following amino acids, Cys/His/Asn, where the group involved in the process of catalysis is sulfur of the cysteine residue (Polgár 1989).

The aspartic peptidase in its active site has two aspartic acid residues that comprise the catalytic dyad. The activity is detected at acidic pH, due to the presence of these residues in the active site. The catalysis involves the presence of the water molecule (Polgár 1989).

The metallo-peptidase has a characteristic which is presence of the binding of metal ions that are involved in the catalysis of the substrate. The catalysis is mediated by the presence of water molecules (Polgár 1989).

Glutamic peptidase has as main residues of the active site, the dyad composed of residues of amino acid, glutamic acid and glutamine. In this kind of peptidase the catalysis is mediated by the presence of water molecules (Fujinaga et al. 2004).

Threonine peptidases has its catalyst residue (threonine) in the region of the N-terminal and shows a similar catalytic mechanism as in serine and cysteine peptidases which are involved in the formation of an intermediate acyl-enzyme (Lbwe et al. 1995).

In the 1960s, more modern methods that involved the studies of peptidases, were driven by the availability of sequences. Once the sequences of trypsin, chymotrypsin and other pancreatic peptidases were determined it became apparent that the proteins were related by a common evolutionary ancestor (Barrett and Rawlings 2007). During the 1990s, methods of nucleotide sequencing led to a rapid increase in the availability of deduction of the amino acid sequence of peptidases. This rich supply of data described the main evolutionary families and clans of peptidases (Barrett and Rawlings 2007).

The peptidase families were assigned on the similarity of the amino acid sequence. Some families were grouped into clans. A peptidase clan was defined as a group of families where there was some evidence of an evolutionary relationship, despite the lack of significant similarities in sequence (Barrett and Rawlings 2007).

Proteases Production by Filamentous Fungi

Microbial Diversity: Isolation and Screening

Microorganisms are prevalent as a source of new enzymes, whereas animals and plants contribute less than 10 percent of the total quantity of enzymes processed. From an economic standpoint, microorganisms offer a practical solution to the challenge of obtaining enzyme activities because they are easy to maintain and show rapid growth. In addition, they can be induced to concentrate on the production of just one of the most desired component. They also have key attributes that can be used in biotechnological processes since they can be isolated, purified, selected and tested as pure cultures from a natural source (Nakayama 1981). Thus, microbial diversity is an important source of genetic resources of biological advancement which improves the quality of biotechnology.

Recent studies of microbial ecology show that the extent of microbial diversity in nature is much greater than previously presumed. Some traditional strategies of isolation and selection of microorganisms in different environments such as soil, polluted areas, animal or plant pathogens, has ensured the development of new enzymes. In the case of proteases, producing fungi may be isolated from different sources. Garden soil was reported to be a preferred source for the isolation of fungi that produce alkaline protease, presumably because of the different biological activities that generate transient alkaline conditions in these environments. These organisms were also isolated from non-alkaline habitats such as neutral and acidic soils and therefore appear to be widespread (Kumar and Takagi 1999).

After isolation, these organisms are grown in specific media that support proteolytic activity using appropriate substrates such as skimmed milk or casein (Kumar and Takagi 1999). Different kinds of media have also been used for the production of proteases including agar nutrient (Joshi and Ball 1993), glucose extract of glucose-asparagine yeast agar (GYA) (Sen and Satyanarayana 1993), agar MYGP (Srinivasan et al. 1983), peptone and yeast extract and glucose (PPYG) (Gee et al. 1980).

The first step in the production of proteases of fungal origin is to identify and acquire the microorganism of interest which may be either a wild type or genetically modified strain. The microorganism to be used for the production of proteases, or other enzymes of interest should have certain characteristics so as to be safe from a biological point of view (*status* GRAS-*generally recognized as safe)*; grow in a simple and inexpensive media to produce minimal undesirable secondary metabolites, excreting the enzyme in large amounts and in a manner in which it is easily separated and purified, being pathogenic free and genetically stable.

The genera *Aspergillus*, *Penicillium* and *Trichoderma* stand out among the filamentous fungi described by the ability to produce proteases. For example, *Aspergillus oryzae* produces acid, neutral and alkaline proteases, which are active at a pH range of 4–11. The production of alkaline proteases has been described for *Aspergillus sojae*, *A. fumigatus*, *A. niger* (Nasuno and Ohara 1971), and *Rhizopus oryzae* (Banerjee and Bhattacharyya 1992a). Acid proteases are produced by *Thermoascus aurantiacus* Miehe, *Aspergillus flavus*, *Rhizomucor pusillus* (Macchione et al. 2008) fungi, and neutral proteases by *Chrysosporium keratinophilum* (Singh 2002).

Protease Overexpression

In recent years with the evolution of molecular biology and genetic engineering, filamentous fungi have been frequently used for the production of proteases. The production of microbial strain enzymes can be improved by enhancing the fermentation conditions, conventional mutagenesis (UV or chemical exposure) or recombinant DNA technology, which selectively generates mutants with high proteolytic activity. The recombinant DNA techniques are important in the construction of genetically modified organisms in order to obtain proteases. The isolation and cloning of the gene encoding a protease is a critical step in the engineering of microbial producers (Gupta et al. 2002), which can be transformed by plasmids that comprise the genome, that provides greater long term stability. For these reasons, filamentous fungi have great potential as hosts for recombinant DNA. In particular, the genus *Aspergillus* has been used successfully for the production of recombinant proteins. *E. coli* and *B. subtilis* are the principal hosts of recombinant DNA.

Mutagenesis by X-ray or UV irradiation and chemical mutagenesis have been used in the past 50 yr to obtain mutant strains with high levels of protease excretion as compared to parent protease. Currently, the spore mutagenesis is often carried out using UV light which is less aggressive to the cell than X-rays. This technique has been applied to isolate various mutant organisms which produce proteases using *A. niger* (Mattern et al. 1992) and *A. nidulans* (Katz et al. 1996). The use of mutagens such as nitrosoguanidine was also described (Kolattukudy et al. 1993, Moralejo et al. 2000).

Chemical-physical Parameters for the Production of Proteases

Growth parameters, whether physical, chemical or nutritional, significantly affect the growth and production of proteases by microorganisms. To this end, many optimization studies are performed in order to increase the yield of these enzymes. This is because the cells are capable of responding to physical and chemical stimuli from the external environment through complex biochemical mechanisms that regulate gene expression and physiology of the organism, and by extension, their performance in the formation of the desired product (Madigan et al. 2000).

Considerable changes in the fermentation media must be made to increase the yield of the protease. Each organism or strain has a particular condition that must be recognized in order to obtain the maximum production of enzymes. Protease production is strongly influenced by physical factors such as pH, temperature, aeration, inoculum density and incubation time, in addition to factors of the component of the culture media and metal ions (Hameed et al. 1999, Puri et al. 2002).

The culture medium used for cultivation of the microorganism which produces protease must contain a source of carbon, nitrogen, and some other required factors necessary for the proper development of the lineage. It is important to consider the need for inducing components that promote catabolite repression.

In recent years, a variety of agricultural or industrial by-products of industrial residues, such as wheat meal, cottonseed meal, soybean hulls, among others, have been studied in order to obtain low-cost substrate for protease production and reduction of environmental pollution (Anwar and Saleemuddin 1998). Proteins present in fibrous materials such as feathers, hooves, horns and hair are also abundantly available in nature and can be converted into protein concentrates of amino acids by the action of microbial proteases (Anwar and Saleemuddin 1998, Busson et al. 1999). Whey, a by-product of industrial waste containing mainly lactose of dairy products and salts has been used as a potential substrate for alkaline protease production.

These industrial residues have as their main components cellulose, hemicellulose, pectin, starch and protein, being characterized as extremely heterogeneous material and serving both as a source of energy and carbon as support for microbial growth (Pandey 2003). Since these substrates have low water activity in which the growth conditions are close to the natural environment of filamentous fungi, their growth and the proteases and other enzymes production are possible.

Besides the composition of agro-industrial wastes, environmental conditions that influence the types of fermentation, such as temperature, pH, water activity, oxygen levels and agitation that significantly affect cell growth and product formation, should be taken into account. In a study by Narahara et al. (1982), who studied the effect of substrate humidity on the activity of proteases and amylases produced by *Aspergillus oryzae*, it was observed that the specific activities decreased significantly in cultures performed with substrates in more humid substrates although these conditions were favorable to cell growth.

Thus, the choice of culture medium is vital for the success of the fermentation process in the choice of the organism. This does not mean that the media which best promotes the development of the microorganism favors the production of proteases.

Carbon sources: carbohydrates are sources for microbial fermentation, although other sources can be used, such as alcohols, alkanes and organic acids. However, growth on glucose or other readily assimilated carbon sources prevents the synthesis of enzymes involved in the use of other substrates such as complex polysaccharides (Wang et al. 2005). This seems to apply to the production of extracellular fungal proteases. Examples of the suppression of glucose and other carbon sources on the levels of secreted proteases have been described. For example, the growth profile and secretion of protease *A. niger* was affected by glucose, although when xylose is in the media the results are similar (Braaksma et al. 2009). In studies using a mycelium of *A. nidulans*, it was observed that when using glucose, lactose, galactose or glycerol, the production of protease was severely repressed (Katz 2000). Compared to *A. oryzae* there was a sharp drop in production of protease when mycelia were transferred to a medium with casein and glucose compared to a medium with only casein (Fukushima et al. 1989).

The production of proteases in filamentous fungi is clearly affected by the presence of glucose. Catabolic repression can be caused by various other carbon sources but glucose is suspected to be the most repressive. In various *Aspergillus* species such as *A. nidulans* (Dowzer and Kelly 1989), *A. oryzae* (Kim et al. 2001), and *A. niger* (Drysdale et al. 1993), the presence of glucose inhibits the transcription of many target genes. Other enzymes such as cellulases, amylases and invertase were also repressed in the presence of glucose as the carbon source.

Nitrogen sources: The growth of filamentous fungi, as well as their ability to produce protease in liquid media in the presence of different nitrogen sources have often been described (Macchione et al. 2008, Djamel el al. 2009). It is very common that the microbial growth is satisfactory but with reduced levels of protease. The presence of peptone, casein, gelatin and meat extract can act as inductors in certain microorganisms, whereas others repress and reduce the production of protease, due to accumulation of nitrogenous catabolites such as amino acids and ammonia.

The nitrogen source can also be provided mainly from by-products of other industries, such as corn, yeast extract, and soybean meal.

The influence of the nitrogen source in the *pepA* and *pepB* gene expression in *Aspergillus niger* was investigated by transferring cells to media with and without ammonia. Cells grown with ammonia as the nitrogen source showed very low levels of protease and low levels of transcripts, however mRNA levels were much higher when cells were grown without ammonia (Jarai and Buxton 1994).

Effect of pH, and temperature: Many fungal genes are regulated by the pH of the environment and this parameter was found to be essential for the expression of proteases. Denilson (2000) reported the effect of pH on the expression of protease genes and of extracellular phosphatises genes of *Aspergillus niger*. In fermentations controlled with *A. niger* at pH 4.0 and 5.0, protease activity was higher at pH 4.0 (Braaksma et al. 2009). The genes of *A. niger*, *pepA* and *pepB*, two encoding extracellular proteases are not expressed in alkaline conditions and suggest that gene regulation occurs predominantly at the level of mRNA (Jarai and Buxton 1994).

A. niger is well known by acidification of the medium as it develops. A system of regulation of pH may be especially important because in addition to the regulatory effect on gene expression, the pH of the culture can also affect the fungal morphology (Whitaker and Long 1973). In *A. nidulans*, pH regulation is mediated primarily by seven genes, *pacC*, *palA*, *palB*, *palC*, *palF*, *palH* and *palI*, where *pacC* plays a key role in the regulation of gene expression influenced by pH (Peñalva and Arst 2002). Genes homologous to *pacC* have been described in *A. fumigatus* and *A. oryzae* (Bignell et al. 2005).

For the production of alkaline proteases, the pH should be maintained above 7.5, even for a long period of fermentation. The pH reduction during fermentation kinetics may indicate information about the production of proteases, such as the beginning and end of production of proteases.

Temperature is another parameter which must be controlled to promote the production of enzymes. Some studies report the importance of production of proteases by thermophilic fungi, as these organisms are known to produce thermostable enzymes. The use of these enzymes has many advantages, especially in the food industry due to the high process temperatures for improving the solubility of reagents, reaction speed, and to prevent contamination by mesophilic microorganisms.

Thermophilic fungi grown at 45°C, such as *Thermoascus aurantiacus* Miehe, *Thermomyces lanuginosus*, *Aspergillus flavus* 1.2, *Aspergillus* sp. 13.33, *Aspergillus* sp. 13.34, *Aspergillus* sp. 13.35, *Rhizomucor pusillus* 13.36 and *Rhizomucor* sp. 13.37 (Macchione et al. 2008) produced significant amounts of protease when grown in different agro-industrial residues.

Metal Ions: The presence of metal ions is required for the fermentation process of the production of proteases. However, the requirement is variable for different proteases (neutral, acidic or alkaline), as well as the exact concentration of the ions. The use of AgNO3 or ZnSO4 showed a significant increase for the protease of *Rhizopus oryzae* (Banerjee and Bhattacharyya 1992b).

The effects of phosphorus and sulfur in the expression of proteases have been reported for some years now. However, there are few studies in filamentous fungi. Recently a strong reduction of the proteolytic activity was observed in *A. nidulans*, when the medium was supplemented with sulfur. The regulatory factors involved in repression by phosphorus metabolism are still unknown.

Industrial Application of Protease

Proteases are extremely important from the industrial point of view and are widely used in the laundry detergents industry and foods such as coagulation of milk for cheese production, meat tenderization, also in the baking industries, pharmaceuticals, leather manufacturing, and in bioremediation processes as well as the recovery and utilization of waste and by-products (Rao et al. 1998, Gupta et al. 2002, Salleh et al. 2006). The various functions performed by proteases in biochemical metabolism and cell signaling processes also stimulated the study and use of the latest generation of protease inhibitors as therapeutic products (Turk 2006).

Detergent Industry

Proteases have a long history of application in the food industry and detergents. In the mid-20th century, Rohn (1913), developed and patented the first method of washing fabrics in detergents containing enzymes and were marketed until the 60s. The formulation was based on a trypsin (a digestive enzyme produced in the pancreas of mammals), but the product was not a great success because the enzyme was not sufficiently active in alkaline medium (pH greater than 9.0) (Rohn 1913 apud Bon 1995). However, in 1959, the Swiss chemist Jaag, who worked for the detergents company Gebrüder Schnyder in Biel, Switzerland, developed a new product called Bio 40, which contained a bacterial protease best suited to industrial purposes.

Proteases are used in all types of detergents including domestic use and those used for the cleaning of contact lenses or dentures (Kumar and Takagi 1999, Gupta et al. 2002, Salleh et al. 2006). Today proteases are found in most types of marketed brands of detergent (Salleh et al. 2006). The ideal detergent containing protease should have specificity for the substrate to facilitate the removal of a variety of stains such as those released by body secretions, and also blood and food debris. Proteases should also possess activity and stability at high pH levels, being compatibable with other chelating and oxidizing agents added to detergents (Cowan et al. 1985, Anwar and Saleemuddin 1998, Kumar and Takagi 1999, Salleh et al. 2006).

In order to have a better performance in the use of the protease in detergent, the pI of the enzyme must match with the pH of the detergent solution (Rao et al. 1998, Gupta et al. 2002, Salleh et al. 2006). A combination of amylase, lipase and cellulase in detergents containing protease increases the efficiency of the compound (Rao et al. 1998, Salleh et al. 2006). The use of these enzymes as additives in detergent allows proteases hidrolisem stains consisting of proteins, amylase and cellulase to be effective against carbohydrates, while the lipases are more effective against fats (Anwar and Saleemuddin 1998). Detergent producers are now the largest consumers of industrial enzymes.

The major proteases used in the detergent industry are serine proteases produced by strains of *Bacillus* sp. Although bacteria are the main microorganisms used in the detergent industry, the production cost is high since it requires expensive methods to remove bacterial contaminants from the final products. But fungal proteases have an advantage over bacterial proteases since mycelia can be easily separated by filtering techniques (Anwar and Saleemuddin 1998).

The fungal alkaline protease *Conidiobolus coronatus* was consistent with commercial detergents used in India while maintaining 43 percent of its activity in the presence of calcium (25mM) and glycine (1M) (Rao et al. 1998). These data imply that protease obtained from *C. coronatus* has potential for use in laundry detergents (Anwar and Saleemuddin 1998).

Leather Industry

The alkaline proteases which have activities of elastase and keratinase may be used in leather processing, especially during immersion, which promotes swelling of the skin, hair removal and maceration (Kumar and Takagi 1999, Gupta et al. 2002). Conventional methods used in leather processing involve toxic chemicals such as sodium sulfide, which increase waste and environmental pollution. With the use of proteases instead of chemicals there is an improvement in the quality of leather and this also reduces environmental pollution (Rao et al. 1998, Gupta et al. 2002, Salleh et al. 2006). The alkaline conditions contribute to the swelling of the skin allowing access, and the action of the proteases promotes the degradation of elastin and keratin, producing a softer leather used mainly for clothing and leather goods. Studies with microorganisms have shown success in the use of alkaline proteases in leather processing from *Aspergillus flavus*, *Streptomyces* sp. and *Bacillus* sp. (Kumar and Takagi 1999, Salleh et al. 2006).

Food Industry

In the food industry proteases have been used in the preparation of soy hydrolysates for tenderizing meat, bakery products and in cheese production, among others.

Bakery

Wheat flour is an important component in bakery processes and contains an insoluble protein, gluten, which determines the dough's characteristics. Proteases from *Aspergillus oryzae* have been used to modify wheat gluten, reducing mixing time and increasing the volume of the final product—bread (Shanley and Walsh 2006, Rao et al. 1998). In baking, proteases are also used in the manufacture of biscuits, because of partial hydrolysis of these proteins to improve product texture. For this process acid proteases produced from fungi are usually used (Salleh et al. 2006).

Meat Tenderizing

The tenderization of muscle protein is extremely necessary in the meat industry. The resistance of meat is due to the presence of collagen, and in older animals is composed of cross-linked collagen that promotes a higher resistance when compared to young animals. Papain, a protease derived from plants is used for the softening process of meat from older animals. A solution containing papain can also be injected into the animal 30 min before slaughter. This facilitates uniform distribution of the proteolytic enzyme in the body. Papain is injected in its oxidized form which is inactive. After slaughter, with the physiological changes in the animal, papain is then converted into its reduced form which is catalytically active (Shanley and Walsh 2006).

Cheese Production

In the dairy industry, bovine rennet is still the most widely used in making cheese. Renin, also known as chymosin, is an important commercial protease used in the manufacture of cheese. This enzyme catalyzes the limited k-casein cleavage, the major protein of milk. Subsequently, the casein micelles precipitate to form curd. Curd is further processed to produce cheese. However with the increase in cheese production there has been a shortage of calf rennet (Ustunol and Hicks 1990, Barbano et al. 1992).

Microorganisms such as *Rhizomucor miehei, R. pusillus, Endothia parasitica, Aspergillus oryzae* and *Irpex lactics* are extensively used for the production of proteases for use as milk coagulants (Neelakantan et al. 1999). An important factor to be considered on the effect of productivity and on sensory characteristics of cheese, such as flavor and texture is the coagulant used, these last two are greatly affected by the proteolytic activity of enzymes.

In the manufacture of cheese, bovine chymosin has long been the preferred enzyme due to high specificity for casein. This specificity improves the the quality of cheese (Salleh et al. 2006). An alternative to calf rennet called "genetic rennet" consisting of pure chymosin has come into the market. The production is made possible by recombinant DNA technology, that allows cloning the gene which encodes for calf chymosin in *Escherichia coli, Saccharomyces cerevisiae, Aspergillus oryzae, Kluyveromyces lactics, A. nidulans, A. niger* and *Trichoderma reesei* (Neelakantan et al. 1999). Another enzyme extract with the power of coagulating milk, already produced industrially, is derived from the fungus *Aspergillus niger* var. *awamori* (Neves-Souza and Silva 2005).

Sweeteners

Aspartame is a dipeptide consisting of L-aspartate and L-phenylalanine, in which the L configuration in both amino acids is responsible for the sweet flavor of aspartame. This non-caloric sweetener was approved by the Food and Drug Administration. The use of enzymatic synthesis of aspartame is preferred for maintenance of stereospecificity as the cost of production by chemical methods is higher. Although proteases are considered hydrolytic enzymes, these enzymes can catalyze synthesis reactions. A preparation from *Bacillus thermoprotyolyticus* is used for the enzymatic synthesis of aspartame. Toya Soda (Japan) and DSM (Holland) are the major producers of industrial aspartame (Rao et al. 1998).

Pharmaceutical Industry

Protease inhibitors

In the pharmaceutical industry, enzyme inhibitors are among the main products highlighting the therapeutic classes (WTN 2009), protease inhibitors are among them. Among the classes of interest are those that inhibit serine, metallo, cysteine, aspartate proteases. The serine proteases act to combat inflammatory conditions such as asthma and rheumatoid arthritis as well as for the creation of anticoagulant drugs that act on the coagulation cascade (Leung et al. 2000).

In the class of metalloproteases, there are several products such as enalapril, captopril and ACE inhibitors (angiotensin-converting enzyme) that are marketed (Zimmer et al. 2009). The ACE inhibitors have annual sales exceeding US$6 billion and have been in the market for over 20 yr, 13 of them are currently approved for clinical use in several clinical trials. They are widely used in the treatment of cardiovascular diseases including hypertension and heart failure (Turk 2006).

The class of aspartate proteases presents the first success in the construction of enzyme inhibitors from computational techniques (Zimmer et al. 2009). The human immunodeficiency virus (HIV) has an aspartic protease which is essential for the retroviral life cycle, and has become a good target for chemotherapy with specific inhibitors (Sandhya et al. 2005).

Other Applications

The proteases produced by *Aspergillus oryzae* species (Luizym and Nortase) have been administered orally to aid digestion and correct certain lytic enzyme deficiencies. It was found that alkaline protease of *Conibiobolus coronatus* is capable of replacing trypsin in cultured animal cells (Rao et al. 1998). Alkaline proteases with collagenase activity produced by *Aspergillus niger* have been used in therapeutic applications. These enzymes hydrolyze various types of collagen, releasing low molecular weight peptides that have been used in treating diseases (Kumar and Takagi 1999). Kudrya and Simonenko (1994) explored the elastolytic activity of *Bacillus subtilis* in elastoterase preparation that was applied in the treatment of burns, purulent wounds and deep abscesses (Kumar and Takagi 1999). Proteases from *Bacillus* sp. are also used as a thrombolytic agent having fibrinolytic activity (Kim et al. 1996, Kumar and Takagi 1999).

There is still much to be discovered in the study of proteases and the use of molecular modeling for the design of more potent inhibitors will certainly contribute to the discovery of new therapeutic approaches for many diseases that affect humans (Junior and De Simone 2001).

Silver is present as a waste product in the processing of photo sensitive material such as photographic and X-ray film approximately 1.5–2.0 percent in weight. The use of chemicals and burning X-ray films are conventional methods for the recovery of silver. These methods result in increased pollution and bad odor. However, a 99 percent recovery of silver using alkaline proteases has been reported. In addition to the recovery of silver from a film made of polyester, retrieval is also possible, using alkaline proteases (Fujiwara et al. 1991).

Conclusions and Future Perspectives

Currently, proteases represent 60 percent of the world market and 40 percent is represented by microbial proteases. The principal industrial enzymes used in the market are amylase, lipase, protease, ligase, phytase, cellulase and xylanase. According to a recent report from Business Communications Company (BCC 2008) the world market for industrial enzymes increased from US$2.2 billion in 2006, with an expected US$2.7 billion in 2012 and US$4.9 billion in 2013. Pharmaceutical enzymes now account for 38.4 percent of special enzymes, and should reach a percentage of 47.2 percent in 2014. The application with the highest growth is predicted in the field of enzyme replacement therapy, which should account for more than half of the demand for enzymes for pharmaceutical use in 2014. Geographically, the United States rank as the fastest growing region, including part of the declining European demand. The segment of enzymes for food and beverages will also have an above-average growth, benefiting both market expansion as well as new applications in the dairy and bakery fields.

Proteases are a class of enzymes with immense physiological and commercial importance. In this context, microorganisms in general and fungi in particular are an inexhaustible source of proteases with a broad variety in the world market, and have been extensively use in the food and detergent industry and also in pharmaceuticals. The fungi represents an invaluable resource for biotechnological innovations and plays an important role in the search for an improved strain used in the industry, and demonstrates the advantages of being easy to handle in relation to growing conditions, and to genetic manipulation.

References

Alves, A.C.V., E. Rogana, C.F. Barbosa and D.L. Ferreira-Alves. 2007. The correction of reaction rates in continuous fluorometric assays of enzymes. J. Biochem. Biophys. Methods 70: 471–479.

Anwar, A. and M. Saleemuddin. 1998. Alkaline proteases: a review. Bioresource Technol. 64: 175–183.

Banerjee, R. and B.C. Bhattacharyya. 1992a. Extracellular alkaline protease of a newly isolated *Rhizopus oryzae.* Biotechnol. Lett. 14: 301–304.

Banerjee, R. and B.C. Bhattacharyya. 1992b. Optimisation of multiple inducers effect on protease biosynthesis by *Rhizopus oryzae.* Bioprocess Eng. 7: 225–228.

Barbano, D.H. and R.R. Rasmussen. 1992. Cheese yield performance of fermentation-produced chymosin and other milk coagulants. J. Dairy Sci. Champaign. 75: 1–2.

Barret, A.J. Classification of peptidases. pp. 1–15. *In:* Alan J. Barret, 1994. Meth. Enzymol., Academic Press, Inc., Califórnia.

Barrett, A.J. and N.D. Rawlings. 2007. Species of peptidases. Biol. Chem. 388: 1151–1157.

BCC-Business Communications Company, Inc. 2008. *In*: Report BIO030E Enzymes for Industrial Applications Norwalk, US.

Bignell, E., S. Negrete-Urtasun, A.M. Calcagno, K. Haynes, H.N. Arsnt, Jr. and T. Rogers. 2005. The *Aspergillus* pH-responsive transcription factor PacC regulates virulence. Mol. Microbiol. 55: 1072–1084.

Bon, E.P.S. 1995. A tecnologia enzimática no Brasil. Enzitec. 1: 9–14.

Braaksma, M., A.K. Smilde., M.J. van der Werf and P.J. Punt. 2009. The effect of environmental conditions on extracellular protease activity in controlled fermentations of *Aspergillus niger*. Microbiol. 155: 3430–3439.

Busson, B., P. Engströn and J. Doucet. 1999. Existence of various structural zones in keratinous tissues revealed by X-ray microdifracction. J. Synchrotron Rad. 6: 1021–1030.

Cowan, D., R. Daniel and H. Morgan. 1985. Thermophilic proteases: properties and potential applications. Trends Biotechnol. 3: 68–72.

Denilson, S.H. 2000. pH Regulation of Gene Expression in Fungi. Fungal Gen. Biol. 29: 61–71.

Djamel, C., T. Ali and C. Nelly. 2009. Acid Protease Production by Isolated Species of *Penicillium.* Eur. J. Science Research 25: 469–477.

Dowzer, C.E. and J.M. Kelly. 1989. Cloning of the *creA* gene from *Aspergillus nidulans*: a gene involved in carbon catabolite repression. Curr. Genet. 15: 457–459.

Drysdale, M.R., S.E. Kolze and J.M. Kelly. 1993. The *Aspergillus niger* carbon catabolite repressor encoding gene, *creA*. Gene. 130: 241–245.

Ekici, O.D., M. Paetzel and R.E. Dalbey. 2008. Unconventional serine proteases: Variations on the catalytic Ser/His/Asp triad configuration. Protein Sci. 17: 2023–2037.

Fretweel, J.F., S.M.K. Ismail, J.M. Cummings and T.L. Selby. 2008. Characterization of a randomized FRET library for peptidase specificity determination. Mol. BioSyst. 4: 862–870.

Fujinaga, M., M.M. Cherney, H. Oyam, K. Oda and M.N.G. James. 2004. The molecular structure and catalytic mechanism of a novel carboxyl peptidase from *Scytalidium lignicolum*. Proc. Natl. Acad. Sci. 101: 3364–3369.

Fujiwara, N., K. Yamamato and A. Masui. 1991. Utilization of thermostable alkaline protease from an alkaliphilic thermophile for the recovery of silver from used x-ray film. J. Fermt. Bioeng. 72: 306–308.

Fukushima, Y., H. Itoh, T. Fukase and H. Motai. 1989. Continuous protease production in a carbon-limited chemostat culture by salt tolerant *Aspergillus oryzae*. Appl. Microbiol. Biotechnol. 30: 604–608.

Garcia-Carreño, F.L. and M.N. Del Toro. 1997. Classification of proteases without tears. Biochem. Edu. 25: 161–167.

Gee, J.M., B.M. Lund, G. Metcalf and J.L. Peel. 1980. Properties of a new group of alkalophilic bacteria. J. Gen. Microbiol. 117: 9–17.

Gupta, R., Q.K. Beg and P. Lorenz. 2002. Bacterial alkaline proteases: molecular approaches and industrial applications. App. Microbiol. Biotechnol. 59: 15–32.

Hameed, A., T. Keshavarz and C.S. Evans. 1999. Effect of dissolved oxygen tension and pH on the production of extracellular protease from a new isolate of *Bacillus subtilis* K2 for use in leather processing. J. Chem. Technol Biotechnol. 74: 5–8.

Hartley, B.S. 1960. Proteolytic enzymes. Annu. Rev. Biochem. 29: 45–72.

Hatanaka, T., Y. Uesugi, J. Arima, H. Usuki and M. Iwabichi. 2009. Biochemical characterization of a novel metalloendopeptidase from *Streptomyces aureofaciens* TH-3 with post-proline hydrolysis activity. Enzyme Microb. Technol. 44: 295–301.

Jarai, G. and F. Buxton. 1994. Nitrogen, carbon, and pH regulation of extracellular acidic proteases of *Aspergillus niger.* Curr. Genet. 26: 238–244.

Joshi, A. and B. Ball. 1994. Extracellular alkaline enzymes of facultative bacteria of $CaCO_3$ kilns near Jabalpur. Indian J. Microbiol. 33: 179–181.

Junior, F.P.S. and S.G. De Simone. 2001. Proteases como alvos de quimioterapia. Biotecnol. Ciência Desenvol. 22: 12–17.

Katz, M.E. 2000. The *Aspergillus nidulans xprF* gene encodes a hexokinase-like protein involved in the regulation of extracellular proteases. Genet. 156: 1559–2000.

Katz, M.E., P.K. Flynn, P.A. VanKuyk and B.F. Cheetham. 1996. Mutations affecting extracellular protease production in the filamentous fungus *Aspergillus nidulans*. Mol. Gen. Genet. 250: 715–724.

Kim, S., J.H.S.S. Whang and K.S. Chae. 2001. Isolation and the nucleotide sequence of the *creA* gene for a carbon catabolite repressor of *Aspergillus oryzae*. Food Sci. Biotechnol. 10: 90–93.

Kim, W.K. Choi, Y. Kim, H. Park, J. Chol, Y. Lee, H. Oh, I. Kwon and S. Lee. 1996. Purification and Characterization of a fibrinolytic enzyme produced from *Bacillus* sp. strain CK 11-4 screened from chungkook-jangs. Appl Environ Microbiol. 62: 2482–2488.

Kolattukudy, P.E., J.D. Lee, L.M. Rogers, P. Zimmerman, S. Ceselski., B. Foz, B. Stein and E.A. Copelan. 1993. Evidence for possible involvement of an elastolytic serine protease in aspergillosis. Infect. Immun. 61: 2357–2368.

Kudrya, V.A. and I.A. Simonenko. 1994. Alkaline serine proteinase and lectin isolation from the culture fluid of *Bacillus subtilis*. Appl. Microbiol. Biotechnol. 41: 505–509.

Kumar, C.G. and H. Takagi. 1999. Microbial alkaline proteases: from a bioindustrial viewpoint. Biotechnol. Adv. 17: 561–594.

Lbwe, J., D. Stock, B. Jap, P. Zwickl, W. Baumeister and R. Huber. 1995. Crystal structure of the 20S proteasome from the Archaeon *T. acidophilum* at 3.4 A resolution. Science 268: 533–539.

Leung, D., G. Abbenante and D.P. Fairlie. 2000. Peptidadse inhibitors: current status and future prospects. J. Med. Chem. 43: 305–341.

Macchione, M.M., C.W. Merheb, E. Gomes and R. Da Silva. 2008. Protease production by different thermophilic fungi. Appl. Biochem. Biotechnol. 146: 223–230.

Madigan, M.T., J.M Martinko and J. Parker. 2000. Brock Biology of Microrganisms. 9th edn. Prentice-Hall, Inc., New Jersey, USA.
Mattern, I.E., J.M. Noort, P. Berg, D.B. Archer, I.N. Roberts and C.A.M.J.J. Hondel. 1992. Isolation and characterization of mutants of *Aspergillus niger* deficient in extracellular proteases. Mol. Gen. Genet. 234: 332–336.
Moralejo, F.J., R.E. Cardoza, S. Guitiérrez, H. Sisniega, I. Faus and J.F. Martin. 2000. Overexpression and lack of degradation of thaumatin in an aspergillopepsin A-defective mutant of *Aspergillus awamori* containing an insertion in the *pepA* gene. Appl. Microbiol. Biotechnol. 54: 772–777.
Nakayama, A. Source of Industrial Microrganisms.pp. 355–410. *In*: H.J. Rehm and G. Reed [eds.]. 1981. Biotechnology Microbial Fundamentals. Verlag Chemic, Weinheim.
Narahara, H., Y. Koyama, T. Yoshida, S. Pichangkura, R. Ueda and H. Taguchi. 1982. Growth and enzyme production in a solid-state culture of *Aspergillus oryzae*. J. Ferment. Technol. 60(4): 311–319.
Nasuno, S. and T. Ohara. 1971. Hyperproduction of proteinase and some hydrolytic enzymes by mutants of *Aspergillus sojae*. Agric. Biol. Chem. 35: 829–835.
Neelakantan S., A.K. Mohanty and J.K. Kaussihik. 1999. Production and use of microbial enzymes for dairy processing. Curr. Sci. 77: 143–148.
Neves-Souza, R.D. and R.S.S.F. Silva. 2005. Estudo de Custo-Rendimento do Processamento de Queijos Tipo Minas Frescal com Derivado de Soja e Diferentes Agentes Coagulantes. Ciência Tecnologia de Alimentos. 25: 170–174.
Pandey, A. 2003. Solid state fermentation. Biochem. Eng. J. 13: 81–84.
Peñalva, M.A. and H.N. Arst. 2002. Regulation of gene expression by ambient pH in filamentous fungi and yeasts, Microbiol. Mol. Biol. 66: 426–446.
Polgár, L. 1989. Mechanism of protease action. CRC Press, Inc. USA, Florida. p. 232.
Polgár, L. 2005. The catalytic triad of serine peptidases. Cell. Mol. Life Sci. 62: 2161–2172.
Puri, S., Q.K. Beg and R. Gupta. 2002. Optimization of alkaline protease production from *Bacillus* sp. using response surface methodology. Curr Microbiol. 44: 286–290.
Rao, M.B., A.M. Tanksale, M.S. Ghatge and V.V. Deshpande. 1998. Molecular and biotechnological aspects of microbial protease. Microbiol Mol Biol. Rev. 62: 597–635.
Rawlings, N.D. and A.J. Barrett. 1993. Evolutionary families of peptidases. Biochem. J. 290: 205–218.
Rawlings, N.D., A.J. Barrett and A. Bateman. 2011. Asparagine peptidase lyases: a seventh catalytic type of proteolytic enzymes. J Biol Chem 286: 38321–38328.
Salleh, A.B., C.N.A. Razak, R.N.Z.R.A. Rahman and M. Basri. Protease: Introduction. pp. 23–39. *In*: A.B. Salleh, R.N.Z.R.A. Rahman. and M. Basri [eds.]. 2006. New Lipases and Proteases, New York.
Sandhya, K.S., M. Nampoothiri and A. Pandey. Microbial Proteases. pp. 165–179. *In*: J.L. Barredo [ed.]. 2005. Microb. Enz. Biotransformations. Humana Press & Totowa, New Jersey, USA.
Schechter, I. and A. Berger. 1967. On the size of the active site in peptidases I Papain. Biochem. Biophys. Res. Commun. 27: 157–162.
Sen, S. and T. Satyanarayana. 1993. Optimization of alkaline protease production by thermophilic *Bacillus licheniformis* S-40. Ind. J. Microbiol. 33: 43–47.
Shanley, N. and G. Walsh. Applied Enzymology. An Overview. pp. 1–15. *In*: B.M. McGrath and G. Walsh [eds.]. 2006. Directory of Therapeutic Enzymes. Taylor & Francis, New York, USA.
Singh, C.J. 2002. Optimization of an extracellular protease of *Chrysosporium keratinophilum* and its potential in bioremediation of keratinic wastes. Mycopathol. 156: 151–156.
Srinivasan, M.C., H. G. Vartak, V.K. Powar and I.I. Sutar. 1983. High activity alkaline protease production by a *Conidiobolus* sp. Biotechnol Lett. 5: 285–288.
Turk, B. 2006. Targeting proteases: successes, failures and future prospects. Nature. 5: 785–799.

Ustunol, Z. and C.L. Hicks. 1990. Effect of calcium addition on yield of cheese manufactured with *Endothia parasitica* protease. J. Dairy Sci. 7: 17–25.
Wang, L., D. Ridgway, T. Gu and M. Moo-Young. 2005. Bioprocessing strategies to improve heterologous protein production in filamentous fungal fermentations. Biotechnol. Adv. 23: 115–129.
Whitaker, A. and P.A. Long. 1973. Fungal pelleting. Proc. Biochem. 8: 27–31.
Wildeboer, D., F. Jeganathan, R.G. Price and R.A. Abuknesha. 2009. Characterization of bacterial peptidases with a panel of fluorescent peptide substrates. Anal. Biochem. 384: 321–328.
Wiseman, A. 1991. Manual de Biotecnologia de las enzimas. Zaragoza: Acribia. p. 444.
WTN. Wisconsin Technology Network. Disponível em: <http://wistechnology.com/articles/4924/>Acesso em abril de 2009.
Zhang, B. 2004. Design of FRET-based GFP probes for detection of peptidase inhibitors. Biochem. Biophys. Res. Commun. 323: 674–678.
Zimmer, K.R., G.L. Borre, D.S. Trentin, C.W. Junior, A.P. Frasson, A.A. Graeff, P. Gomes and A.J. Macedo. 2009. Enzimas microbianas de uso terapêutico e diagnostic clinic. Rev. Liberato. 10: 123–137.

5

Fungal Laccases: Structure, Function and Application

Alessandra Piscitelli,[1,2,*] *Cinzia Pezzella,*[1,2] *Vincenzo Lettera,*[1,2] *Paola Giardina,*[1] *Vincenza Faraco*[1] and *Giovanni Sannia*[1,2]

Introduction

Laccases (EC 1.10.3.2, p-diphenol dioxigen oxidoreductase) are a class of oxidative enzymes which are attracting increasing interest for their potential industrial applications. Laccases belong to the family of blue multicopper oxidases (MCO). Although the presence of laccase-like enzymes has been reported in bacteria and insects, laccases can be mainly divided into two groups: laccases from higher plants and those from fungi. Infact the presence of laccases has been described in almost every fungus examined. They are widely distributed in ascomycetes, duteromycetes, and basidiomycetes. Laccases are also reported in saprophytic ascomycetes of composts, in a few fungi-forming ectomycorrhiza, or orchideoid mycorrhiza, as well as from the species of genera that contain both saprotrophic and mycorrhizal fungi (Dwivedi et al. 2011b). So far, laccase production has not been demonstrated in Zygomycetes and Chytridiomycetes except by a zygomycete from the genus Mucor (Bonugli-Santos et al. 2010).

[1]Department of Chemical Sciences, University of Naples "Federico II" Complesso Universitario Monte S. Angelo, via Cintia, 4 80126 Napoli, Italy.
[2]Biopox s.r.l. via Salita Arenella, 9-80129 Napoli, Italy.
*Corresponding author: apiscite@unina.it

With raising numbers of laccase encoding genes isolated and sequenced, the extreme conservation of four protein regions was confirmed. Many more laccase genes have been cloned from different fungal species on designing degenerative primers from the conserved copper-binding domains. In many instances, more than one gene per species was obtained.

Expression of different patterns of laccase isoenzymes coded by gene families is differentially regulated, depending on the growth conditions and physiological states (Piscitelli et al. 2011). Several studies have indicated that the expression of most fungal laccase genes is affected by metal ions, such as copper and cadmium, and by aromatic compounds.

Although many studies on laccase activity have been conducted, its biological role in fungi is still unclear. Fungal laccases are suggested to carry out a variety of physiological roles including morphogenesis, lignocellulose degradation, soil organic matter cycling, fruiting body formation, different pathways of pigment production, fungal plant-pathogen/host interaction, defense, stress response on diverse environmental challenges and copper and iron homeostasis (Giardina et al. 2010).

This chapter will focus on fungal laccases, describing their structure, function and the state of the art of their engineering. In particular it focuses at laccase applications in different industrial fields.

Structural Characteristics

Most fungal laccases are monomeric glycoproteins of approximately 60–70 kDa with an acidic isoelectric point (pI) around pH 4.0, even if several exceptions exist (Giardina et al. 2010). Their glycosydic moiety generally accounts for 10–25 percent of their molecular weight, rarely for more than 30 percent (De Souza and Peralta 2003, Shleev et al. 2004). Typical metal content of laccases includes one type-1 (T1) copper (Cu1), and one type-2 (T2) and two type-3 (T3) copper ions (Cu2 and Cu3), with Cu2 and Cu3 arranged in a trinuclear cluster (TNC) (Giardina et al. 2009).

Comparison of more than 100 fungal laccase sequences performed by multiple sequence alignments has revealed the presence of four ungapped sequence regions, L1–L4, identified as essential laccase sequence features, that are useful to distinguish the laccases within the broader class of blue multicopper oxidases (MCOs) (Kumar et al. 2003) (Fig. 5.1A).

More in detail, L2 and L4 are in line with the earlier reported copper signature sequences of MCOs, while L1 and L3 are distinctive to laccases. The L1–L4 laccase sequence regions include the 12 amino acid residues acting as copper ligands. Furthermore, superimposition of laccase 3D structures has shown the occurrence of four loop regions, designated loops I, II, III, and IV, and involved in substrate binding (Larrondo et al. 2003) (Fig. 5.1B).

A

Protein	L1	L2	L3	L4
TvL1KYA	^{64}HWHGFFQKGTNWADGPAFINQCPI	^{104}GTFWYHSHLSTQYCDGLRGPF	^{395}HPFHLHGH	^{447}GPWFLHCHIDFHLEAGFAVVF
TvL1GYC	^{64}HWHGFFQAGTNWADGPAFVNQCPI	^{104}GTFWYHSHLSTQYCDGLRGPF	^{395}HPFHLHGH	^{447}GPWFLHCHIDFHLEAGFAIVF
R1G1V10	^{64}HWHGFFQAGTTEMDGPAFVNQCPI	^{104}GTYWYHSHLSTQYCDGLRGAF	^{396}HPFHLHGH	^{446}GPWFLHCHIDWHLEAGLAVVF
CcL1A65	^{64}HWHGLFQRGTNWADGADGVNQCPI	^{104}GTFWYHSHFGTQYCDGLRGPM	^{395}HPFHLHGH	^{446}GPWFFHCHIEFHLMNGLAIVF
LtL2QT6	^{64}HWHGFFQKGTNWADGPAFINQCPI	^{104}GTFWYHSHLSTQYCDGLRGPF	^{394}HPFHLHGH	^{446}GPWFLHCHIDFHLDAGFAVVM
TtL2HRG	^{64}HWHGFFQHGTNWADGPAFVNQCPI	^{104}GTFWYHSHLSTQYCDGLRGPI	^{394}HPFHLHGH	^{444}GPWFLHCHIDFHLEAGFAVVM
Laccase Cons	HWHGxxxxxxxxxDGxxxxxQCPI	GTxWYHSHxxxQYCDGLRGxF	HPxHLHGH	GPWxLHCHIDxHxxxGLxxxL
		L		A F A M F
		I		V E F M
		M		

B

Protein	Substrate binding loops					
	Loop I		Loop II	Loop III	Loop IV	
TvL1KYA	^{154}VAAK--LG--PA--FPLG	^{203}LSCDPNY	^{261}ANP-----NFGNVG----F	^{330}FNFNGTNF--FI	^{384}FPATAAAPGAP	^{454}HIDFHLEAGF
TvL1GYC	^{154}TAAR--LG--PR--FPLG	^{203}ISCDPNY	^{261}ANP-----NFGTVG----F	^{330}FNFNGTNF--FI	^{384}LPATALAPGAP	^{454}HIDFHLEAGF
R1G1V10	^{154}SLST--VLFPNPNKAPPA	^{208}TSCFPNY	^{266}ANP-----SNGRNG----F	^{334}IGRNATTADFTI	^{390}IP-----GGGN	^{453}HIDWHLEAGL
CcL1A65	^{154}IPAP--SI-----QGAAQ	^{202}LSCDPNW	^{260}AQP-----NKGRNGLAGTF	^{333}LGFSGGRF–TI	^{397}VP–AGVLGGP	^{453}HIEFHLMNGL
LtL2QT6	^{154}VAAK--LG--PR--FPKG	^{203}LSCDPNY	^{261}ANP-----NFGTTG----F	^{329}FNFDGTNF--FI	^{383}FPATTAAPGAP	^{453}HIDFHLDAGF
TtL2HRG	^{154}LAAK--VG--SP–VPT-A	^{202}LSCDPNH	^{260}ALP-----NSGTRN----F	^{329}FGFAGGKF--TI	^{383}LPATAAAPGFP	^{451}HIDFHLEAGF

Figure 5.1 Comparison of the laccase signature sequences (A) and of the substrate binding loops (B) of the laccases with known three-dimensional structure: CcL1A65 from *Coprinus cinereus*, TvL1GYC from *Trametes versicolor*, TvL1KYA from *Trametes versicolor*, R1G1V10 from *Rigidoporus lignosus*, LtLQT6 from *Lentinus tigrinus*, TtL2HRG from *Trametes trogii*.

Three-dimensional structures have been so far reported for various basidiomycete laccases such as, Lac-Cc (CcL1A65) from *Coprinus cinereus* (Ducros et al. 1998), LccI (TvL1GYC) (Piontek et al. 2002) and LacIIIb (TvL1KYA) from *Trametes versicolor* (Bertrand et al. 2002), RlL (R1G1V10) from *Rigidoporus lignosus* (Garavaglia et al. 2004), LtL (LtL2QT6) from *Lentinus tigrinus* (Ferraroni et al. 2007), and TtL (TtL2HRG) from *Trametes trogii* (Matera et al. 2008). As far as ascomycete laccases are concerned, the 3D structure of the laccase MaL (MaL1GWO) from *Melanocarpus albomyces* has been solved (Hakulinen et al. 2002), along with the structure of the recombinant enzyme rMaL (rMaL2Q9O), expressed in *Trichoderma reesei* (Hakulinen et al. 2008). Recently, the structure of the ascomycete laccase from *Thielavia arenaria* was also solved (Kallio et al. 2011). The comparison of the 3D structures has shown that all laccases share a similar molecular architecture consisting of three sequentially arranged cupredoxin-like domains. Each domain exhibits a greek key β-barrel topology, which is similar to that of small copper proteins such as azurin and plastocyanin (Murphy et al. 1997) and is also common to all the members of the MCOs family, such as ascorbate oxidase (Messerschmidt et al. 1992) and mammalian ceruloplasmin (Zaitsev et al. 1996). The Cu1 is located in domain 3, whilst the TNC cluster is embedded between domains 1 and 3 and the aminoacidic residues involved in TNC copper coordination are provided by both domains. The structure of basidiomycete laccases is stabilized by two disulfide bridges between domains 1 and 3 and between domains 1 and 2, whereas three disulfide bridges were found in ascomycete structures (Hakulinen et al. 2002, Kallio et al. 2011). A C-terminal protruding tail (13–14 amino acids long) has been found in the deduced amino acidic sequences of ascomycete laccases. This tail is generally cleaved off by proteolysis at a conserved cleavage site to produce the active form of the enzyme. Analyses of the 3D structure of ascomycete laccases have shown this C-terminal extension as a plug obstructing the solvent channel, thus leading to the hypothesis that its cleavage is required to allow the entry of oxygen and the subsequent exit of water molecules.

It is worth noting that the mapping of regions L1–L4 onto the laccases' three-dimensional structure indicated a specific, more or less C-2 symmetric, protein conformational motif characterizing the active site apparatus of the enzymes (Kumar et al. 2003). The observed intra-protein homologies between L1 and L3 and between L2 and L4 at both structure and sequence levels suggest that the quasi C-2 symmetric active site conformational motif may have arisen from a structural duplication event.

Reaction Mechanism

Laccases, as well as the other MCOs, such as CueO, Fet3p and bilirubin oxidase (BOD), couple single-electron oxidations of various substrates to the four-electron reduction of dioxygen to water. The broad range of laccase substrates includes aminophenols, polyphenols, polyamines and aryl diamines.

MCO reactivity involves a minimum of four Cu ions, arranged in a type 1 (T1) Cu and a trinuclear Cu (TNC) site (Solomon et al. 1996). The mononuclear center Cu1, whose potential ranges from +350 to +800 mV, is the primary acceptor of electrons from the reducing substrate (Xu et al. 1996, Klonowska et al. 2002). Electrons are then transferred to the TNC, which is the site of dioxygen reduction.

The electrons extracted at the Cu1 from the reducing substrates are transferred through an intra-molecular electron transfer pathway, to the Cu1 ligand Cys and subsequently split between the two His which coordinate the Cu3a and Cu3b copper ions (Solomon et al. 2008). Two solvent channels provide access to TNC, located in the interior of the protein structure. The first channel points towards the two Cu3 ions on one side of the TNC allowing the molecular oxygen to enter and bind to it. The second channel pointing towards the Cu2 ion on the other side of the cluster permits water molecules produced in the O_2 reduction to move to the bulk solvent (Piontek et al. 2002, Enguita et al. 2004, Ferraroni et al. 2007).

The detailed description of the catalytic events taking place at the three copper atoms in the dioxygen reduction site, is still subject of intense debate. A generally accepted reaction mechanism for O_2 reduction by the TNC involves two two-electron transfers, starting from a fully reduced enzyme (Fig. 5.2).

In the first step, TNC reacts with O_2 to form a 2-electron reduced peroxide, bridging between the type 2 and type 3 Cu, forming the so-called peroxy intermediate (PI) (Solomon et al. 2008). According to other authors, the peroxide in the PI appears to bridge only the Cu3 pair and not to connect the reduced Cu2 (Ferraroni et al. 2007, Matera et al. 2008, Ferraroni et al. 2012). The second two-electron transfer results in cleavage of the O–O bond and formation of a second intermediate, the native intermediate (NI), a fully oxidized species. In the NI intermediate all copper atoms of the cluster are bridged by the product of full O_2 reduction, that remain bound to the trinuclear site as μ_3-oxo and μ_2-hydroxo bridging ligands (Lee et al. 2002). The all-bridged structure of NI is consistent with the rapid $4e^-$ reduction of NI to the fully reduced enzyme, because the μ_3-oxo bridge would allow electron delocalization over the three Cu centers for electron transfer through the cluster. The first step is rate determining, while the second, involving the $2e^-$ reductive cleavage of the O–O bond, is faster. The

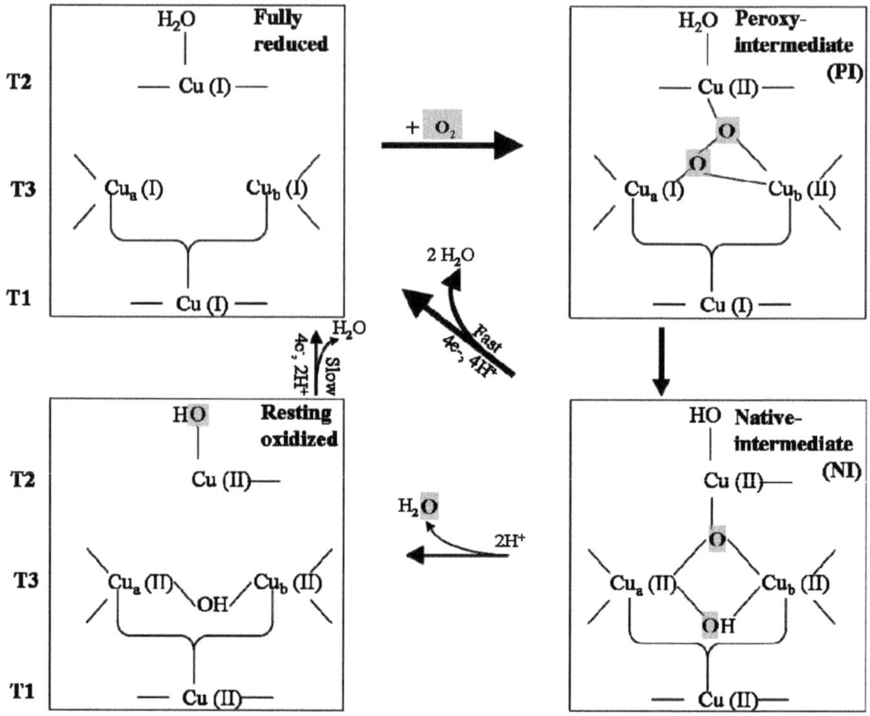

Figure 5.2 Mechanism of O_2 reduction to water by the MCOs. Tight arrows indicate the steps that take place in the catalytic cycle of the MCOs, thin arrows indicate steps that can be experimentally observed but are not part of the catalytic cycle. The peroxy intermediate is a $2e^-$ reduced specie, and the native intermediate is a $4e^-$ reduced specie. From Giardina et al. 2010.

catalytic cycle is completed upon reduction of NI by a total of four electrons, regenerating a fully reduced enzyme (Lee et al. 2002).

Moreover, the NI can decay to the resting oxidized form with a rate that is too slow to be in the catalytic cycle. The resting form, a fully oxidized form as NI; has the Cu2 isolated from the other atoms of the clusters. The decay of NI involves large structural rearrangement of the μ_3-oxo-bridge ligand from the inside to the outside of the cluster. In the absence of reducing substrate, NI slowly decays to the resting enzyme, in which the one remaining O atom of the O_2 is bound as OH^- to Cu2 outside the cluster and the two Cu3 centers are bridged by a OH^- ligand. Thus, the NI appears to be the catalytically relevant fully oxidized form of the enzyme (Lee et al. 2002).

The three Cu centers in the TNC are sequentially lowered by the reducing substrate via the Cu1. According to the mechanism proposed by Yoon and coworkers (Yoon et al. 2007), Cu3a is expected to reduce first, because of the presence of a highly conserved negatively charged residue (D77 in TvL1KYA) near the Cu2 and Cu3b centers, significantly lowers the reduction potentials of these Cu centers below that of Cu3a. Then, Cu2 would likely reduce before Cu3b. Infact reduction of Cu3b before Cu2 would result in protonation of the OH^- bridge, leading to the loss of the Cu2–Cu3b electronic coupling for rapid electron transfer (ET) to Cu2. Further reduction of the remaining Cu3b center would be fast via the Cys–His pathway between Cu1 and Cu3 centers and would be accompanied by dissociation of two water molecules from the TNC.

Mediators

The high applicability of laccases is ascribed to their broad substrate specificity, to the use of molecular oxygen as electron acceptor and to the generation of water as the sole reaction by-product (Paice et al. 1995). However, these green biocatalysts possess a quite low redox potential (≤ 0.8 V), that limit their oxidation of the phenolic compounds (Kawai et al. 1987) whereas non-phenolic substrates cannot be oxidized by laccases directly. Actually, substrates too large to penetrate into the enzyme active site or characterized by high redox potential cannot be oxidized. Nevertheless, laccases are considered to play a major role in the lignin degradation process by some white-rot fungi (Thurston 1994).

Suitable compounds, the so-called mediators, can act as electrons shuttles, enabling laccases to indirectly oxidize large molecules and even non-phenolic substrates (Bourbonnais et al. 1997). Once oxidized by the enzyme, mediators diffuse far away from the enzymatic pocket and can oxidize target compounds that in principle are not substrates of laccase, by mechanisms different from the enzymatic one. An effective redox mediator should be a small-size compound, able to generate stable radicals that do not inactivate the enzyme, its oxidized radical form should have a half-life long enough to permit its diffusion towards the substrate, and its reactivity would allow recycling without degeneration.

The first mediator used in the LMS for pulp delignification has been ABTS (2,20-Azino-bis(3-ethylbenzothiazoline-6-sulfonic acid) (Bourbonnais et al. 1990). Since then about a 100 molecules have been tested for their ability to oxidize lignin or lignin models (Canãs and Camarero 2010).

Xu and coworkers (2000) have shown that the activity and stability of N-O$^\bullet$ radicals seem to be better balanced in comparison with those of a phenoxy radical. Indeed the most effective mediators for lignin degradation has proved to be the N-heterocycles bearing N–OH groups and in particular 1-hydroxybenzotriazole (HBT). However, the application of these enzymatic systems is in part hindered by the economic cost of these artificial mediators and the generation of possible toxic species. For these reasons, many studies have focused on searching alternative "natural" mediators, environmental-friendly and available at low cost (Canãs and

Camarero 2010). The term "natural", not only would indicate the natural origin (lignocellulosic materials, fruits, seeds, etc.) of these compounds, but also their role in nature, being most likely the usual mediators of laccase activities during the biodegradation of lignin polymer by white-rot fungi. Acetosyringone and syringaldehyde, both dimethoxy substituted phenols derived from syringyl lignin units, provide dye decolorization rates higher than those obtained with the powerful HBT mediator (Camarero et al. 2005, Murugesan et al. 2009). The recalcitrant aromatic structures of PAH have similarities with the lignin network. In this case, the outstanding efficiency of *p*-coumaric acid to promote the oxidation of anthracene and benzo[a] pyrene by laccase (upto 100 percent removal), was much better than those found for other natural mediators described elsewhere or ABTS, with close similarity to that of HBT (Canãs et al. 2007).

Production Optimization: State of Art and Future Challenges

Recently, the potential applications of laccases within different industrial fields have been widely reviewed (Rodríguez-Couto and Toca-Herrera 2006, Osma et al. 2010, Jeon et al. 2012, Strong and Claus 2011). At present, the shift from the laboratory to the market is strictly bound to the ability to obtain the enzyme in high yields from microbial processes. Thus, much effort has been spent trying to boost production yields of laccases for industrial applications (Rodríguez-Couto and Toca-Herrera 2007, Majeau et al. 2010). Numerous approaches to solve the problem, such as optimization of medium formulation (Galhaup et al. 2002, Bhattacharya et al. 2011), growth (Dwivedi et al. 2011a) and fermentation (Rodríguez-Couto and Toca-Herrera 2007) conditions, isolation of new fungal strains (Revankar and Lele 2006, Cruz Ramirez et al. 2012), strain improvement by genetic modification (Liu et al. 2010, Lettera et al. 2011, Del Vecchio et al. 2012) have been recently published both alone and appropriately mixed. Furthermore, laccase heterologous expression in suitable hosts has also been extensively explored as an efficient process for producing appreciable amounts of enzyme (Piscitelli et al. 2010).

Reducing the costs of laccase production by optimizing the fermentation medium, the growth and fermentation conditions is steering research for industrial applications. Synthesis and secretion of laccases from fungal sources are strictly influenced by nutrient levels, culture conditions, developmental stage, as well as the addition of a wide range of inducers to cultural media. Variations among both different fungal species and different isoforms in a same strain (Piscitelli et al. 2011) have been reported after the addition of inducers, mainly metals (Palmieri et al. 2000, Galhaup et al. 2002), surfactants (Zhou et al. 2011), and aromatic or phenolic compounds related to lignin or lignin derivatives (Revankar and Lele 2006). However, when analyzing the final price of the produced enzyme, the use of inducers is not always suitable (Osma et al. 2011). In the last few years cultivation methods have also been a subject of intense analysis with the aim of improving laccase production. Growth has been performed i) submerged in broth (SmF), where microorganisms are grown in liquid medium rich in nutrients and with a high oxygen concentration (Palmieri et al. 2000, Galhaup et al. 2002); ii) with fungus immobilized on supports, and a great improvement in the control of the growth rate, in the separation of cells from the liquid medium, and in the possibility to make repeated batch cultures has been obtained (Birhanli and Yesilada 2010); iii) in a variety of vessels/bioreactors designed ad hoc to minimize trouble linked to mycelial growth (e.g., stirred tank, air-lift) (Rodríguez-Couto and Toca-Herrera 2007); iv) or through solid-state fermentations (SSF), where growth occurs in the absence or near the absence of free liquid, on an inert or a natural substrate as a solid support (Neifar et al. 2009), with superior product yields and simpler downstream processing than SmF, but still with several challenges to solve for the scale-up.

Medium optimization methods are usually laborious and time consuming, since they are performed by changing the levels of one independent variable while fixing other variables at a certain level. On the other hand, statistical design of experimentals (DoE) are powerful tools for searching key factors rapidly from a multivariable system and for defining the optimum settings of these factor levels. The use of different statistical designs for medium optimization has been recently used for laccase by different fungal strains (Vasconcelos et al. 2000, Nyanhongo et al. 2002, Trupkin et al. 2003, Neifar et al. 2011a, Bhattacharya et al. 2011). The highest and astonishing production of 54,600 IU/gds (3,412,500 U L−1), from a locally isolated hyperactive strain of *Pleurotus* sp., was obtained through medium optimization by coupling the regression equation with the differential evolution technique (Bhattacharya et al. 2011).

The cost of the enzyme production could be reduced through the use of zero or negative cost substrates, such as tertiary matter, agricultural (Iandolo et al. 2011a, Iandolo et al. 2011b, Karp et al. 2012) and food wastes (Songulashvili et al. 2007) or wastewater from the food or pulp and paper industries (Majeau et al. 2010). Thus, this approach is recently subject of increased research, mainly because some of the potential lignocellulosic wastes may contain significant concentrations of soluble carbohydrates and inducers of enzyme synthesis (Songulashvili et al. 2007), thus also ensuring an efficient production of ligninolytic enzymes. Indeed, a laccase production of up to 110,000 U L−1 was obtained from a *Ganoderma lucidum* strain grown in optimized food waste containing wheat bran (Songulashvili et al. 2007).

Mushrooms provide a rich but relatively untapped pool of biologically active proteins (Erijavec et al. 2012). Only 10 percent of mushroom-forming species are known, thus discovery of novel laccases with different substrate specificities and improved stabilities, as well as of hyper-producing microbes isolated from various environmental samples, is important for industrial applications. Revankar and Lele (2006) reported the isolation of a new strain of white rot fungus WR-1 from tree bark at UICT campus Mumbai (India), able to produce high activities of laccases (124,000 U L−1) during fermentation. The new isolate WR-1 is an efficient and excellent producer of laccases, being able to produce up to 692,000 U L−1 in induced cultures.

Other approaches to increase laccase production include strain improvement through mutagenesis. Classical breeding can be regarded as an ideal approach to address this topic since, unlike genetic transformation and mutagenic treatments, it can generate strains suitable for "natural or safe processes" (Lettera et al. 2011). Recently, starting from two different *Pleurotus ostreatus* varieties, three laccase higher-producing hybrids have been obtained by crossing selected compatible monokaryons (Del Vecchio et al. 2012). The three selected strains increased the productivity of parental strains upto four folds, reaching an expression level of up to 100,000 U L−1. Furthermore, one selected hybrid exhibits a more complex isoenzymatic pattern, pointing out the potential of classical breeding to also differentiate protein expression pattern.

Nevertheless, there are still only a few economical studies regarding the production of laccase enzymes. Recently, Osma and coworkers (2011) provided an economical evaluation of laccase production by the white-rot fungus *Trametes pubescens* cultivated under both SmF and SSF conditions. SSF conditions presented a final cost 50-fold lower than that obtained culturing under SmF conditions at flask scale. In addition, laccase production under SSF conditions in tray bioreactors reduced the final cost of 4-fold, obtaining a final price of 0.04 cent €/U. In comparison with commercially available laccases, this price is more than 10-fold lower than those commercialized by SIGMAeAldrich® (St. Louis, USA) and Jena Bioscience (Jena, Germany). This, however, does not consider other industrial costs related to packaging, storage and distribution of the enzyme.

Recombinant laccase expression in easily cultivable and handling hosts can allow higher productivity in a shorter time than the native laccase production, and reduces the costs of production opening up new commercial opportunities for their industrial uses (Piscitelli et al. 2010). Expression yields of fungal laccases in bacteria are generally very low due to their requirement for post-translational modifications. In contrast to bacterial expression hosts, yeasts are known to express fungal laccases, even though yields are typically lower than 50 mg L−1. The only reported exceptions regarding laccase heterologously expressed in *Pichia pastoris*. Recently, in a *P. pastoris* fed-batch fermentation, a yield of 495 mg L−1 of a highly chloride tolerant laccase from the plant pathogen ascomycete *Botrytis aclada* was obtained (Kittl et al. 2012). The yields of heterologous laccases obtained in filamentous fungi are considerably higher than those obtained in yeasts. The production levels reported for the expression of the ascomycetes *M. albomyces* laccase in *T. reesei* are among the highest heterologous laccase expression levels reported so far, allowing to obtain 230 mg L–1 in shake-flask cultures, 290 mg L–1 in batch fermentations, and 920 mg L–1 in fed-batch fermentation (Kiiskinen et al. 2004). Baker and White (2001) reported a yield of up to 1000 mg L–1 of a *T. versicolor* laccase expressed in the same host. The state of knowledge of parameters controlling/enhancing fungal laccase production is continuously growing, even if it is still contradictory and incomplete. The overall goal of mass scale laccase production relies on the capability of mycologists, genetists, biochemists and engineers to network their skills and to profit from each others.

Laccase Engineering

Biotechnological exploitation of laccases demands industrial-scale production of stable, active and robust enzymes. To promptly reach this aim, protein engineering has been used as a compelling pathway towards efficient biocatalysis (Ayala et al. 2008). Infact the availability of established recombinant expression systems for laccase isoenzymes has allowed their engineering both to widen knowledge of structure function relationships and to improve their features (Piscitelli et al. 2010). Rational and random engineering have been used to generate laccase variants and will be herein surveyed separately. Nonetheless, it is not possible to make a clear distinction between them because very often one approach becomes a valuable tool in aiding interpretation of results obtained through the other approach.

Rational Mutagenesis

Site directed mutagenesis has been performed over the last couple of decades with the aim to give insights in specific open issues (Fig. 5.3).

Furthermore, the growing library of fungal laccase structures, determined by X-ray crystallography, along with spreading of *in silico* models has been essential for designing single-point mutations and deepening structure-function relationships.

T1 Redox Potential

Understanding how the E° of copper sites in proteins is regulated and how its geometric and electronic structure perturbations influence the electron transfer function of a protein still remains one of the major challenges

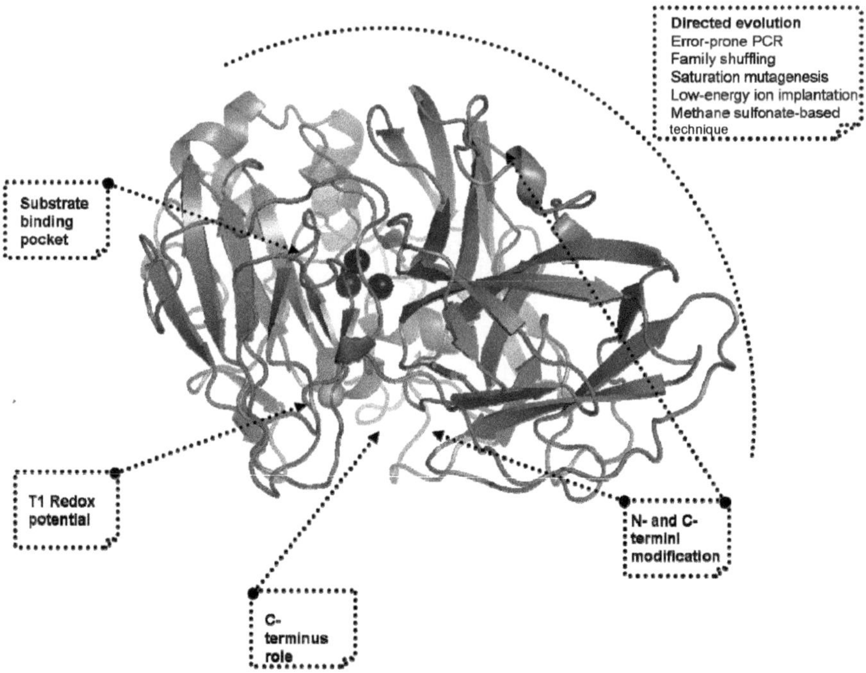

Figure 5.3 Modified laccases with improved stability, activity and specificity for industrial applications.

in the field of metallo-biochemistry. In their pioneering work Xu and coworkers (1998) targeted a pentapeptide segment, believed to be located in the neighborhood of the catalytic copper ions, to perform site directed mutagenesis in order to determine parameters defining the catalytic activity and the redox potentials of these enzymes. Later, the resolution of *C. cinereus* laccase structure (Ducros et al. 1998), determined by X-ray crystallography, boosted the rational choice of candidate residues for site-directed mutagenesis. Mutagenesis experiments in a *Trametes villosa* laccase helped in conceiving that the lack of the T1 Cu fourth axial ligand in laccases is an important factor determining their high values of E° (Xu et al. 1999).

Substrate Binding Pocket

Analysis of high resolution structures of laccases helped to hypothesize many protein substrate interactions, thus many mutants were planned to verify/confirm the role of specific residues.

Madzak and coworkers (2006) modified the well conserved Asp206 residue inducing modifications in catalytic properties of the enzyme and a significant shift of the optimum towards higher pH. Thus significant alterations of the interactions between the reducing substrate and the binding pocket were hypothesized. In the hetero-dimeric laccase POXA3 from *P. ostreatus*, the well conserved Asp is substituted with an Arg residue. Site directed mutants were produced in order to understand the role of this molecular determinant in resolving peculiar properties of *P. ostreatus* laccases (Autore et al. 2009). A significant worsening of catalytic properties was observed along with a decrease of stability when Asp is removed from the substrate binding pocket of another *P. ostreatus* laccase. Based on results obtained from site-directed mutagenesis on the binding pocket of the low-potential ascomyceteous *M. albomyces* Lac1 (MaL), it seems likely that the role of the carboxylic group of its E235 residue is to form a hydrogen bond to the OH-group of a phenolic substrate (such as 2,6-DMP), and this interaction is also important for the catalytic turnover, probably involving the transfer of a proton. The catalytic turnover of phenolic substrates can apparently happen as long as there is a carboxylate in this position (Kallio et al. 2009).

Site-directed mutagenesis has also addressed phenylalanine residues in key positions at the entrance of the binding pocket (Mohamad et al. 2008). These phenylalanines were replaced by smaller-sized but still apolar alanines, and results indicate that these positions contribute to binding and stabilizing bulky hydrophobic aromatic compounds in the substrate binding site (Koschorreck et al. 2008, Galli et al. 2011). The results also highlighted the catalytic importance of residues from the enzymatic pocket that are not directly connected with the electron abstraction process but nevertheless play a critical role at the docking stage (Galli et al. 2011).

C-terminus Role

As mentioned above, C-terminal tail is generally cleaved off by proteolysis at a conserved cleavage site to produce the active form of the enzyme favoring the entry of oxygen and the subsequent exit of water molecules (Hakulinen et al. 2002, Kallio et al. 2011). When Kiiskinen and Saloheimo (2004) studied the expression of *M. albomyces* laccase in *Saccharomyces cerevisiae*, the introduction of a stop codon after the native processing site at the C-terminus gives rise to a six-fold increase in laccase activity. More recently, results obtained with this laccase clearly confirmed the critical role of the last amino acids in its C-terminus. The four C-terminal amino acids of the mature protein penetrate into a tunnel leading towards the trinuclear site, thus there is no space for dioxygen or any other molecules to enter. Furthermore, the C-terminal carboxylate group forms a hydrogen bond with a side chain of a Histidine coordinating to the type 3 copper. In order to analyze the role of the processed C-terminus, site-directed mutants were expressed in *T. reesei* and *S. cerevisiae* (Andberg et al. 2009). Deletion of the last four amino acids dramatically affects enzyme activity. Moreover, the crystal structure of the mutant showed that the mutation of C-terminus clearly affected the trinuclear site geometry.

Whether a similar role of the C-terminal tail is possible among basidiomycete laccases is not yet known.

Gelo-Pujic and coworkers (1999) reported that the redox potential of the laccase from the basidiomycete *T. versicolor* changes when its C terminus is truncated by 11 amino acids, and the barrier to heterogeneous electron transfer is reduced. An unusual C-terminal extension of 16 amino acids has been found in the POXA1b laccase from *P. ostreatus*, although a C-terminal processing of two to four amino acids has been observed for both the native and the heterologously expressed protein in *Kluyveromyces lactis* (Piscitelli et al. 2005). Site directed mutagenesis showed that POXA1b C-terminal tail affects both catalytic performance and stability properties of the enzyme (Autore et al. 2009).

N- and C-termini Modification

In an alternative modification strategy, a *Stachybotrys chartarum* laccase (ascomycete) was linked covalently to peptides that bind to carotenoids, to target the laccase to stained areas of fabric. This resulted in more efficient bleaching on stained fabrics (Janssen et al. 2004). Furthermore, laccase C-terminus was linked to the carbohydrate binding module (CBM) of *Aspergillus niger* cellobiohydrolase B with an improvement in its delignification properties, due to the conferred ability to bind to a cellulosic substrate (Ravalason et al. 2009). Laccases have also been modified with the aim to control their orientation on a solid surface for the development of applications, such as biosensors, or biofuel cells. The formation of a well-ordered monolayer of (C- and N-terminus) histidine-tagged recombinant laccase on gold electrodes by using a short thiol-NTA linker was obtained (Balland et al. 2008).

Random Mutagenesis

There have been many studies in directed evolution of fungal laccases in the last few years, and different strategies have been developed to drive laccase performances towards the desired robustness. The tools used include error-prone PCR (Bulter et al. 2003, Festa et al. 2008, Miele et al. 2010a,b), family shuffling combined with different DNA recombination methods (Bulter et al. 2003, Zumarraga et al. 2007, Zumarraga et al. 2008, Cusano et al. 2009, Matè et al. 2010, Garcìa-Ruiz et al. 2010), saturation mutagenesis (Alcalde et al. 2006), low-energy ion implantation (Hong et al. 2006), methane sulfonate-based techniques (Hu et al. 2007a) (Fig. 5.3).

The first attempt at directed evolution considered heterologous expression of the ascomycete *Myceliophthora thermophila* laccase (MtL) in *S. cerevisiae*. Following 10 iterative rounds of mutagenesis, the laccase yield was raised to about 170-fold and the enzyme showed improved thermostability, as compared with native enzyme activity (Bulter et al. 2003). The most effective mutation (10-fold increase in total activity) adjusts the protein sequence to the different protease specificities of the heterologous host. In another recent study, the MtL mutant was used as a starting point to confer organic co-solvent tolerance (Zumarraga et al. 2007). In this work, five rounds of laboratory evolution were carried out to explore 13,000 clones, using *in vivo* Overlap Extension (IVOE), *In vivo* Assembly of Mutant libraries (IvAM) and error-prone PCR, in combination with *in vivo* DNA shuffling for library creation. The final mutant was fairly active and stable at concentrations as high as 50 percent of organic co-solvents. In a later study, using combinatorial saturation mutagenesis through IVOE, a direct relationship between the C-terminal plug and a conserved tripeptide in the vicinity of the reducing substrate binding site was determined (Zumarraga et al. 2008).

Several directed evolution experiments have been conducted on high redox potential laccase (HRPL) from *P. ostreatus*, producing variants with greater catalytic activity and stability (Festa et al. 2008, Miele et al. 2010a,b). Molecular dynamics simulations were also used to rationalize the functional roles of the principal mutations (Festa et al. 2008). Two other different HRPLs were engineered to enhance their activities and thermostabilities, from basidiomycete PM1 (Matè et al. 2010, Garcìa-Ruiz et al. 2010) and from *Pycnoporus cinnabarinus* (Camarero et al. 2009). As far as PM1 laccase is concerned, eight rounds of evolution were carried out in combination with rational approaches. After screening over 50,000 clones, the total laccase activity was enhanced upto 34,000 fold, the largest improvement ever reported for this kind of system (Matè et al. 2010).

There is also an interesting report on how to engineer chimeric fungal laccases from *Trametes* sp. strain C30 by *in vivo* DNA shuffling (Cusano et al. 2009), where a low redox potential laccase gene was used in all the chimeric libraries to guarantee functional expression. Compared to studies involving mutagenesis, the obtained increase is one of the highest ever observed in a single mutational step, thus confirming how homologous recombination constitute a valuable tool set to study the plasticity of the enzyme.

The aim is that in the near future, new HRPLs engineered by directed evolution and rational approaches can confront the challenges presented by traditional and modern biocatalysis (Matè et al. 2011). The opinion of majority of the scientists worldwide is that there is no ideal laccase to fit all purposes, but there exists a real possibility of designing improved industrial enzymes (Rodgers et al. 2010).

Laccase Potential Exploitation and Industrial Application

Several laccases are candidates for commercialization in several industrial fields. Energy-saving and biodegradable, laccase-based biocatalysts are suitable for the development of highly efficient, sustainable, and ecofriendly industries. However only a few of them are actually present in the market for textiles, food and other industries, because of the high production costs and the still low enzymatic productivity achieved from both natural and recombinant strains (Table 5.1).

Table 5.1 Commercially available laccases.

Company	Product	Application
AB Enzymes	Ecostone LCC10	Denim finishing
Advanced Enzyme Technologies Ltd. (India)	Flavourstar	Brewing
Alfa Kimya Company (Turkey)	Novalite IIS	Denim finishing
Amano Enzyme USA Co. Ltd.	LACCASE M120	Food additive
Americos Industries Inc. (India)	Americos Laccase P	Denim bleaching
	Americos Laccase LTC	Denim bleaching
Apollo Chemical Company (USA)	APCOZYME II-S	Denim finishing and bleaching
BioPox (Italy)	Poxa1b from *P. ostreatus*	Not specified
	PoxA3 from *P. ostreatus*	Not specified
	PoxC from *P. ostreatus*	Not specified
	Mix from *P. ostreatus*	Not specified
	Tailored laccase on demand	Not specified
BioSapien (United Arab Emirates)	Recombinant Not specified laccase expressed in *Aspergillus oryizae*	Denim bleaching and finishing
Chemicals Dyestuffs Ltd. (Hong Kong)	Bleach-cut 3S	Denim bleaching and finishing
CHEMOS GmbH (Germany)	Laccase from *Agaricus bisporus*	Not specified
CHT/BEZEMA (Italy)	DENIMCOL LAC	Denim bleaching
Colotex Biotechnology Co. Ltd. (Hong Kong)	Cololacc BB	Denim bleaching and finishing
Condor Speciality Products (USA)	Hypozyme	Denim deiking
ENZYMES NAVEEN (India)	Not specified laccase	Not specified

Table 5.1 contd....

Table 5.1 contd.

Company	Product	Application
FENKIM KIMYA SANAYI VE TICARET LTD.STI. (Turkey)	Not specified laccase	Denim bleaching
Genencor Inc. (USA)	IndiStar™ Active	Denim finishing
	Primagreen Ecofade LT100	Denim bleaching and shading
HUT (Vietnam)	Not specified laccase	Not specified laccase
Jena Bioscience GmbH (Germany)	Laccase from *T. versicolor*	Not specified
Julich Chiral Solutions GmbH, A Codexis Company (Germany)	Laccase 001	Not specified
Lignozym GmbH (Germany)	Lignozym-process	Pulp bleaching
Metgen (Finland)	Tailored bacterial laccase	Not specified
Nanjing Chemlin Chemical Industry Co.,Ltd. (China)	Not specified laccase	Not specified
Novozyme (Denmark/USA)	DeniliteIITM	Denim finishing
	Novoprime base 268	Denim finishing
	Novozym 51003	Pulp and paper delignification
	Suberase	Cork treatment
Prochimica group (Italy)	Easystone E.DUAL/E.TOP/E. TP5	Denim bleaching
Proenzimas Ltda. (Colombia)	Lacasa Ultratex	Indigo decolourization
Puridet Asia Ltd. (Hong Kong)	Purizyme	Denim bleaching
Season Chemicals Dyestuffs Ltd. (Hong Kong)	Bleach-cut 3S	Denim bleaching
Sigma Aldrich (USA)	Laccase from *A. bisporus*	Not specified
	Laccase from *P. ostreatus*	Not specified
	Laccase from *Rhus vernificera*	Not specified
	Laccase from *T. versicolor*	Not specified
Sunson Industry Group Co., Ltd. (China)	Prozyme LAC	Denim bleaching
Tri-Tex Co. Inc. (Canada)	Trilite II	Indigo decolourization
	Trilite Plus	Indigo decolourization
United States Biological (USA)	Recombinant laccase from *T. versicolor*	Not specified
ZA Biotech (South Africa)	Not specified laccase	Not specified
Zytex Pvt. Ltd. (India)	Zylite	Denim bleaching

Pulp and Paper Industry

The pulp and paper industry uses wood as raw material to produce pulp, paper, board and other products (Kuhad et al. 1997). Laccase integration can improve physical, chemical and mechanical properties (Virk et al. 2012). Primary industrial application of laccases in wood pulp based production involves delignification, bleaching, separation and degradation of lignin in biopulping process, avoiding conventional use of chlorine- or, more recently, oxygen-based chemical oxidants (Luisa et al. 1996). Inspite of the chemical or physical methods, laccases can provide milder and eco-friendly strategies of delignification that also regard the integrity of cellulose. Most laccase applications, however, are fulfilled by means of the LMS (Cadena et al. 2010) that can modify chemical or mechanical properties of lignocellulosic fibers or reduce the refining energy required in thermomechanical pulping. Infact LMS increase wet strength or hydrophobicity of pulp, strength, tensile index, burst index of handsheet (Aracri et al. 2009, Fillat et al. 2010, Moldes and Vidal 2011), integrate antimicrobial or antioxidant proprieties for functional packaging materials (Kudanga et al. 2010).

Moreover, laccases can also be used for coloring cotton cellulose or flax fiber with different colors, depending on the nature of phenol coupled with reaction (Hadzhiyska et al. 2006, Schröder et al. 2007). Conversely, this enzyme can be even used for the opposite process—deinking and decolorization of printed paper (Xu et al. 2009).

Finally, laccases can be used to produce binderless fiberboards and particle boards avoiding artificial adhesives resulting expensive, hazardous, and increasing board mass (Felby et al. 1997, Kharazipour et al. 1997, Felby et al. 2002, Widsten et al. 2003, Felby et al. 2004, Widsten et al. 2004).

Although LMS has been studied extensively, there are still unresolved problems concerning mediator recycling, cost and toxicity. As above mentioned, natural mediators, such as ultra-filtered lignin, envisage some environmental benefits (Canas and Camarero 2010). Thus, laccase-mediated cross-linking of lignocellulose fibers using ultra-filtered lignin results in a twofold increase in wet strength of kraft liner pulp handsheets without losing other critical mechanical properties. The fact that LMS could be easily implemented in the existing bleaching sequences is seen as a major advantage that possibly can lead to a partial replacement of ClO_2 in pulp mills.

Textile Industry

The use of laccases in the textile industry, with or without mediators, is applied in several phases of the whole process (Rodríguez-Couto and Toca-Herrera 2006, Rodríguez-Couto 2012). Infact besides providing new textile dyes, laccases are potentially able to modify dyes changing their property in terms of changing colors and reactivity strength of bonding to textiles. In the last few years, various patents on coloration achieved with laccase have been reported (Kunamneni et al. 2008a). The enzymatic system can be the answer to the increasing demand of the chemical industry to develop eco-friendly processes. Moreover, laccase based biosystems are directly used to modify textile features: the enzyme is successfully used to improve the whiteness in cotton bleaching (Tian et al. 2012) and in biostoning (Paziarloglu et al. 2005), an enzymatic process that substitutes stone washing treatment of the denim textile. Commercially available products for these treatments are reported in Table 5.1. A well-known example of introduction of LMS in the textile industry, is indigo oxidation on denim garments in industrial laundries: the dye chromophore is split to yield two molecules of uncolored compounds (Galante and Formantici 2003).

Evident benefits of laccase integration for the above mentioned applications include chemicals, energy, and saving water, and textile quality improvement. Laccases have also been used to attach functional molecules on textile fibers. For example, laccase-mediated grafting of lauryl gallate on wool resulted in a multifunctional textile material with antioxidant, antibacterial and water repellent properties (Hossain et al. 2009). Moreover, LMS can be used for the anti-shrinking treatment of wool (Lantto et al. 2004). In the post-production treatment, laccases are applied in the cleansing formulation to eliminate the odor on fabrics from several textile products and in detergents to eliminate the odor given off during washing (Aehle et al. 2003, Wolfgang et al. 2005). One of the most interesting applications, however, focuses on the downstream process involving dye removal from the industrial effluents. Many dyes are recalcitrant and toxic to the environment due to the presence of aromatic and metal chlorides (Baughmanet and Perenich 1988, Papic et al. 2004) and stringent legislation regarding the removal of dyes in textile wastewaters require new and more effective methods for dye removal compared to the current chemical, physical and biological processes already applied. Laccases seem an attractive solution for removing dyes as they not only have the potential to degrade dyes of diverse chemical structures, (Forgacs et al. 2004) but can mediate coupling reactions which form the basis for dye removal. A strategy focused on recovering the immobilized enzyme (see below) could be applied.

Food Industry

Food grade laccases are currently of interest in food and beverage industries. Laccases can lead to new functionality, quality improvement, or cost reduction (Brijwani et al. 2010, Osma et al. 2010) because of their reactivity with several alimentary components such as carbohydrates, unsaturated fatty acids, phenols, and thiol-containing proteins.

Laccases can be used in baking. The use of laccases during dough-making is due to their ability to cross-link biopolymers producing increasing strength, stability, and reduced stickiness of the gluten structure. Increased volume, improved crumb structure, and softness of baked products are also reported (Labat et al. 2000, Selinheimo et al. 2006, Renzetti et al. 2010), including the use of low quality flour (Minussi et al. 2002). In addition, laccase can be helpful in increasing the volume and improving crumb structure and softness of the baked products from cereal bread. These baked products, generally gluten-free (Gallagher 2009), are suitable for celiac customers.

Laccases can be applied to some beverage products to enhance or modify their color appearance and their quality. Infact laccases are successfully used for treating polyphenol removal in wines thus avoiding an undesirable alteration in the wine's organoleptic characteristics and color as an alternative to physical-chemical adsorbents (Minussi et al. 2002). An enzymatic formulation (Suberzyme®) for an indirect application in the same sector has been commercialized for oxidative reduction of the cork taint and astringency of bottle cork stoppers, which could negatively affect the aging of wine (Conrad et al. 2000). Treatment with laccases, such as Flavourstar® (Novozyme) (FAO 2004), are also used in the brewing industry to remove the polyphenolic content in beer, thus improving its storage life (Rossi et al. 1988, Mathiasen 1995, McMurrough et al. 1999). Analogously the enzyme could prevent protein-polyphenol haze formation in preparing fruit juice and enhancing its stability (Artik et al. 2004, Ribeiro et al. 2010, Neifar et al. 2011b). Laccases are also used to improve the flavor quality of vegetable oils by eliminating dissolved oxygen avoiding unwanted oxidation (Petersen and Mathiasen 1996). Moreover, laccases are successfully added in tea-based products to enhance their color (Bouwens et al. 1997, Bouwens et al. 1999).

Laccases are also used to pre-treat cacao, before the roasting process in order to improve the flavor and taste of chocolate based products (Takemori et al. 1992, Mensah et al. 2012). The reduction of odors with laccase is documented in patent literature (Tsuchiya et al. 2000) and has been used in some chewing gum formulations. It is also used to perform the cross-link of sugar beet pectin through oxidative coupling to form gels for food ingredients (Norsker et al. 2000, Jung and Wicker 2012).

Pharmaceutical and Cosmetic Industries

Many products generated by laccases are antimicrobial, detoxifying, or active personal-care agents. Laccases can be also used in the synthesis of complex medical compounds such as, antibiotics, sedatives, anti-inflammatory, etc. (Lugaro et al. 1973, Eckenrode et al. 1982, Bhalerao et al. 1994, Kurisawa et al. 2003, Kunamneni et al. 2008b, Mikolasch et al. 2008). Analogously to laccase application in cork treatment for wine bottle packaging, laccases are also used for *in situ* generation of iodine, a reagent widely used as disinfectant (Madhavi and Lele 2009). Some academic research has reported that several laccases have significant antagonistic activity with HIV-1 reverse transcriptase (Wang and Ng 2004) of aceruloplasminemia disease (Harris et al. 2004). These properties, however, will probably never be exploited for a pharmaceutical formulation.

Conversely, laccases find novel applications in cosmetics: dermatological preparations for skin lightening (Golz-Berner et al. 2004), laccase based hair dyes that are less irritant (Lang and Cotteret 1999) or deodorants for personal-hygiene products (Takase et al. 2011).

Organic Synthesis

Recently, increasing interest has been focused on the application of laccases as new biocatalysts in organic synthesis (Riva 2006, Kunamneni et al. 2008c, Witayakran and Ragauskas 2009, Monti et al. 2011). The enzymatic preparation of polymeric polyphenols by the action of laccases has been extensively investigated in the past decades as a viable and non-toxic alternative to the usual formaldehyde-based chemical production of these compounds (Wariishi et al. 2006). Laccase-catalyzed cross-linking reaction of new urushiol analogues for the preparation of "artificial urushi" polymeric films (Japanese traditional coating) was demonstrated (Ikeda et al. 2001). It is also mentioned that laccase induced radical polymerization of acrylamide with or without a mediator (Ikeda et al. 1998). Laccases are also known to polymerize various amino and phenolic compounds (Kunamneni et al. 2008c). Laccases can also be used to synthesize various functional organic compounds including polymers with specific mechanical/electrical/optical properties or pesticides (Keum and Li 2004).

Application of Immobilized Laccases

Enzyme immobilization has widened the scope of laccase application allowing reuse of the biocatalyst, thus benefiting in terms of costs, and especially, improving enzyme performances under optimal process reaction conditions (e.g., higher activity and stability at extreme pHs, elevated temperatures or in organic solvents), a requirement that has often hurdled enzyme application in industrial chemical synthesis. The main benefits provided by laccase immobilization are the increase in thermostability of the enzyme and its resistance to chemical agents such as organic solvents or harsh reaction conditions (extreme pHs) (Duran et al. 2002, Sassolas et al. 2012).

The capacity to retain or recover enzymes also allows biocatalyst separation from the product, thereby permitting continuous processes, and preventing carry-through of protein or activity to subsequent process steps (Brady and Jordaan 2009).

Laccases have been successfully immobilized on various carriers, such as nanoparticles, chitosan, poly (GMA/EGDMA) beads, alginate, and nylon membrane. Physical or chemical interactions may be involved in the binding of the enzyme to the support. The former includes the entrapment of the enzyme in a tridimensional matrix or its encapsulation in a polymer, whereas the latter can occur through adsorption, covalent binding to a matrix, or self-immobilization by cross-linking without the requirement of a support. The advantages and drawbacks can be described for each immobilization method (Illanes et al. 2011), the choice being dependent on the application of laccases that are addressed. Usually the choice of the "ideal" immobilization method is a compromise between the high stabilization achieved by multi-point covalent attachment of the enzyme to the support—at the expense of its activity and of the higher costs of matrix derivatization—and the simpler and easier adsorption and entrapment techniques, where no chemical interaction between laccase and support could affect protein activity, whilst considerable enzyme leakage and limited substrate diffusion toward the catalyst represent the main drawbacks. More recently, the development of carrier-free immobilized biocatalysts has exploited the advantage of the high concentration of active enzyme within the biocatalyst particle, with no additional costs required for the solid matrix.

Laccase immobilization, appears to be a key factor in electrobiochemistry, particularly to design laccase based-biosensors for the detection of phenols in wastewaters or in food industry applications, and in the development of biofuel cells. Moreover, immobilized laccases have been widely used in environmental applications and in the pulp and paper industry. Some interesting examples are described below. Additionally, excellent reviews regarding applications of immobilized laccase can be found in the literature (Duran et al. 2002, Fernandez et al. 2012).

Biofuel Cells

Biological fuel cells generate electrical energy through the oxidation of biomass-based materials catalyzed by microorganisms and/or enzymes. From a sustainable point of view, molecular oxygen is the oxidant of choice used in the cathodic compartment, since it is readily available and a good oxidant. Laccases can reduce oxygen directly to water therefore, they are an attractive alternative to conventional noble metal catalysts for fuel cell cathodes (Farneth and D'Amore 2005), bearing the benefits of renewability, selectivity and low temperature operation. Effective immobilization of laccases on cathodes is essential for the preparation of an efficient and stable fuel cell (Fernandez et al. 2012). The key performance issues for an efficient cathode design are: i) immobilization of the enzyme on the electrode surface in a productive electron-transfer configuration, essential to assure the highest direct electron transfer (DET) between the cathode and the immobilized enzyme; ii) achievement of high enzyme loading without over-dimensioning the electrode; iii) stabilization of the enzyme towards degradation and leaching; iv) availability of the maximum level of molecular oxygen, with reduced mass transfer limitation. Many conducting supports and techniques have been applied to laccase immobilization for biofuel cells. A stable and efficient electrode was obtained by deposition of a *Cerrena unicolor* laccase and a mediator, ABTS, on the surface of a stable voltammetric carbon ceramic electrode (CCE) after encapsulation in two different silicate matrixes (Nogala et al. 2006). In this approach, as in most of the cases, electron conduction from electrode to enzyme involves a mediator, a molecular relay which harvest electrons from the electrode and deliver them to the redox site of the enzyme. An efficient biocatalytic oxygen electrode based on the modification of porous carbon tubes with laccase and ABTS co-immobilized by entrapment within electrochemically polymerized films of polypirrole, was constructed by Servat and coauthors (2007) in order to exploit the porosity of the conducting support for better transportation of dissolved dioxygen solution via convective flow through the pores. In a modified approach, covalent binding of a *T. versicolor* laccase to an aminopolypirrole film showed the best efficiency and an improved stability compared to entrapment techniques (Merle et al. 2008). More recently, the application of three-dimensional hierarchically structured carbon composite materials (carbon microfibers (CMF)/carbon nanotubes (CNT) modified graphite rods) was proposed for the immobilization of a *T. hirsuta* laccase, in order to increase the active electrode surface and to provide a ramified network with open channels for efficient oxygen diffusion and higher oxygen reduction current. When different methods of immobilization on this composite material were compared, covalent binding of enzyme through aldehyde functions introduced by periodate-oxidation of sugar residues at the glycosilation shell of laccases, showed the highest biocatalytic activity with respect to covalent binding by enzyme carboxylic groups or physical adsorption, probably due to a more favorable orientation of the enzyme at the electrode surface (Gutierrez-Sanchez et al. 2012).

Biosensing

Thanks to their ability to catalyze the electron transfer reactions without additional cofactors, laccases have been used in the construction of electrochemical biosensors to detect various toxic phenolic compounds in environmental and food samples, aromatic amines in pharmaceutical formulations (Moccelini et al. 2011), oxygen, and a wide variety of reducing substrates (Shraddha et al. 2011).

Laccase immobilization is a crucial feature in designing the biorecognition part of the biosensor, since the type of immobilization method can affect its performances: factors such as accuracy of measurements, reproducibility, and operational lifetimes are drastically influenced by enzyme stability, whereas biosensor sensitivity can be enhanced by proper enzyme orientation on the transducer surface (Sassolas et al. 2012).

A number of biosensors based on immobilization of laccases to various supports have been reported. In one of the first described examples, fast response and long-term stability were achieved when a laccase from Denilite was immobilized on a silane-modified Pt electrode after subsequent activation with glutaraldehyde (GLU), for the amperometric detection of hydroquinone (sensitivity was 280nA/μM and detection limit of 50nM) and homogentisic acid (sensitivity was 53nA/μM and detection limit of 0.3/μM) (Quan and Shin 2004). When both adsorption and covalent coupling on properly modified graphite electrodes were applied and compared for the construction of a *T. versicolor* laccase based biosensor, covalent immobilization method ensured the highest sensitivities. Moreover, this biosensor displayed its maximum response in the pH range 4.25–5.5, allowing its direct application in wastewaters from agricultural and industrial activities (Portaccio et al. 2006). Carbon based electrodes also exhibited high specific activity and superior cycling properties when employed in a laccase-based biosensor. A large number of phenols could be detected in OMW with high sensitivity (3nA/μM determined towards hydroquinone HQ) when *R. lignosus* laccase was covalently immobilized by carbodiimide chemistry on a modified gold surface with a glassy carbon (GC) electrode as the amperometric transducer (Vianello et al. 2004). Moreover, Mousty et al. (2007) described a laccase biosensor based on the entrapment of the enzyme into redox-active LDH (layered double hydroxide) with a clay colloidal suspension on GC electrodes. In this case, ABTS intercalated within the LDH layers, playing the role of a redox mediator and performing the electrical wiring of laccase. Very low detection limits for dissolved oxygen (0.06–4 nM), azide (5.5 nM), fluoride (6.9 nM) and cyanide (6.2 nM) were observed for this biosensor, allowing its use for the detection of laccase inhibitors (Wang et al. 2008). Recently, the use of nanomaterials (e.g., conducting polymer nanowires, carbon nanotubes, nanoparticles) for the design of biosensing devices has improved the performances of detection platforms, by enhancing the electron transfer between redox centers in the protein and the electrode surface and by offering a larger surface area for laccase immobilization. Infact composite materials have been designed to create a laccase-based biosensor with superior analytical features. In particular, Fe_3O_4 magnetic nanoparticles with core-shells (Fe_3O_4-SiO_2) were amino-modified, and laccase was cross-linked to these nanoparticles by GLU. The obtained bionanoparticles were then immobilized to the surface of a GC electrode. This biosensor showed high sensitivity, a wide range and low cost in manipulation along with good stability and selectivity (linear range for HQ was 0.1 nM–0.1375 mM with a detection limit of 0.01.5 nM) (Zhang et al. 2007). Composite materials based on integration of carbon nanotubes with some conducting polymer (e.g., polyaniline, PANI) have also been exploited. In an interesting example, laccase from *Ganoderma* sp. was covalently immobilized to a gold (Au) electrode modified with manganese dioxide nanoparticles (MnO_2NPs) decorated carboxylated multiwalled carbon nanotubes (cMWCNTs)/PANI composite, for the detection of total phenolic content in tea leaves extract (Rawal et al. 2011). This biosensor was used 150 times over a period of 5 mon and showed good performances in terms of wide linear range, response time, detection limit, reusability and stability.

Biodegradation

The improvement in biocatalyst performances provided by laccase immobilization (higher stability to temperature, extreme pH values or presence of organic solvents), makes their application particularly suited for the bio-treatment of industrial effluents contaminated with xenobiotics, such as phenolic compounds and textile dyes. Moreover, the possibility to employ immobilized enzymes in continuously operated bioreactors, is especially attractive to cope with the huge volume of polluted wastewaters demanding remediation (Peralta-Zamorra et al. 2003).

Decolorization of several dyes has been achieved by means of laccase covalently bound to different supports, such as alumina oxide pellets (Rodríguez-Couto et al. 2007) and controlled porosity-carrier beads (Champagne and Ramsay et al. 2007). Effective dye removal was also carried out by *C. unicolor* laccase covalently immobilized on mesostructured siliceous cellular foams (MCGs) (Rekuc et al. 2009) or on supports functionalized with epoxy groups (Kunamneni et al. 2008a). In the latter case, the degradation of different classes of dyes was reported, along with an improvement of stability toward pH, temperature and storage of the immobilized enzyme.

The entrapment of laccase in alginate-chitosan complex membrane has also been used for bioremediation applications (Palmieri et al. 2005, Phetsom et al. 2009). In an interesting example, a crude laccase preparation from *P. ostreatus* was immobilized in copper-alginate beads attaining high yield of laccase activity, and was then packed in a fixed bed bioreactor employed for the continuous decolorization of RBBR. Under the best conditions, 70 percent RBBR decolorization was achieved even after 20 cycles, although decolorization time exponentially increased after the 10th cycle. The best performance was obtained by decreasing the amount of enzyme loaded and by improving laccase retention using chitosan-coated alginate beads (Palmieri et al. 2005).

The potential of laccase for polycyclic aromatic hydrocarbons (PAHs) remediation was demonstrated for a laccase from *T. versicolor* immobilized on kaolinite, since high degradation percentages were achieved together with higher pH and thermal stability (Hu et al. 2007b). The removal of phenolic compound from olive mill wastewaters (OMWs) was obtained by *Pycnoporus coccineus* laccase immobilized on epoxy-activated resins. Both degradation and polymerization of phenolics were induced by the enzyme. In addition the immobilization resulted in an improvement of thermal and pH stability (Berrio et al. 2007). Entrapment has also been used to immobilize laccases in several works related to phenol degradation (Niladevi and Prema 2008).

Removal of Bisphenol A (BPA), a new emerging pollutant belonging to the Endocrine Disrupting Chemicals (EDC) frequently found in wastewaters, was achieved by a *T. versicolor* laccase covalently bound to nylon membranes. Best results were obtained at low concentration of BPA and non-isothermal conditions, parameters commonly found in polluted waters, thus highlighting the significance of these findings (Diano et al. 2007). Laccases covalently immobilized on polycrylonite (PAN) beads, were employed in a fluidized bed reactor for the degradation of different bisphenols, showing high performances and operational stability (Nicolucci et al. 2011). Likewise fumed silica nanoparticles derivatized by *Coriolopsis polyzona* laccase by a novel developed SAI method (Sorption-Assisted Immobilization), displayed improved stability in real wastewaters (Zimmermann et al. 2011) and proved to be active regarding bisphenol A removal (Hommes et al. 2011). The use of *C. polyzona* laccase cross-linked enzyme aggregates (CLEAs) was also reported for the degradation of some EDCs (Cabana et al. 2007).

Other Applications of Immobilized Laccases

Immobilized laccases find applications in several interesting industrial fields. In textile industry, a laccase from Denilite II S (Novozymes) was immobilized by adsorption on cotton fabrics in order to improve the antimicrobial properties of the fabrics (Ibrahim et al. 2007). As far as the paper and pulp industry is concerned, Gamelas and coworkers (2007) designed a reactor based on confined immobilized laccases for the oxidation of POMs (polyoxometalates), wherein the oxidized products are separated by ultrafiltration, subsequently used in biobleaching of kraft pulp, and then recirculated to the enzymatic reactor in their reduced form. This system optimizes the two steps of the process separately, maximizing both production of POMs and biobleaching of pulp.

Finally, immobilized laccases were used for the modification of milled wood lignin and residual kraft lignin, with the aim to obtain new added-value products from their conversion. In this approach, oxidation and depolymerization of lignin could be achieved in the presence of mediators (Crestini et al. 2010).

Market and Emerging Trends

It is evident that fungal laccases have a promising future for industrial applications, offering a green alternative to several current methods which are environmentally unfriendly and less specific.

Until the last decade this potential was not exploited due to several issues: producing large costs of the enzyme, stability and specific activity for a specific industrial process, and finally, the costs and the eco-friendly of mediators, if required.

As mentioned above, recent efforts to abundantly produce these enzymes in heterologous hosts offer a viable alternative to increase enzyme production in native hosts, while using immobilization techniques may improve enzyme recoverability together with activity and stability. In the recent years, these approaches resulted in the scaling up processes from bench scale to pilot and industry levels.

The first commercial product based on laccase enzyme was launched in 1996 by Novozyme (Novo Nordisk, Denmark) (Table 5.1): DeniliteI$_{TM}$, the first industrial laccase and the first bleaching enzyme acting with the help of a redox mediator molecule. The product exhibited good performance, but handling characteristics were not ideal. Thus, in 1999, Novozyme North America Inc. (USA) launched DeniLiteII$_{TM}$ (Table 5.1) based on a new type of laccase with higher activity than that of DeniliteI$_{TM}$. In 2001, the company Zytex Pvt. Ltd. (India) developed Zylite (Table 5.1), a formulation based on LMS capable of degrading indigo in a very specific way.

Today, the increasing demand of specific and eco-friendly treatments for garment manufactures, together with the improved techniques for natural and recombinant laccase production, have induced the creation of new small companies that offer customized formulations of laccases to target specific garment wet process conditions. As an example, Chemicals Dyestuffs Ltd. (Hong Kong) launched a fully formulated laccase enzyme preparation named Bleach-cut 3S (Table 5.1): varying the dosage and/or treatment time, it is possible to obtain almost all shades, even the lightest ones. This preparation, containing laccase enzyme, redox mediator and buffer system, is extremely efficient in decolorization or bleaching of indigo. Bleach-cut 3S can also be used for clean-up of backstaining and enhancement of denim abrasion. Several other examples can certainly be found within the textile sector. Most of the new companies are located in the Asian continent, where, despite global crises, the textile industry is expanding, probably also due to the less beaurocratic constraints for industrial production and lower wages of laborers in these regions.

Conversely, a few food grade laccases are available on the market: Flavoustar, Suberase and Laccase M120 (Table 5.1) are produced from big companies that can invest in good manufacturing processing and product safety, certifying enzymatic treatment of food and beverages. Several integrations of food grade laccase formulations have begun to be used in different bread preparations in the northern Europe and in North America, though the products are not still available on the global market.

Despite the great potential in delignification processing, less enzymatic formulations are available for pulp and paper industries. This basically depends on two fundamental factors: the high cost of the enzyme and its stability which result in making it prohibitive at a low cost process and the poor available information that could be advertised for a correct and actual integration in an "ancient" and consolidated process flow, such as paper production.

The remaining supply is related to enzymes suitable for specific treatments that are not explicated and require purified and well characterized laccases for laboratory or pilot scale testing.

In this field new companies, such as Metgen and BioPox, are actually focused on offering not only a generic product for all customers, but to develop and produce laccases tailored to the industrial customer needs. Through recombinant expression systems along with enzyme mutagenesis specific parameters can be improved, such as substrate specificity, resistance to temperature or other environmental factors. Nevertheless, the biodiversity of different environments as well as laboratory evolution may provide new laccases, more powerful and specific, to be used in future applications.

Acknowledgments

This work was supported by grants from Ministero dell'Università e della Ricerca Scientifica (Progetti di Rilevante Interesse Nazionale) and by P.O.R. Campania FSE 2007-2013, Project CREMe. Alessandra Piscitelli and

Cinzia Pezzella are recipients of fellowships funded by "Progetto di Ricerca Industriale "Filiere agro-industriali integrate ad elevata efficienza energetica per la messa a punto di processi di Produzione Eco-compatibili di Energia e Bio-chemicals da fonte rinnovabile e per la valorizzazione del territorio" (acronimo EnerbioChem), PON01_01966 CUP B71H11000670005, finanziato nell'ambito del Programma Operativo Nazionale R&C 2007–2013—(D.D. Prot. n. 01/Ric. del 18/01/2010, l'"Invito")".

References

Aehle, W., T.L. Baldwin, G.G. Janssen, C.J. Murray, F.J.C. Van Gastel, H. Wang and D.S. Winetzky. 2003. WO Patent 2003,023,067 A1.

Alcalde, M., M. Zumárraga, J. Polaina, A. Ballesteros and F.J. Plou. 2006. Combinatorial saturation mutagenesis by *in vivo* overlap extension for the engineering of fungal laccases. Comb. Chem. High Throughput Screen. 9: 719–727.

Andberg, M., N. Hakulinen, S. Auer, M. Saloheimo, A. Koivula, J. Rouvinen and K. Kruus. 2009. Essential role of the C-terminus in *Melanocarpus albomyces* laccase for enzyme production, catalytic properties and structure. FEBS J. 276: 6285–6300.

Aracri, E., A. Fillat, J.F. Colom, A. Gutiérrez, J.C. Del Río, A.T. Martínez and T. Vidal. 2010. Enzymatic grafting of simple phenols on flax and sisal pulp fibres using laccases. Bioresour. Technol. 101: 8211–8216.

Artik, N., M. Karhan and G. Aydar. 2004. Effects of polyphenoloxidase (LACCASE) application on clarity stability of sour cherry juice. J. Food Techn. 2: 237–243.

Autore, F., C. Del Vecchio, F. Fraternali, P. Giardina, G. Sannia and V. Faraco. 2009. Molecular determinants of peculiar properties of a *Pleurotus ostreatus* laccase: analysis by site-directed mutagenesis. Enzyme Microb. Technol. 45: 507–513.

Ayala, M., M.A. Pickard and R. Vazquez-Duhalt. 2008. Fungal enzymes for environmental purposes, a molecular biology challenge. J. Mol. Microbiol. Biotechnol. 15: 172–180.

Baker, C.J.O. and T.C. White. 2001. Expression of the laccase IV gene from *Trametes versicolor* in *Trichoderma reesei*. ACS Symp. Ser. 413–425.

Balland, V., C. Hureau, A.M. Cusano, Y. Liu, T. Tron and B. Limoges. 2008. Oriented immobilization of a fully active monolayer of histidine-tagged recombinant laccase on modified gold electrodes. Chemistry. 14: 7186–7192.

Baughman, G.L. and T.A. Perenich. 1988. Fate of dyes in aquatic systems: I. Solubility and partitioning of some hydrophobic dyes and related compounds. Environ. Toxicol. Chem. 7: 183–199.

Berrio, J., F.J. Plou, A. Ballesteros, A.T. Martínez and M.J. Martínez. 2007. Immobilization of *Pycnoporus coccineus* laccase on Eupergit C: stabilization and treatment of olive oil mill wastewaters. Biocatal. Biotransform. 25: 130–134.

Bertrand, T., C. Jolivalt, P. Briozzo, E. Caminade, N. Joly, C. Madzak and C. Mougin. 2002. Crystal structure of a four-copper laccase complexed with an arylamine: insights into substrate recognition insights and correlation with kinetics. Biochemistry. 41: 7325–7333.

Bhalerao, U.T., C. Muralikrishna and B.R. Rani. 1994. Laccase enzyme catalysed efficient synthesis of 3-substituted-1,2,4-triazolo (4,3-b)(4,1,2)benzothiadiazine-8-ones Tetrahedron, 50: 4019–4024.

Bhattacharya, S.S., V.K. Garlapati and R. Banerjee. 2011. Optimization of laccase production using response surface methodology coupled with differential evolution. N. Biotechnol. 28: 31–39.

Birhanli, E. and O. Yesilad. 2010. Enhanced production of laccase in repeated-batch cultures of *Funalia trogii* and *Trametes versicolor*. Biochem. Eng. J. 52: 33–37.

Bonugli-Santosa, R.C., L.R. Durranta, M. da Silva and L. Durães Settec. 2010. Production of laccase, manganese peroxidase and lignin peroxidase by Brazilian marine-derived fungi. 46: 32–37.

Bourbonnais, R. and M.G. Paice. 1990. Oxidation of nonphenolic substrates—an expanded role for laccase in lignin biodegradation. FEBS Lett. 267: 99–102.

Bourbonnais, R., M.G. Paice, B. Freiermuth, E. Bodie and S. Borneman. 1997. Reactivities of various mediators and laccases with kraft pulp and lignin model compounds. Appl. Environ. Microbiol. 63: 4627–4632.

Bouwens, E.M., K. Trivedi, C. Van Vliet and C. Winkel. 1997. Method of enhancing color in a tea based foodstuff. European patent application. EP Patent 760,213 A1.

Bouwens, E.M., K. Trivedi, C. Van Vliet and C. Winkel. 1999. Method of enhancing color in a tea based foodstuff. US Patent 5,879,730 A.

Brady, D. and J. Jordaan. 2009. Advances in enzyme immobilisation. Biotechnol. Lett. 31: 1639–1650.

Brijwani, K., A. Rigdon and P.V. Vadlani. 2010. Fungal Laccases: Production, Function, and Applications in Food Processing. Enzyme Res. 2010: 149748.

Bulter, T., M. Alcalde, V. Sieber, P. Meinhold, C. Schlachtbauer and F.H. Arnold. 2003. Functional expression of a fungal laccase in *Saccharomyces cerevisiae* by directed evolution. Appl. Environ. Microb. 69: 987–995.

Cabana, H., J.P. Jones and S.N. Agathos. 2007. Preparation and characterization of cross-linked laccase aggregates and their application to the elimination of endocrine disrupting chemicals. J. Biotechnol. 132: 23–31.

Cadena, M., T. Vidal and A.L. Torres. 2010. Can the laccase mediator system affect the chemical and refining properties of the eucalyptus pulp? Bioresour. Technol. 101: 8198–8204.

Camarero, S., D. Ibarra, M.J. Martínez and A.T. Martínez. 2005. Lignin-derived compounds as efficient laccase mediators for decolorization of different types of recalcitrant dyes. Appl. Environ. Microbiol. 71: 1775–1784.

Camarero, S., A. Cañas, M.J. Martínez, A.T. Martínez, A. Ballesteros, F.J. Plou, E. Record, M. Asther and M. Alcalde. 2009. High redox potential laccases engineered by directed evolution. International Extension Patent, # PCT/ES2009/070516.

Cañas, A.I. and S. Camarero. 2010. Laccases and their natural mediators: Biotechnological tools for sustainable eco-friendly processes Biotechnol. Adv. 28: 694–705.

Cañas, A., M. Alcalde, F.J Plou, M.J. Martínez, A.T. Martínez and S. Camarero. 2007. Transformation of polycyclic aromatic hydrocarbons by laccase is strongly enhanced by phenolic compounds present in soil. Environ. Sci. Technol. 41. 2964–2971.

Champagne, P. and J.A. Ramsay. 2007. Reactive blue 19 decolouration by laccase immobilized on silica beads. Appl. Microbiol. Biotechnol. 77: 819–823.

Conrad, L.S., W.R. Sponholz and O. Berker. 2000. Treatment of cork with a phenol oxidizing enzyme. US Patent 6152966.

Crestini, C., R. Perazzini and R. Saladino. 2010. Oxidative functionalisation of lignin by layer-bylayer immobilised laccases and laccase microcapsules. Appl. Catal., A. 372: 115–123.

Cruz Ramírez, M.G., J.M. Rivera-Ríos, A. Téllez-Jurado, A.P. Maqueda Gálvez, Y. Mercado-Flores and A. Arana-Cuenca. 2012. Screening for thermotolerant ligninolytic fungi with laccase, lipase, and protease activity isolated in Mexico. J. Environ. Manage. 95 Suppl: S256–59.

Cusano, A.M., Y. Mekmouche, E. Meglecz and T. Tron. 2009. Plasticity of laccase generated by homeologous recombination in yeast. FEBS J. 276: 5471–5480.

De Souza, C.G.M. and R.M. Peralta. 2003. Purification and characterization of the main laccase produced by the white-rot fungus *Pleurotus pulmonarius* on wheat bran solid state medium. J. Basic Microbiol. 43: 278–286.

Del Vecchio, C., V. Lettera, C. Pezzella, A. Piscitelli, G. Leo, L. Birolo and G. Sannia. 2012. Classical breeding in *Pleurotus ostreatus*: a natural approach for laccase production improvement. Biocatal. Biotransform. 30: 78–85.

Diano, N., V. Grano, L. Fraconte, P. Caputo, A. Ricupito, A. Attanasio, M. Bianco, U. Bencivenga, S. Rossi, I. Manco, L. Mita, G. Del Pozzo and D.G. Mita. 2007. Non-isothermal bioreactors in enzymatic remediation of waters polluted by endocrine disruptors: BPA as a model of pollutant. Appl. Catal., B. 69: 252–261.

Ducros, V., A.M. Brzozowski, K.S. Wilson, S.H. Brown, P. Østergaard, P. Schneider, D.S. Yaver, A.H. Pedersen and G.J. Davies. 1998. Crystal structure of the type-2 Cu depleted laccase from *Coprinus cinereus* at 2.2 A° resolution. Nat. Struct. Biol. 5: 310–316.

Durán, N., M.A Rosa, A. D'Annibale and L. Gianfreda. 2002. Applications of laccases and tyrosinases (phenoloxidases) immobilized on different supports: a review. Enzyme Microb. Technol. 31: 907–931.

Dwivedi, P., V. Vivekanand, N. Pareek, A. Sharma and R.P. Singh. 2011a. Co-cultivation of mutant *Penicillium oxalicum* SAU(E)-3.510 and *Pleurotus ostreatus* for simultaneous biosynthesis of xylanase and laccase under solid-state fermentation. N. Biotechnol. 28: 616–626.

Dwivedi, U.N., P. Singh, V.P. Pandey and A. Kumar. 2011b. Structure-function relationship among bacterial, fungal and plant laccases. J. Mol. Catal. B Enzym. 68: 117–128.

Eckenrode, F., W. Peczynska-Czoch and J. Rosazza. 1982. Microbial transformations of natural anitumor agents XVIII: conversions of vindoline with copper oxidases. J. Pharm. Sci. 71: 1246–1250.

Enguita, F.J., D. Marcal, L.O. Martins, R. Grenha, A.O. Henriques, P.F. Lindley and M.A. Carrondo. 2004. Substrate and dioxygen binding to the endospore coat laccase from *Bacillus subtilis*. J. Biol. Chem. 279: 23472–23476.

Erjavec, J., J. Kos, M. Ravnikar, T. Dreo and J. Sabotič. 2012. Proteins of higher fungi—from forest to application. Trends Biotechnol. 30: 259–273.

FAO. 2004. Laccase from Myceliophthora thermophila expressed in Aspergillus oryzae. Chem. Techn. Assess. (Cta).

Farneth, W.E. and M.B. D'Amore. 2005. Encapsulated laccase electrodes for fuel cell cathodes. J. Electroanal. Chem. 581: 197–205.

Felby, C., L.S. Pedersen and B.R. Nielsen. 1997. Enhanced auto-adhesion of wood fibers using phenol oxidases. Holzforschung 51: 281–286.

Felby, C., J. Hassingboe and M. Lund. 2002. Pilot-scale production of fiberboards made by laccase oxidized wood fibers: board properties and evidence for cross-linking of lignin. Enzyme Microb. Technol. 31: 736–741.

Felby, C., L.G. Thygesen, A. Sanadi and S. Barsberg. 2004. Native lignin for bonding of fiber boards-evaluation of bonding mechanisms in boards made from laccase-treated fibers of beech (Fagus sylvatica). Ind. Crop. Prod. 20: 181–189.

Fernández-Fernández, M., M.A. Sanromán and D. Moldes. 2012. Recent developments and applications of immobilized laccase. Biotechnol. Adv. http://dx.doi.org/10.1016/j.biotechadv.2012.02.013.

Ferraroni, M., N.M. Myasoedova, V. Schmatchenko, A.A. Leontievsky, L.A. Golovleva, A. Scozzafava and F. Briganti. 2007. Crystal structure of a blue laccase from *Lentinus tigrinus*: evidences for intermediates in the molecular oxygen reductive splitting by multicopper oxidases. BMC Struct. Biol. 7: 60–72.

Ferraroni, M., I. Matera, A. Chernykh, M. Kolomytseva, L.A. Golovleva, A. Scozzafava and F. Briganti. 2012. Reaction intermediates and redox state changes in a blue laccase from *Steccherinum ochraceum* observed by crystallographic high/low X-ray dose experiments. J. Inorg. Biochem. doi:10.1016/j.jinorgbio.2012.01.011.

Festa, G., F. Autore, F. Fraternali, P. Giardina and G. Sannia. 2008. Development of new laccases by directed evolution: functional and computational analyses. Proteins. 72: 25–34.

Fillat, A., J.F. Colom and T. Vidal. 2010. A new approach to the biobleaching of flax pulp with laccase using natural mediators. Bioresour. Technol. 101: 4104–4110.

Forgacs, E., T. Cserháti and G. Oros. 2004. Removal of synthetic dyes from wastewaters: a review. Environ. Int. 30: 953–971.

Galante, Y. M. and C. Formantici. 2003. Enzyme applications in detergency and in manufacturing industries. Curr. Org. Chem. 7: 1399–1422.

Galhaup, C., H. Wagner, B. Hinterstoisser and D. Haltrich. 2002. Increased production of laccase by the wood-degrading basidiomycete *Trametes pubescens*. Enzyme Microb. Technol. 30: 529–536.

Gallagher, E. 2009. Improving gluten-free bread quality through the application of enzymes. Agro. Food Industry Hi-Tech. 20: 34–37.

Galli, C., P. Gentili, C. Jolivalt, C. Madzak and R. Vadalà. 2011. How is the reactivity of laccase affected by single-point mutations? Engineering laccase for improved activity towards sterically demanding substrates. Appl. Microbiol. Biotechnol. 91: 123–131.

Gamelas, J.A.F., A.S.N. Pontes, D.V. Evtuguin, A.M.R.B. Xavier and A.P. Esculcas. 2007. New polyoxometalatelaccase integrated system for kraft pulp delignification. Biochem. Eng. J. 33: 141–147.

Garavaglia, S., M.T. Cambria, M. Miglio, S. Ragusa, V. Iacobazzi, F. Palmieri, C. D'Ambrosio, A. Scaloni and M. Rizzi. 2004. The structure of *Rigidoporus lignosus* laccase containing a full complement of copper ions, reveals an asymmetrical arrangement for the T3 copper pair. J. Mol. Biol. 342: 1519–1531.

García-Ruiz, E., D. Maté, A. Ballesteros, A.T. Martínez and M. Alcalde. 2010. Evolving thermostability in mutant libraries of ligninolytic oxidoreductases expressed in yeast. Microb. Cell Fact. 9: 17–29.

Gelo-Pujic, M., H.H. Kim, N.G. Butlin and G.T. Palmore. 1999. Electrochemical studies of a truncated laccase produced in *Pichia pastoris*. Appl. Environ. Microbiol. 65: 5515–5521.

Giardina, P., V. Faraco, C. Pezzella, A. Piscitelli, S. Vanhulle and G. Sannia. 2010. Laccases: a never-ending story. Cell Mol. Life Sci. 67: 369–385.

Golz-Berner, K., B. Walze, L. Zastrow and O. Doucet. 2004. Cosmetic and dermatological preparation containing copperbinding proteins for skin lightening. WO Patent 2004017931.

Gutiérrez-Sànchez, C., W. Jia, Y. Beyl, M. Pita, W. Schuhmann, A.L. De Lacey and L. Stoica. 2012. Enhanced direct electron transfer between laccase and hierarchical carbon microfibers/ carbon nanotubes composite electrodes. Comparison of three enzyme immobilization methods Electrochim. Acta http://dx.doi.org/10.1016/j.electacta.2011.12.134.

Hadzhiyska, H., M. Calafell, J.M. Gibert Dagà and T. Tzanov. 2006. Laccase-assisted dyeing of cotton. Biotechnol. Lett. 28: 755–759.

Hakulinen, N., L.L. Kiiskinen, K. Kruus, M. Saloheimo, A Paananen, A. Koivula and J. Rouvinen. 2002. Crystal structure of a laccase from *Melanocarpus albomyces* with an intact trinuclear copper site. Nat. Struct. Biol. 9: 601–605.

Hakulinen, N., M. Andberg, J. Kallio, A. Koivula, K. Kruus and J. Rouvinen. 2008. A near atomic resolution structure of a Melanocarpus albomyces laccase. J. Struc. Biol. 162: 29–39.

Harris, Z.L., S.R. Davis-Kaplan, J.D. Gitlin and J. Kaplan. 2004. A fungal multicopper oxidase restores iron homeostasis in aceruloplasminemia. Blood. 103: 4672–4673.

Hommes, G., C.A. Gasser, C.B. Howald, R. Goers, D. Schlosser, P. Shahgaldian and P.F. Corvini. 2011. Production of a robust nanobiocatalyst for municipal wastewater treatment. Bioresour. Technol. http://dx.doi.org/10.1016/j.biortech.2011.11.129.

Hong, Y., Y. Xiao, H. Zhou, W. Fang, M. Zhang, J. Wang, L. Wu and Z. Yu. 2006. Expression of a laccase cDNA from *Trametes* sp. AH28-2 in *Pichia pastoris* and mutagenesis of transformants by nitrogen ion implantation. FEMS Microbiol. Lett. 258: 96–101.

Hossain, K.H.M.G., M.D. González, G.R. Lozano and T. Tzanov. 2009. Multifunctional modification of wool using an enzymatic process in aqueous-organic media. J. Biotechnol. 141: 58–63.

Hu, M.R., Y.P. Chao, G.Q. Zhang, X.Q. Yang, Z.Q. Xue and S.J. Qian. 2007a. Molecular evolution of *Fome lignosus* laccase by ethyl methane sulfonate-based random mutagenesis *in vitro*. Biomol. Eng. 24: 619–624.

Hu, X., X. Zhao and H. Hwang. 2007b. Comparative study of immobilized *Trametes versicolor* laccase on nanoparticles and kaolinite. Chemosphere. 66: 1618–1626.

Iandolo, D., A. Amore, L. Birolo, G. Leo, G. Olivieri and V. Faraco. 2011a. Fungal solid state fermentation on agro-industrial wastes for acid wastewater decolourization in a continuous flow packed-bed bioreactor. Biores. Technol. 102: 7603–7607.

Iandolo, D., A. Piscitelli, G. Sannia and V. Faraco. 2011b. Enzyme production by solid substrate fermentation of *Pleurotus ostreatus* and *Trametes versicolor* on tomato pomace. Appl. Biochem. Biotechnol. 163: 40–51.

Ibrahim, N.A., M. Gouda, A.M. El-Shafei and O.M. Abdel-Fatah. 2007. Antimicrobial activity of cotton fabrics containing immobilized enzymes. J. Appl. Polym. Sci. 104: 1754–1761.

Ikeda, R., H. Tanaka, H. Uyama and S. Kobayashi. 1998. Laccase-catalyzed polymerization of acrylamide. Macromol. Rapid Commun. 19: 423–425.

Ikeda, R., H. Tanaka, H. Oyabu, H. Uyama and S. Kobayashi. 2001. Preparation of artificial urushi via an environmentally benign process. Bull. Chem. Soc. Jpn. 74: 1067–1073.

Illanes, A., A. Cauerhff, L. Wilson and G.R. Castro. 2011. Recent trends in biocatalysis engineering. Bioresour. Technol. http://dx.doi.org/10.1016/j.biortech.2011.12.050.

Janssen, G.G., T.M. Baldwin, D.S. Winetzky, L.M. Tierney, H. Wang and C.J. Murray. 2004. Selective targeting of a laccase from *Stachybotrys chartarum* covalently linked to a carotenoid-binding peptide. J. Pept. Res. 64: 10–24.

Jeon, J.R., P. Baldrian, K. Murugesan and Y.S. Chang. 2012. Laccase-catalysed oxidations of naturally occurring phenols: from *in vivo* biosynthetic pathways to green synthetic applications. Microb. Biotechnol. 5: 318–332.

Jung, J. and L. Wicker. 2012. Laccase mediated conjugation of sugar beet pectin and the effect on emulsion stability. 28: 168–173.

Kallio, J.P., S. Auer, J. Jänis, M. Andberg, K. Kruus, J. Rouvinen, A. Koivula and N. Hakulinen. 2009. Structure-function studies of a *Melanocarpus albomyces* laccase suggest a pathway for oxidation of phenolic compounds. J. Mol. Biol. 392: 895–909.

Kallio, J.P., C. Gasparetti, M. Andberg, H. Boer, A. Koivula, K. Kruus, J. Rouvinen and N. Hakulinen. 2011. Crystal structure of an ascomycete fungal laccase from *Thielavia arenaria*-common structural features of asco-laccases. FEBS J. 278: 2283–2295.

Karp, S.G., V. Faraco, A. Amore, L. Birolo, C. Giangrande, V.T. Soccol, A. Pandey and C.R. Soccol. 2012. Characterization of laccase isoforms produced by *Pleurotus ostreatus* in solid state fermentation of sugarcane bagasse. Biores. Technol. http://dx.doi.org/10.1016/j.biortech.2012.03.058.

Kawai, S., T. Umezawa, M. Shimada, T. Higuchi, K. Koide, T. Nishida, N. Morohoshi and T. Haraguchi. 1987. Ca–Cb cleavage of phenolic b-1 lignin substructure model compound by laccase of *Coriolus versicolor*. Mokuzai Gakkaishi 33: 792–797.

Keum, Y.S. and Q.X. Li. 2004. Reduction of nitroaromatic pesticides with zero-valent iron. Chemosphere. 54: 255–263.

Kharazipour, A., A. Hüttermann and H.D. Lüdemann. 1997. Enzymic activation of wood fibers as a means for the production of wood composites. J. Adhes. Sci. Technol. 11: 419–417.

Kiiskinen, L.L. and M. Saloheimo. 2004. Molecular cloning and expression in *Saccharomyces cerevisiae* of a laccase gene from the ascomycete *Melanocarpus albomyces*. Appl. Environ. Microb. 70: 137–144.

Kiiskinen, L.L., K. Kruus, M. Bailey, E. Ylosmaki, M. Siika-Aho and M. Saloheimo. 2004. Expression of *Melanocarpus albomyces* laccase in *Trichoderma reesei* and characterisation of the purified enzyme. Microbiol. 150: 3065–3074.

Kittl, R., K. Mueangtoom, C. Gonaus, S.T. Khazaneh, C. Sygmund, D. Haltrich and R. Ludwig. 2012. A chloride tolerant laccase from the plant pathogen ascomycete *Botrytis aclada* expressed at high levels in *Pichia pastoris*. J. Biotechnol. 157: 304–314.

Klonowska, A., C. Gaudin, A. Fournel, M. Asso, J. Le Petit, M. Giorgi and T. Tron. 2002. Characterization of a low redox potential laccase from the basidiomycete *C30*. Eur. J. Biochem. 269: 6119–6125.

Koschorreck, K., S.M. Richter, A. Swierczek, U. Beifuss, R.D. Schmid and V.B. Urlacher. 2008. Comparative characterization of four laccases from *Trametes versicolor* concerning phenolic C-C coupling and oxidation of PAHs. Arch. Biochem. Biophys. 474: 213–219.

Kudanga, T., E.N. Prasetyo, J. Sipilae, G.S. Nyanhongo and G.M. Guebitz. 2010. Enzymatic grafting of functional molecules to the lignin model dibenzodioxocin and lignocellulose material. Enzyme Microb. Technol. 46: 272–280.

Kuhad, R.C., A. Singh and K.E.L. Eriksson. Microorganisms and enzymes involved in the degradation of plant fiber cell wall. *In:* K.E.L. Eriksson [ed.]. 1997. Biotechnology in the pulp and paper industry. Advances in biochemical engineering biotechnology. Springer Verlag. Berlin.

Kumar, S.V.S., P. S. Phale, S. Durani and P. P. Wangikar. 2003. Combined sequence and structure analysis of the fungal laccase family. Biotechnol. Bioeng. 83: 386–394.

Kunamneni, A., F.J. Plou, A. Ballesteros and M. Alcalde. 2008a. Laccases and their applications: A patent review. Recent. Pat. Biotechnol. 2: 10–24.

Kunamneni, A., I. Ghazi, S. Camarero, A. Ballesteros, F.J. Plou and M. Alcalde. 2008b. Decolorization of synthetic dyes by laccase immobilized on epoxy-activated carriers. Process Biochem. 43: 169–178.

Kunamneni, A., S. Camarero, C. García-Burgos, F.J. Plou, A. Ballesteros and M. Alcalde. 2008c. Engineering and applications of fungal laccases for organic synthesis. Microb. Cell Fact. 7: 32.

Kurisawa, M., J.E. Chung, H. Uyama and S. Kobayashi. 2003. Laccase catalyzed synthesis and antioxidant property of poly(catechin) Macromol. Biosci. 3: 758–764.

Labat, E., M.H. Morel and X. Rouau. 2000. Effects of laccase and ferulic acid on wheat flour doughs. Cereal Chem. 77: 823–828.

Lang, G. and J. Cotteret. 1999. Hair dye composition containing a laccase. (L'Oreal, Fr.). WO Patent 9,936,036.

Lantto, R., C. Schänberg and J. Buchert. 2004. Effects of Laccase-Mediator Combinations on Wool. Textile Res. J. 74: 713–717.

Larrondo, L.F., L. Salas, F. Melo, R. Vicuna and D. Cullen. 2003. A novel extracellular multicopper oxidase from *Phanerochaete chrysosporium* with ferroxidase activity. Appl. Eviron. Microbiol. 69: 6257–6263.

Lee, S.K., S.D. George, W.E. Antholine, B. Hedman, K.O. Hodgson and E.I. Solomon. 2002. Nature of the Intermediate Formed in the Reduction of O_2 to H_2O at the Trinuclear Copper Cluster Active Site in Native Laccase. J. Am. Chem. Soc. 124: 6180–6193.

Lettera, V., C. Del Vecchio, A. Piscitelli and G. Sannia. 2011. Low impact strategies to improve ligninolytic enzyme production in filamentous fungi: the case of laccase in *Pleurotus ostreatus*. C R Biol. 334: 781–788.

Liu, Z., D. Zhang, Z. Hua, J. Li, G. Du and J. Chen. 2010. Improvement of laccase production and its properties by low-energy ion implantation. Bioprocess Biosyst. Eng. 33: 639–646.

Lugaro, G., G. Carrea, P. Cremonesi, M.M. Casellato and E. Antonini. 1973. The oxidation of steroid hormones by fungal laccase in emulsion of water and organic solvent. Arch. Biochem. Biophys. 159: 1–6.

Luisa, M., F.C. Goncalves and W. Steiner. Purification and characterization of laccase from a newly isolated wood-decaying fungus. pp. 258–266. *In:* T.W. Jeffries and L.L. Viikari (eds.). 1996. Enzymes for pulp and paper processing. American Chemical Society. Washington, USA.

Madhavi, V. and S.S. Lele. 2009. Laccase properties, use. BioRes. 4: 1694–1717.

Madzak, C., M.C. Mimmi, E. Caminade, A. Brault, S. Baumberger, P. Briozzo, C. Mougin and C. Jolivalt. 2006. Shifting the optimal pH of activity for a laccase from the fungus *Trametes versicolor* by structure-based mutagenesis. Protein Eng. Des. Sel. 19: 77–84.

Majeau, J.A., S.K. Brar and R.D. Tyagi. 2010. Laccases for removal of recalcitrant and emerging pollutants. Bioresour. Technol. 101: 2331–2350.

Maté, D., C. García-Burgos, E. García-Ruiz, A. Ballesteros, S. Camarero and M. Alcalde. 2010. Laboratory evolution of high-redox potential laccases. Chem. Biol. 17: 1030–1041.

Maté, D., E. García-Ruiz, S. Camarero and M. Alcalde. 2011. Directed evolution of fungal laccases. Curr. Genomics. 12: 113–122.

Matera, I., A. Gullotto, S. Tilli, M. Ferraroni, A. Scozzafava and F. Briganti. 2008. Crystal structure of the blue multicopper oxidase from the white-rot fungus *Trametes trogii* complexed with p-toluate. Inorg. Chim. Acta. 361: 4129–4137.
Mathiasen, T.E. 1995. Laccase and beer storage. WO Patent 9,521,240 A2.
McMurrough, I., D. Madigan, R. Kelly and T. O'Rourke. 1999. Haze formation shelf-life prediction for lager beer. Food Techn. 53: 58–63.
Mensah, C.A., N.A. Adamafio, K. Amaning-Kwarteng and F.K. Rodrigues. 2012. Reduced tannin content of laccase-treated cocoa (Theobroma cacao) pod husk. Int. J. Biol. Chem. 6: 31–36.
Merle, G., L. Brunel, S. Tingry, M. Cretin, M. Rolland, K. Servat, C. Jolivalt, C. Innocent and P. Seta. 2008. Electrode biomaterials based on immobilized laccase. Application for enzymatic reduction of dioxygen. Mater Sci. Eng. C. 28: 932–938.
Messerschmidt, A., R. Ladenstein, R. Huber, M. Bolognesi, L. Avigliano, R. Petruzzelli, A. Rossi and A. Finazzi Agro'. 1992. Refined crystal structure of ascorbate oxidase at 1.9 A° resolution. J. Mol. Biol. 224: 179–205.
Miele, A., P. Giardina, E. Notomista, A. Piscitelli, G. Sannia and V. Faraco. 2010a. A semi-rational approach to engineering laccase enzymes. Mol. Biotechnol. 46: 149–156.
Miele, A., P. Giardina, G. Sannia and V. Faraco. 2010b. Random mutants of a *Pleurotus ostreatus* laccase as new biocatalysts for industrial effluents bioremediation. J. Appl. Microbiol. 108: 998–1006.
Mikolasch, A., M. Wurster, M. Lalk, S. Witt, S. Seefeldt, E. Hammer, F. Schauer, W.D. Jülich and U. Lindequist. 2008. Novel beta-lactam antibiotics synthesized by amination of catechols using fungal laccase. Chem. Pharm. Bull. 56: 902–907.
Minussi, R.C., G.M. Pastore and N. Duràn. 2002. Potential applications of laccase in the food industry. Trends Food Sci. Techn. 13: 205–216.
Moccelini, S.K., A.C. Franzoi, I.C. Vieira, J. Dupont and C.W. Scheeren. 2011. A novel support for laccase immobilization: Cellulose acetate modified with ionic liquid and application in biosensor for methyldopa detection. Biosens. Bioelectron. 26: 3549–3554.
Mohamad, S.B., A.L. Ong and A.M. Ripen. 2008. Evolutionary trace analysis at the ligand binding site of laccase. Bioinformation. 2: 369–372.
Moldes, D. and T. Vidal. 2011. New possibilities of kraft pulp biobleaching with laccase and sulfonated mediators. Process. Biochem. 46: 656–660.
Monti, D., G. Ottolina, G. Carrea and S. Riva. 2011. Redox reactions catalyzed by isolated enzymes Chem. Rev. 111: 4111–4140
Mousty, C., L. Vieille and S. Cosnier. 2007. Laccase immobilization in redox active layered double hydroxides: a reagentless amperometric biosensor. Biosens. Bioelectron. 22: 1733–1738.
Murphy, M.E.P., P.F. Lindley and E.T. Adman. 1997. Structural comparison of cupredoxin domains: domain recycling to construct proteins with novel functions. Protein Sci. 6: 761–770.
Murugesan, K., I.H. Yang, Y.M. Kim, J.R. Jeon and Y.S. Chang. 2009. Enhanced transformation of malachite green by laccase of *Ganoderma lucidum* in the presence of natural phenolic compounds. Appl. Microbiol. Biotechnol. 82: 341–350.
Neifar, M., A. Jaouani, R. Ellouze-Ghorbel, S. Ellouze-Chaabouni and M.J. Penninckx. 2009. Effect of culturing processes and copper addition on laccase production by the white-rot fungus *Fomes fomentarius* MUCL 35117. Lett. Appl. Microbiol. 49: 73–78.
Neifar, M., A. Kamoun, A. Jaouani, R. Ellouze-Ghorbel and S. Ellouze-Chaabouni. 2011a. Application of asymmetrical and hoke designs for optimization of laccase production by the white-rot fungus *Fomes fomentarius*. Enzyme Res. 2011: 1–12.
Neifar, M., R. Elleouze-Ghorbel, A. Kamoun, S. Baklouti, A. Mokni, A. Jaouani and S. Ellouze-Chaabouni. 2011b. Effective clarification of pomegranate juice using laccase treatment optimized by response surface methodology followed by ultrafiltration. J. Food Process. Eng. 34: 1199–1219.

Nicolucci, C., S. Rossi, C. Menale, T. Godjevargova, Y. Ivanov, M. Bianco, L. Mita, U. Bencivenga, D.G. Mita and N. Diano. 2011b. Biodegradation of bisphenols with immobilized laccase or tyrosinase on polyacrylonitrile beads. Diano N. Biodegradation. 22: 673–683.
Niladevi, K.N. and P. Prema. 2008. Immobilization of laccase from *Streptomyces psammoticus* and its application in phenol removal using packed bed reactor. World J. Microbiol. Biotechnol. 24: 1215–1222.
Nogala, W., E. Rozniecka, I. Zawisza, J. Rogalski and M. Opallo. 2006. Immobilization of ABTS-laccase system in silicate based electrode for biolectrocatalytic reduction of dioxygen. Electrochem. Commun. 8: 1850–1854.
Norsker, M., M. Jensen and J. Adler-Nissen. 2000. Enzymatic gelation of sugar beet pectin in food products. Food Hydrocol. 14: 237–243.
Nyanhongo, G.S., J. Gomes, G. Gübitz, R. Zvauya, J.S. Read and W. Steiner. 2002. Production of laccase by a newly isolated strain of *Trametes modesta*. Bioresour. Technol. 84: 259–263.
Osma, J.F., J.L. Toca-Herrera and S. Rodríguez-Couto. 2010. Uses of laccases in the food industry. Enzyme Res. 2010: 918761.
Osma, J.F., J.L. Toca-Herrera and S.Rodríguez-Couto. 2011. Cost analysis in laccase production. J. Environ. Manage. 92: 2907–2912.
Paice, M.G., R. Bourbonnais, I.D. Reid, F.S. Archibald and L. Jurasek. 1995. Oxidative bleaching enzymes: a review. J. Pulp Pap. Sci. 21: 280–284.
Palmieri, G., P. Giardina, C. Bianco, B. Fontanella and G. Sannia. 2000. Copper induction of laccase isoenzymes in the ligninolytic fungus *Pleurotus ostreatus*. Appl. Environ. Microbiol. 66: 920–924.
Palmieri, G., P. Giardina and G. Sannia. 2005. Laccase-mediated Remazol Brilliant Blue R decolorization in a fixed-bed bioreactor. Biotechnol. Progr. 21: 1436–1441.
Papic, S., N. Koprivanac, A.L. Bozic and A. Metes. 2004. Removal of some reactive dyes from synthetic wastewater by combined Al(III) coagulation/carbon adsorption process. Dyes Pigments 62: 291–298.
Paziarloglu, N.K., M. Sariisik and A. Telefoncu. 2005. Laccase production by Trametes versicolor and application to denim washing. Proc. Biochem. 40: 1673–1678.
Peralta-Zamora, P., C.M. Pereira, E.R.L. Tiburtius, S.G. Moraes, M.A. Rosa, R.C. Minussi and N. Duran. 2003. Decolorization of reactive dyes by immobilized laccase. Appl. Catal., B Environ. 42: 131–134.
Petersen, B.R. and T.E. Mathiasen. 1996. Deoxygenation of an oil product with a laccase. WO Patent 9,631,133, A1.
Phetsom, J., S. Khammuang, P. Suwannawong and R. Sarnthima. 2009. Copper-alginate encapsulation of crude laccase from *Lentinus polychrous lev.* and their effectiveness in synthetic dyes decolorizations. J. Biol. Sci. 9: 573–583.
Piontek, K., M. Antorini and T. Choinowski. 2002. Crystal structure of a laccase from the fungus *Trametes versicolor* at 1.90-A° resolution containing a full complement of coppers. J. Biol. Chem. 277: 37663–37669.
Piscitelli, A., P. Giardina, C. Mazzoni and G. Sannia. 2005. Recombinant expression of *Pleurotus ostreatus* laccases in *Kluyveromyces lactis* and *Saccharomyces cerevisiae*. Appl. Microbiol. Biotechnol. 69: 428–439.
Piscitelli, A., C. Pezzella, P. Giardina, V. Faraco and G. Sannia. 2010. Heterologous laccase production and its role in industrial applications. Bioeng Bugs. 1: 252–262.
Piscitelli, A., P. Giardina, V. Lettera, C. Pezzella, G. Sannia and V. Faraco. 2011. Induction and transcriptional regulation of laccases in fungi. Curr. Genomics. 12: 104–112.
Portaccio, M., S. Di Martino, P. Maiuri, D. Durante, P. De Luca, M. Lepore, U. Bencivenga, S. Rossi, A. De Maio and D.G. Mita. 2006. Biosensors for phenolic compounds: the catechol as a substrate model. J. Mol. Catal. B: Enzym. 41: 97–102.
Quan, D. and W. Shin. 2004. Amperometric detection of hydroquinone and homogentisic acid with laccase immobilized platinum electrode. Bull. Korean Chem. Soc. 25: 833–837.
Ravalason, H., I. Herpoël-Gimbert, E. Record, F. Bertaud, S. Grisel, S. de Weert, C.A. van den Hondel, M. Asther, M. Petit-Conil and J.C. Sigoillot. 2009. Fusion of a family 1

carbohydrate binding module of *Aspergillus niger* to the *Pycnoporus cinnabarinus* laccase for efficient softwood kraft pulp biobleaching. J. Biotechnol. 142: 220–226.
Rawal, R., S. Chawla and C.S. Pundir. 2011. Polyphenol biosensor based on laccase immobilized onto silver nanoparticles/multiwalled carbon nanotube/polyaniline gold electrode Anal. Biochem. 419: 196–204.
Rekuć, A., J. Bryjak, K. Szymańska and A.B. Jarzebski. 2009. Laccase immobilization on mesostructured cellular foams affords preparations with ultra high activity. Process Biochem. 44: 191–198.
Renzetti, S., C.M. Courtin, J.A. Delcour and E.K. Arendt. 2010. Oxidative and proteolytic enzyme preparations as promising improvers for oat bread formulations: rheological, biochemical and microstructural background. Food Chem. 119: 1465–1473.
Revankar, M.S. and S.S. Lele. 2006. Enhanced production of laccase using a new isolate of white rot fungus WR-1. Process Biochem. 41: 581–588.
Ribeiro, D.S., S.M.B. Henrique, L.S. Oliveira, G.A. Macedo and L.F. Fleuri. 2010. Enzymes in juice processing: a review. Int. J. Food Sci. Techn. 45: 635–641.
Riva, S. 2006. Laccases: blue enzymes for green chemistry. Trends Biotechnol. 24: 219–226.
Rodgers, C.J., C.F. Blanford, S.R. Giddens, P. Skamnioti, F.A Armstrong and S.J. Gurr. 2010. Designer laccases: a vogue for high-potential fungal enzymes? Trends Biotechnol. 28: 63–72.
Rodríguez-Couto, S. 2012. Laccases for Denim Bleaching: An Eco-Friendly Alternative. The Open Textile J. 5: 1–7.
Rodríguez-Couto, S. and J.L. Toca-Herrera. 2006. Laccase in the textile industry. Biotechnol. Mol. Biol. Rev. 1: 117–122.
Rodríguez-Couto, S. and J.L. Toca-Herrera. 2007. Laccase production at reactor scale by filamentous fungi. Biotechnol Adv. 25: 558–569.
Rodríguez Couto, S., J.F. Osma, V. Saravia, G.M. Gübitz and J.L. Toca Herrera. 2007. Coating of immobilised laccase for stability enhancement: a novel approach. Appl. Catal., A Gen. 329: 156–160.
Rossi, M., G. Giovanelli, C. Cantarelli and O. Brenna. 1988. Effects of laccase and other enzymes on barley wort phenolics as a possible treatment to prevent haze in beer. Bulletin de Liaison-Groupe Polyphenols. 14. 85–88.
Sassolas, A., L.J. Blum and B.D. Leca-Bouvier. 2012. Immobilization strategies to develop enzymatic biosensors. Biotechnol. Adv. 30: 489–511.
Schröder, M., N. Aichernig, G.M., Gübitz and V. Kokol. 2007. Enzymatic coating of lignocellulosic surfaces with polyphenols. Biotechnol. J. 2: 334–341.
Selinheimo, E., K. Kruus, J. Buchert, A. Hopia and K. Autio. 2006. Effects of laccase, xylanase and their combination on the rheological properties of wheat doughs. J. Cereal Sci. 43: 152–159.
Servat, K., S. Tingry, L. Brunel, S. Querelle, M. Cretin, C. Innocent, C. Jolivalt and M. Rolland. 2007. Modification of porous carbon tubes with enzymes: application for biofuel cells. J. Appl. Electrochem. 37: 121–127.
Shleev, S.V., O. Morozova, O. Nikitina, E.S. Gorshina, T. Rusinova, V.A. Serezhenkov, D.S. Burbaev, I.G. Gazaryan and A.I. Yaropolov. 2004. Comparison of physico-chemical characteristics of four laccases from different basidiomycetes. Biochimie 86: 693–703.
Shraddha, R. Shekher, S. Sehgal, M. Kamthania and A. Kumar. 2011. Laccase: microbial sources, production, purification, and potential biotechnological applications. Enzyme Res. doi: 0.4061/2011/217861.
Solomon, E.I., U.M. Sundaram and T.E. Machonkin. 1996. Multicopper oxidases and oxygenases. Chem. Rev. 96: 2563–2605.
Solomon, E.I., A.J. Augustine and J. Yoon. 2008. O_2 Reduction to H_2O by the multicopper oxidases. Dalton Trans. 30: 3921–3932.
Songulashvili, G., V. Elisashvili, S.P. Wasser, E. Nevo and Y. Hadar. 2007. Basidiomycetes laccase and manganese peroxidase activity in submerged fermentation of food industry wastes. Enzyme Microb. Technol. 41: 57–61.

Strong, P.J. and H. Claus. 2011. Laccase: a review of its past and its future in bioremediation. Critical Reviews in Environmental Science and Technology. 41: 373–434.

Takase, T., A. Narise and K. Sakurai. 2011. Deodorant composition. Int. Pat. Appl. WO Patent 2,011,105,042.

Takemori, T., Y. Ito, M. Ito and M. Yoshama. 1992. Flavor and taste improvement of cacao nib by enzymatic treatment. JP Patent 04,126,037 A2.

Thurston, C.F. 1994. The structure and function of fungal laccases. Microbiology 140: 19–26.

Tian, L., C. Branford-White, W. Wang, H. Nie and L. Zhu. 2012. Laccase-mediated system pretreatment to enhance the effect of hydrogen peroxide bleaching of cotton fabric. Int. J. Biol. Macromol. 50: 782–787.

Trupkin, S., L. Levin, F. Forchiassin and A. Viale. 2003. Optimization of a culture medium for ligninolytic enzyme production and synthetic dye decolorization using response surface methodology. J Ind Microbiol Biotechnol. 30: 682–690.

Tsuchiya, R., B.R. Petersen and S. Christensen. 2000. Oxidoreductases for reduction of malodor. US Patent 6,074,631 A.

Vasconcelos, A.F.D., A.M. Barbosa, R.F.H. Dekker, I.S. Scarminio and M.I. Rezende. 2000. Optimization of laccase production by *Botryosphaeria* sp. in the presence of veratryl alcohol by the response surface method. Proc. Biochem. 35: 1131–1138.

Vianello, F., A. Cambria, S. Ragusa, M.T. Cambria, L. Zennaro and A. Rigo. 2004. A high sensitivity amperometric biosensor using a monomolecular layer of laccase as biorecognition element. Biosens. Bioelectron. 20: 315–321.

Virk, A.P., P. Sharma and N. Capalash. 2012. Use of laccase in pulp and paper industry. Biotechnol. Prog. 28: 21–32.

Wang, F., C. Guo, H. Liu and C. Liu. 2008. Immobilization of *Pycnoporus sanguineus* laccase by metal affinity adsorption on magnetic chelator particles. J. Chem. Technol. Biotechnol. 83: 97–104.

Wang, H.X. and T.B. Ng. 2004. A novel laccase with fair thermostability from the edible wild mushroom (Albatrella dispansus). Biochem. Biophys. Res. Commun. 315: 450–454.

Wariishi, H., D. Nonaka, S. Nishihashi, T. Hirahashi and K. Ito. 2006. Enzymic preparation of polyphenylne oxides. JP Patent 2,006,280,259 A2.

Widsten, P., J.E. Laine, S. Tuominen and P. Qvintus-Leino. 2003. Effect of high defibration temperature on the properties of medium-density fiberboard (MDF) made from laccase-treated hardwood fibers. J. Adhes. Sci. Technol. 17: 67–78.

Widsten, P., S. Tuominen, P. Qvintus-Leino and J.E. Laine. 2004. The influence of high defibration temperature on the properties of medium-density fiberboard (MDF) made from laccase-treated softwood fibers. Wood Sci. Technol. 38: 521–528.

Witayakran, S. and A. J. Ragauskas. 2009. Synthetic applications of laccase in green chemistry. Adv. Synth. Catal. 351: 1187–1209.

Wolfgang, A., T.M. Baldwin, F.J.C. Van Gastel, G.G. Janssen, C.J. Murray, H. Wang and D.S. Winetzky. 2005. US Patent 2005,058,996 A1.

Xu, F., W. Shin, S.H. Brown, J. A. Wahleithner, U.M. Sundaram and E.I. Solomon. 1996. A study of a series of recombinant fungal laccases and birilubin oxidase that exhibit significant differences in redox potential, substrate specificity and stability Biochim. Biophys. Acta 1292: 303–311.

Xu, F., R.M. Berka, J.A. Wahleithner, B.A. Nelson, J.R. Shuster, S.H. Brown, A.E. Palmer and E.I. Solomon. 1998. Site-directed mutations in fungal laccase: effect on redox potential, activity and pH profile. Biochem. J. 334: 63–70.

Xu, F., A.E. Palmer, D.S. Yaver, R.M. Berka, G.A. Gambetta, S.H. Brown and E.I. Solomon. 1999. Targeted mutations in a *Trametes villosa* laccase, axial perturbations of the T1 copper. J. Biol. Chem. 274: 12372–12375.

Xu, F., J.J. Kulys, K. Duke, K. Li, K. Krikstopaitis, H. J.W. Deussen, E. Abbate, V. Galinyte and P. Schneider. 2000. Redox chemistry in laccase-catalyzed oxidation of N-Hydroxy compounds. Appl. Environ. Microbiol. 66: 2052–2056.

Xu, Q., Y. Fu, Y. Gao and M. Qin. 2009. Performance and efficiency of old newspaper deinking by combining cellulase/hemicellulase with laccase-violuric acid system. Waste Manag. 29: 1486–1490.

Yoon, J., B.D. Liboiron, R. Sarangi, K.O. Hodgson, B. Hedman and E.I. Solomon. 2007. The two oxidized forms of the trinuclear Cu cluster in the multicopper oxidases and mechanism for the decay of the native intermediate. Proc. Natl. Acad. Sci. USA. 104: 13609–13614.

Zaitsev, I., V. Zaitsev, G. Card, K. Moshkov, B. Bax, A. Ralph and P. Lindley. 1996. The nature of the copper centres in human ceruloplasmin. J. Biol. Inorg. Chem. 1: 15–23.

Zhang, Y., G. Zeng, L. Tang, D. Huang, X. Jiang and Y. Chen. 2007. A hydroquinone biosensor using modified core-shell magnetic nanoparticles supported on carbon paste electrode. Biosens. Bioelectron. 22: 2121–2126.

Zhou, M.F., X.Z. Yuan, H. Zhong, Z.F. Liu, H. Li, L.L. Jiang and G.M. Zeng. 2011. Effect of biosurfactants on laccase production and phenol biodegradation in solid-state fermentation. Appl. Biochem. Biotechnol. 164: 103–114.

Zimmermann, Y.S., P. Shahgaldian, P.F. Corvini and G. Hommes. 2011. Sorption-assisted surface conjugation: a way to stabilize laccase enzyme. Appl. Microbiol. Biotechnol. 92: 169–178.

Zumárraga, M., T. Bulter, S. Shleev, J. Polaina, A. Martínez-Arias, F.J. Plou, A. Ballesteros and M. Alcalde. 2007. *In vitro* evolution of a fungal laccase in high concentrations of organic cosolvents. Chem. Biol. 14: 1052–1064.

Zumárraga, M., S. Camarero, S. Shleev, A. Martinez-Arias, A. Ballesteros, F.J. Plou and M. Alcalde. 2008. Altering the laccase functionality by *in vivo* assembly of mutant libraries with different mutational spectra. Proteins. 71: 250–260.

6

Fungal Amylases: Applications and Functional Properties

Tony Marcio da Silva

Introduction

Starch is the second main carbohydrate synthesized by plants via photosynthesis. This substance is entirely composed by β-D-glucose and is also an essential energy source for most organisms. Enzymes involved in the starch degrading process are called amylases or amylolytic enzymes. Amylases are responsible for carbon recycling present in starchy polysaccharides in general, and play an important role in the food industries. According to Glazer and Nikaido (1995), besides the food industry, amylases are used in textile industries for removing the starch paste applied to the yarn in order to increase its resistance to friction during processing. They can also be used in the paper industry for reducing the viscosity of the starch solution applied during its processing. Amylases can also be used in detergent formulas to partly refrain its aggressiveness on wood and porcelain (Gupta et al. 2003).

Department of Biology, Faculty of Philosophy, Sciences and Letters of Ribeirão Preto, University of São Paulo, 14040-901 Ribeirão Preto, SP, Brazil.
E-mail: tonymarcio@gmail.com

Amylases are derived from many different sources, but those derived from microorganisms have received special attention because of their great number of desirable characteristics. Microorganisms such as fungi, bacteria, yeasts and actinomycetes are potential producers of amylase, but those produced by fungi and bacteria have dominated most industry sectors (Pandey et al. 2000).

Starch

Starch is one of the most abundant polymers in nature after cellulose. It is the main carbohydrate synthesized by plants. Starch is a substance of reserve for higher plants. It is composed entirely by D-glucose. It is also an essential energy source for many organisms, including man. This polysaccharide is easily assimilated in the human diet and is a major source of food energy worldwide, since it is found in large amounts in corn, potatoes, wheat and rice, which are the base of the diet for most people. Starch present in vegetables can be hydrolyzed producing glucose, maltose, and oligosaccharide syrups of lower molecular mass, which in turn can be used to produce other chemicals as well as substrates in fermentation processes (Vieille and Zeikus 2001).

Starch is stored in plants as insoluble granules, which are basically composed by two chemically similar polymers : amylose and amylopectin, in a proportion of 25 and 75 percent, respectively. Amylose is essentially made up of linear helical chains consisting of glucose units joined by α-1,4 glycosidic linkages, while amylopectin is formed by a highly branched structure composed by residues of glucose chains connected by α-1,4 glycosidic bonds, from this point ramifications formed by series of 25 glucose residues are connected by α-1, 6 linkages (Laszlo et al. 1986).

Amylolytic Enzymes

Amylases are enzymes able to hydrolyze the α-1,4 glycosidic linkages from the starch molecule. These enzymes are used by plants in order to degrade the starch accumulated in some types of tubers; by animals to digest starch present in the food, and by many prokaryotic and eukaryotic microorganisms which use starch as a carbon source (Peixoto et al. 2003).

Amylases can be divided into two categories according to their action model: as endoamylases and as exoamylases. Endoamylases hydrolyze glycosidic linkages at random within a starch molecule releasing oligosaccharides, and the exoamylases successively hydrolyze the glycosidic linkages from the non-reducing end of the molecule, releasing glucose or maltose (Guzmán-Maldonado and Paredes-Lopez 1995). The action model of some amylolytic enzymes of the amylolytic complex is briefly shown in Fig. 6.1.

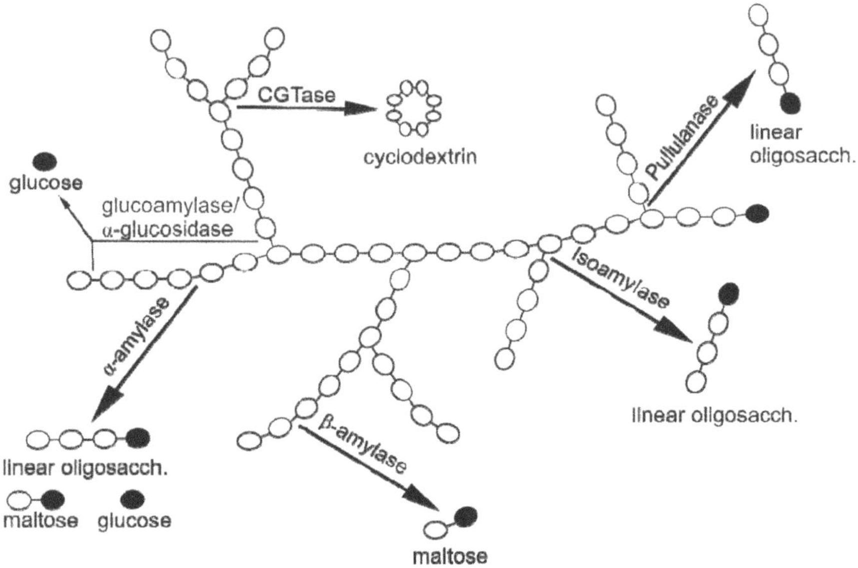

Figure 6.1 Schematic representation of the action model of some amylases. Modified from Bertoldo and Antranikian 2002.

Some amylases that help in the starch converting into compounds of low molecular mass are described below (Vihinen and Mäntsälä 1989, Antranikian et al. 1995):

α-amylases (EC 3.2.1.1 1,4-α-glucan glucanohydrolase): are endoenzymes that hydrolyze the internal α-1,4 glycosidic linkages from amylose and amylopectin, releasing oligosaccharides with different chain lengths. This enzyme can be divided into two categories according to the degree of the substrate hydrolysis: α-amylase of saccharification that hydrolyzes 50–60 percent of starch and α-amylase of liquefaction that hydrolyzes about 30–40 percent. Specificity of α-amylase to α-1,4 glycosidic linkages are not absolute, some α-amylases can degrade some α-1,6 links, however, the reaction rate is less than the α-1,4 links;

β-amylases (EC 3.2.1.2 1,4 α-D-glucan maltohydrolase): are exoenzymes that hydrolyze the penultimate α-1,4 linkage from the non-reducing end of the starch molecule, releasing β-maltose units. Amylose is completely converted into maltose as the conversion ratio of amylopectin to maltose is about 50–60 percent depending on the degree of ramification;

Glucoamylases (GA) (EC 3.2.1.3 1,4-α-D-glucan glucohydrolase): are exoenzymes that hydrolyze α-1, 4 glycosidic linkages, by removing successive units of glucose from the non-reducing end of the chain, releasing molecules of β-D-glucose. This enzyme is also capable of hydrolyzing α-1, 6 and some type of α-1,3 links, but at a much slower speed.

α-D-glycosidase (EC 3.2.1.20 α-D-glycoside glucohydrolase): in general, are exoenzymes of microbial origin. They hydrolyze the α-1,4 and α-1,6 linkages of polysaccharides such as starch, glycogen and olygosaccharides of a short chain, which are formed by the action of other amylases, releasing units of D-glucose from the non-reducing end. Three main types of α-glycosidases are described in literature. Type I has high affinity for aryl glycosides such as p-nitrophenyl glucopyranoside (PNPG) compared to short oligosaccharides. Those representing type II are more active on maltose than aryl glycosides and type III are similar to type II, but they hydrolyze starch and oligosaccharides at the same speed. There are also those that synthesize maltooligosaccharides through maltose in a process known as transglycosylation.

Exo-(1,4)- α-D-glucanases (EC 3.2.1.60 1,4-α-D-glucan maltotetrahydrolase /3.2.1.98 1,4-α-D-glucan maltohexahydrolase): are exoenzymes that instead of releasing successive maltose units, as occurs with β-amylase, they release for example, maltotetraose (EC 3.2.1.60) and are exoenzymes that hydrolyze α-1, 4 glycosidic linkages, by removing successive units of glucose from the non-reducing end of the chain, releasing molecules of β-D-glucose. This enzyme is also capable of hydrolyzing α-1, 6 and some type of α-1, 3 links, but at a much slower speed and maltohexaose as the major products of enzyme action over starch;

Pullulanases (EC 3.2.1.41 α-dextrin-6-glucanohydrolase): are enzymes that hydrolyze the α-1,6 linkages of pullulan and other oligosaccharides. The product of hydrolysis corresponds to maltotetraose, maltotriose and maltose;

Isopullulanases (EC 3.2.1.57 pullulan-4-glucanohydrolase): are enzymes that also hydrolyze α-1,4 linkages of pullulan, but they do not have any activity over starch. Isopanose is the final action product of these enzymes;

Isoamylases (EC 3.2.1.68 glycogen-6-glucanohydrolase): are amylases debranching enzymes that hydrolyze the α-1,6 linkages of amylopectin, glycogen and oligosaccharides, but they do not hydrolyze the α-1,6 linkages of pullulan;

Ciclomaltodextrin D-glucotransferase (CGTase) (EC 2.4.1.19 1,4-α-D-(1,4-D-Glucan) transferase): are exoenzymes which hydrolyze starch through complex reactions forming non-reducing cyclic oligosaccharides, known as cyclodextrins.

Industrial Application

Processing Industry of Starch and Food

Two main steps can be identified in the industrial processing of starch: the liquefaction and the saccharification steps. In a first stage, starch is gelatinized under 110°C until it becomes soluble. The gelatinized starch is subjected to the liquefaction process that is accomplished by the action of a thermostable α-amylase obtained from *Bacillus subtilis*, *B. amyloliquefaciens* and *B. licheniformis*, releasing maltodextrins with low DE (dextrose equivalent) as the final product (Vihinen and Mäntsälä 1989). After complete liquefaction, the material is cooled to a temperature of about 60°C for the saccharification step, which is the complete transformation of the liquefied starch into glucose. This process usually uses glucoamylases from *Aspergillus niger*. Glucose, by the action of glucose isomerase undergoes isomerization and can be converted into fructose. Other amylolytic enzymes as β-amylase and pullulanases, may also be used in the saccharification process, resulting in a maltose syrup (Glazer and Nikaido 1995, Haki and Rakshit 2003).

The use of these enzymes in food processing has been very beneficial, as in addition to their efficiency which may be similar or greater than that of chemical processing, there is no formation of unwanted compounds. In addition, amylolytic enzymes have the advantage of catalyze reactions under moderate conditions; besides, they are natural and non-toxic. Thus making them invaluable to the food industry, in addition most of these enzymes are active at low concentrations (Fenema 1985).

Baking and Brewing

The first enzyme to be used in the baking industry was α-amylase, opening the field of enzymatic activity for other types of enzymes such as proteases, lipases, xylanases, pullulanases, and cellulases, among others. The purpose of using amylolytic enzymes in this industrial sector is to improve the quality of the final product, in such a way as to increase the volume and softness of bread, forming a crispy crust on its surface, in addition to modifying their color making it more attractive to the consumer. It is worth noting that, when the dough is removed from the oven it begins to undergo changes that include hardening, and the bread shell becomes less crunchy (the bread becomes limp and loses its flavor). All these changes are called aging, and one way to delay this process is by adding several enzymes in the process of bread production, since the literature refers to their effect on starch as an anti-aging action. Examples of enzymes that can be used as described previously are: α-amylase, β-amylases and glucoamylases (Pandey et al. 2000).

Another observed effect of adding a mix of amylases during bread manufacture is the fast decomposition of the starchy components used as carbon sources (which may be wheat flour, rye flakes, barley flakes, and cassava flour), releasing sugars readily assimilated by yeast which, via anaerobic metabolism, releases a larger amount of carbon dioxide that in turn, acts by increasing the volume and softness of the bread.

In the brewing industry the raw material usually used is barley (rich in starch) thus, the use of amylolytic enzymes aims to hydrolyze starch to glucose for the subsequent alcoholic fermentation by yeasts. The germination of the barley grain itself, produces the necessary enzymes for the degradation of starch (α and β-amylases).

In addition to barley, breweries use other sources of starch grains that can readily be obtained, such as maize and rice. These cereal grains are more resistant to the action of enzymes in their natural form, because they are boiled before being added to the mixture of malt, therefore, it is feasible that they pass through gelatinization with α-amylase before undergoing the hydration and alcoholic fermentation processes.

Fuel

Amylases have played an important role in the conversion of starch into ethanol simultaneously using, the saccharification and fermentation processes (Singh et al. 1995, Kobayashi et al. 1998, Curvelo-Santana et al. 2010). The application of amylases in ethanol production has partly solved the problems that some countries face regarding the lack of fuel reserves. Glucoamylases and α-amylases of *A. niger* are added into the fermentative medium containing cells of *Saccharomyces cerevisiae* during the process of ethanol production. The starch present in the fermentative medium is converted, through the combined action of amylases, into glucose, which is then converted into ethanol by fermentation (Rajoka et al. 2004).

Detergents and Soap Powder Industries

Currently, enzymes comprise one of the most modern existing ingredients in detergents. However, the first products containing free enzymes that were used caused skin irritation. This problem was solved by using immobilized enzymes in modern formulations. Amylases found in detergents and soap powder act on stains containing starch, such as sauces, fruit, chocolate and others. The starch contained in these materials, usually acts as a glue which holds even patches of proteolytic and lipid sources. Amylases currently sold are produced by Bacillus licheniformis and B. amyloliquefaciens in submerged fermentation and they hydrolyze starch turning it into short chain oligosaccharides which is soluble in water (site: http://www.novozymes.com).

In order to successfully implement these amylases in this industry field, they must show a moderate resistance to alkali pHs, and also work at high temperatures, since many washing machines and dish washers use hot water.

Paper and Textile Industry

Amylases, more specifically the α-amylases, have been used in the paper industry aiming to reduce the viscosity of the starch solution that is used on paper during its processing. The action of α-amylase on the starch layer used on paper coating also contributes to its final quality (Gupta et al. 2003).

During the manufacture of fabrics a considerable amount of starch is introduced in the plant fiber, in order not to break the yarns during the weaving process (the process performed in a device called loom). Therefore, yarn fabrics made from cotton, as well as cotton blended with synthetic fibers are coated with an adhesive called gum, being the most common one, the starch solution, and thus, support the tension to which they are exposed while they are on the loom.

After manufacturing the fabric on the loom, the gum needs to be removed in order to prepare for its completion. Earlier this removal was performed using products such as, strong acids or alkalis. However, for the past few years, the treatment with amylolytic enzymes has been preferred because of their high degree of effectiveness and specific action. Furthermore, enzymes are not considered environmental pollutants, so that the elements of this process are the most accepted.

Fungal Amylases

The application of fungi in processes, where the products are commercially important, has increased over the decades. The use of these microorganisms by humans is not a recent phenomenon, since it is known that ancient Egyptians already used these organisms in fermentation processes for the production of alcoholic beverages (Papagianni 2004).

In general, fungi can be present in many different environments such as soils, decomposing organic materials and plants, among others.

This diversity of environments, has evolutionarily ensured that these organisms have the ability to synthesize a series of enzymes with different characteristics, allowing them to use the organic matter from the medium as a carbon source (Gouka et al. 1997). The ability to synthesize in large-scale, as well as the ease with which they are secreted to the external environment, comprise some of these characteristics (Iwashita 2002). Among these numerous enzymes, amylases appear as one of the most important.

Amylases are derived from many sources, including plants, animals and microorganisms. Microorganisms such as fungi, bacteria, yeast and actinomycetes are potential producers of amylases. However, amylases produced by fungi and bacteria have dominated most of the industrial sectors (Pandey et al. 2000). Among the amylases synthesized by fungi, α-amylases and glucoamylases are those that appear more frequently. Alpha-amylases are produced by many genera of filamentous fungi and yeasts. The α-amylases produced by some bacteria as for example, *Bacillus* sp., *B. amyloliquefaciens* and *B. liqueniformes*, with commercial application are preferred over those produced by fungi and yeasts, because they present some desirable characteristics such as thermoactivity and thermostability and therefore, they can be applied in processes that take place at high temperatures (Pandey et al. 2000). However, bacterial α-amylases do not have the capability of saccharification as some fungal α-amylases. Thus, it is necessary to produce fungal enzymes that exhibit thermostability similar to those of bacteria.

Glucoamylases generally come from many sources such as plants, animals and microorganisms. They effectively catalyze starch saccharification reaction within narrow limits of temperature, which happens because of the catalytic active conformation of glucoamylases used in industrial changes as the temperature increases. This in turn affects their use in industrial processes, where long storage periods at high temperatures are required.

Glucoamylases commercially available are mostly produced by strains of *Aspergillus* and *Rhizopus*. Glucoamylases from *Aspergillus* genus are more thermostable, with maximal activity between 50 and 55°C, being rapidly inactivated at temperatures close to 60°C (Zaha and Zeikus 1989). In addition the ability for large scale production, where some species of *Aspergillus* can synthesize more than one form of the same enzyme with different characteristics compared to those synthesized by bacteria with unique characteristics and properties (Kumar et al. 2004).

Besides the genus *Aspergillus*, species of other genera such as, *Rhizopus, Mucor, Humicola* and *Thermomyces,* among others, appear in the literature as potential amylase producers, especially the amylases produced by *Rhizopus oligosporus DAR 2710, R. microsporus, Rhizopus sp., Mucor pusillus,*

Thermomucor, Humicola grisea, Thermomyces lanuginosus, T. lanuginosus ATCC 34626, *Rhizomucor pusillus* A 13.36 and *Aspergillus niveus* (Somkuti and Steinberg 1980, Tosi et al. 1993, Mishra and Maheshwari 1996, Cruz et al. 1997, Jin et al. 1999, Nguyen et al. 2000, Peixoto et al. 2003, Kaur and Satyanarayana 2004, Silva et al. 2005, Silva et al. 2009, respectively).

Unicellular fungi, usually known as yeast, as well as filamentous fungi, are found around many different ecosystems. This diversity of environments inhabited by these microorganisms has enabled researchers to isolate species with desirable features for industry (Skorupa et al. 2002). The ability to degrade starch by certain species of yeast has been crucial for its application in biotechnological processes as sources of hydrolytic enzymes for ethanol production (Mot 1984).

The production of amylolytic enzymes by yeast, has received much attention due to the small volume in which they are secreted. Even though it is known that these enzymes are produced by most species, among those species known as amylases producers, the following must be highlighted: *Schwanniomyces alluvius, Lipomyces konomenkoae* and *Cryptococcus flavus* (Wilson and Ingledew 1982, Prieto et al. 1995 and Skorupa et al. 2002, respectively).

In recent years, extensive research in molecular biology, along with the development and application of new techniques of recombinant DNA in some species of yeast has led to large-scale production of amylases (Shiraga et al. 2004, Moller et al. 2004).

Production and Purification of Fungal Amylases

Aiming to make the hydrolysis of starch economically viable, many fungi with excellent potential to produce amylases have been isolated in recent years. Some studies have reported production and purification strategies of different types of fungal amylolytic enzymes that are thermostable and thermotolerant, and which can be applied in bioprocesses. Once they are produced, the purification methods used can vary considerably, but most of the purification protocols involve a series of steps. The purification practice of extracellular fungal amylase has been accomplished through the use of classical methods, which involve the separation of the mycelial mass from the fermentation media, by means of centrifugation or ultrafiltration steps. The filtrate, free from mycelium, is concentrated through the precipitation of ammonium sulfate or organic solvents as acetone or ethanol. The enzyme present in the crude extract is then subjected to chromatographic processes, usually ion exchange, affinity and gel filtration columns. The intracellular enzymes are usually obtained through the destruction of the cell walls, and then similar procedures used to purify the extracellular amylase are followed. Some suitable strategies for the purification of fungal amylases are shown in Table 6.1.

Table 6.1 Strategies used for the purification of fungal amylases.

Amylase	Microorganism	Purification techniques used	References
α-amylase	*Aspergillus flavus*	ammonium sulfate precipitation and ion-exchange chromatography	Khoo et al. 1994
α-amylase	*Aspergillus oryzae*	amylopectin affinity, DEAE-Sepharose ion-exchange chromatography and Sephacryl S-200 HR	Chang et al. 1995
glucoamylase	*Rhizopus strains*	80% $(NH4)_2SO4$ precipitation, CM-SephadexC-50 chromatography	Morita et al. 1998
glucoamylase	*Thermomyces lanuginosus*	Ammonium sulfate precipitation; DEAE-Toyopearl; Butyl-Toyopearl; Sephacryl S-300; FPLC Mono-Q	Li et al. 1998
glucoamylase	*Scytalidium thermophilum*	DEAE-cellulose; CM-cellulose	Cereia et al. 2000
α-amylase	*Aspergillus carbonarius*	4 M sucrose concentration, S-Sepharose (fast flow) column, Q-Sepharose (fast flow) column	Okolo et al. 2000
glucoamylase	*Scytalidium thermophilum*	DEAE-cellulose; ultrafiltration; CM-cellulose	Aquino et al. 2001
α-amylase	*Scytalidium thermophilum*	DEAE-cellulose; CM-cellulose; Sepharose 6B	Aquino et al. 2003
glucoamylase	*Thermomucor indicae seudaticae*	Lyophilization and acetone precipitation, SP-sepharose, sephadex G-50	Kumar and Satyanarayana 2003
α-amylase	*Aspergillus flavus* var. Columnaris	Ammonium sulfate precipitation; dialysis; Sephadex G-200	El-Safey and Ammar 2004
glucoamylase	*Penicillium* sp. *X-1*	65% $(NH4)_2SO4$ precipitation	Sun et al. 2007
glucoamylase	*Paecilomyces variotii*	Dialysis; DEAE-cellulose; Sephadex G-100	Michelin et al. 2008
glucoamylase	*Scytalidium thermophilum* 15.1	DEAE-cellulose; CM-cellulose	Ferreira-Nozawa et al. 2008
glucoamylase	*Scytalidium thermophilum* 15.3	CM-cellulose; concanavalin A-Sepharose	Ferreira-Nozawa et al. 2008
glucoamylase	*Aspergillus niveus*	DEAE-fractogel; Concanavalin A-Sepharose	Silva et al. 2009
α-glucosydase	*Aspergillus niveus*	DEAE-fractogel; Sephacryl S-200	Silva et al. 2009
α-amylase	*Penicillium camenberti* PL21	Ammonium sulfate precipitation; dialysis; Sephadex G-100; DEAE-Sepharose CL-6B	Nouadri et al. 2010
α-amylase	*Paecilomyces variotii*	DEAE-cellulose, Sephadex G-100 and electroelution	Michelin et al. 2010
α-amylase	*Penicillium janthinellum*NCIM 4960	Ammonium sulfate precipitation; DEAE-cellulose	Sindhu et al. 2011

Physico-chemical Characterization

Temperature, pH and Ions

In general, many enzymes are denatured and lose their activity at temperatures above 50–60°C. However, thermostable enzymes are of great industrial and biotechnological use because they are more suitable for harsh industrial processes that take place at high temperatures. The advantages of conducting biotechnological processes at elevated temperatures include high reaction rates and increased diffusion coefficient of the substrate due to decreased viscosity, reduced risk of contamination by mesophilic common features, and the facility of handling and transport (Ray and Nanda 1996, Haki and Rakshit 2003).

Thermostable amylases have been isolated from many different sources, including plants, animals and microorganisms. Despite the wide distribution of these enzymes, microbial sources such as fungi and bacteria are most commonly used for industrial production, due to their low cost, shorter length of production and easiness in production and optimization (Prakash and Jaiswal 2010). Amylases with high optimum temperature are industrially desirable because they eliminate the chances of denaturing in bioreactors and the contamination of the solution. Many factors can affect the thermostability of enzymes, including the purity of the preparation, the presence of calcium, the type of substrate, and other stabilizers (Vihinen and Mäntsala 1989). Therefore some approaches such as chemical modifications in the presence of specific ions, immobilization and genetic engineering have been used in an attempt to increase the stability against temperature.

The optimum pH for α-amylases, varies in ranges of 2.0 to 10.5, indicating the evolutionary adaptation to different environments (Vihinen and Mäntsala 1989). Generally, the optimum pH for glucoamylases is between 2.0 and 7.0. The pH stability is also an important feature for the continuous production, as pH changes are frequent in large-scale fermenters.

Many metal ions can activate fungal amylases. The ones that show the best effects are the divalent ions as Ca^{2+}, Ba^{2+}, Mn^{2+}, Mg^{2+} and Fe^{2+}. The metal ions as Zn^{2+}, Cu^{2+}, Pb^{2+}, Hg^{2+}, Cd^{2+}, Ca^{2+}, Mn^{2+}, Ni^{2+}, Fe^{3+}, Ag^{+} and Al^{3+}, have been reported as inhibitors for glucoamylases, while others as Ba^{2+}, Ca^{2+}, Co^{2+}, Cu^{2+}, Mn^{2+}, Zn^{2+}, Na^{2+}, Mg^{2+}, Sn^{2+} and Fe^{2+} have a positive effect on the enzyme, increasing its activity (Kumar and Satyanarayana 2009). Several cations, especially heavy metals, with sulfhydryl reagent groups, EDTA and EGTA inhibit the activity of α-amylase. Since Ca^{2+}, in general, plays a role in the stabilization and activation of α-amylases (Vihinen and Mäntsala 1989). In general thiols, such as β-mercaptoethanol, dithiotreithol (DTT) and cysteine increase the activity and stability of β-amylases. In some cases, enzymes partially inactivated by temperature or other agents may be partly reactivated by the addition of thiols (Ray and Nanda 1996).

Molecular Mass, Isoelectric Point and Glycosylation

The molecular mass of α-amylases is in a range that varies from 10 to 139 kDa, and most of these enzymes are in the range of 50–60 kDa (Vihinen and Mäntsala 1989). For glucoamylases, molecular masses from 26.8 to 250 kDa can be found (see Table 6.2). The Isoelectric point, or pI, is the pH where the enzyme electric charge is zero. The pI values are important for understanding enzyme behavior against different purification procedures at different pHs. The pI value of amylases varies widely, from 3.25 to 10.1. Oligosaccharides present in glycoproteins are responsible for the signaling of cell secretion, maintenance of protein structure and protection against stress caused by temperature, pH and pressure (Kumar and Satyanarayana 2009). The degree of glycosylation varies greatly in amylases, from 1.8 to 80 percent. For example the *A. niger* glucoamylases have 19 percent neutral carbohydrate with an average size of two monosaccharides linked by α-1.2 and α-1.6 glycosidic linkages. Mannose is most often found in *O*-linked serine and threonine residues (Kumar and Satyanarayana 2009).

These days it is well known that protein glycosylation plays extremely important role in cellular functioning. Glycosylation can contribute to protein secretion, as well as to its function, stability and immunogenicity. Concerning the membrane glycoproteins, glycans may mediate cell communication with the extracellular medium. In addition to this, recent studies have shown that N-glycosylation can also have important effects on enzyme activity, function and substrate specificity.

In recent years, many studies have demonstrated the involvement of the systematic removal of putative sites from several glycosylated enzymes in order to identify the precise role glycan has on the regulation of enzyme secretion, activity and substrate specificity. Studies on the removal of N-glycan from glucoamylases produced by *Aspergillus niger,* resulted in the exposure of the protein hydrophobic regions thereby leading to a reduced thermal stability, and a greater aggregation and flexibility of the enzyme; similar to the glucoamylase produced by *Aspergillus awamori*, it was observed that a lower secretion and reduced thermostability; while the amylase produced by *Aspergillus oryzae*, the N-glycosylation had no observed effect over the enzyme secretion, activity or thermostalibity. Complementary studies showed an increase in thermostability when new glycosylation sites were artificially introduced in the glucoamylase produced by *Aspergillus awamori.*

Analysis of the inhibition of α-amylase N-glycans produced by *Cryptococcus flavus* have demonstrated that the non-glycosylated form was

Table 6.2 Biochemical characteristics of amylases produced by fungi.

Fungi	Amylase type	Molecular Mass (kDa)	T (°C) optimum/ stability	pH optimum/ stability	pI	Activator/ inhibitor	Glycosylation (%)	References
Arthrobotrys amenospora	Glucoamylase	44.7	55/–	5.6/–	–	–	–	Norouzian et al. 2000
Aspergillus awamori var. kawachi HF-15	Glucoamylase	250	–/65 (15 min)	–/2.0–8.0 (20h)	–	–	–	Flor and Hyashida 1983
Aspergillus carbonarius	β-amylase	32	40/30–80	6.0/3.0–7.0	–	Co^{2+}/EDTA, Na^{+}, Mg^{2+}, Ca^{2+}, Zn^{2+}, Ba^{2+}, Si^{2+}, Ag^{+}	–	Okolo et al. 2000
Aspergillus terreus (CTC 826)	Glucoamylase	–	60/40–60	4.0/3.0–7.0	–	–	–	Ali and Hossain 1991
A. fumigatus	Glucoamylase	42	65/–	4.5–5.5/–	–	–	23	Silva and Peralta 1998
Aspergillus terreus (NA 170)	Glucoamylase	–	45/30–75	5.0/3.0–7.0	–	–	–	Ghosh et al. 1991
Aspergillus oryzae	Glucoamylase	65	65/<46	4.5/–	4.2	–	–	Hata et al. 1997
Aspergillus tamarii	α-amylase	–	30/50–65	4.5–6.0/4.0–7.0	–	–	–	Moreira et al. 1999
A. niger C	Glucoamylase	70	40/–	4.5/–	–	–	–	Paszczynski et al. 1982
Aspergillus sp. K-27	Glucoamylase	76	45/–	4.0–7.1/–	–	–	–	Abe et al. 1988
A. satoi	Glucoamylase	90	50/–	4.5/–	–	–	18.0	Takahashi et al. 1981
A. oryzae	Glucoamylase	68.4	56/–	4.0–4.5/–	–	–	7.8	Ono et al. 1988

Chalara paradoxa	Glucoamylase	76	50/–	5.0/–	3.8	–	6.8	Monma et al. 1987
Chaetomium thermophium	Glucoamylase	64	65/–	4.0/–	–	–	11.7	Chen et al. 2005
Corticium rolfsii AHU 9627	Glucoamylase	78	60/	4.5/	3.85	–	–	Nagasaka et al. 1998
Thermomyces lanuginosus	Glucoamylase	75	70/–	4.4–5.6/–	4.1–4.3	–	3.27	Nguyen et al. 2000
Trichoderma reesei	Glucoamylase	66	70/–	5.5/–	4.0	–	–	Fagerstrom and Kalkkinen 1995
Curvularia lunata	Glucoamylase	66	50/–	4.0/–	–	–	–	Feng et al. 2007
Cephalosporium eichhorniae	Glucoamylase	26.8	46/–	4.6/–	3.25	–	28	Day 1978
Humicola sp.	Glucoamylase	72.8	55/–	4.7/–	–	–	–	Riaz et al. 2007
Humicola grisea	Glucoamylase	74	60/–	6.0/–	–	–	5	Campos and Felix 1995
H. grisea var. *thermoidea*	Glucoamylase	63	55/–	5.0/–	–	–	1.8	Tosi et al. 1993
Mucor pusillis	α-amylase	–	70/60–65	3.5–4.0/–	–	–	–	Somkuti and Steinberg 1980
Monascus sp.	Glucoamylase	53	–	4.5–4.8/–	–	–	13.5	Yasuda et al. 1989
N. crassa exo I	Glucoamylase	82	60/–	5.4/–	–	–	5.1	Spinelli et al. 1996
Paecilomyces varioti AHU 9417	Glucoamylase	69	37/–	4.5/–	–	–	13	Takeda et al. 1985

Table 6.2 contd....

Table 6.2 contd.

Fungi	**Amylase type**	**Molecular Mass (kDa)**	**T (°C) optimum/ stability**	**pH optimum/ stability**	**pI**	**Activator/ inhibitor**	**Glycosylation (%)**	**References**
Paecilomyces varioti	Glucoamylase	86.5	55/–	5.0/–	3.5	–	27.5	Michelin et al. 2008
Penicillium sp. X-1	Glucoamylase	–	65/<70	6.5/3.0–8.0	–	Mn^{2+}, Ca^{2+}, Li^{2+}, Fe^{3+}, Cu^{2+} /Mg^{2+}, Ni^{2+}, Zn^{2+}	–	Sun et al. 2007
Rhizopus sp.	Glucoamylase	74	–	4.5/–	–	–	9.2	Takahashi et al. 1978
Scytalidium thermophilum	Glucoamylase	86	60/–	6.5/–	8.4	–	25.5	Cereia et al. 2000
Scytalidium thermophilum	Glucoamylase	75	70/–	5.5/–	8.3	–	9.8	Aquino et al. 2001
Thermomyces lanuginosus F1	Glucoamylase	70	70/–	6.0/–	4.0	–	–	Odibo and Ulbrich-Hofmann 2001
Thermomucor indicae-seudaticae	Glucoamylase	42	60/half-life of 12h at 60°C	7.0/–	8.2	Co^{2+}, Fe^{2+}, Ag^{+}, Ca^{2+}, Cu^{2+}, Mg^{2+}, trealose /Hg^{2+}, Zn^{2+}, Ni^{2+}, Mn^{2+}, EDTA	–	Kumar and Satyanarayana 2003

more sensitive to proteolytic degradation when compared to the glycosylated native form (Barros et al. 2009). Studies on N-glycans inhibition, carried out by our group have found that N-linked glycans of α-glucosydase produced by *Aspergillus niveus* had no significant influence on the enzyme production and secretion, and on its catalytic properties. However, the deglycosylated form was more sensitive to proteolytic degradation (Silva et al. 2010).

The analysis of *O*-glucansmakes it possible to verify some interesting data. A glucoamylase produced by *Aspergillus niveus* is highly *O*-glycosylated and its catalytic activity is strongly dependent of this process. The *O*-glycans structures of glucoamylases produced by *Aspergillus niveus* are highly variable, as well as their function on the structures, this makes it necessary for specific analysis of each glucoamylase.

The amount of glycosylation and its distribution in amylase varies among different fungi, a factor that can influence the function of other amylases studied. In recent decades, several studies have elucidated the role of glycosylation in enzyme secretion, stability and intracellular trafficking, and also its action in the regulation of enzyme activity and substrate specificity. However, many of the functions that the glycosylation exerts on the enzyme activity remains unknown, and further studies are required.

For the past few decades, the use of fungi in bioprocesses has gained importance due to the production of many enzymes with different physicochemical characteristics and an excellent potential for industrial application (Table 6.2).

Future Prospects

Amylases are among the most important enzymes used in industrial processes for the conversion of starch into dextrins and low molecular weight saccharides. As these enzymes have some peculiar characteristics, their commercial use has been restricted to those obtained from a few species of fungi and bacteria. This fact should encourage researchers aiming to obtain strains which produce enzymes that meet the current needs of this emerging market. Research leading to the discovery of strains that produce amylase with these characteristics along with the use of recombinant DNA techniques and protein engineering is a field with enormous future prospects.

Acknowledgment

This chapter was supported by grants from Fundação de Amparo à Pesquisa do Estado de São Paulo (FAPESP).

References

Abe, J.I., K. Nakajiima, H. Nagano, S. Hizkuri and K. Obata. 1988. Properties of the raw starch digesting amylase of Aspergillus sp. K-27: a synergistic action of glucoamylase and α-amylase. Carbohydr. Res. 175: 85–92.

Ali, S. and Z. Hossain. 1991. Characteristics for glucoamylase from *Aspergillus terreus*. J. App. Bacteriol. 71: 144–46.

Antranikian, G., A. Rudiger, F. Canganella, M. Klingeberg and A. Sunna. 1995. Biodegradation of polymers at temperatures up to 130-degrees-c. J. Macromol. Sci. Pure. Appl. Chem. 32: 661–669.

Aquino, A.C.M.M., J.A. Jorge, H.F. Terenzi and M.L.T.M. Polizeli. 2001. Thermostable glucose-tolerant glucoamylase produced by thermophilic fungus *Scytalidium thermophilum*. Folia Microbiol. 46(1): 11–16.

Aquino, A.C.M.M., J.A. Jorge, H.F. Terenzi and M.L.T.M. Polizeli. 2003. Studies on a thermostable α-amylase from the thermophilic fungus *Scytalidium thermophilum*. Appl. Microbiol. Biotechnol. 61: 323–28.

Barros, M.C., S.R. Nascimento, M.H. Ramada, A.S. Galdino, L.M de Moraes and C.J. Ulhoa. 2009. The influence of N-glycosylation on biochemical properties of Amy1, an alpha-amylase from the yeast *Cryptococcus flavus*. Carbohydr Res. 344: 1682–1686.

Bertoldo, C. and G. Antranikian. 2002. Starch-hydrolyzing enzymes from thermophilic archaea and bacteria. Current Opinion in Chemical Biology. 6: 151–160.

Campos, L. and C.R. Felix. 1995. Purification and charcterization of a glucoamylase from *Humicola grisea*. Appl. Environ. Microbiol. 61: 2436–38.

Cereia, M., H.F. Terenzi, J.A. Jorge, L.J. Greene, J.C. Rosa and M.L.T.M. Polizeli. 2000. Glucoamylase activity from the thermophilic fungus *Scytalidium thermophilum*: biochemical and regulatory properties. J. Basic. Microbiol. 40: 83–92.

Chang, C.T., M.S. Tang and C.F. Lin. 1995. Purification and properties of alpha-amylase from *Aspergillus oryzae* ATCC 76080. Biochem. Mol. Biol. Int. 36(1): 185–93.

Chen, J., D.C. Li, Y.Q. Zhang and Q.X. Zhou. 2005. Purification and characterization of a thermostable glucoamylase from *Chaetomium thermophilum*. J. Gen. Appl. Microbiol. 51: 175–81.

Cruz, R., E.L. Sousa, E.H.E. Hoffmanm, M.Z.B. Bellini, V.D. Cruz and C.R. Vieira. 1997. Relationship between carbon source, production and patter action of α-amylase from *Rhizopus* sp. Rev. Microbiol. 28: 101–105.

Curvelo-Santana, J.C., D.D. Ehrhardt and E.B.Tambourg. 2010. Otimização da produção de álcool de mandioca. Ciênc. Tecnol. Aliment., Campinas 30(3): 613–617.

Day, D.F. 1978. A thermophilic glucoamylase from *Cephalosporium eichhorniae*. Curr. Microbiol. 1: 181–84.

El-Safey, E.M. and M.S. Ammar. 2004. Purification and characteraization of α-amylase isolated from *Aspergillus falvus* var. columnaris. Ass. Univ. Bull. Environ. Res. 7(1): 93–100.

Fagerstrom, R. and N. Kalkkinen. 1995. Characterization, subsite mapping and partial amino acid sequence of glucoamylase from the filamentous fungus *Trichoderma reesei*. Biotechnol. Appl. Biochem. 21: 223–31.

Fenema, O.R. 1985. Introdución a la Ciência de los Alimentos. Barcelona. Ed. Reveré, S.A. 1: 445.

Feng, B., Hu. W, B.P. Ma, Y. Wang, H. Huang, S. Wang and X. Qian. 2007. Purification, characterization, and substrate specificity of a glucoamylase with steroidal saponin-rhamnosidase activity from *Curvularia lunata*. App. Microbiol. Biotechnol. 76: 1329–1338.

Feng, P.H., S. Berensmeier, K. Buchholz and P.J. Reilly. 2002. Production, purification, and characterization of *Thermoanaerobacterium thermosaccharolyticum* glucoamylase. Starch/Starke. 54: 328–37.

Ferreira-Nozawa, M.S., J.L. Rezende, L.H.S. Guimarães, H.F. Terenzi, J.A. Jorge and M.L.T.M. Polizeli. 2008. Mycelial glucoamylases produced by the thermophilic fungus *Scytalidium thermophilum* strains 15.1 and 15.8: purification and biochemical characterization. Braz. J. Microbiol. 39: 344–52.

Flor, P.Q. and S. Hayashida. 1983. Production and Characteristics of Raw Starch-Digesting Glucoamylase from a Protease-Negative, Glycosidase-Negative *Aspergillus awamori* var. *kawachi* Mutant. Appl. Environ. Microbiol. 45(3): 905–12.

Ghosh, A., B. Chatterjee and A. Das. 1991. Purification and characterization of glucoamylases of *Aspergillus terreus* NA-170 mutant. J. Appl. Bacteriol. 71: 162–69.

Glazer, A.N. and H. Nikaido. 1995. Microbiol Biotechnology, W.H. Freeman and Company, New York.

Gouka, R.J., P.J. Punt and C.A.M.J.J. Van Den Hondel. 1997. Efficient production of secreted proteins by *Aspergillus*: progress, limitations and prospects. Appl. Microbiol. Biotechnol. 47: 1–11.

Gupta, R., P. Gigras, H. Mohapatra, V.G. Kumar and B. Chauhan. 2003. Microbiol α-amylases: a biotecnological perspective. Process Biochem. 38: 1599–1616.

Guzmán-Maldonado, H. and O. Paredes-López. 1995. Amylolytic enzymes and products derived from starch: A review. Crit. Rev. Food Sc. Nut. 35: 373–403.

Haki, D.G. and K.S. Rakshit. 2003. Developments in industrially important thermostable enzymes: Bioresour. Technol. 89: 17–34.

Hata, Y., H. Ishida, Y. Kojima, E. Ichikawa, A. Kawato and K. Suginami. 1997. Comparison of two glucoamylases produced by *Aspergillus oryzae* in solid-state culture (koji) and in submerged culture. J. Ferment. Bioeng. 84: 532–37.

Iwashita, K. 2002. Recent studies of protein secretion by filamentous fungi. J. Biosci. Bioeng. 94: 530–535.

Jin, B., H.J. Leeuwen, B. Patel, H.W. Doelle and Q. YU. 1999. Production of fungal protein and glucoamilase by *Rhizopus oligosporus* from starch processing wasterwater. Process Biochem. 34: 59–65.

Kaur, P. and T. Satyanarayana. 2004. Production and saccharification by a thermostable and neutral glucoamylase of a thermophilic mould *Thermomucor indicae-seudaticae*. J. Microbiol. Biotechnol. 20: 419–425.

Khoo, S.L., A.A. Amirul, M. Kamaruzaman, N. Nazalan and M.N. Azizan. 1994. Purification and characterization of alpha-amylase from *Aspergillus flavus*. 39(5): 392–398.

Kobayashi, F., T. Sawada, Y. Nakamura, M. Ohnaga, M. Godliving and T. Ushiyama. 1998. Sacaharification and alcohol fermentation in starch solution of stream-explodet potato. Appl. Bioch. Biotechnol. 69: 177–189.

Kottwitz, B., H. Upadek and G. Carrer. 1995. Applications and benefits of enzymes in detergents. Chim. Ogg. 12: 21–24.

Kumar, R.K., S.K. Venkatesh and U.S. Kumar. 2004. Evidence that cleavage of the precursor enzyme by autocatalysis caused secretion of multiple amylases by *Aspergillus niger*. FEBS Letters. 557: 239–242.

Kumar, S. and T. Satyanarayana. 2003. Purification and Kinetics of a Raw Starch-Hydrolyzing, Thermostable, and Neutral Glucoamylase of the Thermophilic Mold *Thermomucor indicae-seudaticae*. Biotechnology Progress. 19: 936–944.

Kumar, S. and T. Satyanarayana. 2009. Overproduction of Glucoamylase by a Deregulated Mutant of a Thermophilic Mould *Thermomucor indicae-seudaticae*. Appl. Biochem. Biotechnol. 158: 113–125.

Laszlo, H., L.M. Basso and C.M.L. Coelho. 1986. Química de Alimentos, Nobel, São Paulo.

Li, D.C., Y.J. Yang, Y.L. Peng and C.Y. Shen. 1998. Purification and characterization of extracellular glucoamylase from the thermophilic *Thermomyces lanuginosus*. Mycol. Res. 102: 568–72.

Michelin, M., R. Ruller, R.J. Ward, L.A.B. Moraes, J.A. Jorge, H.F. Terenzi and M.L.T.M. Polizeli. 2008. Purification and biochemical characterization of a thermostable extracellular glucoamylase produced by the thermotolerant fungus *Paecilomyces variotii*. J. Ind. Microbiol. Biotechnol. 35: 17–25.

Michelin, M., T.M. Silva, V.M. Benassi, S.C. Peixoto-Nogueira, L.A. Moraes, J.M. Leão, J.A. Jorge, H.F. Terenzi and M.L.T.M. Polizeli. 2010. Purification and characterization of a thermostable α-amylase produced by the fungus *Paecilomyces variotii*. Carbohydr. Res. 345(16): 2348–53.

Mishra, R. and R. Maheshwari. 1996. Amylases of thermophilic fungus *Thermomyces lanuginosus*. Their purification, properties, action on starch and response to heat. J. Biosc. 21: 653–672.

Moller, K., M.Z. Sharif and L. Olsson. 2004. Production of fungal α-amylase by *Saccharomyces kluyveri* in glucose-limited cultivations. J. Biotechnol. 111: 311–318.

Monma, M., K. Mikuni, H. Ishigami and K. Kainuma. 1987. Purification of the glucoamylase components of *Chalara paradoxa* by affinity chromatography and chromatofocusing. Carbohydr. Res. 159: 255–61.

Moreira, F.G., F.A. Lima, S.R.F. Pedrinho, V. Lenartoviez, C.M.S. Souza and R.M. Peralta. 1999. Production of amylases by *Aspergillus tamari*. Rev. Microbiol. 30: 157–62.

Morita, H., M. Matsunaga, K. Mizuno and Y. Fujio. 1998. A comparison of raw starch-digesting glucoamylase production in liquid and solid cultures of Rhizopus strains. J. Gen. Appl. Microbiol. 44: 211–216.

Mot, R., K. Andries and H. Verachtert. 1984. Comparative study of starch degradation and amylase production by Ascomycete yeast species. Appl. Microbiol. 5: 106–118.

Nagasaka, Y., K. Kurosawa, A. Yokota and F. Tomita. 1998. Purification and properties of the raw-starch-digesting glucoamylases from *Corticium rolfsii*. Appl. Microbiol. Biotechnol. 50: 323–30.

Nguyen, D.G., J.M. Rezessy-Szabó and A. Hoschke. 2000. Optimisation of composition of media for the production of amylolytic enzymes by *Thermomyces lanuginosus* ATCC 34626. Food Technol. Biotechnol. 38: 229–234.

Norouzian, D., K. Rostami, I.D. Nouri and M. Saleh. 2000. Subsite mapping of purified glucoamylases I, II, III produced by *Arthrobotrys amerospora* ATCC 34468. World J. Microbiol. Biotechnol. 16: 155–61.

Nouadri, T., Z. Meraihi, D.D. Shahrazed and B. Leila. 2010. Purification and characterization of the α-amylase isolated from *Penicillium camemberti* PL21. African Journal of Biochemistry Research. 4(6): 155–162.

Odibo F.J.C. and R. Ulbrich-Hofmann. 2001. Thermostable α-amylase and glucoamylase from *Thermomyces lanuginosus* F1. Acta Biotechnol. 21: 141–153.

Okolo, B.N., F.S. Ire, L. Ezeogu, C.U. Anyanwu and F.J.C. Odibo. 2000. Purification and some properties of a novel raw starch-digesting amylase from *Aspergillus carbonarius*. J. Sci. Food Agric. 81: 329–36.

Ono K., S. Shigeta and S. Oka. 1988. Effective purification of glucoamylase in Koji, a solid culture of *Aspergillus oryzae* on steamed rice, by affinity chromatography using an immobilized acarbose (BAY g-542). Agric. Biol. Chem. 52: 1707–1714.

Pandey, A., P. Nigam, C.R. Soccol, V.T. Soccol, D. Singh and R. Mohan. 2000. Advances in microbial amylases. Rev. Biotechnol. Appl. Biochem. 31: 135–152.

Papagianni, M. 2004. Fungal morphology and metabolite production in submerged mycelial processes. Review. Biotechnol. Adv. 22: 189–259.

Paszczynski, A., I. Miedziak, J. Lobarzewski, J. Kochmanska and J. Trojanowski. 1982. A simple method of afinity chromatograpy or the purification of glucoamylase obtained from *Aspergillus niger* C. FEBS Lett. 149: 63–66.

Peixoto, S.C., J.A. Jorge, H.F. Terenzi and M.L.T.M. Polizeli. 2003. *Rhizopus microsporus var. rhizopodiformis*: a thermotolerant fungus with potential for production of thermostable amylases. Int. Microbiol. 6: 269–273.

Prakash, O. and N. Jaiswal. 2010. α-amylase: an ideal and representative of thermostable enzymes. Appl. Biochem. Biotechnol. 160: 2401–2414.

Prieto, J.A., B.R. Bort, J. Martinez, F. Randez-Gil, C. Buesa and P. Sanz. 1995. Purification and characterization of a new α-amylase of intermediate thermal stability from the yeast *Lipomyces kononenkoae*. Biochem. Cell. Biol. 73: 41–49.

Rajoka, M.I., A. Yasmin and F. Latif. 2004. Kinetics of enhanced ethanol productivity using raw starch hydrolyzing glucoamilase from *Aspergillus niger* mutant produced in solid state fermentation. Appl. Microbiol. 39: 13–18.

Ray, R.R. and G. Nanda. 1996. Microbial β-amilases: biosynthesis, charachteristics and industrial aplications. Crit. Rev. Microbiol. 22(3): 181–99.

Riaz, M., R. Perveen, M.R. Javed, H. Nadeem and M.H. Rashid. 2007. Kinetic and thermodynamic properties of novel glucoamylase from *Humicola* sp. Enz. Microb. Technol. 41: 558–64.

Shiraga, S., M. Kawakami and M. Ueda. 2004. Construction of combinatorial library of starch-binding domain of *Rhizopus oryzae* glucoamylase and screening of clones with enhanced activity by yeast display method. J. Mol. Catalysis B: Enz. 28: 229–234.

Silva, B.W. and W.M. Peralta. 1998. Purification and characterization of a thermostable glucoamylase from *Aspergillus fumigatus*. Can. J. Microbiol. 44: 493–97.

Silva, T.M., D. Attili-Angelis, A.F.A. Carvalho, R. Silva and E. Gomes. 2005. Production of sacharogenic and dextrinogenic amylases by *Rhizomucor pusillus* A 13.36. J. Microbiol. 43: 561–568.

Silva, T.M., A. Maller, A.R.L. Damásio, M. Michelin, R.J. Ward, I.Y. Hirata, J.A. Jorge, H.F. Terenzi and M.L.T.M. Polizeli. 2009a. Properties of a purified thermostable glucoamylase from *Aspergillus niveus*. J. Ind. Microbiol. Biotechnol. 36: 1439–1446.

Silva, T.M., M. Michelin, A.R.L. Damasio, A. Maller, F.B.R. Almeida, R. Ruller, R.J. Ward, J.C. Rosa, J.A. Jorge, H.F. Terenzi and M.L.T.M. Polizeli. 2009b. Purification and biochemical characterization of a novel *a*-glucosidase from *Aspergillus niveu*. Antonie van Leeuwenhoek. 96: 569–578.

Silva, T.M., F.B. Almeida, A.R.L. Damásio, A. Maller, M. Michelin, J.A. Jorge, E.S. Hanna, M.C. Roque-Barreira, H.F. Terenzi and M.L.T.M. Polizeli. 2010. Tunicamycin inhibition of N-glycosylation of alpha-glucosidase from *Aspergillus niveus*: partial influence on biochemical properties. Biotec. Lett. 32: 1449–1455.

Sindhu, R., G.N. Suprabha and S. Shashidhar. 2011. Purification and characterization of α-amylase from *Penicillium janthinellum* (NCIM 4960) and its application in detergent industry. Biotechnol. Bioinf. Bioeng 1(1) 25–32.

Singh, D., J.S. Dahiya and P. Nigam. 1995. Simultaneous raw starch hydrolysis and ethanol fermentation by glucoamylase from *Rhizocolonia solini* and *Saccharomyces cerevisae*. J. Basic. Microbiol. 35: 117–121.

Skorupa, P.N., L.M.P. Moraes and F.A.G. Torres. 2002. Characterization of new variant of the yeast *Cryptococcus flavus* isolated from the biodversity of Brasilian Cerrado. Abstracts from the 23rd meeting on Genetics of Microorganims, Pirenópolis, GO, Brazil.

Somkuti, A.G. and H.D. Steinberg. 1980. Thermoacidophilic extracellular α-amylase of *Mucor pusillus*. Devel. Ind. Microbiol. 21: 327–337.

Spinelli, B.B.L., M.L.T.M. Polizeli, H.F. Terenzi and J.A. Jorge. 1996. Biochemical characterization of glucoamylase from the hyper producer exo-1 mutant strain of *Neurospora crassa*. FEMS Microbiol. Lett. v. 138, p. 173–177.

Sun, H., X. Ge and W. Zhang. 2007. Production of a novel raw-starch-digesting glucoamylase by *Penicillium sp*. X-1 under solid state fermentation and its use in direct hydrolysis of raw starch. World J. Microbiol. Biotechnol. 23: 603–613.

Takahashi, T., N. Inokuchi and M. Irie. 1978. Purification and some properties of three forms of glucoamylase from a *Rhizopus* Species. The Journal Biochemistry. 84: 1183–1194.

Takeda, Y., H. Matsui, M. Tanida, S. Tani and S. Chib. 1985. Purification and substrate specificity of glucoamylase of *Paecilomyces varioti* AHU9417. Agric. Biol. Chem. 49: 1633–1641.

Tosi, L.R.O., H.F. Terenzi and J.A. Jorge. 1993. Purification and characterization of an extracellular glucoamylase from the thermophilic *Humicola grisea* var. *thermoidea*. Can. J. Microbiol. 39: 846–52.
Vieille, C. and G.J. Zeikus. 2001. Hiperthermophilic enzymes: Sources, uses and molecular mechanisms for termostability. Microbiol. Mol. Biol. 65: 1–43.
Vihinen, M. and P. Mäntsälä. 1989. Microbial amylolytic enzymes. Crit. Rev. Biochem. Mol. Biol. 24: 329–418.
Wilson, J.J. and W.M. Ingledew. 1982. Isolation and characterization of *Schwanniomyces alluvius* amylolytic enzymes. Appl. Environ. Microbiol. 44: 301–307.
Yasuda, M., M. Kuwae and H. Matsushita. 1989. Purification and properties of two forms of glucoamylase from *Monascus* sp. No. 3403. Agric. Biol. Chem. 53: 274–249.
Zaha, C.B. and G. Zeikus. 1989. Microbial glucoamylases: biochemical and biotechnological features. Starch Staerke. 41: 57–64.

7

Regulation of Plant Cell Wall Degrading Enzymes Formation in Filamentous Fungi

Roberto N. Silva

Introduction

Energy production and its use directed towards sustainability, environmental quality and safety is among the great challenges at the present time. In recent years, there have been great incentives to study alternative fuels that are not derived from oil. New technologies are being developed to answer this challenge in several centers of biotechnological research. In particular, the production of bioethanol from lignocellulosic residues—a promising alternative source of fuel production—is receiving increasing attention. The amount of energy that originates from the sun is stored as carbon via photosynthesis by plants (2.0×10^{18} Btu/year) is approximately 10 times higher than the annual energy demand worldwide (Kumar et al. 2008).

Department of Biochemistry and Immunology, School of Medicine, University of São Paulo, 14049-900, Ribeirão Preto, SP, Brazil.
E-mail: rsilva@fmrp.usp.br

The vegetal biomass is formed by the combination of carbon dioxide in the atmosphere and water during photosynthesis with chlorophyll, which produces carbohydrates, and solar energy stored in the chemical bonds of the structural components of the biomass. Chemically, the biomass is primarily composed of cellulose, hemicellulose, and pectin, followed by lignin and proteins. Approximately 1.3×10^{10} tons (in dry mass) of vegetal biomass are produced per year worldwide, which can generate approximately 180 million tons of lignocellulosic matter per year, which, if transformed into energy, could supply two-thirds of the energy demand worldwide (Demain et al. 2005). The main sources of energy in the 21st century will probably be of biological origin and produced through biotechnological processes. Biomass currently comprises 1 percent of the worldwide electric energy; however, it is calculated that in 20 yr, it will comprise 30 percent of the total energy consumption.

Lignocellulosic biomass is easy to find, but converting it to ethanol is not. Although the substitution of gasoline for cellulosic ethanol can reduce global warming (Farrel et al. 2006), this process remains expensive on the industrial scale. To produce ethanol from cellulose and hemicellulose, it is necessary to promote intense hydrolysis of its structures to attain fermentable sugars. The most studied and used techniques include acid and enzymatic hydrolysis, the latter being the cleanest and most promising. However, degradation of the plant cell wall components is a complex process that involves the synergistic action of a great number of extracellular enzymes that are produced in nature by fungi and bacteria as energy for their own survival. The development of efficient enzymes that are capable of processing the plant biomass is one way to thoroughly exploit the cellulosic and hemicellulosic fractions of lignocellulosic residues. Thus, it is important to elucidate the biochemistry and physiology of the fungi that produce the enzymes which are capable of degrading plant cell walls, as well as understanding how expression of the genes coding for cellulases, hemicellulases, pectinases, and ligninases is regulated. The most studied filamentous fungi in respect to the production of enzymes that degrade plant biomass are *Trichoderma reesei* (*Hypocrea jecorina*), Aspergillus niger, Penicillium, and the ligninolytic white-rot fungus Phanerochaete. The following sections will provide a review on the structure and composition of the plant cell wall, its degradation, and the production of enzymes that degrade plant biomass, focusing on the systems of regulation involved in the synthesis of these enzymes in filamentous fungi.

Structure and Composition of Plant Cell Wall

The plant cell wall is a complex structure that provides several functions, such as mechanical support, transport, and protection against pathogenic organisms. It is formed from polysaccharides, including cellulose, hemicellulose, and pectin, proteins and phenol compounds (lignin polymers). Morphologically, the plant cell wall is divided in three distinct regions: (1) the middle lamella, which is rich in pectic substances, (2) the primary wall, which is composed of cellulose microfibrils (~30%), hemicelluloses (~30%), pectins (~30%), and structural proteins (~10%), and (3) the secondary wall, which is composed of three layers (S1, S2, and S3). Each of the secondary wall layers differs in thickness, composition, and orientation of the cellulose microfibrils (Raven et al. 2003). Layers S1 and S2 are rich in hemicellulose, while S3 has more cellulose (Fig. 7.1).

Cellulose is an insoluble, crystalline homopolymer formed from glucose units linked by type β-1,4 glycosidic bonds, and comprises a significant part of the biomass on Earth. The smallest unit of cellulose is cellobiose, which is composed of two units of glucose (Fig. 7.2). In nature, the degree of polymerization (DP) of glucose chains ranges from approximately 10,000 to 15,000 glucose units, depending on the origin (Sjöström 1981).

Cellulose has a relatively simple chemical structure, but it is not easy to degrade. Its filaments are joined together by forming crystalline microfibrils, which are stabilized by hydrogen and *van der Waals* bonds, to form highly insoluble structures. In addition to these crystalline regions, cellulose may

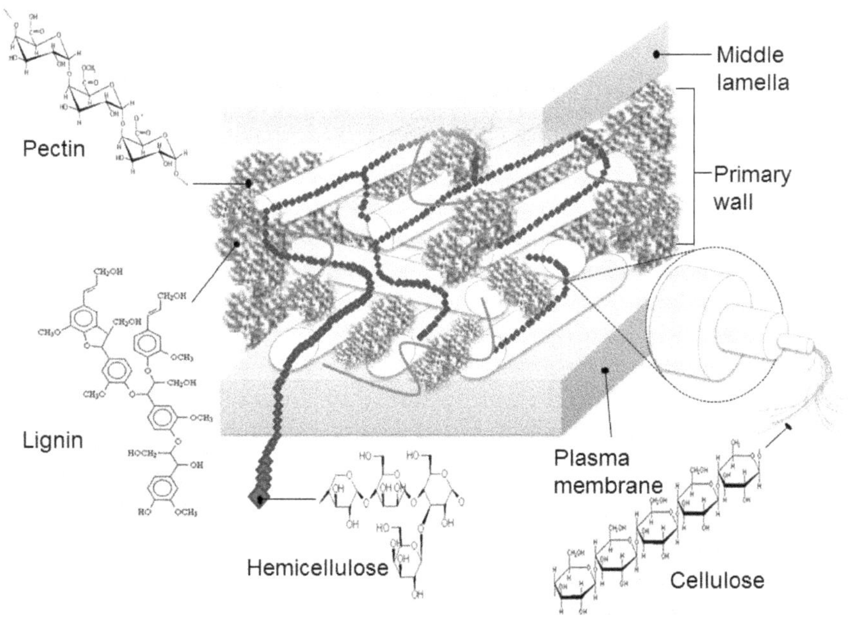

Figure 7.1 Schematic representation of the plant cell wall and composition in terms of cellulose, hemicellulose, lignin and pectin.

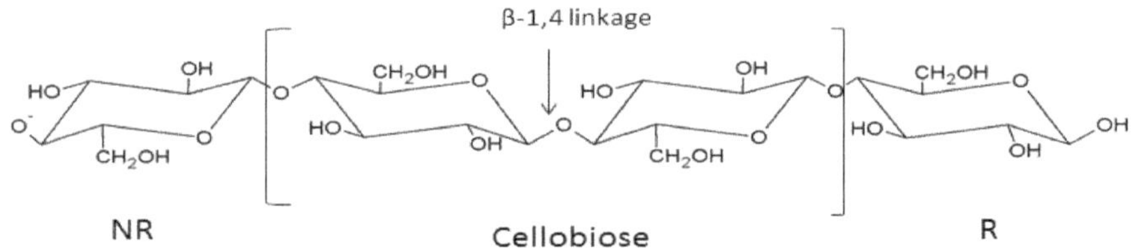

Figure 7.2 Residues of cellobiose in cellulose chain. Non Reducing end (NR), Reducing end (R).

also contain less-ordered regions named amorphous or paracrystalline regions (Miettinen-Oinonen 2004).

Hemicellulose is the second most abundant polysaccharide in nature, and has a heterogeneous composition of several sugar units, such as D-xylose, D-mannose, D-glucose, D-galacturonic acid, D-glucoronic acid, D-galactose, L-arabinofuranose and L-rhamnose. Hemicellulose is generally formed by two–six different sugars that are linked by β-1,4 bonds (de Vries et al. 2002). The structure and biosynthetic pathways of hemicelluloses are widely known, and alternative terms, such as glycan "cross-linking," have been proposed for their correct designation (Vergara and Carpita 2001). However, many authors still use the term "hemicelluloses." Hemicelluloses are classified according to the main sugars they contain, including glucans, xylans, mannans, galactans, and galacturans). Other sugars may also be present in small amounts, including α-L-rhamnose and α-L-fucose, and some sugar with hydroxyl groups may be replaced by acetyl groups.

The main hemicellulose polymer in cereals is xylan. It is located between lignin and cellulose fibers, forming covalent bonds with lignin and hydrogen bonds with cellulose fibers (Carpita and Gibeaut 1993). Xylan is a heteropolysaccharide formed by a main chain of D-xylopyranose residues linked by β-1,4 bonds to side chains of α-L-arabinofuranose residues (usually called arabinoxylans) and D-glucuronic acid (usually called glucoxylans). Xylan may also contain D-galactopyranose residues linked by β-1,5 bonds, L-arabinose or xylose linked by β-1,4 bonds, or acetyl groups attached at positions two or three of D-xylose or feruloyl, or p-coumaroyl groups attached at position 5 of L-arabinose (Fig. 7.3). These various conformations demonstrate the highly heterogeneous character of this group of polysaccharides (de Vries and Visser 2001).

Another hemicellulose polymer usually found in soft and hard wood is (galacto)glucomannan, which consists of a backbone of D-mannose and D-glucose residues linked by β-1,4 bonds with side groups of D-galactose. Soft woods mainly contain galactoglucomannan, while hard woods mainly contain glucomannan (Fig. 7.4). The water solubility of galactoglucomannan depends on the amount of D-galactose present in the structure.

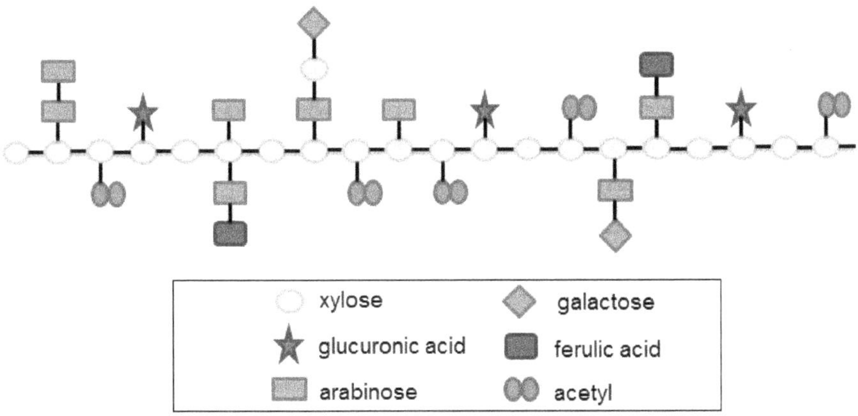

Figure 7.3 Schematic representation of xylan. Adapted from: de Vries and Visser (2001).

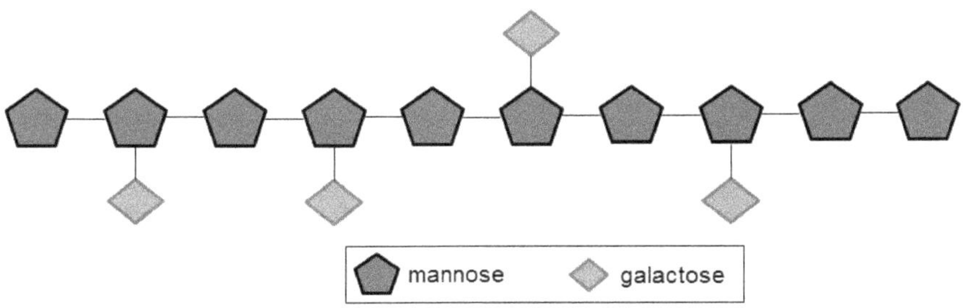

Figure 7.4 Schematic representation of galactomannan.

Xyloglucans are additional hemicellulose polymers that are found in the cell wall of dicotyledons and some monocotyledons. They consist of a backbone of D-glucose linked by β-1,4 bonds replaced by D-xylose. L-Arabinose and D-galactose residues can also be attached to xylose and L-fucose attached to galactose. Two main types of xyloglucans have been identified in plant cell walls. XXXG-type xyloglucans consist of repeated units of three D-glucopyranose residues linked by β-1,4 bonds, replaced with D-xylopyranose by a α-1,6 bond, and separated by a non-replaced glucose residue. XXGG-type xyloglucans consist of two xylose-replaced glucose residues that are separated by two non-replaced glucose residues (Fig. 7.5). Xyloses can be covered with galactose units (XLXG, XXLG, XLLG) and fucose units (XXFG). Xyloglucans interact with cellulose microfibrils through hydrogen bonds and are responsible for the structural integrity of the cellulose network.

Pectins are the least abundant polysaccharides in plant cell walls but are the most complex. The backbone of pectin consists mainly of α-1,4-linked D-galacturonic acid residues that may be methylated or acetylated. This

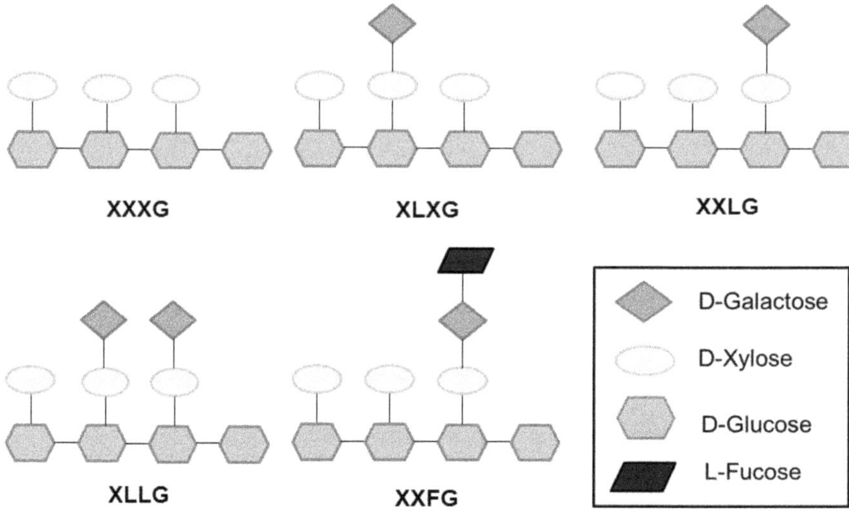

Figure 7.5 Schematic structures of repeating units of xyloglucan structures XXXG, XLXG, XXLG, XLLG and XXFG.

is known as the "smooth" or "hairy" region and consists of two different structures of D-xylose replaced by galacturonan and rhamnogalacturonan to which long chains of arabinan and galactan are linked through rhamnose residues. Pectins can also be classified into three main groups, homogalacturonan (linear polymer), xylogalacturonan (β-1,3-linked D-xylose branches), and rhamnogalacturonan (a complex structure with alternating L-rhamnose and D-galacturonic acid residues) (Fig. 7.6) (Caffall and Mohnen 2009, van den Brink and De Vries 2011).

In contrast to cellulose and hemicellulose, which have well-known and defined structures, lignins are amorphous, tri-dimensional biopolymers with complex molecular structures that depend on variables like plant species and age. In nature, lignin is embedded in a matrix composed of carbohydrates, cellulose and hemicellulose. It is difficult to establish the exact constitution of lignin, because it presents itself as a complex network of phenylpropanoid units linked by various types of links and also because they suffer structural changes during isolation processes. Lignin monomers are comprised of units of guaiacyl (lignin type G) of the precursor trans-coniferyl alcohol, units of syringyl (lignin type S), units of trans-sinapyl alcohol, and units of hydroxyphenyl (lignin type H) of the precursor trans-p-coumaric (Fig. 7.7).

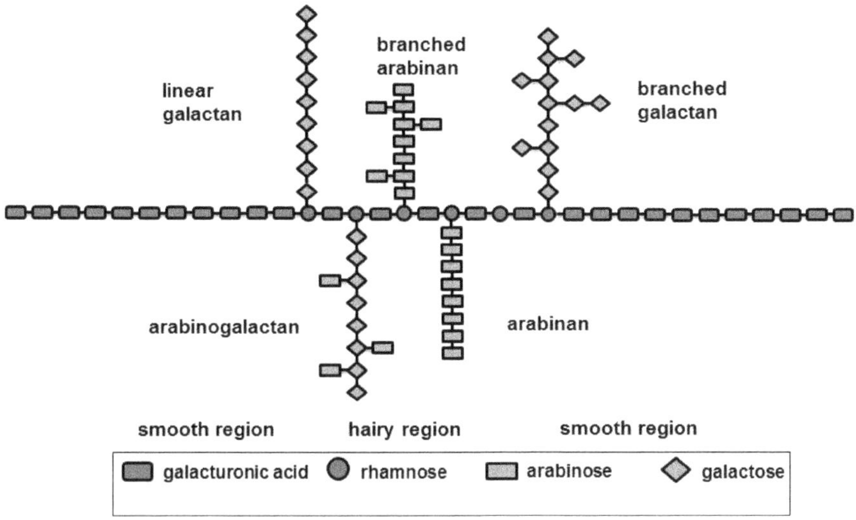

Figure 7.6 Schematic presentation of complex pectin structure.

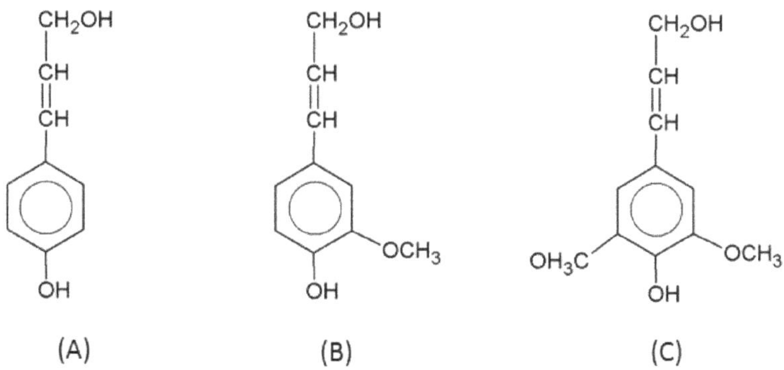

Figure 7.7 Chemical structure of lignin monomers. p-Coumaryl alcohol (A), Coniferyl alcohol (B), Sinapyl alcohol (C).

Enzymatic Degradation of the Plant Cell Wall

Cellulose Degradation

A wide variety of fungi, such as *Aspergillus* spp., *Trichoderma reesei*, *Trichoderma viride*, *Penicillium pinophilum*, *Phanerochaete chrysosporium*, *Fusarium solani*, *Talaromyces emer*sonii, *Neurospora crassa*, and *Rhizopus oryzae*, produce cellulolytic enzymes that hydrolyze cellulose (Bhat and Bhat 1997, Murashima et al. 2002). However, the cellulolytic complex secreted by

T. reesei is the most studied and its mode of action has been well documented (Miettinen-Oinonen 2004).

T. reesei is the most prominent degradative species of the genus *Trichoderma*. It was originally isolated in the South Pacific (Solomon Islands) during World War II, where it was found to degrade cotton fabrics, tents, and belts. Since then, this fungus has been recognized as a potential source of industrial cellulases.

The cellulolytic system of this fungus is constituted by at least three different types of enzymes: exoglucanases (cellobiohydrolases EC 3.2.1.91), endoglucanases (EC 3.2.1.4), and β-glucosidases (EC 3.2.1.21), which occur in different isoforms (Saloheimo et al. 1994).

T. reesei secretes at least eight types of endoglucanases (CEL7B, CEL5A, CEL5B, CEL12A, CEL61A, CEL45A, CEL74A, and CEL61B) that differ in molecular mass and isoelectric point (pI) (Penttila et al. 1986, Saloheimo et al. 1988, Ward et al. 1993, Saloheimo et al. 1994, Foreman et al. 2003). CEL7B is an endoglucanase produced in large amounts by *T. reesei* and accounts for roughly 5 percent of the total protein released into the culture medium, while CEL5A accounts for up to 0.5 percent, and the remaining endoglucanases are minor components (Saloheimo et al. 2003).

Two types of cellobiohydrolases (CEL7A and CEL6A) have been identified in the cellulolytic complex secreted by *T. reesei* (Teeri 1997). These exoglucanases are produced in larger quantities than other cellulolytic enzymes of the complex. CEL7A comprises about 60 percent of the total protein released to the culture medium, while CEL6A comprises up to 20 percent (Miettinen-Oinonen 2004).

β-glucosidases, also called cellobiases, hydrolyze cellobiose generated by cellobiohydrolase and endoglucanases in glucose. Strictly speaking, β-glucosidases are not true cellulases, because they act on soluble substrates. Nevertheless, their contribution is important to the efficiency of cellulose hydrolysis by removing cellobiose, a potent competitive inhibitor of cellobiohydrolases, from the reaction medium (Miettinen-Oinonen 2004). Since the *T. reesei* genome was sequenced (Martinez et al. 2008), seven β-glicosidases (CEL1A, CEL1B, CEL3A, CEL3B, CEL3C, CEL3D, and CEL3E) have been described. Among them, CEL1A is able to transglycosylate as well as hydrolyze cellobiose (Saloheimo et al. 2002a).

Cellulases produced by *T. reesei*, as well as those produced by most filamentous fungi, are composed of two domains that are functionally and structurally distinct: the catalytic domain, which constitutes most of the protein, and a substrate-binding domain called the carbohydrate-binding module (CBM). The two domains are connected by a highly glycosylated sequence ("linker").

Besides classical cellulases, new types of cellulases have recently been discovered in *T. reesei*, including swollenin (SWOI). SWOI is a protein that has no catalytic activity but modifies cellulose by expanding its crystalline. Swollenin has an expansin-like domain in the N-terminal region, which is normally found in plants during their growth phase. It has been suggested that SWOI makes cellulose fibers more accessible for attack by cellulases (Saloheimo et al. 2002b) and are transcriptionally regulated when cellulose is present (Verbeke et al. 2009). The properties and structural organization of cellulases produced by *T. reesei* are reported in Table 7.1.

The enzymes of *T. reesei*'s cellulolytic complex attack the substrate in a synergistic manner, increasing the speed with which soluble products are formed as compared to those of isolated enzymes (Miettinen-Oinonen 2004). Endoglucanases hydrolyze internal bonds in the most accessible (amorphous) regions of cellulose to generate new free ends. Cellobiohydrolases can penetrate the crystalline regions of microfibrils to degrade the cellulose polymer from the ends generated by endoglucanases. β-glucosidases complement the degradation by cleaving the cello-oligosaccharides and cellobiose generated by cellobiohydrolases (Teeri 1997). Cellobiohydrolases can act very rapidly on ends generated by endoglucanases, indicating that the enzymes are distributed along the cellulose fiber. There is also synergism between different subtypes of cellobiohydrolases; some enzymes that attack reducing ends (CBHI/CEL7A), and others attack non-reducing ends (CBHII/CEL6A) of cellulose fibers (Miettinen-Oinonen 2004). The degree of synergism varies according to the type of cellulose used. In the presence of crystalline cellulose, there is a high degree of endo-exo synergism, which decreases in amorphous cellulose and is very low or absent in soluble derivatives of cellulose, such as carboxymethylcellulose (CMC) (Heikinheimo and Buchert 2001) (Fig. 7.8).

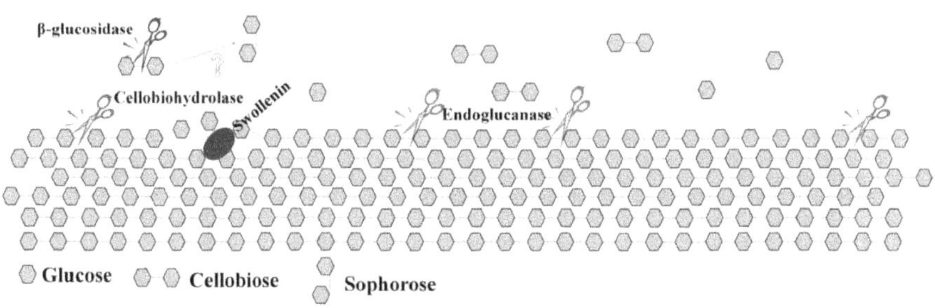

Figure 7.8 A schematic view of cellulose degradation and sophorose formation by transglucosylating.

Table 7.1 Properties and structural organization of cellulases produced by *Trichoderma reesei* (adapted from Miettinen–Oinonen 2004).

Family GH	Enzyme	Amino acids	Molecular mass (kDa)	pI	Structural organization	Ref.
1	CEL1A	466	52,24	5,33*	--	Takashima et al. 1999, Saloheimo et al. 2002a
1	CEL1B	484	55	5.55*	--	Foreman et al. 2003
3	CEL3A	744	78,4	6.38*	--	Barnett et al. 1991
3	CEL3B	874	93,94	5,73*	--	Foreman et al. 2003
3	CEL3C	833	90,71	5,38*	--	Foreman et al. 2003
3	CEL3D	700	77	6,02*	--	Foreman et al. 2003
3	CEL3E	765	83	6,04*	--	Foreman et al. 2003
5	CEL5A	397	48	5,5	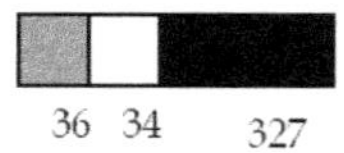	Saloheimo et al. 1988
5	CEL5B	438	46,8	4,32*	--	Foreman et al. 2003
6	CEL6A	447	50–58	5,1–6,3	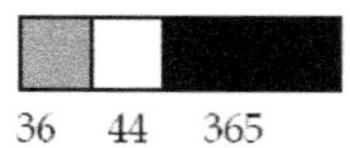	Teeri et al. 1987
7	CEL7B	437	50–55	4,6		Penttilä et al. 1986
7	CEL7A	497	59–68	3,5	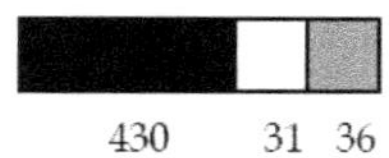	Shoemaker et al. 1983, Teeri et al. 1983
12	CEL12A	218	25	7,4	--	Ward et al. 1993, Okada et al. 1998

45	CEL45A	225	23	2,8–3	 166 23 36	Saloheimo et al. 1994
61	CEL61A	326	37	5,29*	 233 56 37	Saloheimo et al. 1997
61	CEL61B	249	26,8	7,62*	--	Martinez et al. 2008
74	CEL74A	838	75–105	4,1–4,3	 810 28	Foreman et al. 2003

■ Catalytic domain
□ Linker
▒ Cellulose binding domain (CBD)
*determined on the basis of the amino acid sequence

Hemicellulose Degradation

Although hemicelluloses are relatively complex polysaccharides, their enzymatic degradation is well understood. Xylan represents the main class of hemicellulose. Due to its heterogeneity and complexity, xylan's complete hydrolysis requires a variety of enzymes that act cooperatively to convert it into xylose units. Its hydrolysis occurs through the action of endo-enzymes that internally cleave the main chain, exo-enzymes that liberate monomeric sugars, and accessory enzymes that cleave side chains of polymers or oligosaccharides. The catalytic modules of hemicellulases may be glycosyl hydrolases (GHS) that hydrolyze glycosidic bonds or carbohydrate esterases (CES) that hydrolyze acetate ester bonds or ferulic acid side groups.

Arabinofuranosidases (EC 3.2.1.55) hydrolyze the non-reducing terminal group of arabinofuranose of arabinans, arabinoxylans, and arabinogalactans. Glucuronidases (EC 3.2.1) hydrolyze type α-1,2 glycosidic bonds between xylose and glucoronic acid. Complete hydrolysis of xylans requires the action of esterases, which break the bond between xylose and acetic acid (acetyl-xylan esterase—EC 3.1.1.6) and bonds between arabinofuranose residues and phenolic acid (phenolic acid esterase) and coumaric acid (acid p-coumaric esterase—EC 3.1.1). Endoxylanases hydrolyze type β-1,4 glycosidic bonds inside the xylan chain to produce xylo-oligosaccharides, which are converted in xylose by β-xylosidase. These enzymes act cooperatively to convert xylan in xylose units (Subramaniyan and Prema 2002).

The chemical bonds hydrolyzed by hemicellulases and the classification of these enzymes into families are listed in Table 7.2.

Among the fungi that degrade hemicellulose, the most studied belong to the genus *Aspergillus*. Already, 20 genes that encode endoxylanases and β-xylosidases have been reported (van den Brink and de Vries 2011). Next, the fungus *T. reesei* has been reported to have four genes that encode xylanases, designated xynl to xyn4, and 10 genes that encode hemicellulases (Foreman et al. 2003). In addition, xylanase genes have been cloned from other fungi, such as *Penicillium, A. bisporus*, and *Magnaporthe grisea* (Wu et al. 1995, Kimura et al. 2000, Emami and Hack 2002, Alcocer et al. 2003). A degradation scheme for three hemicelluloses, xylan, galacto(gluco)mannan, and xyloglucan are shown in Fig. 7.9 as an example of a hemicellulolytic system.

Table 7.2 Hemicellulases and accessory enzymes and their classification into families (glycosyl hydrolase [GH] and carbohydrate esterase [CE]).

Enzyme	**Substrates**	**EC**	**Family**	**Ref.**
β-1,4-Endoxylanase	β-1,4-xylan	3.2.1.8	**GH** 5, 8, 10, 11, 43	Krengel and Dijkstra 1996
β-D-xylosidase	β-1,4-xylo-oligomers xylobiose	3.2.1.37	**GH** 3, 39, 43, 52, 54	van Peij et al. 1997
α-L-Arabinofuranosidase	α-arabinofuranosyl (1→2) xylo-oligomers (1→3) α-1,5-arabinan	3.2.1.55	**GH** 3, 43, 51, 54, 62	Flipphi et al. 1993
Endo-α-1,5-arabinase	α-1,5-arabinan	3.2.1.99	**GH** 43	Bauer et al. 2006
α-Glucuronidase	4-O-methyl-α-acid glucuronic xylo-oligomers (1→2)	3.2.1.139	**GH** 67	de Vries et al. 2002
Endo-β-1,4-mannanase	Β-1,4-mannan	3.2.1.78	**GH** 5, 26	Ademark et al. 1998
Exo-β-1,4-manosidase	β-1,4-manno-oligomers mannobiose	3.2.1.25	**GH** 1, 2, 5	Ademark et al. 2001
α-galactosidase	α-galactopyranose (1→4) manno-oligomers	3.2.1.22	**GH** 4, 27, 36, 57	den Herder et al. 1992, de Vries et al. 1999a
β-glucosidase	β-glucopyranose (1→4) mannopyranose	3.2.1.21	**GH** 1, 3	Fedorova et al. 2008
Endo-galactanase	β-1,4-galactan	3.2.1.89	**GH** 53	Christgau et al. 1995
Acetylxylan esterase	2- or 3-*O*-acetyl xylan	3.2.1.72	**CE** 1, 2, 3, 4, 5, 6, 7	van Peij et al. 1998a
Acetylmannan esterase	2- or 3-O-acetyl mannan	3.1.1.6	**CE** 1	de Vries et al. 1997
Ferulic and p-coumaric acid esterases		3.1.1.73	**CE** 1	de Vries et al. 1997, de Vries et al. 2002

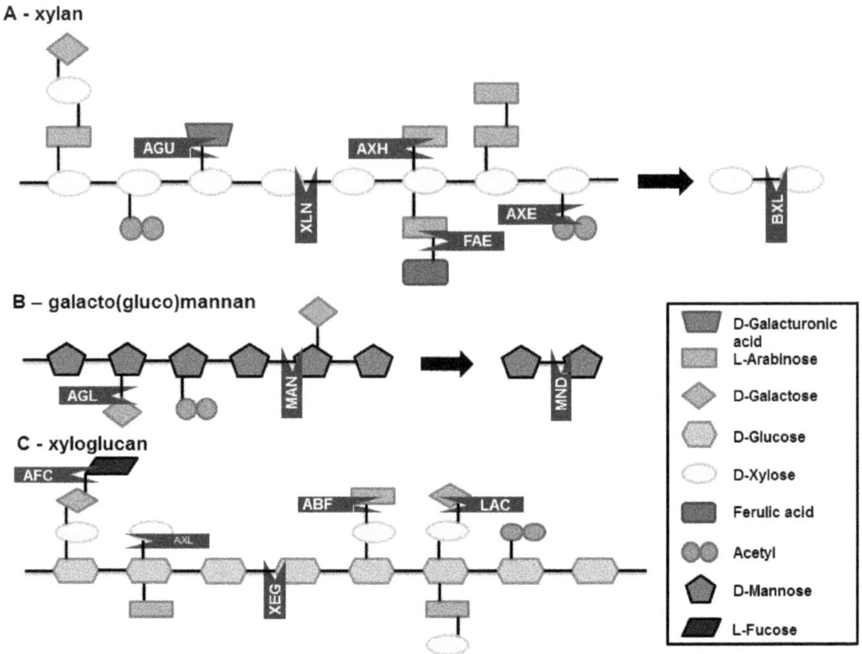

Figure 7.9 Schematic hydrolysis and structure of three hemicelluloses, xylan (A), galacto(gluco) mannan (B), and xyloglucan (C), ABF (α- arabinofuranosidase), AFC (α-fucosidase), AGL (α-1,4-galactosidase), AGU (α-glucuronidase), AXE (acetyl(xylan) esterase), AXH (arabinoxylan α-arabinofuranohydrolase), AXL (α-xylosidase), BXL (β-1,4-xylosidase), FAE (feruloyl esterase), LAC (β-1,4-galactosidase), MAN (β-1,4-endomannanase), MND (β-1,4-mannosidase), XEG (xyloglucan-active β-1,4-endoglucanase), XLN (β-1,4-endoxylanase). Adapted from: van den Brink and de Vries (2011).

Pectin Degradation

Pectinases form a group of enzymes that degrade pectic substances by hydrolyzing glycosidic bonds along the carbonic chain. Pectinolytic enzymes can be divided into three broad groups (Sakai et al. 1993, Uenojo and Pastore 2007): (1) protopectinases, which solubilize protopectin to form pectin, (2) esterases, which catalyze esterification of pectin by removing methyl ester groups, and (3) depolymerases, which catalyze hydrolytic cleavage of α-1,4 glycosidic bonds. Table 7.3 shows the classification and examples of pectinases produced by fungi.

Pectin degradation is a complex process that requires at least two classes of enzymes: glycoside hydrolases (GH 28, 78, and 105) and polysaccharide lyases (PL 1, 3, 4, 9, and 11). Glycosil hydrolases endo- and exo-polygalacturonases cleave α-1,4 glycosidic bonds between galacturonic acids of the backbone of "smooth" regions. "Hairy" regions are attacked by

Table 7.3 Classification of pectinolytic enzymes produced by fungi.

Type of pectinase	EC	Sigla	Common name	Ref.
De-esterifying	3.1.1.11	PMGE	pectin esterase	Khanh et al. 1991
Hydrolase	3.2.1.15	endo-PG	polygalacturonase	Bussink et al. 1991
	3.2.1.67	exo-PG 1	polygalacturonase	Hara et al. 1984
	3.2.1.82	exo-PG 2	polygalacturonase	Hara et al. 1984
Lyase	4.2.2.2	endo-PGL	pectate lyase	Benen et al. 2000
	4.2.2.9	exo-PGL	pectate lyase	Benen et al. 2000
	4.2.2.10	endo-PMGL	pectin lyase	Kusters-Van et al. 1992
	4.2.2.10	exo-PMGL	pectin lyase	Machida et al. 2005

endo- and exo-rhamnogalacturonase, which cleave α-1,2-glycosidic bonds formed between D-galacturonic acid and L-rhamnose residues, and by other poorly characterized enzymes (xylogalacturonase, α-rhamnosidases, glucuronyl hydrolases, and rhamnogalacturonan hydrolases) (reviewed by van den Brink and de Vries 2011).

Pectin and pectato lyases cleave α-1,4 bonds of D-galacturonic acid residues on the inside of "smooth" regions of pectin (Lombard et al. 2010). Rhamnogalacturonan lyases differ substantially in their structure from pectin and pectato lyases, and act preferentially from four residues of L-rhamnose at the reducing end. Their action is regulated by the presence of acetyl groups in the backbone of rhamnogalacturonan; therefore, they must work in cooperation with rhamnogalacturonan acetyl-esterases for effective degradation (de Vries and Visser 2001).

The complete degradation of the pectin structure requires accessory enzymes to remove the side chains and make the main chain accessible to the pectinolytic enzymes. These accessory enzymes include alpha-arabinofuranosidases, β-galactosidases and β-xylosidases, which are also used for the degradation of hemicellulose and endo- and exo-arabinanases, β-endogalactanases, and esterases that are specific for pectin degradation (Martens-Uzunova and Schaap 2009). A schematic mechanism of the action of pectinases and accessory enzymes is presented in Fig. 7.10.

Lignin Degradation

Fungi degrade lignin by secreting enzymes collectively known as "ligninases." Ligninolytic fungi cannot use lignin as their only source of carbon and energy. However, lignin degradation is necessary for the fungus to access cellulose and hemicellulose, promote hydrolysis, and consequently use these polymers as a carbon and energy source. In the vast majority of fungi studied, ligninolysis occurs during secondary metabolism (i.e., under conditions of limited nutrient sources) (Kirk and Farrell 1987). Ligninases

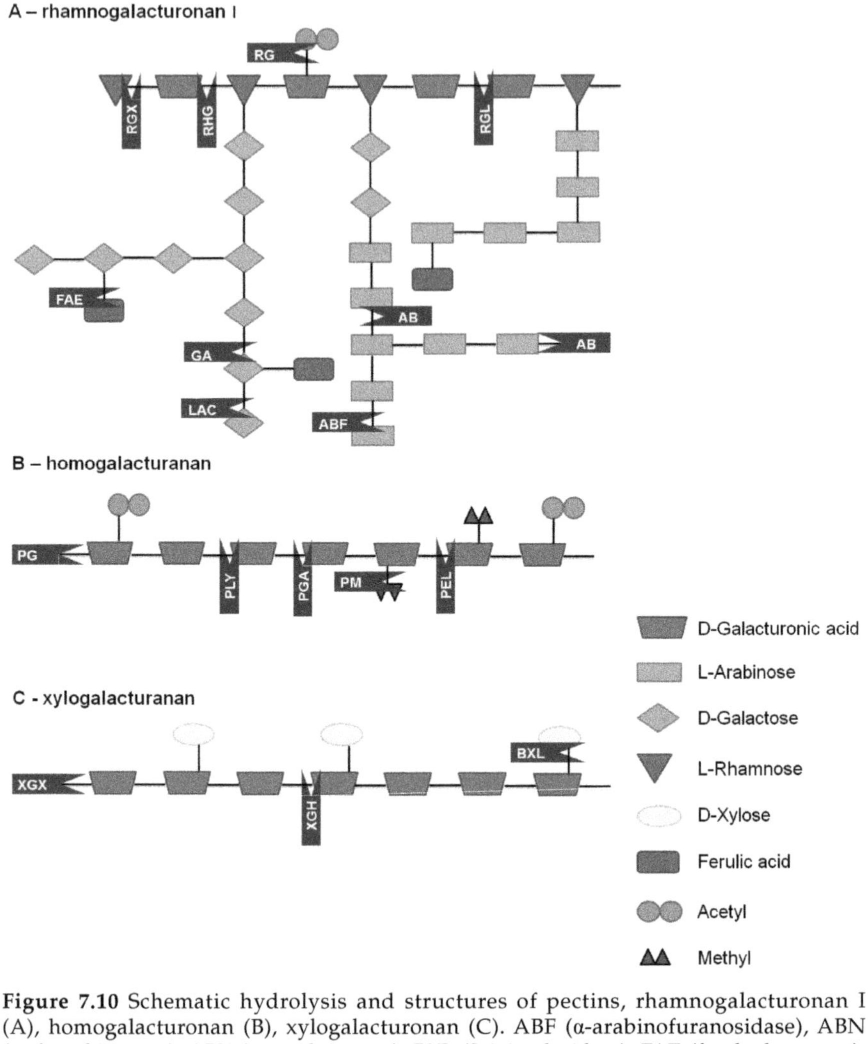

Figure 7.10 Schematic hydrolysis and structures of pectins, rhamnogalacturonan I (A), homogalacturonan (B), xylogalacturonan (C). ABF (α-arabinofuranosidase), ABN (endoarabinanase), ABX (exoarabinanase), BXL (β-1,4-xylosidase), FAE (feruloyl esterase), GAL (β-1,4-endogalactanase), LAC (β-galactosidase), PEL (pectin lyase), PLY (pectate lyase), PGA (endopoly-galacturonase), PGX (exo-polygalacturonase), PME (pectin methyl esterase), RGAE (rhamnogalacturonan acetyl esterase), RGL (rhamnogalacturonanlyase), RHG (endorhamnogalacturonase), RGX (exorhamnogalacturonase), XGH (endoxylo-galacturonase), XGX (exoxylogalacturonase). Adapted from: van den Brink and de Vries (2011).

are classified as phenol oxidases (laccase) (EC 1.10.3.1), heme peroxidases, lignin peroxidase (LiP) (EC 1.11.1.14), and manganese peroxidases (MnP) (EC 1.11.1.13). In addition, two other enzymes participate in delignification, glyoxal oxidase and glucose-2-oxidase (EC 1.1.3.10), which generate hydrogen peroxide that is acted upon by peroxidase. Table 7.4 shows the classification of ligninolytic enzymes produced by fungi and their general reactions.

White-rot fungi are the most abundant wood degraders in nature. Lignin degradation by white-rot fungi has been studied for more than half of a century. Studies show that *P. chrysosporium* produces several isoenzymes of LiP and MnP. However, it is incapable of producing laccase (Singh and Chen 2008). The genome of *P. chrysosporium* contains 10 LiP and 5 MnP genes (Martinez et al. 2004). Genes that encode for LiP have been well characterized not only in *P. chrysosporium*, but also in other white-rot fungi, such as *P. radiata* (Saloheimo et al. 1989), *Bjerkandera adusta* (Kimura et al. 1991), *T. versicolor* (Gold and Alic 1993), *Pleurotus eryngii* (Ruiz-Duenas et al. 1999), and *Coriolus versicolor* (Wang et al. 2008). Moreover, studies have shown that fungal degradation of wood by enzymes is preceded by non-enzymatic degradation, where low-molecular-weight oxidant compounds initiate changes in the lignin structure (Tanaka and Enoki 1999).

Table 7.4 Classification of ligninases and their general reactions.

Type of enzyme	EC	General reaction	Ref.
Laccase	1.10.3.1	4 benzenediol + O_2 = 4 benzosemiquinone + 2 H_2O	Temp et al. 1999
Lignin peroxidase (LiP)	1.11.1.14	(3,4-dimethoxyphenyl) methanol + H_2O_2 = 3,4-dimethoxybenzaldehyde + 2 H_2O.	Gaskell et al. 1991
Manganese peroxidase (MnP)	1.11.1.13	2 Mn^{2+} + 2 H^+ + H_2O_2 = 2 Mn^{3+} + 2 H_2O	Godfrey et al. 1990
Glucose-2-oxidase	1.1.3.10	D-glucose + O_2 = 2-dehydro-D-Glucose + H_2O_2	Nishimura et al. 1996

Production of Enzymes that Degrade Plant Cell Wall by Filamentous Fungi

Production of Cellulases

Since the 1960s, studies have investigated the mechanisms of cellulase production by the fungus *T. reesei* (Mandels and Reese 1960, Gritzali and Brown 1979, Kubicek et al. 1988, Seiboth et al. 1992, Carle-Urioste et al. 1997, Ilmén et al. 1997b).

It was postulated that cellulases are adaptive enzymes and that the induction of the production of these enzymes requires oligosaccharides liberated from cellulose itself. This hypothesis is supported by the fact that cellulases are induced by several disaccharides, including cellobiose, lactose, sophorose, and 1,5-lactone-δ-cellobiono (reviewed by Schmoll and Kubicek 2003).

Today, there are three different models for explaining the mechanism of induction. One of them, proposed in the 1970s, suggests that the initial production of these oligosaccharides from cellulose is attributed to low constitutive levels of cellulases found in these fungi (Gritzali and Brown 1979). Another hypothesis, presented by Kubicek and collaborators, is based on the observation that only conidia and not mycelia can grow in the presence of cellulose in the culture medium. The conidia of *Trichoderma* spp. contain a series of different enzymes capable of hydrolyzing several types of polysaccharides. This is an alternative explanation for the initial degradation of cellulose (Kubicek et al. 1988). Seiboth et al. (1992), using mutants for several cellulase genes, supported Kubicek's hypothesis and showed the importance of CEL6A and CEL5A for the induction of cellulase. Consequently, constitutive expression of CEL6A and CEL5A bound to the surface of conidia acts synergistically in the initial degradation of cellulose molecules, thereby generating the inducers of cellulase biosynthesis and promoting the expression of cellulase genes and fungal growth in cellulose. A third hypothesis, proposed by Ilmen et al. (1997b), is based on the observation that all cellulases were significantly expressed after total consumption of glucose in the culture medium. However, this mechanism of derepression in cellulase production is not yet clarified and seems to be independent of a lack of carbon source or catabolic repression.

Whatever the true mechanism underlying the initial attack on the cellulose molecule, the final product must be a low molecular weight inducer that is captured by fungal cells. The identity of this compound remains unknown. However, several oligosaccharides have been demonstrated to induce cellulase in laboratorial experiments, such as (1) cellobiose (β-D-Gluc-(1→4)-D-Gluc 4-O-β-D-Glucopyranosyl-D-glucose), which is the major product from cellulose hydrolysis, (2) sophorose (β-D-Gluc-(1→2)- D-Gluc 2-O-β-D-Glicopyranosyl-D-glucopyranose), which induces the expression of cellulase genes to a high level and has thus been considered a natural inducer of cellulase production in *T. reesei* (Sternberg and Mandels 1979), and (3) lactose (1,4-O-β-D-galactopyranosyl-D-glucose), which is not a normal product of cellulose degradation but is an important inducer of cellulase and is a soluble carbon source used for economical production of cellulase and heterologous proteins under the control of cellulase promoters by *T. reesei* (Morikawa et al. 1995). In addition, several other oligosaccharides, such as D-cellobiono-1,5-lactone, laminaribiose, gentiobiose, xylobiose, and L-sorbose, in cultured *T. reesei* induce the expression of cellulases (Vaheri et al. 1979, Durand et al. 1988, Margolles-Clark et al. 1997, Nogawa et al. 2001). It has also been reported that L-arabitol and different xylans can induce the expression of *cbh1* in *T. reesei* (Margolles-Clark et al. 1997).

Several models have been proposed to explain the production of cellulases by *T. reesei*. Upon the sequencing of *T. reesei's* genome and techniques for studying gene expression, new components have been added to the proposed models. In summary, *T. reesei* produces cellulases in the presence of cellulose as shown in Fig. 7.11 (model proposed by Suto and Tomita 2001).

Immediately upon contacting cellulose, cellobiohydrolase present on the surface of conidia spores degrades the cellulose, producing cello oligosaccharides. These are then hydrolyzed to glucose and transglycosylated by a constitutive β-glucosidase, thereby forming sophorose (1). Germination of conidia spores proceeds as follows. Glucose acts as a carbon source, and sophorose acts as an inducer; both enter the cell (2). Sophorose induces cellulases, which are secreted into the extracellular medium (3). Cellulases then degrade cellulose, increasing the amounts of cello oligosaccharides and glucose (4). Glucose is then assimilated. Cello oligosaccharides are hydrolyzed to glucose and transglycosylated to form more sophorose (5) in a reinitiating step (3). This cycle continues until no cellulose remains in the medium.

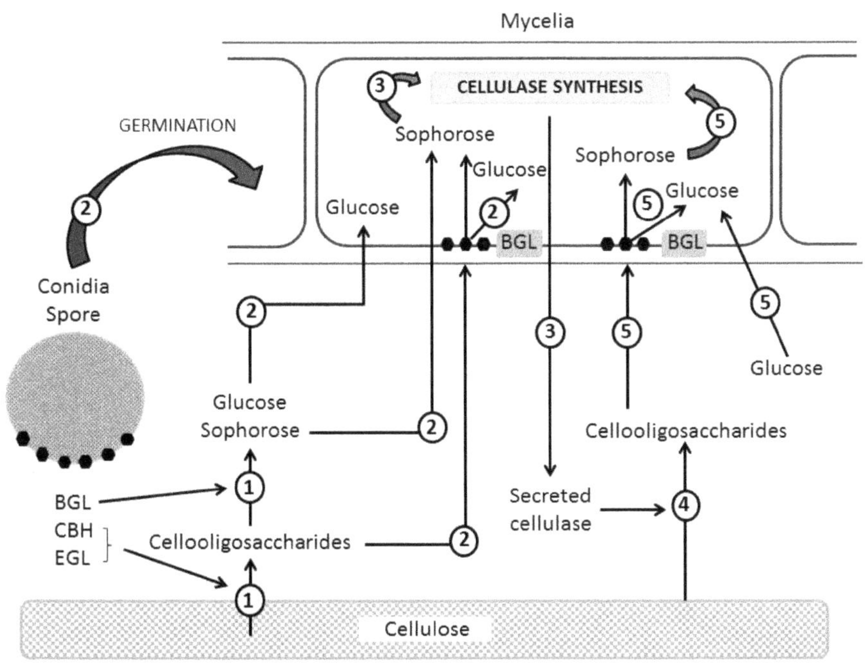

Figure 7.11 Model of cellulase induction in *T. reesei*. The numbers 1 to 5 correspond to steps of induction and are explained in the text. BGL (β-glucosidase), CBH(cellobiohydrolase), EGL (endoglucanases). Adapted from: Suto and Tomita (2001).

Production of Xylanases

Xylanases are synthetized by mesophilic and thermophilic fungi (Smith et al. 1991). However, the xylanolytic system has been studied in more detail in mesophilic species, including the fungi *Aspergillus* and *T. reesei*. Regarding cellulases, xylanase production is initiated by physical contact between constitutive xylanases released by the fungus and the inducer (Polizeli et al. 2005). It has been reported that the optimal culture conditions for the production of xylanases are achieved when the fungus is cultured in a medium containing xylan or mixtures of plant cell wall polymers (Zeilinger et al. 1996). Constitutive xylanases hydrolyze xylan and release small amounts of β-D-xylopyranosyl oligosaccharides, such as xylobiose and xylotriose. These oligo saccharides are then transported into the cell with the aid of β-xyloside permeases, thereby triggering the expression of genes in the xylanolytic system. While permease activity in the induced cell decreases in the presence of glucose, it is very effective in the presence of xylanolytic inducers (Polizeli et al. 2005) (Fig. 7.12).

In *Aspergillus* species, the production of xylanolytic enzymes may be induced by D-xylose, xylobiose, and the heterodisaccharide D-glucose-β-1,2-D-xylose (Bisaria and Mishra 1989). However, xylanotic enzymes are not induced by monomeric substrates, such as glucose and galactose,

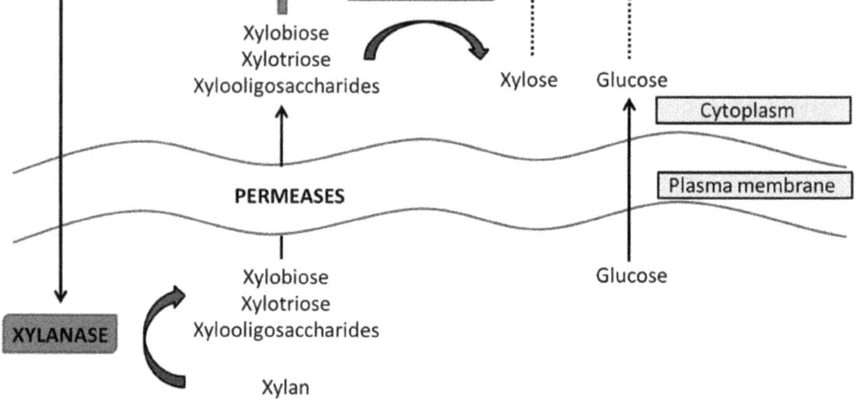

Figure 7.12 Hypothetical model of xylanolytic regulation. Adapted from: Polizeli et al. (2005).

or polymeric substrates, such as cellulose and pectin. Additionally, in *A. niger*, D-xylose also induces the expression of several other genes encoding enzymes that cleave the side chain, such as α-glucuronidase (*aguA*), acetylxylan esterase (*axeA*), and feruloyl esterase (*faeA*) (de Vries and Visser 1999b, de Vries et al. 2002). In *T. reesei*, on the other hand, xylobiose induces genes that encode enzymes related to xylan hydrolysis (*xyn1*, *xyn2*, and β-xilosidase [*bxl1*]) as well as genes encoding α-galactosidase (*agl1* and *agl2*) and glucuronidase (*glr1*), which cleave the side chain. However, low concentrations of xylose only induce expression of *xyn1*, while *xyn2* has a partially constitutive expression and is induced by xylobiose and sophorose. This demonstrates that in *T. reesei* the two main xylanases are not co-regulated (Mach et al. 1996, Zeilinger et al. 1996). Studies have shown that the regulation of xylanolytic genes expression is not conserved among fungal species (de Graaff et al. 1994). Nevertheless, a general model for substrate recognition and a mechanism for induction of xylanolytic enzymes were proposed for *T. reesei* (Zeilinger et al. 1998). The gene *xyn2* is expressed at a low level and encodes an enzyme that attacks the xylan polymer, causing the release of oligosaccharides and inducing endoxylanases I and II and β-xylosidase 1.

Production of Pectinases

The production of pectinolytic enzymes is induced by pectins of beetroot (Bussink et al. 1992) and apple (Suykerbuyk et al. 1997), polygalacturonic acid (Ruttkowski et al. 1991), and low concentrations of galacturonic acid. Furthermore, some reports have suggested constitutive production of pectinolyic enzymes (Parenicova et al. 2000). In the presence of glucose, enzyme production is reduced to basal levels. Unlike the regulation of fungal (hemi-)cellulose production, the molecular mechanism underlying the induction of the pectinolytic system is not well described. It has been best studied in *A. niger* (de Vries et al. 2002). Several factors, such as carbon source, pH of the culture medium, cAMP concentration, catabolic repression, and amount of available oxygen, may be regulated by the expression of fungal pectinases. The expression of several genes is controlled by catabolic repression mediated by cAMP in the presence of glucose and catabolic products of pectin. Environmental factors, including osmolarity, temperature, oxygen and nitrogen availability, lack of iron, and DNA-damaging agents, differentially affect expression of pectinases (Hugouvieux-Cotte-Pattat et al. 1996). Other factors and specific binding regions for transcription factors have been identified in the promoter region of some pectinases, such as the HAP2/3/4 complex (see "Regulation of cellulase and hemicellulose gene transcription" below). Further, a hexanucleotidic sequence (CCCTGA) was detected in the promoter of several genes encoding pectin lyase (Benen et al. 1996). However, the function of these elements has not been investigated.

Production of Ligninases

The production and regulation of the expression of ligninolytic enzymes has been best studied in *P. chrysosporium*. The expression of manganese peroxidase (*mnp*) and lignin peroxidase (*lip*) is difficult to study in ligninolytic fungi, because they have several homologous genes but different regulatory mechanisms. In general, the expression of ligninolytic genes is triggered by depletion of nutrients, such as nitrogen, carbon or sulfur. As such, the expression of most ligninolytic genes is considered to be a stress response to the depletion of those nutrients. LiPs are regulated by catabolic repression, while MnPs are not unrepressed by carbon limitation (Bonnarme et al. 1991). In *P. chrysosporium*, the expression of *mnp* is activated by depletion of nitrogen in the culture medium and by the presence of Mn (II) (Gold and Alic 1993). However, Mn (II) exerts this effect when nitrogen is limited. Genetic regulatory elements for these enzymes are not yet fully elucidated, but studies have identified putative metal-response elements in the promoter region of MnP (Gettemy et al. 1998). Expression of *mnp* is also temporarily induced by a temperature shock in cultures with limiting amounts of nitrogen, and sequences corresponding to the consensus temperature-shock element ($CN_2GAAN_2TTCN_2G$) were found in the promoter of *mnp* (Mayfield et al. 1994). Expression of laccases, on the other hand, does not depend on nutrient depletion. Laccases are constitutively expressed in many basidiomycota fungi (Scheel et al. 2000) and are the first enzymes to attack lignin, since its degradation product induces the expression of laccase and other ligninolytic enzymes (Smith et al. 1998). Constitutive expression of laccases is also increased by the presence of inducers, such as 2,5-xylidine, in fungi, such as *Trametes villosa* (Yaver et al. 1996), *T. versicolor* (Collins and Dobson 1997), and *A. bisporus* (Smith et al. 1998).

Regulation of Cellulase and Hemicellulase Gene Transcription

Fungi can respond to different types of chemical and physical stimuli, including nutrients, osmolarity, pH, and light (Bölker 1998). Therefore, fungi can flexibly adapt and quickly respond to environmental changes in nutrient composition by regulating gene expression according to demand. For instance, in a complex medium, *T. reesei* chooses the best carbon and energy source and negatively regulates the expression of genes required for degradation of other, less favorable carbon sources (Seiboth et al. 2011).

Similarly, *A. niger* regulates the induction of hemicellulose expression based on polysaccharide recognition (Hasper et al. 2000). At present, the mechanisms underlying regulation of cellulase and hemicellulase are best studied in *A. niger* and *T. reesei,* which share some similarities as well as important differences (Stricker et al. 2008).

Regardless of the nature of the cellulose inducer, cellulase production is regulated at the transcriptional level in a coordinated manner that depends on the available carbon and light source (Gyalai-Korpos et al. 2010). Transcription of the main cellulase components (CEL7A, CEL6A, CEL7B, CEL5A, CEL12A, CEL61A, CEL45A, CEL61B, CEL74A) is induced by cellulose and a variety of disaccharides, including lactose, β-cellobiono-1,5-lactone, cellobiose, and sophorose (Kubicek et al. 1993, Ilmen et al. 1997b, Nogawa et al. 2001, Foreman et al. 2003). In contrast, significant levels of cellulases are not produced when fructose, glucose or glycerol are used as carbon sources (Ilmen et al. 1997b). On the other hand, hemicellulases, particularly xylanases (*xyn1* and *xyn2*) in *T. reesei* and *A. niger,* are induced in the presence of cellulose, xylan, sophorose, xylobiose, and L-arabitol.

An analysis of the promoter regions of the two most highly expressed cellulase genes (*cbh1* and *cbh2*) revealed that their expression is both positively and negatively regulated (reviewed by Schmoll and Kubicek 2003). Saloheimo et al. (2000) and Aro et al. (2001) isolated and identified two genes that regulate transcription of cellulase genes. The first, *ace1* (activator of cellulase gene expression), encodes the protein ACEI, which binds to DNA and contains three Cys(2)-His(2)-type zinc-finger motifs. The second, *ace2*, encodes the protein ACEII, which also binds DNA and belongs to the class of binuclear zinc family (Zn(II)Cys6) that is unique to fungi (Aro et al. 2001).

In vitro, ACEI binds to the promoter region of *cbh1* at eight sequences of AGGCAAA and some sequences of AGCA are preceded by relatively AT-rich regions. Identical sequences were identified in promoters of other cellulase genes, such as *egl1* (1 binding site) and *cbh2* (3 binding sites), and in the promoters of xylanase genes, *xyn1* and *xyn2* (1 binding site each) (Fig. 7.13) in *T. reesei*. This supports the hypothesis of coordinated regulation of the cellulolytic system of *T. reesei* (reviewed by Schmoll and Kubicek 2003). Deletion of *ace1* in *T. reesei* delayed fungal growth in cellulose-containing culture medium; nevertheless, sophorose and cellulose expression was increased by about 2–30 fold. Deletion of *ace1* in *T. reesei* also increased the expression of xylanase genes, *xyn1* and *xyn2*. These results indicate that ACEI acts as a repressor of cellulase and xylanase gene expression (Aro et al. 2001). However, other fungi display different cellulase and xylanase gene expression, suggesting that ACEI may have a more general role in regulating their expression as opposed to acting only as a repressor (Aro et al. 2005).

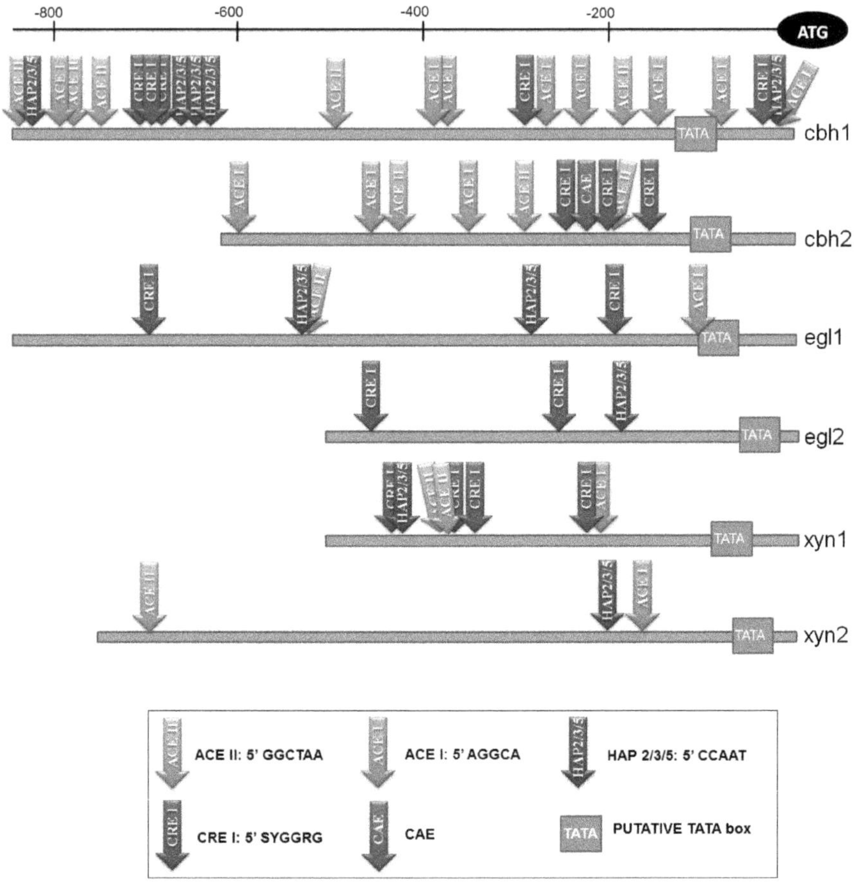

Figure 7.13 Schematic representation of putative transcription factor binding sites within the cellulase and xylanase promoters of *T. reesei*.

ACEII, on the other hand, can bind to the same region as factor XYR1 [GGC(T/A)$_4$]. Although the two factors bind to sequences, the sequence similarity of the Zn(II)Cys6 domain is reduced compared to other proteins with the same domain. Accordingly, studies suggest that ACEII requires phosphorylation and dimerization to bind this motif (Stricker et al. 2006). At least five ACEII binding sites have been identified in the promoter region of *cbh1*, three in *cbh2*, one in *egl1*, two in *xyn1*, and one in *xyn2* (Fig. 7.13). Some of these sites are located near the binding region of the catabolic repressor, CREI. This region is flanked by four CCAAT sequences, to which the HAP2/3/5 complex binds, and by an ACEI binding site (Aro et al. 2001).

Deletion of *ace2* in *T. reesei* decreased mRNA levels of *cbh1*, *cbh2*, *egl1*, and *egl2* and decreased the activity of cellulases by 30–70 percent when the fungus was grown in cellulose-containing medium (Aro et al. 2001). However, induction of cellulase by sophorose was not affected, suggesting different mechanisms in sophorose's regulatory function (Schmoll and Kubicek 2003). It was also reported that ACEII binds to the promoter of *cbh2*. In contrast to deletion of the binding motif in the promoter of *cbh1*, deleting the binding motif in the promoter of *cbh2* reduced its induction by cellulose and by sophorose (Zeilinger et al. 1998). ACEII deletion in *T. reesei* caused a 30–45 percent reduction in the expression of *xyn2* but did not significantly affect the expression of *xyn1* under conditions of induction by cellulose. Deletion of both *ace1* and *ace2* similarly affected the expression of cellulases and xylanases, suggesting a coordinated regulation during induction by cellulose. Importantly, no ortholog of *ace2* has been identified in *Aspergillus* genomes, indicating that *Aspergillus* and *Trichoderma* differ in transcriptional regulation underlying cellulose and hemicellulose hydrolysis (Stricker et al. 2008).

The first transcription factor that activates cellulase and xylanase expression in filamentous fungi was isolated from *A. niger* (van Peij et al. 1998b). XlnR has a binuclear zinc group with a DNA-binding domain and binds *in vitro* to the sequence GGCTAATAA in the promoters of xylanase-encoding genes. XlnR transcriptionally activates xylanases (*xlnA*, *xlnB*, and *xlnC*) as well as endoglucanases (*eglA* and *eglB*) and cellobiohydrolases (*cbhA* and *cbhB*), demonstrating the shared regulation of hemicellulases and cellulases. A loss of function mutation in *xlnR* reduced or eliminated xylose- and xylan-induced expression of genes that are not regulated by XlnR. XlnR in *A. niger* also regulates transcription of genes that encode intracellular enzymes involved in D-xylose catabolism, including D-xylose reductase (*xyrA*) (Hasper et al. 2000). These data indicate that in *A. niger*, there is a connection between the regulation of intra- and extracellular enzymes. XlnR has also been isolated from other fungi, such as *A. oryzae* (Marui et al. 2002), *A. kawachii* (AB064658), *A. nidulans* (AJ272537), *T. reesei* (AF479644), *Penicillium funiculosum*, and *P. purpurogenum* (Kroon et al. 2000, Chavez et al. 2002)

In *T. reesei* (*H. jecorina*), an ortholog termed xylanase regulator 1 (*xyr1*) binds to the consensus sequence GGCTAA. XYR1 differs from XlnR in just two amino acids and has 49 percent sequence homology (Stricker et al. 2006, Rauscher et al. 2006). XYR1 is the main transcriptional activator of genes that encode hydrolases, such as, *xyn1*, *xyn2*, *bxl1*, *cbh1*, *cbh2*, *egl1*, and *bgl1*. Deletion of *xyr1* eliminates induction of cellulases in the presence of cellulose and sophorose, highlighting its essential role in induction. Further, XYR1 regulates the expression of other genes that encode xylanases, β-xylosidases, and some α-L-arabinofuranosidases (Stricker et al. 2006). This pattern was also observed in *A. niger*. Specifically, loss of xlnR function reduced or ablated the induction of the above-mentioned genes in the presence of xylose or xylan. Conversely, overexpression of xlnR increased the expression of some of these genes (van Peij et al. 1998a, Gielkens et al. 1999).

Another feature of the regulation of cellulase and hemicellulase expression is catabolic repression (CCR). In the presence of glucose, several genes necessary for the use of other carbon sources are repressed. CCR is an important regulatory mechanism that represses genes encoding enzymes required for using complex carbon sources when there is a high availability of simple sugars, like glucose (Gancedo 1998). In filamentous fungi, CCR is mediated by the transcription factors CREA and CREI. Besides the well known catabolic repressors in fungi (e.g., MIGl in yeast and CreA in *Aspergillus*), others have been isolated and characterized in *H. jecorina* (Strauss et al. 1995, Ilmén et al. 1996), *Humicola grisea* (Takashima et al. 1999), and *Sclerotinia sclerotiorum* (Vautard-Mey et al. 1999).

Under conditions of high glucose in *T. reesei*, sophorose cannot induce gene expression of cellulases. Similarly, addition of glucose to the media of *T. reesei* that is already induced by sophorose reduces the expression of *cbh1*, *cbh2*, *egl1*, *egl2*, and *egl5* (Ilmén et al. 1997b). However, alone, this data do not prove that glucose is a direct repressor, because glucose could prevent cellular entry of the inducer by inhibiting the disaccharide permease (Kubicek et al. 1993, Vankuyk et al. 2004). Glucose's direct repressive effect was established through the use of mutant repressor genes and the analysis of cellulase promoters (Takashima et al. 1996 and Ilmén 1997a). *cre1* encodes a catabolic repressor, CREI (Strauss et al. 1995), which is homologous to CreA from *A. nidulans* (Dowzer and Kelly 1991). This protein has a consensus sequence at its binding site, SYGGRG. CRE binding sites were identified in the promoters of *cbh1*, *xyn1*, *cbh2*, *egl1*, and *egl2* of *T. reesei* (Fig. 7.13) and in the promoter of *xlnA* of *A. nidulans*. The number of binding sites varied between the promoters, suggesting tight regulation by this protein. Mutations in CRE binding sites affected the expression of these genes even in the presence of glucose (Takashima et al. 1996). CREA also regulated the expression of hemicellulase genes, *xlnB*, *xnlD*, *faeA*, *aguA*, *afbA*, *afb*, and *abnA* in *A. niger*. Expression of these genes was regulated by the balance between CREA and the activator XlnR (de Vries and Visser 1999b).

Cziferszky et al. (2003) performed a more detailed study of the mechanism by which CREI is activated in the presence of glucose during the induction of the cellulolytic system in *T. reesei*. The authors showed that CREI is a phosphoprotein with a phosphorylation motif at Ser_{241}, containing an acidic region as a target for phosphorylation. Binding of CREI on to its target DNA site in *T. reesei* seems to involve a casein kinase that phosphorylates a serine in the acidic region of CREI (Cziferszky et al. 2003). Interestingly, the most commonly used *T. reesei* strain and publically known cellulase hyperproductor, RutC-30, was originally isolated after it was selected for growth in cellobiose in the presence of 2-deoxi-glucose

(Montenecourt and Eveleigh 1979). It has a mutated *cre1* gene, with only 20 percent of the wild-type coding region, and thus forms a defective protein without the C-terminus. A Northern blot analysis confirmed that this strain expresses some cellulases, such as *cbh1*, in the presence of even low concentrations of glucose (Zeilinger et al. 2003). Recently, Seidl et al. (2008) reported that RutC-30 had lost an approximately 85-kb genomic fragment during the mutagenesis process, which resulted in the loss of 29 genes, including those related to protein transport and primary metabolism. An analysis of *cre1* deletion mutants demonstrated that catabolic derepression was not sufficient to increase cellulase and xylanase production, suggesting that overexpression of these enzymes depends on inducers (Nakari-Setälä et al. 2009). Recent studies have also shown that CREI is involved in other cellular processes of *T. reesei*, such as growth, protein transport, nitrogen capture, chromatin remodeling, transcriptional mediation, and development (Portnoy et al. 2011).

Other transcription factors involved in the expression of cellulase genes in *T. reesei* were identified by the electrophoretic mobility shift assay (EMSA) using mycelia extracts in conditions with or without sophorose induction and several fragments of the promoter of *cbh2*. Zeilinger et al. (1998) detected a nucleotide sequence (ATTGGTAATA-) named *cbh2*-activating element (CAE) that binds to protein complexes (Fig. 7.13). *In vivo* "footprinting" experiments showed that protein binding to CAE occurred under conditions of induction and non-induction. EMSA analysis with mutations in specific regions of CAE revealed that protein binding required an intact copy of any of the two adjacent motifs: CCAAT in the template strand and GTAATA in the coding strand. Simultaneous mutations in these motifs led to a loss of binding of protein complexes (Zeilinger et al. 1998).

The proteins that bind to GTAATA motifs have not yet been fully identified (Zeilinger et al. 2001). However, EMSA experiments using antibodies have revealed that Hap-type proteins can bind the CCAAT motif. Zeilinger et al. (2001) cloned *hap2*, *hap3*, and *hap5* of *T. reesei*, which encode proteins with homologous regions to HapB, HapC, and HapE of *Aspergillus* as well as known Haps of other organisms. Hap2, Hap3, and Hap5 were all required to bind CAE of the promoter of *cbh2* in *T. reesei in vitro*. Therefore, the CCAAT motif in the template strand of the CAE is bound in a complex in *T. reesei*, similar to the Hap complex in other organisms (reviewed by Schmoll and Kubicek 2003). CCAAT sequences have been identified in 50 regions of about 30 percent of eukaryotic genes. The first-described CCAAT binding complex, designated as the Hap complex, consists of the proteins Hap2, Hap3, Hap4, and Hap5 and was identified in *S. cerevisiae* (Pinkham and Guarente 1985, Olesen and Guarente 1990, McNabb et al. 1995). In *S. cerevisiae*, the Hap complex is needed for several functions, including positive regulation of respiratory genes involved in oxidative phosphorylation during growth in non-fermentable carbon sources (Gancedo 1998, Marui et al 2002). Abrahão-Neto et al. (1995) reported that mitochondrial activity is necessary for maximal expression of cellulase genes in *T. reesei*. Carraro et al. (1998) showed that heterologous transcription of genes controlled by the promoter of *cbh1* in *S. cerevisiae* also depends on mitochondrial activity, suggesting the presence of promoter sequences that bind regulatory proteins.

The transcriptional regulators of cellulase genes are known to bind their targets in a constitutive fashion; however, their exact actions are unknown (Schmoll and Kubicek 2003). The Hap2/3/5 complex can regulate gene transcription by chromatin remodeling (Narendja et al. 1999). Zeilinger et al. (2003) investigated changes in chromatin structure of the *cbh2* promoter during induction of the cellulase gene. Under conditions of induction or catabolic repression, changes in the structure of chromatin were observed. During repression, formation of nucleosomes occurred near the TATA box, thus preventing the interaction of the ACEII regulators with the Hap2/3/5 complex. In contrast, under conditions of induction, the TATA box was accessible. Therefore, these results suggested that both the Hap2/3/5 complex and ACEII are involved in forming nucleosomes in the promoter of *cbh2* and respond to the absence of nucleosomes by activating cellulase gene expression (Zeilinger et al. 2003). Studies suggest that in the presence of two more components in this process, an unknown protein joins the RNA polymerase II complex to the Hap2/3/5 complex and a protein with a repeating WD domain, which is abundant in the presence of cellulose and lactose in *T. reesei* and absent in a mutant lacking cellulase (Schmoll et al. 2005). WD-domain-containing proteins have a great variety of functions, including signal transduction, synthesis and processing of RNA, chromatin assembly, vesicular transport, cytoskeleton assembly, cell-cycle regulation, and apoptosis (Schmoll et al. 2005).

Mutations in the CCAAT motif of promoters or deletion of subunits generally reduce the expression of HAP-regulated genes, both at basal levels and in response to specific stimuli. This indicates that the CCAAT motif cooperates with others specific elements to induce transcription. In the promoter of the *cbh2* gene of *T. reesei*, mutation of one element of the CCAAT sequence reduced the activity of the *cbh2* promoter by approximately 20 percent upon induction by sophorose (Zeilinger et al. 1998). Zeilinger et al. (1998) also proposed a role for the CCAAT element in regulating the xylanase promoters, *xyn1* and *xyn2*. Mutation both of the CCAAT element or the GTAATA motif reduced the transcription of *cbh2*, whereas simultaneous mutation of both ablated transcription of *cbh2* (Zeilinger et al. 1998). Zeilinger et al. (2001) showed that both basal transcription and induction of *xyn2* depended on binding of the Hap2/3/5 complex and ACEII (Aro et al. 2001) to an undecameric motif (GGGTAAATTGG), termed xylanase-activating element (XAE). In contrast to *cbh2*, a mutation in the CCAAT sequence of the XAE in the *xyn2* promoter of *T. reesei* slightly increased transcription of *xyn2* in the presence of glycerol and xylan, suggesting different regulatory mechanisms (Wurleitner et al. 2003). No studies have investigated the role of HAP complexes in the regulation of genes encoding enzymes that degrade polysaccharides in *Aspergillus* (de Vries and Visser 2001). However, the involvement of HAP-like complexes in the regulation of xylanases and cellulases in *T. reesei* suggests that these systems may be regulated by HAP-like complexes in *Aspergillus.*

The physiological processes that trigger secretion of macromolecules and metabolites may depend on adaptation to environmental pH, as is the case of penicillin secretion by the fungus *A. nidulans* (Espeso et al. 1993). Enzymatic degradation of cellulose and hemicellulose by filamentous fungi may occur in varying conditions, including changes in pH. Regulation by pH was first described by Nahas et al. (1982). They found that secretion of Pi-repressible acid and alkaline phosphatases by *N. crassa* depended on the pH of the culture medium. Specifically, acid and alkaline phosphatases were preferentially secreted in acidic and alkaline media, respectively. This adaptive response to environmental pH is a universal phenomenon in many cellular processes and is essential for the efficiency of contractible elements (actin, myosin, and microtubules) in the control of the cell cycle and cell proliferation. This adaptive response also contributes to the pathogenicity of some fungi (reviewed in detail by Martinez-Rossi et al. 2011). PacC is the main factor controlling pH-dependent expression in *Aspergillus*. PacC has a dual function: at an alkaline pH, it activates the expression of alkaline-related genes and represses the expression of acid-induced genes (Tilburn et al. 1995). PacC has a Zn-finger (Cys2His2)-type DNA-binding domain and binds to promoter sites containing the hexanucleotide GCCARG (Caddick et al. 1986, Tilburn et al. 1995, Espeso et al. 1997). When the fungus is in an alkaline pH, PacC is proteolytically processed, and the proteolyzed form acts as an activator of expression of alkaline-induced genes (i.e., those that are preferentially expressed in alkaline medium) and as an inhibitor of expression of acid-induced genes (i.e., those that are preferentially expressed in acidic medium). The *pal* (*A, B, C, F, H* and *I*) genes participate in this signaling cascade in alkaline pH, promoting two proteolytic cleavages required for the activation of the PacC protein. Three forms of PacC, with molecular weights of 72 kDa ($PacC^{72}$), 53 kDa ($PacC^{53}$), and 27 kDa ($PacC^{27}$), are present in protein extracts (Orejas et al. 1995, Diez et al. 2002). Studies in *A. nidulans* showed that the system that regulates pH via PacC also controls other activities, including post-translational modifications of secreted enzymes and synthesis of enzymes responsible for energy production (Nozawa et al. 2003a, Nozawa et al. 2003b). Studies with PacC mutants showed that two genes coding for xylanases in *A. nidulans* (*xlnA* and *xlnB*) are regulated by PacC. *xlnB* is expressed under acidic pH conditions, while *xlnA* is expressed under alkaline conditions. The same study also showed that in *A. nidulans*, the expression of *abfB* that encodes α-L-arabinofuranosidase is also regulated by PacC. PacC homologues have also been found to mediate pH regulation in *A. niger* (MacCabe et al. 1996), *Penicillium chrysogenum* (Suarez and Penalva 1996), and *T. harzianum* (Moreno-Mateos et al. 2007). However, there are few detailed studies of PacC in the regulation of enzymes that degrade plant cell walls.

Cell Signaling

In *Aspergillus* the cell signaling pathways underlying the production of enzymes that degrade the plant cell wall are not well known. These pathways have been better characterized in *T. reesei*. Although previous studies have shown that cellulase genes are induction dependent, neither the nature of the inducer nor their signaling pathways have been fully elucidated (Schmoll and Kubicek 2003). In contrast to the amount of data available on the architecture of the promoters of hydrolases in *T. reesei*, there are few studies on its signal transduction mechanisms (Mach and Zeilinger 2003). Wang and Nuss (1995) demonstrated the involvement of a G protein in *cbh1*'s induction by cellulose. Sestak and Farkas (1993) reported that the efficacy of cellulase induction doubled in the presence of dibutyryl-cAMP, a membrane permeable form of cAMP, or phophodiesterase inhibitors, which degrade cAMP. These data suggested cellulase inducers may use involve G proteins and cAMP for signaling. Sequencing of *T. reesei*'s genome identified three G-alpha protein subunits, one G-beta protein subunit, and one G-gamma protein subunit (Schmoll 2008). In *T. reesei*, the potential roles of two G-proteins, GNA1 and GNA3, in the expression of cellulase genes were investigated. Studies with mutants expressing constitutively active GNA3, as well as anti-sense *gna3* mutants revealed that this G-protein positively regulates cellulase gene expression under conditions of constant light but not in the dark (Schmoll et al. 2009). This suggests that the signaling pathways underlying production of cellulases may be more complex than once imagined. A null mutant for GNA1 (Δ*gna1*) exhibited a lack of cellulase gene expression in the presence of cellulose in the light, while cellulase expression was increased in the dark (Seibel et al. 2009). These data indicated that these two G-alpha subunits differ in function and may respond to different signals. Other components of the cAMP signaling pathway were recently analyzed in *T. reesei*. The results showed that both adenyl-cyclase and protein kinase A are involved in light-modulated cellulase gene expression when induced by lactose (Schuster et al. 2012). Cellulase gene expression is totally dependent on inducers and is influenced by light; this finding prompted a new study of the factors that increase cellulase production in industrial settings, where light being the new factor to be considered. In *N. crassa*, about 7 percent of the genome may be regulated by light signaling pathways (Lewis et al. 2002). In *T. reesei*, the ENVOY protein is a photoreceptor blue-light ortholog (VIVID) of *N. crassa*, which has been thoroughly studied and is involved in the regulation of the system of cellulase, carbon source, G-protein, and light (Schmoll 2008). Few studies have elucidated cell signaling pathways involved in the production of hemicellulases. In *T. reesei*, signaling through the calcium/calmodulin system may be involved in some steps of endoxylanase II production both at the levels of transcription and secretion (Mach and Zeilinger 2003). Studying these signaling pathways will be facilitated by knowledge of the *T. reesei* genome (Martinez et al. 2008) and by new functional genomics tools (Schuster et al. 2012). Specifically, it will be possible to construct mutants for each gene whose function one wishes to evaluate. This information will help develop new metabolic engineering strategies for obtaining lineages capable of overproducing cellulases and hemicellulases.

Conclusions and Future Perspectives

The increase in our knowledge of the biochemical machinery used by microorganisms to degrade biomass can lead to the development of biologically based processes with the potential to convert biomass on an industrial scale. The development of low-cost enzymatic systems is essential for making the conversion of biomass to biofuel economically viable. To date, a large amount of information on production regulation, mechanisms of action, and enzyme diversity in filamentous fungi has been obtained. However, much remains to be learned regarding the mechanisms of induction and the signaling pathways involved in the synthesis of these enzymes that degrade the plant cell wall. Through the development of novel molecular biology techniques over the past 15 yr, some transcription factors and their binding sites have been identified and characterized in certain fungi, particularly in *T. reesei* and *A. niger*. Thus, there has been a substantial increase in the knowledge of regulating the production of cellulases and hemicellulases in particular. Another important achievement in this area was the sequencing of fungi genomes (see http://www.broadinstitute.org/scientific-community/science/projects/fungal-genome-initiative/fungal-genome-initiative; http://genome.jgi.doe.gov/programs/fungi/index.jsf), which are being used in comparative genomics and cloning studies, in the identification of new genes for degrading biomass, and as a precious databank for transcriptomics and proteomics. New tools of functional genomics, which offer the possibility of deleting the entire genome of a fungus, for example *Neurospora crassa* and *T. reesei*, are being developed in several research laboratories.

In parallel to this, analysis of the enzymatic content secreted by these fungi has also received great attention. Based on the enormous progress made in electrophoresis, chromatography, and mass spectrometry techniques in recent years, preliminary studies in the analysis of the proteins secreted by fungi—the so-called secretome—have been published. However, only a limited number of studies on the secretome of filamentous fungi ascomycetes and basidiomycetes that degrade biomass, either for its complexity or for industrial interest. At present, with the advancement in software, algorithms, and experimental methods, proteomics is the ideal tool for the study of the secretome of fungi, the analysis of the expression profile, and the identification of new enzymes. When analyzed together, data generated on a large scale from genomics, transcriptomics, metabolomics, and proteomics—the so-called "omics" science—will be able to supply considerably useful tools for understanding the regulation of the production of enzymes that degrade plant cell walls in addition to making it possible to better produce these enzymes in systems of industrial fermentation, thus contributing to the production of bioethanol produced from vegetal biomass on an industrial scale.

Acknowledgments

I thank Professor Valdirene Neves Monteiro, PhD (University of Goias State, Brazil) for critical revision of the text. This work was supported by The State of São Paulo Research Foundation (FAPESP) (proc. 2010/15683-8).

References

Abrahao-Neto, J., C.H. Rossini, S. El-Gogary, F. Henrique-Silva, O. Crivellaro and H. El-Dorry. 1995. Mitochondrial functions mediate cellulase gene expression in *Trichoderma reesei*. Biochem. 34: 10456–10462.

Ademark, P., A. Varga, J. Medve, V. Harjunpaa, T. Drakenberg, F. Tjerneld and H. Stalbrand. 1998. Softwood hemicellulose-degrading enzymes from *Aspergillus niger*: purification and properties of a β-mannanase. J. Biotechnol. 63: 199–200.

Ademark, P., R.P. de Vries, H. Stålbrand and J. Visser. 2001. Cloning and characterisation of genes encoding a β-mannosidase and an α-galactosidase from *Aspergillus niger* involved in galactomannan degradation. Eur. J. Biochem. 268: 2982–2990.

Alcocer, M.J., C.S. Furniss, P.A. Kroon, M. Campbell and D.B. Archer. 2003. Comparison of modular and non-modular xylanases as carrier proteins for the efficient secretion of heterologous proteins from *Penicillium funiculosum*. Appl. Microbiol. Biotechnol. 60: 726–732.

Aro, N., A. Saloheimo, M. Ilmén and M. Penttilä. 2001. ACEII, a novel transcriptional activator involved in regulation of cellulase and xylanase genes of *Trichoderma reesei*, J. Biol. Chem. 276: 24309–24314.

Aro, N., T. Pakula and M. Penttilä. 2005. Transcriptional regulation of plant cell wall degradation by filamentous fungi. FEMS Microbiol. Rev. 29(4): 719–39.

Barnett, C., R. Berka and T. Fowler. 1991. Cloning and amplification of the gene encoding an extracellular β-glucosidase from *Trichoderma reesei*: evidence for improved rates of saccharification of cellulosic substrates. Biotechnology 9: 562–567.

Bauer, S., P. Vasu, S. Persson, A.J. Mort and C.R. Somerville. 2006. Development and application of a suite of polysaccharide-degrading enzymes for analyzing plant cell walls. Proc. Natl. Acad. Sci. USA. 103: 11417–11422.

Benen, J., L. Parenicova, M. Kusters-van Someren, H. Kester and J. Visser. Molecular genetic and biochemical aspects of pectin degradation in *Aspergillus*, p. 331–346. *In:* J.Visser and A.G.J. Voragen [eds.]. 1996. Pectins and pectinases, vol. 14. Elsevier Science, Amsterdam, The Netherlands.

Benen, J.A.E., H.C.M. Kester, L. Parenicova and J. Visser. 2000. Characterization of *Aspergillus niger* pectate lyase A. Biochem. 39: 15563–15569.

Bhat, M. and S. Bhat. 1997. Cellulose degrading enzymes and their potential industrial applications. Biotechnol. Adv. 15: 583–620.

Bisaria, V.S. and S. Mishra. 1989. Regulatory aspects of cellulase biosynthesis and secretion. CRC Crit. Rev. Biotechnol. 9: 61–103.

Bolker, M. 1998. Sex and crime: heterotrimeric G proteins in fungal mating and pathogenesis. Fungal Genet. Biol. 25: 143–156.

Bonnarme, P., J. Perez and T.W. Jeffries. Regulation of Ligninase Production in White-Rot Fungi. *In:* G.F. Leathanl and M.E. Himmel [eds.]. 1991. Enzymes in biomass conversion: Proceedings of a symposium, 1990 April 22– 27, Boston, MA. ACS symposium series 460. American Chemical Society, Washington, DC.

Bussink, H.J., K.B. Brouwer, L.H. de Graaff, H.C. Kester and J. Visser. 1991. Identification and characterization of a second polygalacturonase gene of *Aspergillus niger*. Curr. Genet. 20: 301–307.

Bussink, H.J.D., F.P. Buxton, B.A. Fraaye, L.H. de Graaff and J. Visser. 1992. The polygalacturonases of *Aspergillus niger* are encoded by a family of diverged genes. Eur. J. Biochem. 208: 83–90.

Caddick, M.X., A.G. Brownlee and H.N. Arst, Jr. 1986. Regulation of gene expression by pH of the growth medium in *Aspergillus nidulans*. Mol. Gen. Genet. 203: 346–353.

Caffall, K.H. and D. Mohnen. 2009. The structure, function, and biosynthesis of plant cell wall pectic polysaccharides. Carbohydr. Res. 344(14): 1879–1900.

Carle-Urioste, J.C., J. Escobar-Vera, S. El-Gogary, F. Henrique-Silva, E. Torigoi, O. Crivellaro, A. Herrera-Estrella and H. El-Dorry. 1997. Cellulase induction in *Trichoderma reesei* by cellulose requires its own basal expression. J. Biol. Chem. 272: 10169–10174.

Carpita, N.C. and D.M. Gibeaut. 1993. Structural models of primary cell walls in flowering plants: consistency of molecular structure with the physical properties of the walls during growth. Plant J. 3: 1–30.

Carraro, D.M., J.R. Ferreira, Junior, R. Schumacher, G.G. Pereira, C.P. Hollenberg and H. El-Dorry. 1998. A region of the cellobiohydrolase I promoter from the filamentous fungus *Trichoderma reesei* mediates glucose repression in *Saccharomyces cerevisiae*, dependent on mitochondrial activity, Biochem. Biophys. Res. Commun. 253: 407–414.

Chavez, R., K. Schachter, C. Navarro, A. Peirano, C. Aguirre, P. Bull and J. Eyzaguirre. 2002. Differences in expression of two endoxylanase genes (xynA and xynB) from *Penicillium purpurogenum*. Gene 293: 161–168.

Christgau, S., T. Sandal, L.V. Kofod and H. Dalboege. 1995. Expression cloning, purification and characterization of a beta-1,4-galactanase from *Aspergillus aculeatus*. Curr. Genet. 27: 135–141.

Collins, P. and A.W. Dobson. 1997. Regulation of laccase gene transcription in *Trametes versicolor*. Appl. Environ. Microbiol. 63: 3444–3450.

Cziferszky, A., B. Seiboth and C.P. Kubicek. 2003. The Snf1 kinase of the filamentous fungus *Hypocrea jecorina* phosphorylates regulation-relevant serine residues in the yeast carbon catabolite repressor Mig1 but not in the filamentous fungal counterpart Cre1. Fungal Genet. Biol. 40: 166–175.

de Graaff, L.H., H.C. Van den Broeck, A.J. Van Ooijen and J. Visser. 1994. Regulation of the xylanase-encoding *xlnA* gene of *Aspergillus tubigensis*. Mol. Microbiol. 12: 479–490.

de Vries, R.P. and J. Visser 2001. *Aspergillus* enzymes involved in degradation of plant cell wall polysaccharides. Microbiol. Mol. Biol. Rev. 65: 497–522.

de Vries, R.P., B. Michelsen, C.H. Poulsen, P.A. Kroon, R.H.H. van den Heuvel, C.B. Faulds, G. Williamson, J.P.T.W. van den Hombergh and J. Visser. 1997. The *faeA* genes from *Aspergillus niger* and *Aspergillus tubigensis* encode ferulic acid esterases involved in the degradation of complex cell wall polysaccharides. Appl. Environ. Microbiol. 63: 4638–4644.

de Vries, R.P., H.C. van den Broeck, E. Dekkers, P. Manzanares, L.H. de Graaff and J. Visser. 1999a. Differential expression of three α-galactosidase genes and a single β-galactosidase gene from *Apergillus niger*. Appl. Environ. Microbiol. 65: 2453–2460.

de Vries, R.P. and J. Visser. 1999b. Regulation of the feruloyl esterase (*faeA*) gene from *Aspergillus niger*. Appl. Environ. Microbiol. 65: 5500–5503.

de Vries, R.P., J.A. Benen, L.H. de Graaff and J. Visser. Plant cell wall degrading enzymes produced by *Aspergillus*. pp. 263–279. *In:* H.D. Osiewacz [ed.]. 2002. The Mycota Springer Verlag, Berlin.

Demain, A.L., M. Newcomb and J.H. David Wu. 2005. Cellulase, clostridia, and ethanol microbiol. Mol. Biol. Rev. 69: 124–154.

den Herder, I.F., A.M. Mateo Rosell, C.M. van Zuilen, P.J. Punt and C.A.M.J.J. van den Hondel. 1992. Cloning and expression of a member of the *Aspergillus niger* gene family encoding α-galactosidase. Mol. Gen. Genet. 233: 404–410.

Diez, E., J. Alvaro, E.A. Espeso, L. Rainbow, T. Suarez, J. Tilburn, H. N.Arst and M.A. Peñalva. 2002. Activation of the *Aspergillus* PacC zinc finger transcription factor requires two proteolytic steps. EMBO J. 21: 1350–1359.

Dowzer, C.E.A. and J.M. Kelly. 1991. Analysis of the *creA* gene, a regulator of carbon catabolite repression in *Aspergillus nidulans*, Mol. Cell. Biol. 9: 5701–5709.

Durand, H., H. Clanet and G. Tiraby. 1988. Genetic improvement of *Trichoderma reesei* for large scale cellulase production. Enzyme. Microbiol. Technol. 10: 341–346.

Emami, K. and E. Hack. 2002. Conservation of *XYN11A* and *XYN11B* xylanase genes in *Bipolaris sorghicola*, *Cochliobolus sativus*, *Cochliobolus heterostrophus*, and *Cochliobolus spicifer*. Curr. Microbiol. 45: 303–306.

Espeso, E.A., J. Tilburn, H.N. Arst Jr. and M.A. Peñalva. 1993. pH regulation is a major determinant in expression of a fungal penicillin biosynthetic gene. EMBO J. 12: 3947–3956.

Espeso, E.A., J. Tilburn, L. Sanchez-Pulido, C.V. Brown, A. Valencia, H.N. Arst Jr. and M.A. Pen˜alva. 1997. Specific DNA recognition by the *Aspergillus nidulans* three zinc finger transcription factor PacC. J. Mol. Biol. 274: 466–480.

Farrel, A.E., R.J. Plevin, B.T. Turner, A.D. Jones, M. O'Hare and D.M. Kammem. 2006. Ethanol can contribute to energy and environmental goals. Science 311: 506–508.

Fedorova, N.D., N. Khaldi, V.S. Joardar, R. Maiti, P. Amedeo, M.J. Anderson, J. Crabtree, J.C. Silva, J.H. Badger, A. Albarraq, S. Angiuoli, H. Bussey, P. Bowyer, P.J. Cotty, P.S. Dyer, A. Egan, K. Galens, C.M. Fraser-Liggett and W.C. Nierman. 2008. Genomic islands in the pathogenic filamentous fungus *Aspergillus fumigatus*. PLoS Genet. 4: E1000046–E1000046.

Flipphi, M.J., J. Visser, P. van der Veen and L.H. de Graaff. 1993. Cloning of the *Aspergillus niger* gene encoding alpha-L-arabinofuranosidase Appl. Microbiol. Biotechnol. 39(3): 335–340.

Foreman, P., D. Brown, L. Dankmeyer, R. Dean, S. Diener, N. Dunn-Coleman, F. Goedegebuur, T. Houfek, G. England, A. Kelley, H. Meerman, T. Mitchell, C. Mitchinson, H. Olivares, P. Teunissen, J. Yao and M. Ward. 2003. Transcriptional regulation of biomass-degrading enzymes in the filamentous fungus *Trichoderma reesei*. J. Biol. Chem. 278: 31988–31997.

Gancedo, J.M. 1998. Yeast carbon catabolite repression. Microbiol. Mol. Biol. Rev. 62: 334–361.

Gaskell, J., E. Dieperink and D. Cullen. 1991. Genomic organization of lignin peroxidase genes of *Phanerochaete chrysosporium*. Nucleic Acids Res. 19: 599–603.

Gettemy, J.M., B. Ma, M. Alic and M.H. Gold. 1998. Reverse transcription-PCR analysis of the regulation of the manganese peroxidase gene family. Appl. Environ. Microbiol. 64: 569–574.

Gielkens, M.M., E. Dekkers, J. Visser and L.H. de Graaff. 1999. Two cellobiohydrolase-encoding genes from *Aspergillus niger* require D-xylose and the xylanolytic transcriptional activator XlnR for their expression. Appl. Environ. Microbiol. 65: 4340–4345.

Godfrey, B.J., M.B. Mayfield, J.A. Brown and M.H. Gold. 1990. Characterization of a gene encoding a manganese peroxidase from *Phanerochaete chrysosporium*." Gene 93: 119–124.

Gold, M.H. and M. Alic. 1993. Molecular biology of the lignindegrading basidiomycete *Phanerochaete chrysosporium*. Microbiol. Rev. 57: 605–622.

Gritzali, M. and R.D. Brown. 1979. The cellulase system of *Trichoderma*: relationships between purified extracellular enzymes from induced or cellulose-grown cells, Adv. Chem. Ser. 181: 237–260.

Gyalai-Korpos, M., G. Nagy, Z. Mareczky, A. Schuster, K. Réczey and M. Schmoll. 2010. Relevance of the light signaling machinery for cellulase expression in *Trichoderma reesei* (*hypocrea jecorina*). BMC Res. Notes 3: 330.

Hara, T., J.Y. Lim, Y. Fujio and S. Ueda. 1984. Purification and some properties of exo-polygalacturonase from *Aspergillus niger* cultured in the medium containing Satsuma mandarin peel. Nippon Shokuhin Kogyo Gakkaishi 31: 581–586.

Hasper, A.A., J. Visser and L.H. de Graaff. 2000. The *Aspergillus niger* transcriptional activator XlnR, which is involved in the degradation of the polysaccharides xylan and cellulose, also regulates D-xylose reductase gene expression. Mol. Microbiol. 36: 193–200.

Heikinheimo, L. and J. Buchert. 2001. Synergistic effects of *Trichoderma reesei* cellulases on the properties of knitted cotton fabric. Textile Research Journal 71(8): 672–677.

Hugouvieux-Cotte-Pattat, N., G. Condemine, W. Nasser and S. Reverchon. 1996. Regulation of pectinolysis in *Erwinia chrysanthemi*. Annu. Rev. Microbiol. 50: 213–257.

Ilmén, M., C. Thrane and M. Penttilä. 1996. The glucose repressor gene cre1 of *Trichoderma*: isolation and expression of a full-length and a truncated mutant form. Mol. Gen. Genet. 251: 451–460.

Ilmén, M., T. Nakari-Setälä, M. Tenkanen and M. Penttilä. 1997a. cDNA cloning of a *Trichoderma reesei* cellulase and demonstration of endoglucanases activity by expression in yeast. Eur. J. Biochem. 249: 584–591.

Ilmén, M., A. Saloheimo, M.L. Onnela and M. Penttilä. 1997b. Regulation of cellulase expression in the filamentous fungus *Trichoderma reesei*, Appl. Environ. Microbiol. 63: 1298–1306.

Khanh, N.Q., E. Ruttkowski, K. Leidinger, H. Albrecht and M. Gottschalk. 1991. Characterization and expression of a genomic pectin methyl esterase-encoding gene in *Aspergillus niger*. Gene 106: 71–77.

Kimura, Y., Y. Asada, T. Oka and M. Kuwahara. 1991. Molecular analysis of a *Bjerkandera adusta* lignin peroxidase gene. Appl. Microbiol. Biotechnol. 35: 510–514.

Kimura, I., N. Yoshioka, Y. Kimura and S. Tajima. 2000. Cloning, sequencing and expression of an α-L-arabinofuranosidase from *Aspergillus sojae*. J. Biosci. Bioeng. 89: 262–266.

Kirk, T.K. and R.L. Farrell. 1987. Enzymatic "combustion": the microbial degradation of lignin. Annu Rev Microbiol. 41: 465–505.

Krengel, U. and B.W. Dijkstra. 1996. Three-dimensional structure of Endo-1,4-beta-xylanase I from *Aspergillus niger*: molecular basis for its low pH optimum. J. Mol. Biol. 263: 70–78.

Kroon, P.A., G. Williamson, N.M. Fish, D.B. Archer and N.J. Belshaw. 2000. A modular esterase from *Penicillium funiculosum* which releases ferulic acid from plant cell walls and binds crystalline cellulose contains a carbohydrate binding module. Eur. J. Biochem. 267: 6740–6752.

Kubicek, C.P., G. Mühlbauer, M. Grotz, E. John and E.M. Kubicek-Pranz. 1988. Properties of the conidial-bound cellulase system of *Trichoderma reesei*, J. Gen. Microbiol. 134: 1215–1222.

Kubicek, C.P., R. Messner, F. Gruber, M. Mandels and E.M. Kubicek-Pranz. 1993. Triggering of cellulase biosynthesis by cellulose in *Trichoderma reesei*: involvement of a constitutive, sophoroseinducible and glucose-inhibited ß-diglucoside permease, J. Biol. Chem. 268: 19264–19268.

Kumar, R., S. Sompal and V.S. Om. 2008. Bioconversion of lignocellulosic biomass: biochemical and molecular perspectives. J. Ind. Microbial. Biotechnol. 35(5): 377–391.

Kusters-van, S., M. Flipphi, L.H. de Graaff, H. van den Broeck, H. Kester, A. Hinnen and J. Visser. 1992. Characterisation of the *Aspergillus niger pelB* gene: structure and regulation of expression. Mol. Gen. Genet. 234: 113–120.

Lewis, Z.A., A. Correa, C. Schwerdtfeger, K.L. Link, X. Xie, R.H. Gomer, T. Thomas, D.J. Ebbole and D. Bell-Pedersen. 2002. Overexpression of white collar-1 (WC-1) activates circadian clock-associated genes, but is not sufficient to induce most light-regulated gene expression in *Neurospora crassa*. Mol. Microbiol. 45: 917–931.

Lombard, V., T. Bernard, C. Rancurel, H. Brumer, P.M. Coutinho and B. Henrissat. 2010. A hierarchical classification of polysaccharide lyases for glycogenomics. Biochem. J. 432(3): 437–444.

MacCabe, A.P., J.P.T.W. van den Hombergh, J. Tilburn, H.N.J. Arst and J.Visser. 1996. Identification, cloning and analysis of the *Aspergillus niger* gene *pacC*, a wide domain regulatory gene responsive to ambient pH. Mol. Gen. Genet. 250: 367–374.

Mach, R. and S. Zeilinger. 2003. Regulation of gene expression in industrial fungi: *Trichoderma*. Appl. Microbiol. Biotechnol. 60: 515–522.

Mach, R.L., J. Strauss, S. Zeilinger, M. Schindler and C.P. Kubicek. 1996. Carbon catabolite repression of xylanase I (*xyn1*) gene expression in *Trichoderma reesei*. Mol. Microbiol. 21: 1273–1281.

Machida, M., K. Asai, M. Sano and T. Tanaka et al. 2005. Genome sequencing and analysis of *Aspergillus oryzae*. Nature. 438 (7071): 1157–1161.

Mandels, M.E.T. and E.T. Reese. 1960. Induction of cellulase in fungi by cellobiose. J. Bacteriol. 79: 816–826.

Margolles-Clark, M., M. Ilme'n and M. Penttilä. 1997. Expression patterns of 10 hemicellulase genes from filamentous fungus *Trichoderma reesei* on various carbon sources. J. Biotechnol. 57: 167–179.

Martens-Uzunova, E.S. and P.J. Schaap. 2009. Assessment of the pectin degrading enzyme network of *Aspergillus niger* by functional genomics. Fungal Genet. Biol. 46(Suppl 1): S170–S179.

Martinez, D., L.F. Larrondo, N. Putnam, M.D. Gelpke, K. Huang, J. Chapman, K.G. Helfenbein, P. Ramaiya, J.C. Detter, F. Larimer, P.M. Coutinho, B. Henrissat, R. Berka, D. Cullen and D. Rokhsar. 2004. Genome sequence of the lignocellulose degrading fungus *Phanerochaete chrysosporium* strain RP78. Nat. Biotechnol. 22: 695–700.

Martinez, D., R.M. Berka and B. Henrissat et al. 2008. Genome sequencing and analysis of the biomass-degrading fungus *Trichoderma reesei* (syn. Hypocrea jecorina). Nat. Biotechnol. 26(10): 119.

Martinez-Rossi, N.M., G.F. Persinoti, N.T.A. Peres and A. Rossi. 2011. Role of pH in the pathogenesis of dermatophytoses. Mycoses. doi: 10.1111/j.1439-0507.2011.02162.x.

Marui, J., A. Tanaka, S. Mimura, L.H. de Graaff, J. Visser, N. Kitamoto, M. Kato, T. Kobayashi and N. Tsukagoshi. 2002. A transcriptional activator, AoXlnR, controls the expression of genes encoding xylanolytic enzymes in *Aspergillus oryzae*. Fungal Genet. Biol. 35: 157–169.

Mayfield, M.B., B.J. Godfrey and M.H. Gold. 1994. Characterization of the *mnp2* gene encoding manganese peroxidase isozyme 2 from the basidiomycete *Phanerochaete chrysosporium*. Gene 142: 231–235.

McNabb, D.S., Y. Xing and L. Guarente. 1995. Cloning of yeast HAP5: a novel subunit of a heterotrimeric complex required for CCAAT binding. Genes Dev. 9: 47–58.

Miettinen-Oinonen, A. 2004. *Trichoderma reesei* strains for production of cellulases for the textile industry. Espoo VTT Publications 550.

Montenecourt, B.B. and D.E. Eveleigh. 1979. Selective screening methods for the isolation of high yielding cellulase mutants of *Trichoderma reesei*, Adv. Chem. Ser. 181: 289–301.

Moreno-Mateos, M.A., J. Delgado-Jarana, A.C. Codon and T. Benitez. 2007. pH and Pac1 control development and antifungal activity in *Trichoderma harzianum*. Fungal Genetics and Biology 44(12): 1355–1367.

Morikawa, Y., T. Ohashi, O. Mantani and H. Okada. 1995. Cellulase induction by lactose in *Trichoderma reesei* PC-3-7, Appl. Microbiol. Biotechnol. 44: 106–111.

Murashima, K., T. Nishimura, Y. Nakamura, J. Koga, T. Moriya, N. Sumida, T. Yaguchi and T. Kono. 2002. Purification and characterization of new endo-1,4-β-D-glucanases from *Rhizopus oryzae*. Enzyme Microb. Technol. 30: 319–326.

Nahas, E., H.F. Terenzi and A. Rossi. 1982. Effect of carbon source and pH on the production and secretion of acid-phosphatase (EC3.1.3.2) and alkaline-phosphatase (EC3.1.3.1) in *Neurospora crassa*. J. Gen. Microbiol. 128: 2017–2021.

Nakari-Setälä, T., M. Paloheimo, J. Kallio, J. Vehmaanperä, M. Penttilä and M. Saloheimo. 2009. Genetic modification of carbon catabolite repression in *Trichoderma reesei* for improved protein production. Applied and Environmental Microbiology 75(14): 4853–4860.

Narendja, F.M., M.A. Davis and M.J. Hynes. 1999. AnCF, the CCAAT binding complex of *Aspergillus nidulans*, is essential for the formation of a DNase I-hypersensitive site in the 50 region of the *amdS* gene. Mol. Cell. Biol. 19: 6523–6531.

Nishimura, I., K. Okada and Y. Koyama. 1996. Cloning and expression of pyranose oxidase cDNA from *Coriolus versicolor* in *Escherichia coli*. J. Biotechnol. 52: 11–20.

Nogawa, M., M. Goto, H. Okada and Y. Morikawa. 2001. L-Sorbose induces cellulase gene transcription in the cellulolytic fungus *Trichoderma reesei*. Curr. Genet. 38: 329–334.

Nozawa, S.R., M.S. Ferreira-Nozawa, N.M. Martinez-Rossi and A. Rossi. 2003a. The pH-induced glycosylation of secreted phosphatases is mediated in *Aspergillus nidulans* by the regulatory gene pacC-dependent pathway. Fungal Genet Biol. 39: 286–295.

Nozawa, S.R., G.S. May, N.M. Martinez-Rossi, M.S. Ferreira-Nozawa, J. Coutinho-Netto, W. Maccheroni, Jr. and A. Rossi. 2003b. Mutation in a calpain-like protease affects the posttranslational mannosylation of phosphatases in *Aspergillus nidulans*. Fungal Genet. Biol. 38: 220–227.

Okada, H., K. Tada, T. Sekiya, K. Yokoyama, A. Takahashi, H. Tohda, H. Kumagai and Y. Morikawa. 1998. Molecular characterization and heterologous expression of the gene encoding a low-molecular-mass endoglucanase from *Trichoderma reesei* QM9414. Appl. Environ. Microbiol. 64: 555–563.

Olesen, J.T. and L. Guarente. 1990. The HAP2 subunit of yeast CCAAT transcriptional activator contains adjacent domains for subunit association and DNA recognition: model for the HAP2/3/4 complex. Genes Dev. 4: 1714–1729.

Orejas, M., E.A. Espeso, J. Tilburn, S. Sarkar, H.N. Arst and M.A. Peñalva. 1995. Activation of the *Aspergillus* PacC transcription factor in response to alkaline ambient pH requires proteolysis of the carboxy-terminal moiety. Genes Dev. 9: 1622–1632.

Parenicova, L., J.A.E. Benen, H.C.M. Kester and J. Visser. 2000. *pgaA* and *pgaB* encode two constitutively expressed endopolygalacturonases of *Aspergillus niger*. Biochem. J. 345: 637–644.

Penttilä, M., P. Lehtovaara, H. Nevalainen, R. Bhikhabhai and J. Knowles. 1986. Homology between cellulase genes of *Trichoderma reesei*: complete nucleotide sequence of the endoglucanase I gene. Gene 45: 253–263.

Pinkham, J.L. and L. Guarente. 1985. Cloning and molecular analysis of the HAP2 locus: a global regulator of respiratory genes in *Saccharomyces cerevisiae*. Mol. Cell. Biol. 5: 3410–3416.

Polizeli, M.L., A.C. Rizzatti, R. Monti, H.F.Terenzi, J.A. Jorge and D.S. Amorim. 2005. Xylanases from fungi: properties and industrial applications. Appl. Microbiol. Biotechnol. 67(5): 577–91.

Portnoy, T., A. Margeot, V. Seidl-Seiboth, S. Le Crom, F. Ben Chaabane, R. Linke, B. Seiboth and C.P. Kubicek. 2011. Differential regulation of the cellulase transcription factors XYR1,

ACE2, and ACE1 in *Trichoderma reesei* strains producing high and low levels of cellulase. Eukaryotic Cell 10(2): 262–271.

Rauscher, R., E. Würleitner, C. Wacenovsky, N. Aro, A.R. Stricker, S. Zeilinger, C.P. Kubicek, M. Penttilä and R.L. Mach. 2006. Transcriptional regulation of *xyn1*, encoding xylanase I, in *Hypocrea jecorina*. Eukaryot Cell 5: 447–456.

Raven, P.H., F.E. Ray and E.E. Susan . 2003. Biology of Plants. 6th edn. W.H. Freeman and Company/Worth Publishers, New York, NY.

Ruiz-Duenas, F.J., M.J. Martinez and A.T. Martinez. 1999. Molecular characterization of a novel peroxidase isolated from the ligninolytic fungus *Pleurotus eryngii*. Mol. Microbiol. 31: 223–235.

Ruttkowski, E., N.Q. Khanh, F.-J. Wientjes and M. Gottschalk. 1991. Characterisation of a polygalacturonase gene of *Aspergillus niger* RH5344. Mol. Microbiol. 5: 1353–1361.

Sakai, T., T. Sakamoto, J. Hallaert and E. Vandamme. 1993. Pectin, pectinase and protopectinase: production, properties, and applications. Adv. Appl. Microbiol. 39: 213.

Saloheimo, M., P. Lehtovaara, M. Penttilä, T.T. Teeri, J. Stahlberg, G. Johansson, G. Petterson, M. Claeyssens, P. Tomme and J. Knowles. 1988. EGIII, a new endoglucanase from *Trichoderma reesei*: the characterization of both gene and enzyme. Gene 63: 11–21.

Saloheimo, M., V. Barajas, M.L. Niku-Paavola and J.K. Knowles. 1989. A lignin peroxidase-encoding cDNA from the white rot fungus *Phlebia radiata*: characterization and expression in *Trichoderma reesei*. Gene 85: 343–351.

Saloheimo, A., B. Henrissat, A. Hoffren, O. Teleman and M. Penttilä. 1994. A novel, small endoglucanase gene, egl5, from *Trichoderma reesei* isolated by expression in yeast. Mol. Microbiol. 13: 219–228.

Saloheimo, M., T. Nakari-Setala, M. Tenkanen, M. Penttilä et al. 1997. cDNA cloning of a *Trichoderma reesei* cellulase and demonstration of endoglucanase activity by expression in yeast. Eur. J. Biochem. 249(2): 584–591.

Saloheimo, A., N. Aro, M. Ilmen and M. Penttila. 2000. Isolation of the *ace1* gene encoding a Cys(2)–His(2) transcription factor involved in regulation of activity of the cellulase promoter *cbh1* of *Trichoderma reesei*. J. Biol. Chem. 275: 5817–5825.

Saloheimo, M., J. Kuja-Panula, E. Ylösmäki, M. Ward and M. Penttilä. 2002a. Enzymatic properties and intracellulase loalization of the novel *Trichoderma reesei* β-glucosidase BGLII (Cel1A). Appl. Environ. Microbiol. 68: 4546–4553.

Saloheimo, M., M. Paloheimo, S. Hakola, J. Pere, B. Swanson, E. Nyyssönen, A. Bhatia, M. Ward and M. Penttilä. 2002b. Swollenin, a *Trichoderma reesei* protein with sequence similarity to plant expansins, exhibits disruption activity on cellulosic materials. Eur. J. Biochem. 269: 4202–4211.

Saloheimo, M., T. Pakula, M. Arvas, M. Valkonen and M. Penttilä. 2003. Cellular responses to secretion stress in *Trichoderma reesei*. XXII Fungal Genetics Conference, March 18–23, 2003, Asilomar, Abstract 422.

Scheel, T., M. Hofer, S. Ludwig and U. Holker. 2000. Differential expression of manganese peroxidase and laccase in white-rot fungi in the presence of manganese or aromatic compounds. Appl. Microbiol. Biotechnol. 54: 686–691.

Schmoll, M. 2008. The information highways of a biotechnological workhorse—signal transduction in *Hypocrea jecorina*. BMC Genomics. 20: 9: 430.

Schmoll, M. and C.P. Kubicek. 2003. Regulation of *Trichoderma* cellulase formation: lessons in moleclar biology from an industrial fungus, Acta. Microbiol. Immunol. Hung. 50: 125–145.

Schmoll, M., L. Franchi and C.P. Kubicek. 2005. Envoy, a PAS/LOV domain protein of *Hypocrea jecorina* (Anamorph *Trichoderma reesei*), modulates cellulase gene transcription in response to light. Eukaryot Cell 4: 1998–2007.

Schmoll, M., A. Schuster, R.D.N. Silva and C.P. Kubicek. 2009. The G-alpha protein GNA3 of *Hypocrea jecorina* (anamorph *Trichoderma reesei*) regulates cellulase gene expression in the presence of light. Eukaryot Cell. 8: 410–20.

Schuster, A., K.S. Bruno, J.R. Collett, S.E. Baker, B. Seiboth, C.P. Kubicek, M. Schmoll et al. 2012. A versatile toolkit for high throughput functional genomics with *Trichoderma reesei*. Biotechnology for Biofuels 5: 1.

Seibel, C., G. Gremel, S.R. do Nascimento, A. Schuster, C.P. Kubicek and M. Schmoll. 2009. Light-dependent roles of the G-protein alpha subunit GNA1 of *Hypocrea jecorina* (anamorph *Trichoderma reesei*). BMC Biol. 3, 7: 58.

Seiboth, B., R. Messner, F. Gruber and C.P. Kubicek. 1992. Disruption of the *Trichoderma reesei cbh2* gene coding for cellobiohydrolase II leads to a delay in the triggering of cellulase formation by cellulose, J. Gen. Microbiol. 138: 1259–1264.

Seiboth, B., C. Ivanova and V. Seidl-Seiboth. Trichoderma reesei: A Fungal Enzyme Producer for Cellulosic Biofuels. *In:* Marco Aurélio dos Santos Bernardes [eds.]. 2011. Biofuel Production-Recent Developments and Prospects ISBN 978-953-307-478-8. (InTech, Available from: http://www.intechopen.com/books/biofuel-production-recent-developments-and-prospects/trichoderma-reeseia-fungal-enzyme-producer-for-cellulosic-biofuels)

Seidl, V., C. Gamauf, I.S. Druzhinina, B. Seiboth, L. Hartl and C.P. Kubicek. 2008. The *Hypocrea jecorina* (*Trichoderma reesei*) hypercellulolytic mutant RUT C30 lacks a 85 kb (29 gene-encoding) region of the wild-type genome. BMC Genomics 9: 327.

Sestak, S. and V. Farkas. 1993. Metabolic regulation of endoglucanase synthesis in *Trichoderma reesei*: participation of cyclic AMP and glucose-6-phosphate, Can. J. Microbiol. 39: 342–347.

Shoemaker, S., V. Schweickart, M. Ladner, D. Gelfand, S. Kwok and K. Myambo. 1983. Molecular cloning of exo-cellobiohydrolase derived from *Trichoderma reesei* strain L27. Biotechnology 1: 691–696.

Singh, D. and S. Chen. 2008. The white-rot fungus *Phanerochaete chrysosporium*: conditions for the production of lignin-degrading enzymes. Appl Microbiol Biotechnol. 81: 399–417.

Sjöström, E. 1981. p. 49. *In:* Wood Chemistry Fundamentals and Applications. Academic Press, New York.

Smith, D.C., K.M. Bhat and T.M. Wood. 1991. Xylan-hydrolysing enzymes from thermophilic and mesophilic fungi. World J. Microbiol. Biotechnol. 7: 475–484.

Smith, M., A. Shnyreva, D.A. Wood and C.F. Thurston. 1998. Tandem organization and highly disparate expression of the two laccase genes *lcc1* and *lcc2* in the cultivated mushroom *Agaricus bisporus*. Microbiology 144: 1063–1069.

Sternberg, D. and G.R. Mandels. 1979. Induction of cellulolytic enzymes in *Trichoderma reesei* by sophorose, J. Bacteriol. 139: 761–767.

Strauss, J., R.L. Mach, S. Zeilinger, G. Hartler, G. Stoffler, M. Wolschek and C.P. Kubicek. 1995. *Cre1*, the carbon catabolite repressor protein from *Trichoderma reesei*. FEBS Lett. 376: 103–107.

Stricker, A.R., K. Grosstessner-Hain, E. Würleitner and R.L. Mach. 2006. Xyr1 (xylanase regulator 1) regulates both the hydrolytic enzyme system and D-xylose metabolism in *Hypocrea jecorina*. Eukaryot Cell 5: 2128–2137.

Stricker, A.R., R.L. Mach and L.H. de Graaff. 2008. Regulation of transcription of cellulases- and hemicellulasesencoding genes in *Aspergillus niger* and *Hypocrea jecorina* (*Trichoderma reesei*). Appl. Microbiol. Biotechnol. 78: 211–220.

Suarez, T. and M.A. Penalva. 1996. Characterization of a *Penicillium chrysogenum* gene encoding a PacC transcription factor and its binding sites in the divergent *pcbAB–pcbC* promoter of the penicillin biosynthetic cluster. Mol. Microbiol. 20: 529–540.

Subramaniyan, S. and P. Prema. 2002. Biotechnology of Microbial Xylanases: Enzymology, Molecular Biology, and Application. Critical Reviews in Biotechnology 22(1): 33–64.

Suto, M. and F. Tomita. 2001. Induction and catabolite repression mechanisms of cellulase in fungi. J. Biosci. Bioeng. 92(4): 305–11.

Suykerbuyk, M.E.G., H.C.M. Kester, P.J. Schaap, H. Stam, W. Musters and J. Visser. 1997. Cloning and characterization of two rhamnogalacturonan hydrolase genes from *Aspergillus niger*. Appl. Environ. Microbiol. 63: 2507–2515.

Takashima, S., A. Nakamura, M. Hidaka, H. Masaki and T. Uozumi. 1996. Cloning, sequencing, and expression of the cellulase genes of *Humicola grisea* var. thermoidea. J. Biotechnol. 50: 137–147.

Takashima, S., A. Nakamura, M. Hidaka, H. Masaki and T. Uozumi. 1999. Molecular cloning and expression of the novel fungal β-glucosidase genes from *Humicola grisea* and *Trichoderma reesei*. J. Biochem. 125: 728–736.

Tanaka, H., S. Itakura and A. Enoki. 1999. Hydroxyl radical generation by an extracellular low-molecular-weight substance and phenol oxidase activity during wood degradation by the white-rot basidiomycete *Trametes versicolor*. J. Biotechnol. 75: 57–70.

Teeri, T. 1997. Crystalline cellulose degradation: new insight into the function of cellobiohydrolases. Tibtech. 15: 160–167.

Teeri, T., I. Salovuori and J. Knowles. 1983. The molecular cloning of the major cellulase gene from *Trichoderma reesei*. Biotechnology 1: 696–699.

Teeri, T., P. Lehtovaara, S. Kauppinen, I. Salovuori and J. Knowles. 1987. Homologous domains in *Trichoderma reesei* cellulolytic enzymes: gene sequence and expression of cellobiohydrolase II. Gene 51: 43–52.

Temp, U., U. Zierold and C. Eggert. 1999. Cloning and characterization of a second laccase gene from the lignin-degrading basidiomycete *Pycnoporus cinnabarinus*. Gene 236: 169–177.

Tilburn, J., S. Sarkar, D.A. Widdick, E.A. Espeso, M. Orejas, J. Mungroo, M.A. Peñalva and H.N. Arst, Jr. 1995. The *Aspergillus* PacC zinc finger transcription factor mediates regulation of both acid- and alkaline-expressed genes by ambient pH. EMBO J. 14: 779–790.

Uenojo, M. and G.M. Pastore. 2007. Pectinolytic enzymes. Industrial applications and future perspectives. Quim. Nova 30(2): 388–394.

Vaheri, M., M. Leisola and V. Kauppinen. 1979. Transglycosylation products of cellulase system of *Trichoderma reesei*. Biotechnology 1: 696–699.

van den Brink, J. and R.P. de Vries. 2011. Fungal enzyme sets for plant polysaccharide degradation. Appl. Microbiol. Biotechnol. 91: 147–1492.

Vankuyk, P.A., J.A. Diderich, A.P. MacCabe, O. Hererro, G.J. Ruijter and J. Visser. 2004. *Aspergillus niger mstA* encodes a high-affinity sugar/H+ symporter which is regulated in response to extracellular pH. Biochem. J. 379: 375–383.

van Peij, N.N., J. Brinkmann, M. Vrsanska, J. Visser and L.H. de Graaff. 1997. beta-Xylosidase activity, encoded by *xlnD*, is essential for complete hydrolysis of xylan by *Aspergillus niger* but not for induction of the xylanolytic enzyme spectrum. Eur. J. Biochem. 245(1): 164–173.

van Peij, N.N.M.E., M.M.C. Gielkens, R.P. de Vries, J. Visser and L.H. de Graaff. 1998a. The transcriptional activator *xlnR* regulates both xylanolytic and endoglucanase expression in *Aspergillus niger*. Appl. Environ. Microbiol. 64: 3615–3619.

van Peij, N.NM.E., J. Visser and L.H. de Graaff. 1998b. Isolation and analysis of *xlnR*, encoding a transcriptional activator coordinating xylanolytic expression in *Aspergillus niger*. Mol. Microbiol. 27: 131–142.

Vautard-Mey, G., P. Cotton and M. Fevre. 1999. Expression and compartmentation of the glucose repressor CRE1 from the phytopathogenic fungus *Sclerotinia sclerotiorum*. Eur. J. Biochem. 266: 252–259.

Verbeke, J., P. Coutinho, H. Mathis, A. Quenot, E. Record, M. Asther and S. Heiss-Blanquet. 2009. Transcriptional profiling of cellulase and expansin-related genes in a hypercellulolytic *Trichoderma reesei*. Biotechnology Letters 31(9): 1399–1405.

Vergara, C.E. and N.C. Carpita. 2001. β-D-glycan synthases and the *CesA* gene family: lessons to be learned from the mixed-linkage (1→3,1→4)β-D-glucan synthase. Plant. Mol. Biol. 47: 145–160.

Wang, L., W. Yan, J. Chen, F. Huang and P. Gao. 2008. Function of the iron-binding chelator produced by *Coriolus versicolor* in lignin biodegradation. Sci. China. C. Life. Sci. 51: 214–221.

Ward, M., S. Wu, J. Dauberman, G. Weiss, E. Larenas, B. Bower, M. Rey, K. Clarkson and R. Bott. 1993. Cloning, sequence and preliminary structural analysis of a small high pI endoglucanase (EGIII) from *Trichoderma reesei*. pp. 153–158. *In:* P. Suominen and T. Reinikainen [eds.]. 1993. Proceedings of the Second Tricel Symposium on *Trichoderma* cellulases and other hydrolases. Espoo, vol. 8. Foundation for Biotechnical and Industrial Fermentation Research, Helsinki, Finland.

Wang, P. and D.L. Nuss. 1995. Induction of a *Cryphonectria parasitica* cellobiohydrolase I gene is suppressed by hypovirus infection and regulated by a GTP-binding-protein-linked signaling pathway involved in fungal pathogenesis. Proc. Natl. Acad. Sci. USA. 92: 11529–11533.

Wu, S.C., S. Kauffmann, A.G. Darvill and P. Albersheim. 1995. Purification, cloning and characterization of two xilanases from *Magnaporthe grisea*, the rice blast fungus. Mol. Plant. Microbe Interact. 8: 506–514.

Wurleitner, E., L. Pera, C. Wacenovsky, A. Cziferszky, S. Zeilinger, C.P. Kubicek and R.L. Mach. 2003. Transcriptional regulation of *xyn2* in *Hypocrea jecorina*. Eukaryot. Cell 2: 150–158.

Yaver, D.S., F. Xu, E.J. Golightly, K.M. Brown, S.H. Brown, M.W. Rey, P. Schneider, T. Halkier, K. Mondorf and H. Dalboge. 1996. Purification, characterization, molecular cloning, and expression of two laccase genes from the white rot basidiomycete *Trametes villosa*. Appl. Environ. Microbiol. 62: 834–841.

Zeilinger, S., R.L. Mach, M. Schindler, P. Herzog and C.P. Kubicek. 1996. Different inducibility of expression of the two xylanase genes *xyn1* and *xyn2* in *Trichoderma reesei*. J. Biol. Chem. 271: 25624–25629.

Zeilinger, S., R.L. Mach and C.P. Kubicek. 1998. Two adjacent protein binding motifs in the *cbh2* (cellobiohydrolase II-encoding) promoter of the fungus *Hypocrea jecorina* (*Trichoderma reesei*) cooperate in the induction by cellulose. J. Biol. Chem. 273: 34463–34471.

Zeilinger, S., A. Ebner, T. Marosits, R. Mach and C.P. Kubicek. 2001. The *Hypocrea jecorina* Hap 2/3/5 protein complex binds to the inverted CCAAT-box (ATTGG) within the *cbh2* (cellobiohydrolase II gene) activating element, Mol. Genet. Genomics 266: 56–63.

Zeilinger, S., M. Schmoll, M. Pail, R.L. Mach and C.P. Kubicek. 2003. Nucleosome transactions on the *Hypocrea jecorina* (*Trichoderma reesei*) cellulase promoter *cbh2* associated with cellulase induction. Mol. Genet. Genomics 270: 46–55.

8

Fungal Phytases: Production, Properties and Biotechnological Application

Alexandre Maller[a,*] and
Simone de Carvalho Peixoto-Nogueira[b]

Introduction

The use of enzymes is an old practice, considering that the date of its origin has been lost in time. Said and Pietro (2010) reported that during thousands of years, such biological catalysts have been applied as crude preparations of vegetable, animal or microbial origin, without the essential nature of an enzyme to be fully understood. The first example of the application of crude enzymatic preparations is believed to have been in the production of cheese, a process in which the catalyst used was a macerated calf's stomach. Such reports were present in the oldest civilizations, such as ancient Egypt (Said and Pietro 2010).

Later during the 19th century, the first studies on enzymes, which, infact, corresponded to studies on the industrial application of microorganisms, once that the enzymes (truly "responsible" for the process) were not isolated

Department of Biology, University of São Paulo (USP), Ribeirão Preto, SP 14040-901 Brazil.
[a]E-mail: alemaller@yahoo.com.br
[b]E-mail: peixoto_nogueira@yahoo.com.br.
*Corresponding author

from their producers. It is also pointed out that upto the referred time, nothing was known about enzymes. When their existence was "noted", separating them from the synthesizing organisms was believed to be impossible (Said and Pietro 2004).

According to Glazer and Nikado (1995), at the beginning of the 20th century, the first large-scale enzymatic production, of amylases (α-amylases and glucoamylases) took place, and these were applied in the manufacturing processes of sugar and alcohol, and were the first synthesizing catalysts in this process. This was possible because there was a better understanding on the existence of enzymes, how to produce and extract them from their sourcing organisms. Henceforth this knowledge increased and these days many other catalysts are being produced, among them phytases.

Phytases correspond to enzymes capable of hydrolyzing phytate, the main form of phosphorus storage in vegetables, predominantly found in cereal grains, legumes and oilseeds (Pandey et al. 2001, Vats and Banerjee 2004). These enzymes can be widely used in animal nutrition providing a better use of the phosphate present in feeds.

One of the earliest reports on phytases was by Suzuki et al. (1907) and reviewed by Kida (1909) in a study on the production of phosphoric acid from rice bran, but the first reports about the production of phytases from microbial origin date back to the 70's, and were sourced by a filamentous fungus called *Rhizopogonluteolus* (Theodorou 1971). Since then, many reports have been published on phytase dosage methodologies (Harland and Harland 1980, Zimmermann et al. 2002a), production (Bogar et al. 2003, Vats and Banerjee 2004, Salmon et al. 2012), purification (Pandey et al. 2001, Xu et al. 2012), studies on heterelogous expression (Promdonkoy et al. 2009, Farhat-Khemakhem et al. 2012), immobilization (Correia et al. 2008) and crystallization (Yao et al. 2012).

The possibility for the application of phytases in animal nutrition led to a greater interest in studying more about this catalyst. We present a compiled discussion on important points such as, the biochemical properties of phytic acid and phytases, enzymatic production, biotechnological application of these catalysts and future perspectives.

Biochemical Properties of Phytic Acid and Phytases

The chemical formula of phytic acid is $C_6H_6[OPO(OH)_2]$. It is constituted by an inositol molecule (hexahydroxy-cyclohexane) forming ester linkage with six inorganic phosphate molecules (Fig. 8.1). Thus phytic acid may also be called inositol hexakis-phosphate, *myo*-inositol 1-2-3-4-5-6 hexakis-phosphate, phytate or IP6 (Ali et al. 2010). Phytate is the main form of phosphorus storage in vegetables, consisting from 1 to 5 percent of the weight in cereals, legumes, oilseeds and nuts. In the forage, one-third of

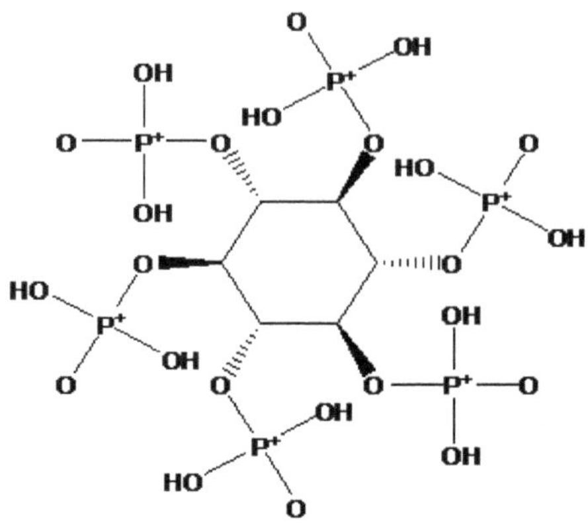

Figure 8.1 Chemical structure of phytic acid.

the phosphorus is in the form of digestible inorganic phosphorus, whereas two-thirds are in the form of organic phosphorus, as phytate. Phytate is a mixture of phytic acid salts (inositol hexaphosphoric acid) (Greiner et al. 2001, Vats and Banerjee 2005). In grains, phytic acid is found mainly in the form of sodium, calcium or potassium phytate, acting as a phosphate reserve source. The phytate molecule is distributed in different components of the grain, being mostly concentrated in the aleurone layer and in the pericarp of wheat and rice, respectively. Corn contains the highest concentration of phytic acid in its germ tube, with an approximate level of 6.5 percent (Fukuji et al. 2008).

The chemical structure of the phytic acid presents 12 dissociable hydrogen atoms in its highly ionizable orthophosphate groups, whose substitution provides the molecule with a chelating property with polyvalent metals, specially divalent and trivalent cations, such as calcium, iron, zinc and manganese, and amine groups of some basic amino-acids. Such complexes are insoluble in a neutral pH and close to the small intestine pH, characterizing phytate as an anti-nutritional factor by reducing the biological availability of minerals and proteins (Ludke et al. 2002).

However, phytic acid has been described as an anti-oxidizing agent, due to its ability to inhibit the formation of hydroxyl radical (*OH) through the formation of chelation with iron, making them catalytically inactive. This way, the addition of small amounts of phytic acid in the diet reduces the oxidative transformations, with a reduction in the degradation of ascorbic acid and a reduction in the peroxidation in chicken meat. Besides, this molecule acts synergistically with vitamin E, preventing the lipid peroxidation (Fukuji et al. 2008). Some animal studies show that phytic acid has inhibitory action of the neoplasic growth in different kinds of cancer, such as breast, colon, liver, prostate, rhabdomyosarcoma and skin. The action mechanism proposed indicates a natural increase in the natural killer cells (CDK), alteration in the signal transduction, stimulation of the cell differentiation genes and anti-oxidizing activity. Some beneficial effects of phytic acid in human health could be the decrease of cholesterol and triglyceride, and the prevention of cancer and heart diseases (Fox and Eberl 2002).

Phytases are enzymes that hydrolyze the phytate ester linkage releasing inositol and inorganic phosphate (Fig. 8.2). The International Union of Biochemistry and Molecular Biology (IUBMB), consulting IUPAC-IUB, Articulation Commission for Biochemical Naming (JCBN), listed two kinds of phytases:

3-phytase (EC 3.1.3.8): Systematic name: *myo*-inositol hexakis phosphate-3-phosphohydrolase; hydrolyzes the ester linkage in the third position of the *myo*-inositol hexakis phosphate in D-*myo*-Ins-1,2,4,5,6-pentakisphosphate and orthophosphate (Lassen et al. 2001);

6-phytase (EC 3.1.3.26): Systematic name: *myo*-inositol hexakis phosphate-6-phosphohydrolase; hydrolyzes the ester linkage in the sixth position of the *myo*-inositol hexakis phosphate in D-*myo*-Ins-1,2,3,4,5-pentakisphosphate and orthophosphate. The subsequent ester linkages in the substrate are hydrolyzed under different rates (Lassen et al. 2001).

Mullaney and Ullah (2003) and Vats and Banerjee (2005) describe three classes of phytases based on their molecular properties. Those include members of the histidine acid phosphatase (HAP), β-helix phytase (BPP) and purple acid phosphatase (PAP), besides the phytases with non-HAP active site.

Histidine acid phosphatase: Members of this group possess a motif conserved in the active site, RHGXRXP and the dipeptide catalytically active HD, which are unique for the class of acid phosphatases of high molecular mass and hydrolyze the phosphomonoester in a two-step mechanism: a

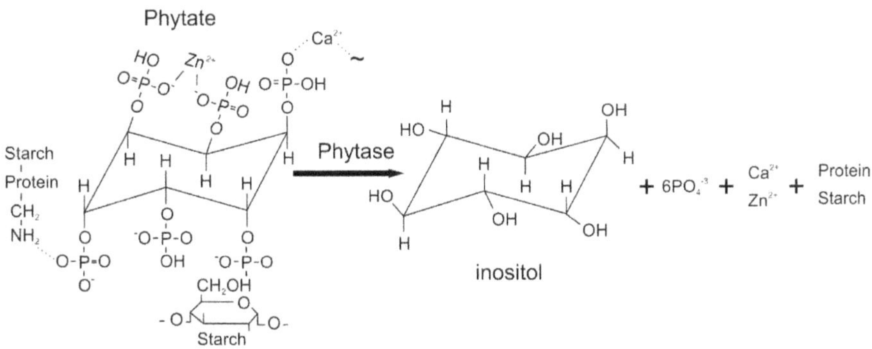

Figure 8.2 Hydrolysis of the phytate-protein-starch-ions complex through the phytase in inositol, phosphate and other elements (adapted from Yao et al. 2012).

nucleophilic attack of the histidine present in the active site of the enzyme in the phosphoester linkage of the phytic acid and protonation of the group through the residue of aspartic acid of the dipeptide HD (Lei and Porres 2003).

β-propeller phytase: A tridimensional model of these enzymes shows a basic shape resembling a six-propeller (Ha et al. 2000), whereby no other known phosphatase presents this kind of structure. Besides, the dependence of Ca^{2+} for thermostability and additional catalytic activity distinguishes this enzyme from the other subdivisions of phytases (Oh et al. 2001). BPPs possess two linkage sites to phosphate (Shin et al. 2001). During the enzymatic reaction, the hydrolysis of the substrate occurs in the "cleavage site" and in the adjacent "affinity site", which increases the linkage affinity to the substrate. The calcium ions facilitate the linkage, creating a favorable electrostatic ambient (Mullaney and Ullah 2003, Greiner et al. 2007).

Purple acid phosphatase or cysteine phytases: A representative of this class is GmPhy (EC 3.1.3.2), which possesses in the active site a purple acid phosphatase motif (PAP) (Mullaney and Ullah 2003). This phytase is characterized by seven conserved residues in the five conserved motifs—**D**X**G**, G**D**XX**Y**, G**NH**(D/E), VXX**H** and G**H**X**H**—involved in the coordination of the dimetal nuclear center (Hegeman and Grabau 2001, Li et al. 2002, Yao et al. 2012). This phytase has a low ability to degrade phytate. This can be explained as soybean germination requires a constant breakdown of phytate over several days (Mullaney and Ullah 2003). A structural model of another PAP, kidney bean purple acid phosphatase (KBPAP), is available in NCBI as number 1KBP. The amino acid sequence comparison between GmPhy and KBPAP revealed 30 percent of homology and both enzymes showed purple acid phosphatase active site motif. Searches in genomic databases reveal sequences connected to PAPs, which are present in plants, mammals, fungi and bacteria (Mullaney and Ullah 2003).

Phytases with non-HAP active site: These are enzymes shown in literature as phytases that do not possess the motif RHGXRXP and catalytic dipeptide HD, found in the active site of HAPs and, this way, possessing the different requirements for their catalytic action. However, some non-HAP phytases have contrasting biochemical characteristics. The enzyme from *Bacillus subtilis* VTT E-68013 requires Ca^{2+} for activity and stability (Kerovuo et al. 1998, Vats and Banerjee 2005), but a 3-phytase from *Klebsiella terrigena* was reported as not being metaloenzyme (Greiner et al. 1997, Vats and Banerjee 2005). Hence, characteristics of phytases with non-HAP active site are similar to PAP and further studies should be conducted to better characterize this class of phytase.

Besides, literature also reports three distinct classes of phytases based on the proteic sequences and the order of phosphate dephosphorylation (*myo*-inositol-hexaphosphate phosphohydrolase), EC 3.1.3.8, EC 3.1.3.26 and EC 3.1.3.72, happening in the positions three, six and five of phosphate, respectively. These phytases consecutively hydrolyze the inositol hexakisphosphate in much lower esters, including inositol monoesters (George et al. 2007).

The lack of extracellular phytases of vegetable origin and the relative abundance of microorganisms in the earth (Vats and Banerjee 2004) indicate that these microorganisms are the key to cycling of phosphate inositol phosphorus in soils. In fact, extracellular phytases of bacteria and fungi can enhance the absorption of phosphorus in plants when they are present in the rhizosphere (Richardson et al. 2001, Idriss et al. 2002).

Research on phytase has a great economical appeal since its discovery up to its commercialization in Europe in 1994 by Gist-Brocades, which emphasized not only its use, but also its ability to produce the enzyme economically. In the late 20th century, the annual sales of phytases as an additive of animal feed were estimated to be US$ 500 million. The recent growth in the market of phosphate additives for animal feed led to a critical commercial development of phytases, which are extensively spread throughout nature (Vats and Banerjee 2005).

Microbial Phytase Production

Great technological and commercial advances have been reached with the addition of phytases in animal feed to enhance the use of the ingredients present in the feed, aiming at better animal development and a smaller environmental contamination with phosphate residues (Haefner et al. 2005, Rao et al. 2009, Pereira et al. 2010, Slominski 2011). In literature there are several strategic studies that target obtaining organisms capable of producing significant quantities of enzymes intended for the manufacture of enzymatic products, that are commercially and economically viable, and could be used as additives in animal food to improve phosphate absorption. Among such studies, we highlight works on the biological prospects of microorganisms that produce different enzymes (Guimarães et al. 2006) and, among them, work on biological prospects of phytases (Ries and Macedo 2009).

From the data on the biological prospects of phytases, reports were found that several organisms produce high levels of phytases, such as bacteria (Zhang et al. 2011), fungi (Xu et al. 2012), yeasts (Ries and Macedo 2009) and plants (Afify et al. 2011). However the production of phytases after fermented processes with wild microorganisms has not been economically viable, either due to the low recovery of the product or to the presence of undesirable molecules, such as mycotoxins that are usually produced by several species of filamentous fungi.

Despite the fact that there are several works in literature on the production of phytases out of wild microorganisms (Nagashima et al. 1999, Mullaney et al. 2000, Bogar et al. 2003, Marounek et al. 2009, Xu et al. 2012), the necessity of producing phytases free from mycotoxins, as well as the need to obtain significant amounts of this product, studies on the heterologous expression appear as a key tool for the economically viable and large-scale production of microbial phytases (Berka et al. 1998, Rodriguez et al. 2000, Yin et al. 2007, Shah et al. 2009, Johnson et al. 2010).

The heterologous expression of phytases most commonly found in literature occurs in bacteria and yeasts. In the case of bacteria, it is usually reported that the *E. coli* phytase enhances the expression in other microorganisms such as yeast *Pichia pastoris* (Chen et al. 2004). There are also reports in literature on the heterologous expression of phytases of fungal origin (from *Aspergillus ficuum*) in *Lactobacillus casei*. In this study, the *Aspergillus ficuum* phytase was cloned in the pIAβ8 vector, followed by the transformation of *L. casei* through electroporation (Zuo et al. 2010).

The main advantage of expressing phytases in *Lactobacillus* is this microorganism is already widely used as a probiotic, helping in growth and fattening, allowing a better resistance against infectious agents, keeping the equilibrium of the intestinal microflora and stimulating the immune system of monogastric animals (Sissons 2006). Therefore the use of a probiotic that offers the above mentioned advantages, associated to a greater use of phytic acid present in food (once it expresses *A. ficuum* phytases) seems to be quite advantageous from a biotechnological point of view (Zuo et al. 2010).

Similar heterologous expression of phytases in organisms which were already commonly used as probiotics—was developed in *Bacillus subtilis* 168. In this work, the gene *phy*US417 was cloned in the pMSP3535 vector, following the transformation of cells of *B. subtilis*. The use of this mutant microorganism also has great potential due to its synergic activity along the gastrointestinal tract of monogastric animals (Farhat-Khemakhem et al. 2012).

The main advantage of using phytases as an animal supplement is the presence of phytic acid in the food, as well as the reduction in the levels of phosphorus excretion in feces (Costa et al. 2007). On the other hand, among the advantages of using heterologous expression of phytases in bacteria, is its fast growth, that results in excellent levels of enzymatic production in a short period of time. The possibility for such enzymes to be expressed by bacteria usually used as probiotics, also highlights the advantage that expression of phytases occuring in the gastrointestinal tract of monogastric animals, offers a better nutritional use of the food (due to the presence of the phytase that hydrolyzes the phytic acid), including other advantages on the use of probiotics, already widely described in literature (Sissons 2006).

Although there are a number of advantages to heterologous expression of phytases in bacteria, at times this is not the best alternative when expressing genes of eukaryotes. Even though there are several works in literature on the expression of phytases of fungi in bacteria (as is the case of heterologous expression of phytases of *Aspergillus ficuum* in *Lactobacilluscasei* —Zuo et al. 2010), in many cases, one should consider other eukaryotes as the expression system. This occurs when the enzyme to be expressed presents many post translational modifications (for example, sugar addition). In the case of heterologous expression in yeasts, generally the use of methylotrophic yeasts, i.e., organisms whose induction of the target gene (in this case, phytase) occurs in the presence of methanol, when the promoter generally used is the *AOX1* (Chen et al. 2004).

Since the beginning of the use of yeasts like *Pichia pastoris* or *Saccharomyces cerevisae* as expression systems, some advantages over the prokaryotic systems were observed: (1) as a microorganism present in foods, *S. cerevisiae* was shown to be highly acceptable for the production of proteins, showing a suitable post translational folding, decreasing the formation of aggregates, therefore rendering biologically active polypeptides; (2) both yeasts present an excellent secretion system, through which one can observe an initial purity of the product in culture media that is considerably free of proteins (Romanos et al. 1992, Kim et al. 2006).

By comparing the expression in *S. cerevisiae* with the expression in *P. pastoris*, one can see that in *Saccharomyces* there are some disadvantages over the expression in *Pichia*, for example, the generation of hyperglycosylation, that at first could represent the advantage of increasing the enzymatic thermostability; however, in some cases, it results in a decrease in the activity levels (Watanabe et al. 2009, Fugthong et al. 2010).

Since the late 80's and early 90's, literature has shown *P. pastoris* as an interesting expression system (Vedvick 1991) for the production of high levels of proteins in large-scale fermentative processes, whose synthesis of recombinant proteins presents some advantages over *S. cerevisae* or expression systems of prokaryotes.

Among these advantages, the following can be highlighted: (1) the fast growth rate that (yet lower than the bacterial growth) is still possible to render a fermentation with high cellular density; (2) high levels of productivity in culture media which are exempt from proteins, resulting in a product with a relative degree of purity, without the inconvenient cost of purification processes; (3) elimination of the problematic contamination with endotoxins and contamination through bacteriophages; (4) easy genetic handling with expression vectors already well characterized in literature, as is the case of vector pPICZαA (Xuan et al. 2009); (5) absence of known human pathogenicity; (6) correct occurrence of several post translational modifications including peptide folding, glycosylations, methylations, acylations, proteolytic adjustment and addressing to cellular compartments; (7) the capacity to modify proteins to be secreted after purifying the crude extract of the cultivation media; (8) this yeast (*P. pastoris*) can grow in a pH range of 3.3 to 7.0, a very interesting property, once that the pH of the cultivation medium can be adjusted in order to minimize the degradation of the proteins secreted (Cregg et al. 2000, Li et al. 2007).

In spite of the advantages of working with yeasts, there are still some proteins of filamentous fungi that do not present a correct processing or that are still produced with lower activity levels, even using *P. pastoris* instead of *S. cerevisae* as an expression system. The development of molecular biology techniques has opened new paths for the use of filamentous fungi as an expression system for the production of homologous and heterologous proteins (Nevalainen et al. 2005). Such studies began particularly with species of the genus *Aspergillus* (Pontecorvo et al. 1953) and, more recently, were also extended to the genus *Trichoderma* (Ramos et al. 2006), and both genus are being extensively used for large-scale production of proteins.

At first, works on the production of heterologous proteins in *Aspergillus* were initiated with *Aspergillus nidulans*, and those species have been studied since 1940 (Pontecorvo et al. 1953). Throughout the last decade, other species, especially *Aspergillus niger* (Kaplan et al. 2011), *Aspergillus awamori* (Rozhkova et al. 2011) and *Aspergillus oryzae* (Uchima et al. 2011), have been more frequently used, since high levels of recombinant proteins may be obtained with these species (Gouka et al. 1997). Another expression mode which is commonly used is *Aspergillu snidulans* A773, an auxotrophic mutant for uracil and uridine (Damásio et al. 2011).

The expression of both homologous and heterologous fungal protein in filamentous fungi is usually very efficient and the production levels are as grams per liter. Strains of *Aspergillus* and *Trichoderma* may secrete homologous proteins with an amount of upto 30 g/L in fermentation processes (e.g., cellulases, Durand et al. 1988, or glucoamylases, Finkelstein et al. 1989). As the example of a heterologous fungal protein, spartic proteinase of *Rhizomucor miehei* was secreted with a yielding of up to 3 g/L for a strain of transforming *A. oryzae* in controlled fermentation (Christensen et al. 1988).

There are still only a few reports about enzymes of certain filamentous fungi expressed in other fungi; however, such studies have been increasing, especially the heterologous expression of enzymes that hydrolyze the cellular wall (Damásio et al. 2011). There are also reports about the heterologous expression of phytases in fungi (Miettinen-Oinonen et al. 1997, Berka et al. 1998).

The greatest and most important advantage of expressing proteins of fungal origin using filamentous fungi as expression systems is that the processing of the *introns* as well as the addition of sugars during glycosylation occurs practically without differences in comparison to the wild strain (Damásio et al. 2011). In economical terms, the characteristic of each expression system varies according to the end product volume and to the process. For industrial processes involving large volumes, the enzymes must be produced in an amount of grams per liter that makes it a competitive alternative in the market. Therefore, for the production of enzymes at an effective cost in industries, there is the need of a serious choice of the best expression system (Bergquist et al. 2002).

Biotechnological Application of Phytases in Animal Nutrition

These days, it is known that the greatest technological advances in animal nutrition are the discoveries of novel ingredients associated to the adjustments of the nutritional requirements. The enhancement of the efficiency in using low-quality raw materials, as well as their wide use, will certainly produce benefits for animal feeding in the next few years, especially for the formulation of animal feeds.

A formulation means a mixture of several components. This way, the concepts of ingredient and nutrient are important. Ingredients are the source of nutrients and can be divided into protein (soybean bran, cotton and sunflower, meat meal, feather and fish, extruded soybean and milk amino-acids), energy (corn, sorghum, wheat, barley, rice, animal fat and vegetal oil) and fiber (pasture, hay, silage, wheat bran, soybean, rice hull and citrus pulp) (Portal Educação 2008).

Nutrients are the active components of the ingredients, and participate in the biochemical process of animal tissue formation, such as amino-acids, vitamins, minerals and bioactive compound. The target of formulating animal feeds is to combine the foods in a way that they provide the correct amounts of nutrients that the animal needs to grow, keep healthy, produce tissue or mate. The animals express demands of specific levels for each nutrient. Normally, one food alone is not able to meet all the demands of the animal at the same time. Hence, a food can have a lot of protein, but little energy, or a lot of vitamin but little amino-acid (Portal Educação 2008). Therefore, it is necessary to combine several foods for the animals so that the end result is a more balanced and efficient compound from a nutritional point of view.

A food additive can be defined as every substance intentionally added to the food aiming at conserving, intensifying or modifying its properties, as long as it does not comprise its nutritional value. In this way, the use of enzymes in the feeding of poultry and cattle has significantly increased in the last years. In the case of poultry, such an increase is primarily because of the high costs of the traditional raw materials and the search for alternative ingredients such as barley, oatmeal, rice, wheat and its by-products, among others. As for cattle, the use of enzymes is because of the fear of diseases like foot-and-mouth disease and the mad-cow disease, forcing importers not to buy meat that has been fed with foods containing food additive of animal origin, such as bone meal, for instance, increasing the demand for the meat of animals fed with diets which consist 100 percent of vegetables (Campestrini et al. 2005).

The enzymatic additives do not possess a direct nutritional function, but they help in the digestive process facilitating the digestibility of the nutrients present in the diet. Such additives have been incorporated into the animal foods aiming at enhancing their performance and, thus, their profitability (Araújo et al. 2008).

Enzymes have been available in the market of animal production since the 90's, with promising results. Their use has also increasingly become a routine practice with a good benefit/cost relation, as is the case of diets based on wheat and barley (Campestrini et al. 2005).

Phosphorus is an important nutrient and limited for biological systems in the environment of the Earth. Paradoxically, a lot of soils contain a relatively large amount of phosphorus that, if totally available, would exceed requirements for the growth of plants and soil microorganisms. However, much of this phosphorus is in the form which is not directly available for use. Up to 80 percent of this compound occurs in the organic form (Richardson et al. 2005), where phytate constitutes the largest fraction (50 percent) (Turner et al. 2002). This way, there is a justification for the incorporation of phytase in the sector producing poultry, swine and fish, which is to improve the availability of phosphorus, as well as minerals, amino-acids and energy.

Nevertheless, many producers supplement the feed with inorganic phosphorus, aiming at reaching the zootechnical requirements of this compound for the feeding of poultry and swine. In an excessive supplementation of phosphorus, this compound, together with the non-digestible phytate, is excreted by the animals and is then deposited on the soil as manure. The growth of cattle density in many regions in Europe resulted in an excess of manure application in the soil. This application was greater than necessary for plantations, resulting in the accumulation of phosphorus (Lan et al. 2002). Such accumulation may pose global ecological problems when this compound reaches rivers, causing the flourishing of cyanobacteria, eutrophication of surface waters, as well as the hypoxia and death of marine animals and, in the long run, phosphate leaching in deep water may also be expected (Mallin 2000, Naqvi et al. 2000).

Due to these problems, there is considerable international interest in enzymes that degrade phytate, in order to promote the biological availability of phosphorus and the reduction in the excretion of this compound in intensive cattle areas.

On the other hand, some countries have soils with low levels of phosphorus, thus requiring its import at a very high cost for feed supplementation. For example, this compound is in low supply in the semi-arid region of northeastern Brazil, which may be a result of the origin material as well as the strong interaction of the phosphate with the soil, a phenomenon known as adsorption, sorption or phosphorus fixation (Corrêa et al. 2004). Besides, studies have demonstrated that the factors influencing the adsorption of phosphorus in most of Brazil are the mineralogy grade of the clay fraction, pH, changeable aluminum and organic matter (Moreira et al. 2006).

Generally, the degradation of phytate through phytase may occur in the digestive tract and/or in the food before consumption. Some cereals like rye, wheat and barley are rich in intrinsic phytases, whereas others like corn contain little or no phytase activity. Vegetable phytase is generally active in feedings, yet its use in animal feeding is limited. This occurs because its content is highly variable in one feed. Besides, the formation of food pellets in temperatures above 70°C, results in the enzyme's partial inactivation. Additionally, the biological efficacy of the cereal phytases is only 40 percent, compared to the microbial phytase produced by species of *Aspergillus* (Zimmermann et al. 2002b).

The phytase produced by microorganisms in the digestive tract may be very efficient in degrading phytate, as was demonstrated by the almost complete availability of vegetal phosphorus in ruminants (Rodehutscord 2001). However, the microbial ecosystem in monogastric animals is located mainly in the large intestine. One can assume that most of the phosphate coming out of the phytate is not absorbed, but excreted after release by the microorganisms.

In order to reduce the excretion of phosphorus by monogastric animals, there was a need to supplement feeds with microbial phytases. Thus in the late 80's, the first phytase product became commercially available. Depending on the diet, and on the type and level of supplementation, phosphorus excretion could be reduced between 25 and 50 percent (Ravindran et al. 1999).

This supplement for non-ruminants also showed benefits for the absorption of other nutrients, such as improvement of calcium availability. Chickens with suitable phosphorus supplementation had an increase in the calcium, based not only on one direct effect (that is, the calcium present in the phytate complex), but also on one indirect effect, highlighting the benefits of that compound. There was also an increase in the availability of magnesium and several traces of elements like zinc, copper, iron and manganese (Schoener and Hoppe 2002).

Besides improving the availability of minerals and traces of elements, the microbial phytase is also capable of increasing protein digestibility. That was demonstrated in swines, chickens, egg-laying hens, turkeys and ducks. The benefits of microbial phytase under proteins and amino-acids may be explained by the degradation of protein-phytate complex and protein-mineral-phytate. Protein-phytate complexes can be formed in the intestine after feeding, as well as complexes between phytate and supplementary free amino-acids, that could be prevented, in parts, with the addition of phytase. Besides, phytate is known for inhibiting the activity of proteolytic enzymes, diminishing the protein digestion even more (Petersen 2009).

Phytate also possesses the capacity to bind to starch, inhibiting the action of amylases. This way, the phytase activity would act in the degradation of the phytate-starch complex, enabling the starch to be cleaved by amylases. Hence, starch breakdown would lead to the release of glucose and consequently the use of energy by monogastric animals. As a consequence, phytase activity would have an indirect effect in the increase of energy for monogastric animals (Ravindran et al. 1999).

The world animal feed production in 2005 reached 626 million tons, that is 2 percent higher than the preceding year. Thanks to the positive role of the more representative countries. The 10 main world producers in 2005 were: USA, European Union, China, Brazil, Mexico, Japan, Canada, Russia, South Korea and India. Brazil had the highest relative increase, which was 30.1 percent between 2000 and 2005 and 7.8 percent in 2005, compared to the preceding year (Nogueira 2006).

The phosphorus source most used by the producers to complement animal feed is bi-calcium phosphate. These days, the cost of the phosphate is around R$ 0.90/kg. The use of 500 units of phytase reduces the inclusion of 6 to 7 kg of phosphate/ton of feed. At present with the cost of phytase at R$ 2.00 per ton of feed, there is the saving of R$ 6.30 of phosphate. To this difference, the value of another raw material should be added, in this case, corn, which would cost R$ 2.80. The amount saved with the supplementation is 3 percent using phytase, besides the additional benefit in the animals' performance enhancement, available at http://www.portaldoagronegocio.com.br/conteudo.php?id=32848. In this way, research on phytases produced by microorganisms is fully justifiable.

Due to such environmental and economic reasons, phytases produced from microorganisms are increasingly attracting commercial interest.

Future Perspectives and Conclusions

Historically, in order to obtain phytases in the industry, microorganisms are used more than vegetable culture tissues and they can be produced either in submerged or solid-state fermentation cultures using agroindustrial residues (wheat straw, sugar cane bagasse, etc.) as carbon and phosphate sources. This enzyme has been widely researched by the biotechnological industry as there are many reports about its importance as a food additive for monogastric animals, especially pigs and poultry. The addition of phytase in these animals' diets may provide an improvement of the phosphate absorption reducing the need of phosphorus supplementation (which is very expensive) and also reducing phosphorus in feces secretion (which may cause soil pollution and, consequently, water contamination).

These days, attention is focused on studies on increasing phytase production and, in this scenario; studies on the heterologous expression of these enzymes in bacteria, yeasts and fungi are highlighted. With the aim of improving the production of enzymes, the heterologous expression of these enzymes in bacteria from the genus *Lactobacillus*, commonly used as prebiotics providing high animal growth, muscle mass gain and greater resistance to diseases added by better phosphorus absorption because of the phytase synthesized by the transformed bacteria have been investigated a lot. Therefore, providing phytase synthesis with the same microorganism used as a prebiotic may result in a product with better efficiency and at a lower cost.

Another important focus of the study has been the immobilization of phytase in order to increase its stability. But the greatest emphasis has been given to studies involving the immobilization of phytase in bacteria of the genus *Lactobacillus*, again due to its wide use as a prebiotic. However, further research is necessary to improve the efficacy of the phytase supplementation in animal diets, which will enable its large biotechnological application.

References

Afify, A.E.M.R., H.S. El-Beltagi, S.M. El-Salam and A.A. Omran. 2011. Bioavailability of iron, zinc, phytate and phytase activity during soaking and germination of white sorghum varieties. PLoS One 6(10): 505–512.

Ali, M., M.N. Shuja, M. Zahoor and I. Qadri. 2010. Phytic acid: How far have we come? Afr. J. Biotechnol. 9: 1551–1554.

Anonymous. 2009. Utilização da fitase nas rações para suínos garante benefícios ao animal e meio ambiente. Accessed on December 02, 2009 at http://www.portaldoagronegocio.com.br/conteudo.php?id=32848.

Araujo, D.M., J.H.V. Silva, E.C. Miranda, J.A. Araujo, F.G.P. Costa and E.N.M. Teixeira. 2008. Farelo de trigo e complexo enzimático na alimentação de poedeiras semipesadas na fase de produção. Rev. Bras. Zoot. 37: 843–848.

Bergquist, P., V. Te'o, M. Gibbs, A. Cziferszky, F.P. Faria, M. Azevedo and H. Nevalainen. 2002. Expression of xylanase enzymes from thermophilic microorganisms in fungal hosts. Extremophiles 6: 177–184.

Berka, R.M., M.W. Rey, K.M. Brown, T. Byun and A.V. Klotz. 1998. Molecular characterization and expression of a phytase gene from the thermophilic fungus *Thermomyces lanuginosus*. Appl. Environm. Microbiol. 64(11): 4423–4427.

Bogar, B., G. Szakacs, A. Pandey, S. Abdulhameed, J.C. Linden and R.P. Tengerdy. 2003a. Production of Phytase by *Mucorracemosus* in Solid-State Fermentation. Biotechnol. Prog. 19: 312–319.

Bogar, B., G. Szakacs, J.C. Linden, A. Pandey and R.P. Tengerdy. 2003b. Optimization of phytase production by solid substrate fermentation. J. Ind. Microbiol. Biotechnol. 30: 183–189.

Campestrini, E., V.T.M. Silva and M.D. Appelt. 2005. Utilização de enzimas na alimentação animal. Rev. Elet. Nutritime. 2: 254–267.

Chen, C.C., P.H. Wu and C.T. Huang. 2004. A *Pichia pastoris* fermentation strategy for enhancing the heterologous expression of an *Escherichia coli* phytase. Enz. Microbial. Technol. 35(4): 315–320.

Christensen, T., H. Woeldike, E. Boel, S.B. Mortensen, K. Hjortshoej, L. Thim and M.T. Hansen. 1988. High level expression of recombinant genes in *A. oryzae*. Biotechnol. 6: 1419–1422.

Corrêa, J.C., M. Mauad and C.A. Rosolem. 2004. Fósforo no solo e desenvolvimento de soja influenciados pela adubação fosfatada e cobertura vegetal. Pesquisa Agropecuária Brasileira. 39: 1231–1237.

Correia, I., S. Aksu, P. Ada, J.C. Pessoa, R.A. Sheldon and I.W.C.E. Arends. 2008. Vanadate substituted phytase: Immobilization, structural characterization and performance for sulfoxidations. J. Inorg. Biochem. 102: 318–329.

Costa, F.G.P., P.A. Brandão, J.S. Brandão and J.H.V. Silva. 2007. Efeito da enzima fitase nas rações de frangos de corte, durante as fases pré-inicial e inicial. Ciênc. Agrotec. 31(3): 865–870.

Cregg, J.M., J.L. Cerrghino, J. Shi and D.R. Higgins. 2000. Recombinant protein expression in *Pichia pastoris*. Mol. Biotechnol. 16(1): 23–52.

Damásio, A.R.L., T.M. Silva, F.B.R. Almeida, F.M. Squina, D.A. Ribeiro, A.F.P. Leme, F. Segato, R. Prade, J.A. Jorge, H.F. Terenzi and M.L.T.M. Polizeli. 2011. Heterologous expression of an *Aspergillus niveus* xylanase GH11 in *Aspergillus nidulans* and its characterization and application. Process Biochem. 46: 1236–1242.

Durand, H., M. Clanet and G. Tiraby. 1988. Genetic improvement of *Trichoderma reesei* for large scale cellulase production. Enzyme Microb. Technol. 10: 341–345.

Farhat-Khemakhem, A., M. Ben Farhat, I. Boukhris, W. Bejar, K. Bouchaala, R. Kammoun, E. Maguin, S. Bejar and H. Chouayekh. 2012. Heterologous expression and optimization using experimental designs allowed highly efficient production of the PHY US417 phytase in *Bacillus subtilis* 168. AMB Express 2(1): 10.

Finkelstein, D.B., J. Rambosek, M.S. Crawford, C.L. Soliday, P.C. McAda and J. Leach. Protein secretion in *Aspergillus niger*. pp. 295–300. *In:* C.L. Hershberger, S.W. Queener and G. Hegeman [eds.]. 1989. Genetics and Molecular Biology of Industrial Microorganisms American Society for Microbiology, Washington, DC, USA.

Fox, C.H. and M. Eberl. 2002. Phytic acid (IP6), novel broad spectrum anti-neoplastic agent: a systematic review. Complement. Ther. Med. 10: 229–234.

Fugthong, A., K. Boonyapakron, W. Sornlek, S. Tanapongpipat, L. Eurwilaichitr and K. Pootanakit. 2010. Biochemical characterization and *in vitro* digestibility assay of *Eupenicillium parvum* (BCC17694) phytase expressed in *Pichia pastoris*. Protein Expr. Purif. 19: 818–856.

Fukuji, T.S., D.L. Ferreira, A.L. Soares, C.E.C. Prete and E.I. Ida. 2008. Ácido fítico de híbridos de milho e alguns produtos industrializados. Acta Sci. Agron. 30: 31–35.

George, T.S., R.J. Simpson, P.J. Gregory and A.E. Richardson. 2007. Differential interaction of *Aspergillus niger* and *Peniophora lycii* phytases with soil particles affects the hydrolysis of inositol phosphates. Soil Biol. Biochem. 39: 793–803.

Glazer, A.N. and H. Nikaido. 1995. Microbiol Biotechnology, W.H. Freeman and Company, New York. Gouka, R.J., P.J. Punt and C.A.M.J.J. van den Hondel. 1997. Efficient production of secreted proteins by *Aspergillus*: progress, limitations and prospects. Appl. Microbiol. Biotechnol. 47: 1–11.

Greiner, R., E. Haller, U. Konietzny and K.D. Jany. 1997. Purification and characterization of a phytase from *Klebsiella terrigena*. Arch. Biochem. Biophys. 341: 201–206.

Greiner, R., N. Carlsson and M.L. Alminger. 2001. Stereospecificity of *myo*-inositol hexakisphosphatedephosphorylation by a phytate-degrading enzyme of *Escherichia coli*. J. Biotech. 84: 53–62.

Greiner, R., B.L. Lim, C.W. Cheng and N.G. Carlsson. 2007. Pathway of phytate dephosphorylation by beta-propeller phytases of different origins. Can. J. Microbiol. 53: 488–495.

Guimarães, L.H.S., S.C. Peixoto-Nogueira, M. Michelin, A.C.S. Rizzatti, V.C. Sandrim, F.F. Zanoelo, A.C.M.M. Aquino, A.B. Junior and M.L.T.M. Polizeli. 2006. Screening of filamentous fungi for production of enzymes of biotechnological interest. Braz. J. Microbiol. 37: 474–480.

Ha, N.C., B.C. Oh, S. Shin, H.J. Kim, T.K. Oh, Y.O. Kim, K.Y. Choi and B.H. Oh. 2000. Crystal structures of a novel, thermostable phytase in partially and fully calcium-loaded states. Nat. Struct. Biol. 7: 147–153.

Haefner, S., A. Knietsch, E. Scholten, J. Braun, M. Lohscheidt and O. Zelder. 2005. Biotechnological production and applications of phytases. Appl. Microbiol. Biotechnol. 68(5): 588–597.

Harland, B.F. and J. Harland. 1980. Fermentative reduction of phytate rye, white and whole breads. Cereal Chem. 57(3): 226–229.

Hegeman, C.E. and E.A. Grabau. 2001. A novel phytase with sequence similarity to purple acid phosphatases is expressed in cotyledons of germinating soybean seedlings. Plant Physiol. 126: 1598–1608.

Idriss, E.E., O. Makarewicz, A. Farouk, K. Rosner, R. Greiner, H. Bochow, T. Richter and R. Borriss. 2002. Extracellular phytase activity of *Bacillus amyloliquefaciens* FZB45 contributes to its plant-growth-promoting effect. Microbiology 148: 2097–2109.

Johnson, S.C., M. Yang and P.P. Murthy. 2010. Heterologous expression and functional characterization of a plant alkaline phytase in *Pichia pastoris*. Protein Expr. Purif. 74(2): 196–203.

Kaplan, O., K. Bezouška, O. Plíhal, R. Ettrich, N. Kulik, O. Vaněk, D. Kavan, O. Benada, A. Malandra, O. Sveda, A.B. Veselá, A. Rinágelová, K. Slámová, M. Cantarella, J. Felsberg, J. Dušková, J. Dohnálek, M. Kotik, V. Křen and L. Martínková. 2011. Heterologous expression, purification and characterization of nitrilase from *Aspergillus niger* K10. BMC Biotechnol. 6: 11–12.

Kerovuo, J., M. Lauraeus, P. Nurminen, N. Kalkkinen and J. Apajalahti. 1998. Isolation, characterization, molecular gene cloning and sequencing of a novel phytase from *Bacillus subtilis*. Appl. Environ. Micro. 64: 2079–2085.

Kida, Y. 1909. Influence of Rice Bran upon the Manurial Value of Phosphoric Acid Contained in Oil Cakes. Jour. Coll. Agric. 1: 367–379.

Kim, T., E.J. Mullaney, J.M. Porres, K.R. Roneker, S. Crowe, S. Rice, T. Ko, A.H. Ullah, C.B. Daly, R. Welch and X.G. Lei. 2006. Shifting the pH profile of *Aspergillus niger* PhyA phytase to match the stomach pH enhances its effectiveness as an animal feed additive. Appl. Environ. Microbiol. 72(6): 4397–4403.

Lan, G.Q., N. Abdullah, S. Jalaludin and Y.W. Ho. 2002. Optimization of carbon and nitrogen sources for phytase production by *Mitsuokella jalaludinii*, a new rumen bacterial species. Lett. Appl. Microbiol. 35: 157–161.

Lassen, S.F., J. Breinholt, P.R. Ostergaard, R. Brugger, A. Bischoff, M. Wyss and C.C. Fuglsang. 2001. Expression, gene cloning, and characterization of five novel phytases from four basidiomycete fungi: *Peniophoralycii*, *Agrocybepediades*, a *Ceriporia* sp., and *Trametespubescens*. Appl. Environ. Microbiol. 67: 4701–4707.

Lei, X.G. and J.M. Porres. 2003. Phytase enzymology, applications, and biotechnology. Biotechnol. Lett. 25: 1787–1794.

Li, D., H. Zhu, K. Liu, X. Liu, G. Leggewie, M. Udvardi and D. Wang. 2002. Purple acid phosphatases of *Arabidopsis thaliana* comparative analysis and differential regulation by phosphate deprivation. J. Biol. Chem. 277: 27772–27781.

Li, P., A. Anumanthan, X.G. Gao, K. Ilangovan, V.V. Suzara, N. Duzgunes and V. Renugopalakrishnan. 2007. Expression of recombinant proteins in *Pichia pastoris*. Appl. Biochem. Biotechnol. 142(2): 105–124.

Ludke, M.C.M.M., J. López and J.V. Ludke. 2002. Fitase em dietas para suínos em crescimento: (I) impacto ambiental. Ciência Rural 32: 97–102.

Mallin, M.A. 2000. Impacts of industrial animal production on rivers and estuaries. Am. Sci. 88: 26–37.

Marounek, M., N. Brenová, O. Suchorská and J. Mrázek. 2009. Phytase Activity in Rabbit Cecal Bacteria. Folia Microbiol. 54 (2): 111–114.

Miettinen-Oinonen, A., T. Torkkeli, M. Paloheimo and H. Nevalainen. 1997. Over expression of the *Aspergillus niger* pH 2.5 acid phosphatase gene in a heterologous host *Trichoderma reesei*. J. Biotechnol. 58(1): 13–20.

Moreira, F.L.M., F.O.B. Mota, C.A. Clemente, B.M. Azevedo and G.V. Bomfim. 2006. Adsorção de fósforo em solos do Estado do Ceará. Revista Ciência Agronômica 37: 7–12.

Mullaney, E.J. and A.H. Ullah. 2003. The term phytase comprises several different classes of enzymes. Biochem. Biophys. Res. Commun. 312: 179–184.

Mullaney, E.J., C.B. Daly, K. Sethumadhavan, E. Rodriquez, X.G. Lei and A.H.J. Ullah. 2000. Phytase Activity in *Aspergillus fumigatus* Isolates. Biochem. Bioph. Res. Comm. 275: 759–763.

Nagashima, T., T. Tange and H. Anazawa. 1999. Dephosphorylation of phytate by using the *Aspergillus niger* phytase with a high affinity for phytate. Appl. Environm. Microbiol. 65(10): 4682–4684.

Naqvi, S.W.A., D.A. Jayakumar, P.V. Narvekar, H. Naik, V.V.S.S. Sarma and W. D'Souza. 2000. Increased marine production of N2O due to intensifying anoxia on the Indian continental shelf. Nature 408: 346–349.

Nevalainen, K.M.H., V.S.J. Te'o and P.L. Bergquist. 2005. Heterologous protein expression in filamentous fungi. Trends Biotechnol. 23(9): 468–474.

Nogueira, S.J. 2006. Mercado mundial de produtos para alimentação animal e a inserção do Brasil. Informações Econômicas 36.

Oh, B.C., B.S. Chang, K.H. Park, N.C. Ha, H.K. Kim, B.H. Oh and T.K. Oh. 2001. Calcium-dependent catalytic activity of a novel phytase from *Bacillus amyloliquefaciens* DS11. Biochemistry 40: 9669–9676.

Pandey, A., G. Szakacs, C.R. Soccol, J.A. Rodriguez-Leon and V.T. Soccol. 2001. Production, purification and properties of microbial phytases. Bioress. Technol. 77: 203–214.

Pereira, A.A., O.M. Junqueira, J.C. Alva, S. Sgavioli, M.F.F.M. Praes and D.N. Júnior. 2010. Utilização de rações de poedeiras comerciais formuladas com fitase e níveis de proteína bruta sobre a excreção de fósforo, nitrogênio e cálcio. Ars. Veterin. 26(3): 178–183.

Petersen, S.T. 2009. The potential ability of swine nutrition to influence environmental factors positively. J. Anim. Sci. 88: E95–E101.

Pontecorvo, G., J.A. Roper, L.M. Hemmons, K.D. MacDonald and A.W.J. Bufton. 1953. The genetics of *Aspergillus nidulans*. Adv. Gene 5: 141–239.

Portal Educação. 2008. Conceitos para formulação de ração: ingredientes e nutrientes. Accessed on September 23, 2011 at http://www.portaleducacao.com.br/veterinaria/artigos/2510/conceitos-para-formulacao-de-racao-ingredientes-e-nutrientes.

Promdonkoy, P., K. Tang, W. Sornlake, P. Harnpicharnchai, R.S. Kobayashi, V. Ruanglek, T. Upathanpreecha, M. Vesaratchavest, L. Eurwilaichitr and S. Tanapongpipat. 2009. Expressionand characterization of *Aspergillus thermostable* phytases in *Pichia pastoris*. FEMS Microbiol. Lett. 290: 18–24.

Ramos, A.S., F.S. Chambergo, E.D. Bonaccorsi, A.J. Ferreira, N. Cella, A.K. Gombert, A. Tonso and H. El-Dorry. 2006. Oxygen- and glucose-dependent expression of Trhxt1, a putative glucose transporter gene of Trichoderma reesei. Biochem. 45 (26): 8184–8192.

Rao, D.E., K.V. Rao, T.P. Reddy and V.D. Reddy. 2009. Molecular characterization, physicochemical properties, known and potential applications of phytases: An overview. Crit. Rev. Biotechnol. 29(2): 182–198.

Ravindran, V., S. Cabahug, W.L. Bryden and P.H. Selle. The influence of microbial phytase on the bioavailability of protein and energy in broiler chickens. pp. 573–584. *In*: M.B. Coelho and E.T. Kornegay [eds.]. 1999. Phytase in animal nutrition and waste management. BASF, Mexico.

Richardson, A.E., P.A. Hadobas, J.E. Hayes, C.P. O'Hara and R.J. Simpson. 2001. Utilization of phosphorus by pasture plants supplied with *myo*-inositol hexaphosphate is enhanced by the presence of soil microorganisms. Plant and Soil. 229: 47–56.

Richardson, A.E., T.S. George, M. Hens and R.J. Simpson. Utilization of soil organic phosphorus by higher plants. pp. 165–184. *In*: B.L. Turner, E. Frossard and D. Baldwin [eds.]. 2005. Organic Phosphorus in the Environment. CABI Publishing, Wallingford.

Ries, E.F. and G.A. Macedo. 2009. Progressive screening of thermostable yeasts for phytase production. Food Sci. Biotechnol. 18(3): 655–660.

Rodehutscord, M. 2001. Current phosphorus evaluation systems for livestock in Germany. Lohmann-Inf 25: 1–8.

Rodriguez, E., E.J. Mullaney and X.G. Lei. 2000. Expression of the *Aspergillus fumigatus* phytase gene in *Pichia pastoris* and characterization of the recombinant enzyme. Biochem. Bioph. Res. Comm. 268: 373–378.

Romanos, M.A., C.A. Scorer and J.J. Clare. 1992. Foreign gene expression in yeast: a review. Yeast, 8(6): 423–88.

Rozhkova, A.M., A.S. Sereda, N.V. Tsurikova, A.K. Nurtaeva, M.V. Semenova, L.V. Rimareva, E.A. Rubtsova, I.N. Zorov, O.A. Sinitsyna and A.P. Sinitsyn. 2011. Creation of a heterologous gene expression system on the basis of *Aspergillus awamori* recombinant strain. Prikl. Biokhim. Mikrobiol. 47(3): 308–3017.

Said, S. and R. Pietro. 2004. Enzimas como agents biotecnológicos. Primeira edição. Ed. Legis Summa. Ribeirão Preto, SP, Brazil.

Said, S. and R. Pietro. 2010. Enzimas de interesse industrial e biotecnológico. Segunda edição. Ed. Eventos

Salmon, D.N., L.C. Piva, R.L. Binati, C. Rodrigues, L.P. Vandenberghe, C.R. Soccol and M.R. Spier. 2012. A bioprocess for the production of phytase from *Schizophyllum commune*: studies of its optimization, profile of fermentation parameters, characterization and stability. Bioprocess Biosyst. Eng. 35: 1067–1079.

Schoener, F.J. and P.P. Hoppe. The effects of phytase in poultry nutrition. *In*: J.M. Mcnab and K.N. Boorman [eds.]. 2002. Poultry feedstuffs: supply, composition and nutritive value. CAB Publishing, Wallingford.

Shah, P., K. Bhavsar, S.K. Soni and J.M. Khire. 2009. Strain improvement and up scaling of phytase production by *Aspergillus niger* NCIM 563 under submerged fermentation conditions. J. Ind. Microbiol. Biotechnol. 36: 373–380.

Shin, S., N.C. Ha, B.C. Oh, T.K. Oh and B.H. Oh. 2001. Enzyme mechanism and catalytic property of beta propeller phytase. Structure 9: 851–858.

Sissons, J.W. 2006. Potential of probiotic organisms to prevent diarrhoea and promote digestion in farm animals—A review. J. Sci. Food Agric. 49(1): 1–13.

Slominski, B.A. 2011. Recent advances in research on enzymes for poultry diets. Poultry Sci. 90: 2013–2023.

Suzuki, U., K. Yoshimura and M. Takaishi. 1907. Ueber ein enzym "Phytase" das "Anhydro-oxy-methylen diphosphorsaure" spaltet. Tokyo Imper. Univ. Coll. Agric. Bu. 7: 503–512.

Theodorou, C. 1971. The phytase activity of the mycorrhizal fungus *Rhizopogon luteolus*. Soil Biol. Biochem. 3: 89–90.

Turner, B.L., M.J. Paphazy, P.M. Haygarth and I.D. McKelvie. 2002. Inositol phosphates in the environment. Philos. Trans. R. Soc. Lond. 357: 449–469.

Uchima, C.A., G. Tokuda, H. Watanabe, K. Kitamoto and M. Arioka. 2011. Heterologous expression and characterization of a glucose-stimulated β-glucosidase from the termite Neotermes koshunensis in *Aspergillus oryzae*. Appl. Microbiol. Biotechnol. 89(6): 1761–1771.

Vats, P. and U.C. Banerjee. 2004. Production studies and catalytic properties of phytases (*myo*-inositol hexakis phosphate phosphohydrolases): an overview. Enzyme Microb. Technol. 35: 3–14.

Vats, P. and U.C. Banerjee. 2005. Biochemical characterisation of extracellular phytase (*myo*-inositol hexakisphosphatephosphohydrolase) from a hyper-producing strain of *Aspergillus niger* van Teighem. J. Ind. Microbiol. Biotechnol. 32: 141–147.

Vedvick, T.S. 1991. Gene expression in yeast: *Pichiapastoris*. Curr. Opinion Biotechnol. 2(5): 742–745.

Watanabe, T., H. Ikeda, K. Masaki, T. Fujii and H. Iefuji. 2009. Cloning and characterization of a novel phytase from wastewater treatment yeast *Hansenula fabianii* J640 and expression in *Pichia pastoris*. J. Biosc. Bioeng. 108(3): 225–230.

Xu, L., G. Zhang, H. Wang and T.B. Ng. 2012. Purification and characterization of phytase with a wide pH adaptation from common edible mushroom *Volvariella volvacea* (Straw mushroom). Indian J. Biochem Biophys. 49(1): 49–54.

Xuan, N.T., M.T. Hang and V.N. Thanh. 2009. Cloning and over expression of an *Aspergillus niger* XP phytase Gene (phyA) in *Pichia pastoris*. World Acad. Sci. Eng. Technol. 56: 750–753.

Yao, M.Z., Y.H. Zhag, W.L. Lu, M.Q. Hu, W. Wang and A.H. Liang. 2012. Phytases: crystal structures, protein engineering and potential biotechnological applications. J. Appl. Microbiol. 112(1): 1–14.

Yin, Q.Q., Q.H. Zheng and X.T. Kang. 2007. Biochemical characteristics of phytases from fungi and the transformed microorganism. Anim. Feed Sci. Technol. 132: 341–350.

Zhang, R., P. Yang, H. Huang, T. Yuan, P. Shi, K. Meng and B. Yao. 2011. Molecular and biochemical characterization of a new alkaline β-propeller phytasefrom the insect symbiotic bacterium *Janthinobacterium* sp. TN115.Appl. Microbiol. Biotechnol. 92(2): 317–25.

Zimmermann, B., H.J. Lantzsch, U. Langbein and W. Drochner. 2002a. Determination of phytase activity in cereal grains by direct Incubation. J. Anim. Physiol. Anim. Nutr. 86: 347–352.

Zimmermann, B., H.J. Lantzsch, R. Mosenthin, F.J. Schoener, H.K. Biesalski and W. Drochner. 2002b. Comparative evaluation of the efficacy of cereal and microbial phytases in growing pigs fed diets with marginal phosphorus supply. J. Sci. Food Agric. 82: 1298–1304.

Zuo, R., J. Chang, Q. Yin, L. Chen, Q. Chen, X. Yang, Q. Zheng, G. Ren and H. Feng. 2010. Phytase gene expression in *Lactobacillus* and analysis of its biochemical characteristics. Microbiol. Res. 165(4): 329–335.

9

Mannan-Degrading Enzyme System

Andrezza Furquim da Cruz

Introduction

The use of enzymes has been exploited by humans for thousands of years in a direct way via the use of crude enzyme preparations derived from animals or vegetables, or in an indirect way via the use of enzymatic action resulting in microbial growth on certain substrates (Neidleman 1992).

Enzymes with potential biotechnological applications can originate from plants (papain, bromelain, α-amylase and β-amylase), animal (pepsin, pancreatin, lipases and esterases) and microbes (xylanase, glucose isomerase, glucose oxidase, catalase, α-amylase, glucoamylase, ciclomaltodextrin-D-glycosiltransferase (CGTases), pectinases, cellulases, lipases, proteases, pullulanase, uricase, phosphatases and lignases (Said and Pietro 2002).

Hemicellulose is the second most abundant heteropolymer in nature. It is present in cell walls of plants and is closely linked to cellulose and lignin forming lignocellulosic biomass (Harris and Stone 2008). It makes up to 25–30 percent of the total wood dry weight. The two most important hemicellulases are hetero-1,4-β-D-xylan and hetero-1,4-β-D-mannans

Departamento de Biologia da Faculdade de Filosofia, Ciências e Letras de Ribeirão Preto, Universidade de São Paulo. Av. Bandeirantes, 3900 Monte Alegre, Ribeirão Preto, São Paulo, Brasil.
E-mail: andrezzafcruz@gmail.com

(Chauhan et al. 2012). The xylans are widely found in grasses and hardwoods, while the mannans are found in softwood and specialized structures such as seeds and fruits (Scheller and Ulvskov 2010, Chauhan et al. 2012).

Due to the structural complexity of hemicellulose, several enzyme systems are required for effective breakage of the cell wall. Hemicellulases are classified in accordance with actions carried out on a separate substrate. Thus, they can be classified as arabinoxylans, acetilglucuronoxylan, xylan, xiloglucan, arabinan, arabinogalactans, galactomannans, glucomannans and galactoglucomannans (Moreira et al. 2011). After xylanases, mannanases are the second most important group of enzymes on hemicellulose hydrolysis. Hence, the mannan-degrading enzymes play an important role in supplying energy to the organisms which produce these enzymes (Moreira and Filho 2008). Furthermore, the hydrolysis of β-1,4-D-mannopyranoside into β-1,4-mannan by these enzymes has allowed their application in pharmaceuticals, food, feed, oil, textiles and pulp and paper industries (Coughlan et al. 1993, Chauhan et al. 2012).

The aim of this chapter is present an overview of the mannan structure and mannan degrading enzymes, as well as, the microorganisms which produces these enzymes. The chapter also discusses some bacterial and fungal cloning and heterologous expression systems that can levels up the enzyme production to grams per liter with high purity, ideal for industrial enzyme production, and many other industrial applications.

Mannan Structure

Mannans and heteromannans are complex biopolymers closely related to cellulose and lignin (Puls and Schuseil 1993) and widely distributed in nature as part of the hemicellulose fraction of hardwoods and softwoods, seeds of leguminous plants, and beans (Capek et al. 2000, Handford et al. in 2003, Buckeridge et al. 2000, Dhawan and Kaur 2007).

Mannans are divided into two principal groups according to their structure, depending on whether the β-1,4-linked backbone contains only D-mannose residues (mannans) or mannose and D-glucose residues (glucomannans). These groups can be subdivided according to the number of α-1,6 galactose linked on the side groups. If there is more than 5 percent (w/w) of D-galactose in the linear backbone chain, they are denominated galactomannans (Fig. 9.1). Sometimes, mannose and glucose on the backbone chain are acetylated at C-2 or C-3 (Moreira and Filho 2008, Van Zyl et al. 2010, Chauhan et al. 2012).

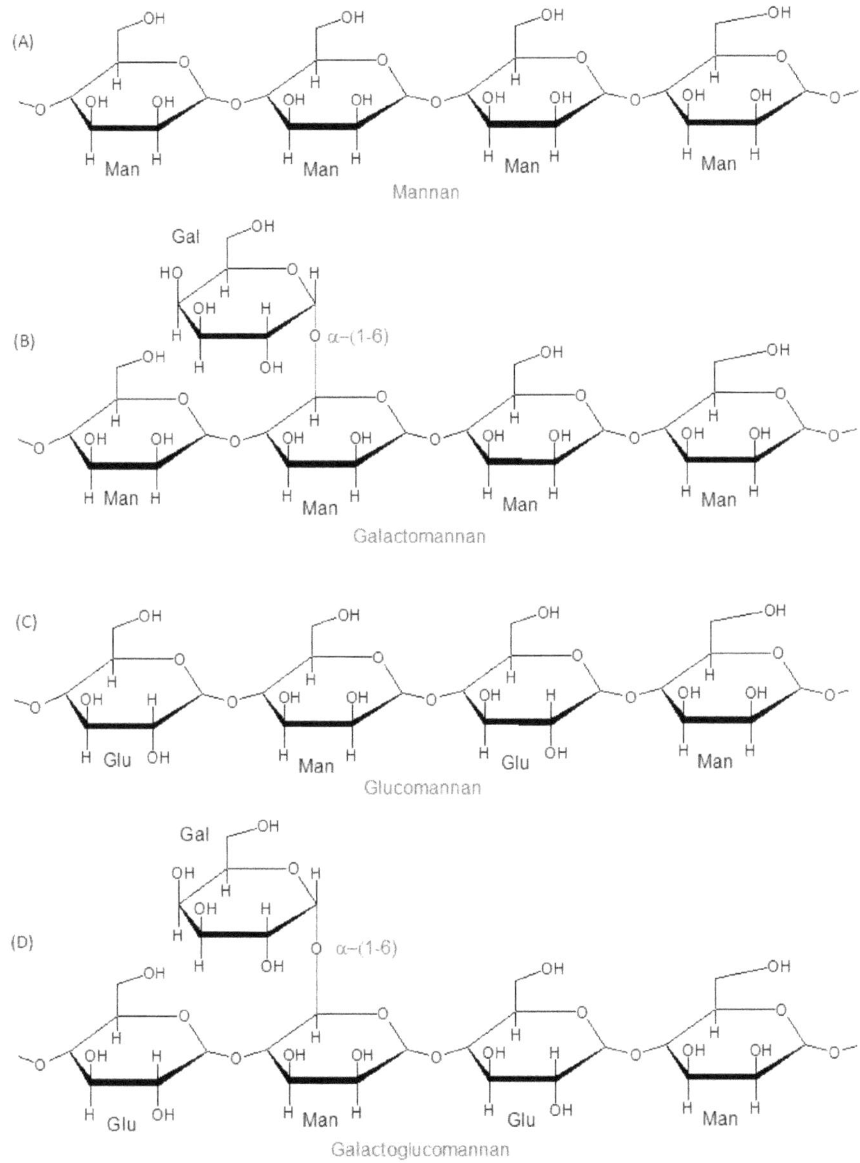

Figure 9.1 General structure of mannans and heteromannans. A) Mannan structure, a main chain of residues, B) galactomannan, an α-1,6 linked galactose residues coupled to some β-1,4 linked mannose (Man) residues C) Glucomannan, a main chain of β-1,4 linked mannose and glucose residues, D) Galactoglucomannan, β-1,4 linked mannose and glucose residues with α-1 ,6 linked galactose residues coupled to some mannose residues.

The mannose and galactose ratio (M:G) is the key to determine the amount of intermolecular association called hyperentanglement (Fig. 9.2B). Without galactose side chains, the mannan backbone aggregates due the intermolecular interactions between unbranched portions of heteromannans (Dhawan and Kaur 2007).

Hardwood mannans are composed by β-1,4-mannopyranose and glucopyranose units, while softwood contains two types of acetylated galactoglucomannans with 1:3:1 and 1:4:0:2 glucose, mannose and galactose

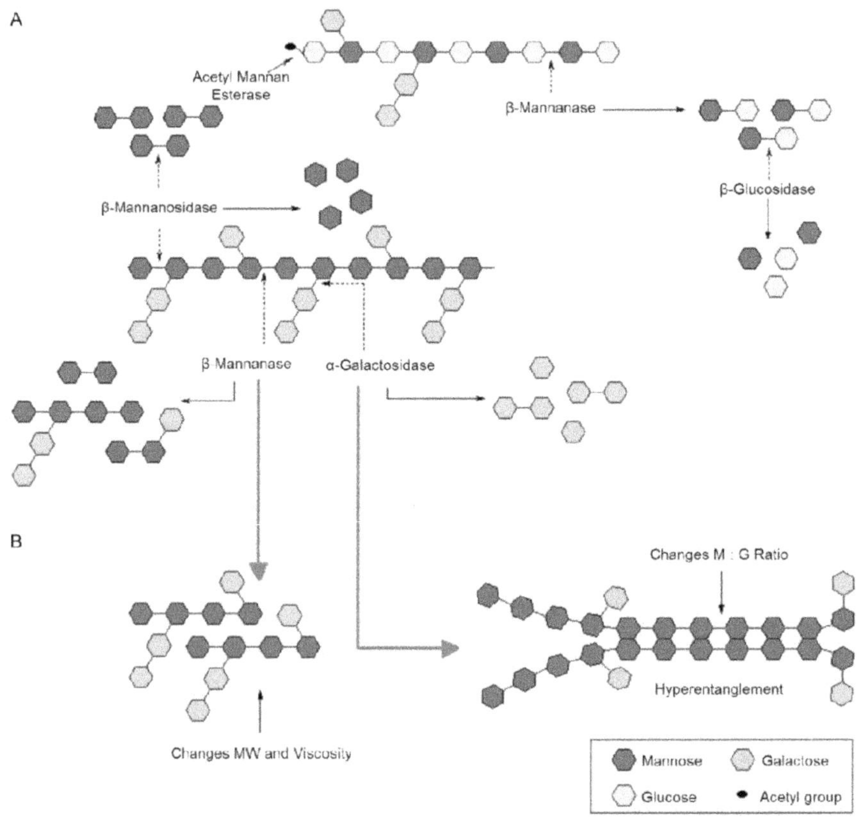

Figure 9.2 (A) Structure of different forms of mannan and enzymes required for their hydrolysis.The structure of mannan is hydrolyzed by β-mannanase, while α-galactosidase and acetyl mannan esterase release galactose and acetyl groups respectively. The products generated by the action of β-mannanase, mannose and glucomannose oligosaccharides, are then hydrolyzed by β-mannosidase and β-glucosidase to generate the monosaccharides mannose, glucose and galactose. (B) Enzymatic action of galactomannans. The β-1,4 linked polymannose chain replaced with α-1,6 galactose and the glycosidic links recognized by β-mannanase and α-galactosidase determining the hyperentanglement.

Color image of this figure appears in the color plate section at the end of the book.

ratios, respectively (Lundqvist et al. 2002). In leguminous seeds, water-soluble galactomannan represents 20 percent of the biomass and is the main source of carbohydrate storage (Dhawan and Kaur 2007).

Mannan-degrading Enzymes System

Mannanases are glycosyl hydrolases which degrade mannans and heteromannans (Dhawan and Kaur 2007). Glycosyl hydrolase (EC 3.2.1.-) are a large group of enzymes that hydrolyze bonds between two or more carbohydrates or between a carbohydrate and no carbohydrate mixture. Their nomenclature is based on substrate specificity and occasionally in their molecular mechanism; this classification does not reflect the structural characteristics of these enzymes. These enzymes are grouped into families based on their amino acid sequences and similarity. The mannanases are classified into families 5, 26 and 113 of GHs according to the Carbohydrate-Active Enzymes Database—CAZy (http://www.cazy.org/) classification (Cantarel et al. 2009). The GH5 family contains bacterial and eukaryotes endo-β-1,4-mannanases whereas families GH26 and 113 contain prokaryote endo-β-1,4-mannanases (Dilopkmol 2010). Generally, most of mannanases show modular organization and contains double-domain proteins with discrete catalytic and non-catalytic modules (Warren 1996).

Due to the complexity of the hemicellulose chemical structure, several enzymes are necessary for their complete degradation (Dilopkimol 2010). Three groups of enzymes are involved in the hydrolysis process for the degradation of mannans: 1,4-β-D-mannan mannohydrolase (β-mannanases —EC 3.2.1.78), 1,4-β-D-mannopyroside hydrolase (beta-mannosidase—EC 3.2.1.25) and 1,4-β-D-glycosyl glucohydrolase (β-glucosidase EC 3.2.1.21). Enzymes such as acetyl mannan esterase (EC 3.2.1.6) and α-galactosidase (EC 3.2.1.22) are required for removing constituents of the side chain of sugars which can be coupled to the structure of the mannans, thereby creating more sites for the subsequent hydrolysis (Hägglund 2003, Dhawan and Kaur 2007, Moreira and Filho 2008).

Mannan-degrading enzymes are responsible for the internal cleavage of β-1,4 linked in the mannan backbone chain including glucomannans, galactomannans and galactoglucomannans, whereas β-mannosidases cleaves β-1,4-mannoside, releasing mannose on the mannan and mannan oligosaccharide non-reducing ends. The β-glucosidases hydrolyze β-1,4-Glc-linked in the terminal portion of glucomannoligosaccharides released by mannanases glucomannans and galactoglucomannans hydrolysis. The α-galactosidases catalyze the hydrolysis of side chains of galactoglucomannans of galactomannans α-1,6-Gal residues, whereas acetyl mannan esterases catalyze the release of galatoglucomannans groups (Fig. 9.2A) (Visser and de Vries 2001, Dhawan and Kaur 2007, Moreira and Filho 2008).

The galactose (Gal) residues replacement and the pattern of distribution of Glc residues (β-D-glucose or β-D-galactose) in the main chain steeply affect mannans degradation. The GH5 mannanases complete hydrolysis generates mannobiose and mannotriose as final products (Dhawan and Kaur 2007).

The active sites of GH5 mannanases have at least five sub sites capable of accommodating substrates of carbohydrate residues in the backbone chain. The hydrolysis seems to require the binding and at least four of these sites. The crystal structure of these mannanases indicates an open cleft-shaped active site with strictly conserved catalytic glutamate residues present on β-strands 4 and 7, respectively (van Zyl et al. 2010).

β-mannanases use double-displacement reactions resulting in the retention of anomeric configuration. This double mechanism involves a nucleophile attack at the anomeric center acid-catalyzed displacement of the leaving group, leaving a covalent glycosyl-enzyme acylal intermediate. The water attacks the anomeric center of the intermediary in the base-catalyzed process to obtain the product and release the enzyme in the original state (Fig. 9.3A-D). Depending on the enzyme, the attacked water can be replaced by reactive donor molecules (e.g., fluorine derivatives) or by high concentrations of oligosaccharides donor molecules which result in a transglycosylation reaction and not hydrolysis (Fig. 9.3E and 9.3F). This transglycosylation reaction has been reported in cell wall modification of plants, seed germination and fruit ripening (Schröder et al. 2006, 2009) and produces molecules that are not mannanases natural substrates such as 6^1-α-D-galactosyl-β-D-mannotriose (van Zyl et al. 2010).

Microorganisms typically produce many enzymes that degrade mannans (Xu et al. 2002). This fact may be due to several factors such as differential processing of mRNA and posttranslational modifications or due to the requirements of binding and degrading substrates of varying complexity (Ferreira and Filho 2004).

In this context, mannanases have been isolated from a wide range of organisms including bacteria, fungi, plants and animals. Among bacteria, β-mannanases have been found in aerobic, anaerobic and different extremophiles as thermophiles, halophiles and psychrophiles. Most β-mannanases are extracellular, however, some of them are found attached to cells. In plants, activity of β-mannanases has been correlated with seed germination and in some cases, also with fruit ripening (Dilopkmol 2010).

The regulation and expression of β-mannanases are not well understood as it occurs with xylanases and cellulases. Nevertheless, as these enzymes are well-represented in the fungal kingdom, their regulation mimics other hemicellulases such as xylanase. Different strains of basidiomycetes have

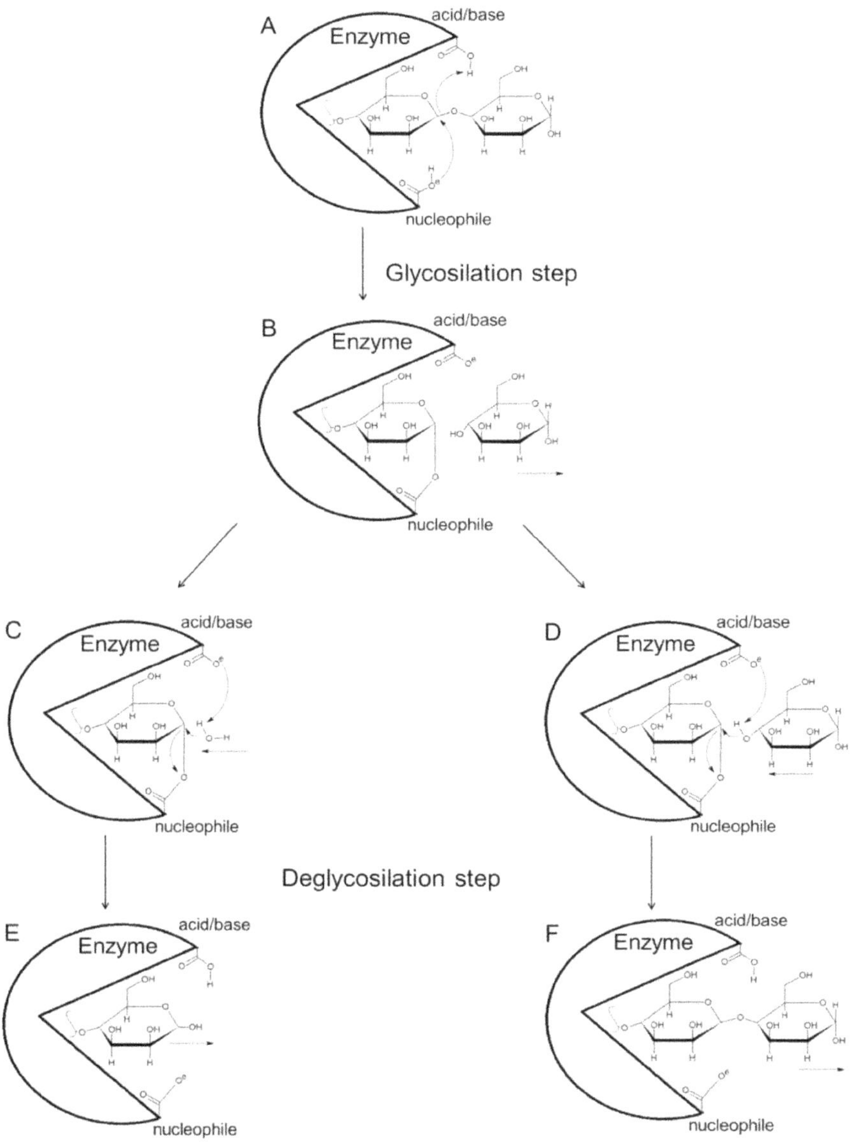

Figure 9.3 Mechanisms of retaining glycosyl hydrolases for mannanases. The glycosylation results from the formation of an oligosaccharide (A + B). Depending on enzyme conditions, the water molecule (C + D) can attack the oligosaccharide to release it or a new mono- or oligosaccharide (E + F) can attack to release a new oligosaccharide. The produced molecule relieves the enzyme during its original protonation stage and a new hydrolysis or transglycosylation reaction can be catalyzed.

been shown to secrete β-mannanases at levels higher than xylanase and endoglucanases. Some strains of *Sclerotium rolfsii* strains were shown to secrete more than seven types of the enzyme (Haltrich et al. 1994, Großwindhager et al. 1999). Interestingly, the production of β-mannanases in *S. rolfsii* strains is better induced by cellulose rather than by manno-oligosaccharides and mannans, showing that these enzymes could have common mechanisms of β-xylanases and endoglucanases induction (Sachslehner et al. 1998).

β-mannanases are also commonly found as part of the repertoire of hemicellulases of ascomycetes such as *Trichoderma* spp. and *Aspergillus* spp., yeast and thermophilic fungi (van Zyl et al. 2010).

The production of mannanases is mainly by submerged fermentation. However, solid fermentation production has been shown already (Chauhan et al. 2012). Microbial mannanases are extra-cellular and inducible by many substrates, including locust bean gum (LBG), konjac powder and wheat bran. Moreover, some factors such as, time, temperature, pH, carbon and nitrogen sources, inorganic salts, agitation and dissolved oxygen concentration should also be considered during the production of mannan-degrading enzymes (Moreira and Filho 2008).

For a better understanding of molecular and biochemical studies of mannan-degrading enzymes, it is necessary that these enzymes be purified to determine their primary amino acid sequences and structures (Sa-Pereira et al. 2003, Moreira and Filho 2008). Several methods have been used in the purification process of mannanases including precipitation with salts (Gübitz et al. 1996, Ademark et al. 1998, Jiang et al. 2006), ion exchange chromatography (Gübitz et al. 1996, Ademark et al. 1998, Ferreira and Filho 2004, Magalhães and Milagres 2009), gel filtration (Ademark et al. 1998, Vieira et al. 2007), ultrafiltration (Gübitz et al. 1996, Jørgensen et al. 2003, Ferreira and Filho 2004, Vieira et al. 2007), high performance liquid chromatography (Jøegensen et al. 2003), hydrophobic interaction chromatography (Sachslehner and Haltrich 1999) and dialysis (Zakaria et al. 1998). The method of choice is made empirically and confirmed by experiments in laboratory scale (Asenjo and Patrick 1990, Weingartner 2010).

Table 9.1 shows some properties of mannan-degrading enzymes produced by different microorganisms. In general, bacterial mannanases have optimum activity at alkaline pH ranges and fungal mannanases at acid intervals (3.0–7.0) (Dhawan and Kaur 2007, Chauhan et al. 2012). The optimum temperature ranges from 37 to 70°C (Eneyskaya et al. 2009), and thermostability ranges from 10 min to 8 hr at different temperatures. They also differ at isoelectric point and kinetic parameters. Once bacterial mannanases are more thermostable then fungal mannanases, they are more suitable for industrial applications such as pulp and cellulose biobleaching (Chauhan et al. 2012).

Table 9.1 Characteristics of mannan degrading enzymes produced by different microorganisms.

Microorganism	Carbon source/ (Fermentation condition)	Temp. optima (°C) of activity	Thermostability	pH optima of activity	Molecular weight of protein (Kda)	Reference
Acinetobacter sp. ST 1-1	CM/ LBG/ (SF)	40	30 min (50°C)	6.0	nr	Titapoka et al. 2008
Bacillusamyloliquefaciens I0A1	GM /(SF)	50	NR	7.0	nr	Mabrouk and Ahwany 2008
Bacillus circulans M-21	GG (SF)	50	4h (40°C)	7.0	33.4	Mou et al. 2011
Klebsiella oxytoca CW23	KM (SF)	70	NR	9.6	nr	Lin et al. 2007
Bacillus sp. MSJ-5	KP (SF)	50	1h (35–65°C)	5.5	40.5	Zhang et al. 2009
Bacillus sp. MG-33	WB (SSF)	65	2h (50–60°)	6.5	nr	Meenakshi et al. 2010
Bacillus subtilis SUT1	BIM (SF)	NR	NR	NR	NR	Rattanasuk and Cairns 2009
Bacillus subtilis strain (CD-3, CD-6, CD-9, CD-10, CD-23, CD-25)	BFM (SF)	45–65	30 min (70°C)	6–6.5	NR	Bo et al. 2009
Cellulosimicrobium sp. HY-13	LGB (SF)	50	1h (37°C)	7.0	34.9	Kim et al. 2011
Chryseobacteriumindologenes	BIM (SF)	NR	NR	NR	NR	Rattanasuk and Cairns 2009
Klebsiellaoxytoca CW23	CM/ LBG (SF)	50	30 min (50°C)	7.0	NR	Titapoka et al. 2008
Paenibacillus sp. MSL-9	GG (SF)	40	NR	8.0	NR	Manjula et al. 2010
Paenibacillus sp. DZ3	LB with glucomannan (SF)	60	1h (60°C)	6.0	39	Chandra et al. 2011
Aspergillus nigergr	LBG (SF)	65	8h (60°C)	5.5	NR	Kote et al. 2009
Aspergillus flavusgr	LBG (SF)	60	6h (60°C)	6.0	NR	Kote et al. 2009

Table 9.1 contd....

Table 9.1 contd.

Microorganism	Carbon source/ (Fermentation condition)	Temp. optima (°C) of activity	Thermostability	pH optima of activity	Molecular weight of protein (Kda)	Reference
Aspergillus niger LW-1	PP/LBG (SSF)	NR	NR	NR	NR	Zhang et al. 2008
Aspergillus niger ATCC 20114	PEPTONE (SF)	50	NR	4.0	NR	Mohamad et al. 2011
Aspergillus niger FTCC 5003	PKC (SF)	NR	NR	NR	NR	Aziz et al. 2008
Aspergillus niger FTCC 5003	PKC (SF)	50	NR	5.5	NR	Abdeshanian et al. 2009
Aspergillus niger USM F-4	Molasses/PKC (SSF)	60	NR	4.0	NR	Rashid et al. 2010
Aspergillus niger	GG (SF)	50, 70	6h (60°C)	3.5	NR	Norita et al. 2010
Aspergillus oryzae NRRL 3448	LBG (Static culture)	55	15 min (55°C)	5.5	18	Fattah et al. 2009
Penicillium occitanis Pol 6	CSFL	40	30 min (50°C)	4.0	41	Blibech et al. 2010, 2011
Scopulariopsis candida LMK 004	LBG (SF)	50	3h (30–40°C)	5.0	28	Mudau and Setati 2008
Scopulariopsis candida LMK 008	LBG (SF)	40	3h (30–40°C)	6.0	NR	Mudau and Setati 2008
Trichoderma reesi	NR	37	NR	3.5	NR	Eneyskaya et al. 2009
Streptomyces sp. PG-08-03	GG (SF)	75	NR	8.0		Bhoria et al. 2009

CM—Copra meal; LBG—Locust bean gum; GM—Galactomannan; GG—guargum; KM—Konjac mannan; KP—Konjac powder; WB—wheat bran; BIM—Bacterial isolation médium; BMF—bacterial fermentation médium; LB—Luria Broth; PP—potato peel; PKC—Palm kernel cake; CSFL—carob seed flour liquid médium; SF—Submerged fermentation; SSF—solid state fermentation; NR—not reported. Modified from Chauhan et al. 2012.

Heterologous Production of β-mannanases

Recombinant DNA technology has enabled the development of several studies using gene cloning and manipulation of mannanases and other hemicellulases with well-known properties in homologous or heterologous hosts, in order to increase the levels of enzyme production via genetic manipulation and select mutant organisms capable of producing great hemi-cellulolytic activity for commercial applications.

Gene cloning is a fast-processing technology that has been fundamental to improve our understanding of the structure-function relationship of genetic systems. Currently, over 50 percent of the most important industrial enzymes are produced by genetically modified microorganisms (Dhawan and Kaur 2007, Chauhan et al. 2012).

Heterologous expression in fungi has been efficiently produced by both fungal and bacterial mannanases in submerged culture. High levels of enzyme expression in heterologous fungal systems associated with the ease of purification of the supernatant substantially reduce the costs of industrial enzymes. The fungal post-translational machinery is generally sufficient for the expression of genes from other organisms which require protein glycosylation, proteolytic processing and formation of disulfide bonds (Nevalainen 2005).

Genes encoding mannanases cloned from different species have been expressed in *Escherichia coli* in most cases. However, some genes have been expressed in other hosts such as *Bacillus megaterium* (Summpunn et al. 2011), *Brevibacillus brevis* (Zhou et al. 2011), *Pichia pastoris* (Qiao et al. 2010), *Kluyveromyces cicerisporous* (Pan et al. 2011). Most fungal mannanases have been expressed in fungal hosts like *P. pastoris* (Duruksu et al. 2009, Cai et al. 2011a) and *Aspergillus* sp. (Petrus et al. 2009).

In order to produce the enzyme on a commercial scale, the enzymes must reach amounts of grams per liter (van Zyl et al. 2010). Heterologous expression in fungi and bacteria has allowed the mannanase production to increase in many cases (Qiao et al. 2008, Duruksu et al. 2009). Table 9.2 shows some heterologous mannanases produced by different microorganisms.

Applications of Mannanases

It is estimated that the production of enzymes will grow 6.8 percent per annum upto 2015. In percentage terms, the industrial catalysts represented 60 percent of the world market of enzymes in 2009. It is expected that by the year 2014 there will be an increase of 42 percent in this market (Freedonia Group Inc. 2009). Currently, the market for industrial enzymes is classified in three application areas: technical, food and animal feed (Moreira and

Table 9.2 Heterologous mannanases produced by different microorganisms' hosts.

Origin Microorganism	Host	Cultivation conditions	Thermostability	pH stability	Family (GHs)	Reference
Bacillus circulans CGMCC 1554	*E. coli*	Tryptone, SF, 37°C, 48 h, 240 rpm	1h (50°C)	7.6	NR	Yang et al. 2009
Bacillus sp. N 16-5	*Pichia pastoris*	SCM, SF, 30°C, 200 rpm	2h (60°C)	10.0	NR	He et al. 2008
Bacillus sp. N 16-5	*Kluyveromyces cicerisporus*	Glucose, SF, 30°C, 144 h	10 min (60°C)	9.56.0	5	Pan et al. 2011
Bacillus subtilis	*E. coli*	LB Induced by IPTG, SF, 37°C	NR	6.0	NR	Yamabhai et al. 2008
Bacillus subtilis MA-139	*P. pastoris*	YPDM, 28°C, 72 h, 250 rpm	30 min (40°C)	6.0	NR	Qiao et al. 2010
Bacillus subtilis HB002	*Sacharomyces cerevisiae*	YPDM, 96 h	1h (70°C)	6.0	NR	Yang et al. 2009
Bacillus subtilis BCC 41051	*E. coli*	GLMM, 180 rpm, 50°C, 36 h	30 min (60°C)	7.0	NR	Summpunn et al. 2011
Bacillus subtilis B23	*Brevibacillus brevis*	NR	NR	6.8	26	Zhou et al. 2011
Cellulomicrobium sp. HY-13	*E. coli*	LB, SF, 30°C, 7 h	15 min (50°C)	6.0	5	Kim et al. 2011
Caldicellulosirupator Rt 8B.4	*E. coli*	LB, SF, 32°C, 225 rpm	NR	5.5	26	Sunna 2010
Clostridium josui	*E. coli*	BMC, 45°C, 48–168 h	NR	6.5	5	Sakka et al. 2010
Clostridium cellulovorans	*E. coli*	NR	NR	7.0	26	Jeon et al. 2011
Paenibacillus polymyxa	*E. coli*	NR	NR	6.0	26	Cho et al. 2008
Paenibacillus sp. BME-14	*E. coli*	LB induced by IPTG, 37°C	7 min (70°C)	4.5	26	Fu et al. 2009

Pantoeae agglomerans A021	*E. coli*	LB induced by IPTG, 37°C, SF, 150 rpm	15-60min (50°C)	6.0	26	Wang et al. 2010
*Vibrio*sp strain MA-138	*E. coli*	LB induced by IPTG, SF, 25°C, 24 h, pH 7	NR	7.0	5	Tanaka et al. 2009
Thermatogao petrophila RKU-1	*E. coli*	LB, SF, 27°C, 16 h, 200 rpm	NR	NR	5	Santos et al. 2010
Agaricusbisporus CEL4	*S. cerevisiae*	NR	Mainly cell wall bound	NR	5	Tang et al. 2001
A. bisporus CEL4	*P. pastoris*	SC	NR	NR	5	Tang et al. 2001
Aspegillus aculeatus Man1	*Y. lipolytica*	Batch, SCM, 50h	NR	NR	5	Roth et al. 2009
Aspegillus aculeatus Man1	*A. niger*	SCM, 200rpm, 7d	75°C	24h, (pH 3–6), BSA	5	Qiao et al. 2008
Aspegillus aculeatus Man1	*S. cerevisiae*	SCM, 4d	24h (50°C)	NR	5	Setati et al. 2001
Aspegillus aculeatus	*A. oryzae*	NR	60–70°C	NR	5	Christgar et al. 1994
Aspergillus aculeatus MRC 11624	*A. niger*	Glucose, SF, 30°C, 168h, 220 rpm	5h (60°C)	NR	5	Petrus et al. 2009
Aspergillus aculeatus VN	*A. niger*	SMM, SF, 30°C, 150 rpm	350min (65°C)	NR	5	Pham et al. 2010
Aspergillus fumigatus	*P. pastoris*	Glucose, SF, 72 h, 155 rpm	3h (70°C)	3h, pH 5, 70°C	5	Duruksu et al. 2009
Aspergillus niger	*P. pastoris*	2% methanol, 96 h, 28°C	80°C	56h, pH 4, 70°C	5	Do et al. 2009
Aspergillus sulphureus	*P. pastoris*	GCM, SF, 28°C, 48 h, 250 rpm	1h (40°C)	1h, pH 2.4, 40°C	5	Chen et al. 2007
Aspergillus sulphureus	*E. coli*	LB/IPTG, SF, 37°C, 12 h	30 min (50°C)	30min, pH 2.4, 50°C	NR	Chen et al. 2008

Table 9.2 contd....

Table 9.2 contd.

Origin Microorganism	Host	Cultivation conditions	Thermostability	pH stability	Family (GHs)	Reference
Aspergilus terreus	*P. pastoris*	3d after induction	55°C	NR	5	Huang et al. 2007
Bispora sp.	*P. pastoris*	GCM, 30°C, 48 h	1h (60°C)	1h, pH 0.5–11, 37°C	5	Luo et al. 2009
Chaetomium sp. *CQ31*	*P. pastoris*	BMGYM/BMMYM induced by methanol, SF, 30°C, 120h, 270 rpm	30 min (55°C)	30 min, pH 5–11, 50°C	5	Katrolia et al. 2012
Mytilus edulis	*P. pastoris*	Fermenter, continuous, SD, 10d, 17°C	50°C	pH 4–9.5	5	Xu et al. 2002
Orpinomyces sp. ManA	*S. cerevisiae*	SCM continuous	NR	24h, pH 3.8–9.5, RT,	5	Ximenes et al. 2005
Penicillium pinophilum C1	*P. pastoris*	SCM induced by metanol, SF, 96 h, 30°C	1h (50°C)	1h, pH 3–10, 37°C	5	Cai et al. 2011a
Penicillium sp. C6	*P. pastoris*	YPDM, SF, 48 h, 30°C	1h (50°C)	1h, pH 4–9, 37°C	5	Cai et al. 2011b
Phialophora sp. P13	*P. pastoris*	SCM, SF, 30°C, 48 h	2h (55°C)	2h, pH 1–7, 37°C	5	Zhao et al. 2010
Phanerochaete chrysosporium Man 5D	*A. niger*	SC	2h (60°C)	3h, pH 4–9.5, RT	5	Benech et al. 2007
Streptomyces sp. S27	*E. coli*	LBG, SF, 42°C, 36 h	1h (50°C)	1 h, pH 5–9, 37°C	5	Shi et al. 2011
Streptomyces thermolilacinus	*E. coli*	NR	NR	NR	5	Kumagai et al. 2011

Trichoderma reesei	Tobacco Chloroplast	MSM	NR	pH 3–7	5	Agrawal et al. 2011
Trichoderma reesei Man1	*P. pastoris*	NR	NR	NR	5	Wei et al. 2005
Trichoderma reesei Man1	*S. cerevisiae*	SD, 200 rpm, 48 h	NR	NR	5	Stalbrand et al.1995
Trichoderma reesei Man5A	*T. reesei*	SD, 160 rpm, 48 h	NR	NR	5	Hagglund et al. 2003
Xanthomonas campestris	*E. coli*	LBG	10 min (35°C)	NR	5	Hsiao et al. 2010

LB—Luria Broth; SCM—Synthetic Complex medium; SF—Submerged fermentation; SSF—solid state fermentation; IPTG—Iso Propyl thio galactoside; YPDM—Yeast peptone dextrose medium; GLMM—GLM medium; BMC—ball millded cellulose; SMM; Solid mineral medium; GCM- Glycerol complex medium; BMGYM—buffered glycerol complex medium; BMMYM—Buffered metanol complex medium; MSM—Murashige and Skoog medium; NR—not reported.

Filho, 2008). In this context, mannanases have many applications such as, improve the quality of food and feedstuffs and bleaching of cellulose pulp and paper industry.

The biobleaching of cellulose pulp of softwoods is the most important and potential application of mannanases. The extraction of lignin from wood fibers is an essential step in the bleaching of pulp. Under alkaline conditions, the pre-treatment allows the hydrolysis of hemicelluloses covalently bound to lignin and facilitates its removal, but the process causes a problem of environmental pollution due to the release of chlorinated compounds (Dhawan and Kaur 2007, Chauhan et al. 2012). Mannanases and their enzymes are capable of selectively cleaving mannan components of the pulp without affecting its quality and are useful in the bleaching process without the presence of chlorine compounds in paper pulp, increasing the brightness and eliminating the use of hydrogen peroxide in the process. A large amount of the pulp is produced from softwood. This pulp contains 15–20 percent of galactomannans whose constituents are excellent candidates for the enzymatic biobleaching of pulp, since they have become economically viable (Gübitz et al. 1997, Benech et al. 2007, Dhawan and Kaur 2007, Chauhan et al. 2012). Once the bleaching is performed at high temperatures and basic conditions, there is a strong demand for mannanases with suitable properties for such industrial conditions (Yanhe et al. 2004).

Cellulose material processing for textile into clothing material involves several steps such as the wire fabric construction or mesh, subsequent dyeing and finishing. Mannanase is useful in the enzymatic washing process and gum removal during the cellulosic material preparation to improve the dyeing material (Palackal et al. 2007).

Beta-mannan is a polysaccharide commonly used in feed ingredients such as soybean meal, guar meal (GM), copra meal (CM), palm kernel meal (PKM) and sesame meal (SM). All these ingredients have a high amount of fiber, low taste, essential amino acids and high viscosity coupled with anti-nutritional factors such as mannan, galactomannan, xylan and arabinoxilana. These oligosaccharides comprise the weight gain and feed conversion, as well as the water and glucose intestinal absorption. The incorporation of mannanases into diets assists on cell wall β-mannan breaking and releasing the encapsulated ingredients increases the intestinal villi in the duodenum and jejunum, leading to an increase in the absorption area and viscosity digestion decreases (Adibmoradi and Mehri 2007, Chauhan et al. 2012).

The use of beta-mannanases from different species in broiler feed allows facilitating the breaking of sugars which can be useful in the feed industry. Because the majority of mannanases is in the form of endo-β-mannanases, it is speculated that manno-oligosaccharides, mannotriose and mannobiose as well as a small amount of mannans are generated when beta-mannanases are included in the animals' diets. Since only mannose residues are absorbed by the intestine, these components become unattractive because they do not produce energy for the animal. To overcome this problem, the inclusion of the bacterium genus *Bifidiobacteria* supplies these components to provide carbohydrates as a food source for bacteria in cecum-suppressing pathogens binding to mannooligossacaride (Lyons and Jacques 2002).

The role of mannanases activated on gastric conditions in animal feed has been demonstrated by Li et al. (2008) and Cai et al. (2011a). Some such as cyprinoids and carp fish do not have stomachs, leading to trypsin and pepsin release directly into the gastrointestinal tract with a neutral pH. Mannanases that are activated at neutral pH and resistant to pepsin proteolysis are allowed to be used in fish feed industries (Li et al. 2008). Cai et al. (2011a) showed that in a constructed gastric fluid system, mannanases were resistant to metal ions and proteases such as pepsin and trypsin besides releasing significant quantities of LBG-reducing sugars, making the enzyme a strong candidate for use in animal feed.

Another application of the mannanases is related to the hydrolysis of galactomannans of coffee. It reduces the viscosity of the extracts, allowing their concentration by low-cost procedures such as evaporation (Moreira and Filho 2008). Mannan is the major polysaccharide of these extracts and is responsible for its viscosity, which negatively affects the technological processing of instant coffee. The coffee mannan is isolated from green beans by delignification, acid wash and subsequent alcoholic precipitation for separation of polysaccharides. When mannanases hydrolyze these polysaccharides, it leads to a low viscosity, which also increases the reducing sugars such as mannotetraose, mannotriose and mannobiose (Dhawan and Kaur 2007).

Mannans are usually found in gums and used as thickening agents in ice cream, hair gel, shampoo, conditioners, dressings and toothpaste. As mannans have a high affinity with cellulose fibers, staining containing mannan is difficult to remove (Chauhan et al. 2012). The β-mannanases were introduced in 2000 as agents against the reappearance of spots during the washes (Kirk et al. 2002). These enzymes cleaves β-1,4 linked mannans by hydrolysis breaking gum polymers into small pieces of more water-soluble carbohydrates which remain free and are siphoned during washing, and thereby prevent the deposit of these fragments again during the process of washing clothes. It also can be used in health and beauty products, sanitation, contact lens cleaners and hard surface (Bettiol et al. 2000).

In the conception of bioenergy, the plant biomass is a rich source of lignocellulolytic components that can be converted into various products, especially bioethanol, microbial and enzyme system of these organisms which hydrolyze these substrates and are responsible for the efficient conversion of lignocellulose (Filho 2008). Thus, the mannan-degrading system associated to other enzymes may be used synergistically to convert the lignocellulose carbohydrates into fermentable sugars (pentose and hexose), since the saccharification process is based on the production of sugars (Moreira and Filho 2008).

Furthermore, mannanases can be used to improve the flow of oil and gas into wells drilled. During drilling, the stream is carried out to the well stream stimulated by expulsion through the slits opened in the surrounding rock. The process is done by flooding the wells with the guar gum polymer and particles of sand which close the well leading to pressure in the rock until its fracture. The polymer carries the sand through the fractures sustaining the cracks to the oil and gas flow (Politz et al. 2000). To facilitate the flow, the polymer is diluted. Mannanases can hydrolyze guar gum at high temperatures, making its introduction into the system somewhat plausible (Dhawan and Kaur 2007).

Mannanases can also be used in copra oil extraction, rich in galactomannans (Moreira and Filho 2008) replacing the use of the manual process as well as producing polycyclic aromatic hydrocarbons (PAH) which facilitate the loss of protein quality and leave the product susceptible to contamination by aflatoxins and rancidity by oxidative process while minimizing the need to refine the oil, providing coconut oil, an aromatic protein containing liquids that can be used as a basis for beverages and isolated coconut protein (Chen and Diosady 2003, Dhawan and Kaur 2007).

Besides the processes mentioned above, mannanases can also be used in the manufacture of pharmaceutical tablets (Fu et al. 2006), drugs for disorders of the urinary tract (Van Zyl et al. 2010), reduction of plasma cholesterol and body fat without reducing the use of the protein and increasing the fecal excretion volume (Takeno et al. 1990), intestinal syndrome treatment (Parisi et al. 2002) and in the treatment of acute diarrhea in children (Alam et al. 2000), where they may promote the release of short chain fatty acids into the small intestine, maintaining the balance of salt and water, if children are supplemented with oral rehydration (Chauhan et al. 2012).

Conclusions

Cellulases and xylanases have been extensively studied and shown to possess significant industrial applications. However, little attention is focused on mannan-degrading enzymes. The complexity of the structure of mannans is reflected in the enzymes that act on the main and side chains required for its hydrolysis. With the increase of heterologous strains producing these enzymes, their application has been gaining importance. The increasing number of microorganisms genomes described reveals a new system of degradation of mannans and a new field of action of these enzymes for commercial use. In order to make feasible the commercialization of these enzymes, it needs to be produced in large scale. In this context, genetic manipulation and understanding of the molecular and biochemical mechanisms using heterologous microorganisms are gaining interest in the academic and industrial areas. The exploitation of microorganisms that produce mannanases for biotechnological applications, such as paper mills, food, food products, among other biofuels, is considered to have a promising future.

References

Abdeshanian, P., N. Samat, A.A. Hamid and W.M.W. Yusoff. 2009. Utilization of palm kernel cake for production of β-mannanase by *Aspergillus niger* FTCC 5003 in solid state fermentation using an aerated column bioreactor. J. Microbiol. Biotechnol. 37: 103–109.

Ademark, P., A. Varga, J. Medve, V. Harjunpää, T. Drakenberg, F. Tjerneld and H. Stålbrand. 1998. Softwood hemicellulose-degrading enzymes from *Aspergillus niger*: purification and properties of a β-mannanase. J. Biotechnol. 63: 199–210.

Adibmoradi and M. Mehri. 2007. Effects of β-mannanase on broiler performance and gut morphology. 16th European Symposium on Poultry Nutrition, Stasburg, France, pp. 471–47.

Agrawal, P., D. Verma and H. Daniell. 2011. Expression of *Trichoderma reesei* β-mannanase in tobacco chloroplasts and its utilization in lignocellulosic woody biomass hydrolysis. PLoS One 6(12): e29302. doi: 10.1371/journal.pone.0029302.

Alam, N.H., R. Meier, H. Schneider, S.A. Sarker, P.K. Bardhan, D. Mahalanabis, G.J. Fuchs and N. Gyr. 2000. Partially hydrolyzed guar gum supplemented oral rehydration solution in the treatment of acute diarrhea in children. J. Ped. Gas.Nut. 31: 503–507.

Asenjo, J.A. and I. Patrick. Large-scale protein purification. pp. 1–28. *In*: E.L.V. Harris and S. Angal [eds.]. 1990. Protein purification applications: a practical approach. IRL, Oxford.

Aziz, S.A., L.G.A. Ong, M.A. Hassan and M.I.A. Karim. 2008. Production parameters optimization of mannanse production from *Aspergillusniger* FTCC 5003 using palm kernel cake as carbon source. Asi. J. Biochem. 3(5): 297–307

Benech, R.O., X. Li, D. Patton, J. Powlowski, R. Storms, R. Bourbonnais, M. Paice and A. Tsang. 2007. Recombinant expression, characterization, and pulp prebleaching property of a *Phanerochaete chrysosporium* endo-β-1,4-mannanase. Enzyme Microb. Technol. 41: 740–747.

Bettiol, J.L.P., J.P. Boutique, L.M.P. Gualco and J.P. Johnston. 2000. Nonaqueous liquid detergent compositions comprising a borate releasing compound and a mannanase. Patent EP1059351.

Bhoria, P., G. Singh and G.S. Hoondal. 2009. Optimization of mannanase production from *Streptomyces* sp. PG-08-03 in submerged fermentation. Bioresources 4(3): 1130–1138.

Blibech, M., R.E. Ghorbel, I. Fakhfakh, P. Ntarima, K. Piens, A.B. Bacha and S.E. Chaabouni. 2010. Purification and characterization of a low molecular weight of β-mannanases from *Penicillium occitanis* Pol6. App. Biochem. Biotechnol. 160: 1227–1240.

Blibech, M., R.E. Ghorbel, F. Chaari, I. Dammak, F. Bhiri, M. Neifar and S.E. Chaabouni. 2011. Improved mannanase production from *Penicillium occitanis* by fed-batch fermentation using acacia seeds. ISRN Microbiol. doi: 10.5402/2011/938347.

Bo, X., D. Lei, T. Xiang-hua, L. Jun-jun, M. Yue-lin and Y. Yun-juan. 2009. Characterization of 6 *Bacillus subtilis* β-mannanases and their genes. Afri. J. Biotechnol. 8(18): 4316–4324.

Buckeridge, M.S., H. Pessoa dos Santos and M.A. Tiné. 2000. Mobilisation of storage cell wall polysaccharides in seeds. Plant Physiol. Biochem. 38: 141–156.

Cai, H., P. Shi, H. Luo, Y. Bai, H. Huang, P. Yang and B. Yao. 2011a. Acidic β-mannanase from *Penicillium pinophilum* C1: cloning, characterization and assessment of its potential for animal feed application. J. Biosci. Bioeng. 112: 551–557. doi: 10.1016/j.jbiosc.2011.08.018.

Cai, H., P. Shi, H. Huang, H. Luo, Y. Bai, P. Yang, K. Meng and B. Yao. 2011b. An acidic β-mannanase from *Penicillium* sp. C6: gene cloning and over-expression in *Pichia pastoris*. World J. Microbiol. Biotechnol. 27: 2813–2819.

Cantarel, B.L., P.M. Coutinho, C. Rancurel, T. Bernard, V. Lombard and B. Henrissat. 2009. The Carbohydrate-Active EnZymes database (CAZy): an expert resource for Glycogenomics. Nucleic Acids Res. 37: 233–8.

Capek, P., M. Kubacková, J. Alföldi, L. Bilisics, D. Lisková and D. Kákoniová. 2000. Galactoglucomannan from the secondary cell wall of *Picea abies* L. Karst. Carbohydr. Res. 329(3): p. 635–45.

Chandra, M.R.S., Y.S. Lee, I.H. Park, Y. Zhou, K.K. Kim and Y.L. Choi. 2011. Isolation, purification and characterization of a thermostable β-mannanase from *Paenibacillus* sp. DZ3. J. Korean Soc. App. Biol. Chem. 54(3): 325–331.

Chauhan, P.S., N. Puri, P. Sharma and N. Gupta. 2012. Mannanases: microbial sources, production, properties and potential biotechnological applications. Appl. Microbiol. Biotechnol. 93(5): p. 1817–30.

Chen, B.K. and L.L. Diosady. 2003. Enzymatic aqueous processing of coconuts. Int. Appl. Sci. Engg. 1: 55–61.

Chen, X., Y. Cao, Y. Ding, W. Lu and D. Li. 2007. Cloning, functional expression and characterization of *Aspergillus sulphureus* betamannanase in *Pichiapastoris*. J. Biotechnol. 128(3): 452–461.

Chen, X., W. Lu, Y. Cao and D. Li. 2008. Prokaryotic expression, purification and characterization of *Aspergillus sulphureus* β-mannanase and site directed mutagenesis of the catalytic residues. Appl. Biochem. Biotechnol. 149: 139–144.

Cho, K.M., R.K. Math, S.Y. Hong, S.M.A. Islam, J.O. Kim, S.J. Hong, H. Kim and H.D. Yun. 2008. Changes in the activity of the multifunctional β-glycosyl hydrolase (Cel44C-Man26A) from *Paenibacillus polymyxa* by removal of the C-terminal region to minimum size. Biotechnol. Lett. 30: 1061–1068.

Christgar, S., S. Kauppinen, L.V. Kofod and H. Dalbogea. 1994. Expression cloning, purification and characterization of a β-1,4-mannanase from *Aspergillus aculeatus*. Biochem. Mol. Biol. Int. 33: 917–25.

Coughlan, M.P., M.G. Tuohy, E.X.F. Filho, J. Puls, M. Claeyssens, M. Vrsanská and M. Hughes. Enzymological aspects of microbial hemicellulases with emphasis on fungal systems. pp. 53–84. *In*: M.P. Coughlan [ed.]. 1993. Hemicellulose and hemicellulases. Portland, London.

de Vries, R.P. and J. Visser. 2001. Aspergillus enzymes involved in degradation of plant cell wall polysaccharides. Microbiol. Mol. Biol. Rev. 65(4): 497–522. Dhawan, S. and J. Kaur. 2007. Microbial mannanases: an overview of production and applications. Crit. Rev. Biotechnol. 27(4): 197–216.

Dilokpimol, A. 2010. Production and Characterisation of Glycoside Hydrolases from GH3, GH5, GH10, GH11 and GH61 for Chemo-Enzymatic Synthesis of Xylo- and Mannooligosaccharides. Thesis.Published.Enzyme and Protein Chemistry.Technical University of Denmark, Denmark. Do, B.-C., T.T. Dang, J.G. Berrin, D. Haltrich, K.A. To, J.C. Sigoillot and M. Yamabhai. 2009. Cloning, expression in *Pichia pastoris*, and characterisation of a thermostable GH5 endo-1,4-β-mannosidase from *Aspergillus niger* BK01. Microb. Cell. Fact. 8: 59.

Duruksu, G., B. Ozturk, P. Biely and Z.B. Ogel. 2009. Cloning, expression and characterization of endo-β-1,4-mannanase from *Aspergillus fumigates* in *Aspergillus sojae* and *Pichia pastoris*. Biotechnol. Prog. 25: 271–276.

Eneyskaya, E.V., G. Sundqvist, A.M. Golubev, F.M. Ibatullin, D.R. Ivanen, K.A. Shabalin, H. Brumer and A.A. Kulminskaya. 2009. Transglycosylating and hydrolytic activities of the β-mannosidase from *Trichodermareesei*. Biochimie. 91: 632–638.
Fattah, A.A.F., A.M. Hashem, A.M.S. Ismail and E.M.A. Refai. 2009. Purification and some properties of β-mannanase from *Aspergillus oryzae* NRRL 3448. J. App. Sci. Res. 5(12): 2067–2073.
Ferreira, H.M. and E.X.F. Filho. 2004. Purification and characterization of a β-mannanase from *Trichoderma harzianum* strain T4. Carbohydr. Polym. 57: 23–29.
Filho, E.X.F. Hemicellulases and biotechnology. pp. 165–176. *In:* S.G. Pandalai [ed.]. 1998. Recent research developments in microbiology. ResearchSignpost, Trivandrum, India. Fu, X., H. Huang, P. Liu, L. Lin, G. Wu, C. Li, C. Feng and H. Hong. 2009. Cloning and characterization of a novel mannananses from *Paenibacillus* sp. BME-14. J. Microbiol. Biotechnol. 20(3): 518–524.
Fu, Y., S.H. Jeong, J. Kim, J.A. Callihan, K. Park and C.M. Pai. 2006. Mannosebased fast dissolving tablets. Patent US20060134195A1.
Großwindhager, C., A. Sachslehner, B. Nidetzky and D. Haltrich. 1999. Endo-β-1,4 mannanase is efficiently produced by *Sclerotiumrolfsii* under depressed conditions. J. Biotechnol. 67: 189–203.
Gübitz, G.M., M. Hayn, M. Sommerauer and W. Steiner. 1996. Mannan degrading enzymes from *Sclerotium rolfsii*: characterization and synergism of two endo-β-mannanases and β-mannosidase. Biores. Technol. 58: 127–135.
Gübitz, G.M., T. Lischnig, D. Stebbing and J.N. Saddler. 1997. Enzymatic removal of hemicellulose from dissolving pulps. Biotechnol. Lett. 19: 491–495.
Hagglund, P., T. Eriksson, A. Collen, W. Nerinckx, M. Claeyssens and H. Stalbrand. 2003. A cellulose-binding module of the *Trichoderma reesei* β-mannanase Man5A increases the mannan-hydrolysis of complex substrates. J. Biotechnol. 101: 37–48.
Haltrich, D., B. Laussamayer, W. Steiner, B. Nidetzky and K.D. Kulbe. 1994. Cellulolytic and hemicellulolytic enzymes of *Sclerotium rolfsii*: Optimization of the culture medium and enzymatic hydrolysis of lignocellulosic material. Bioresour. Technol. 50(1): 43–50.
Handford, M.G., T.C. Baldwin, F. Goubet, T.A. Prime, J. Miles, X. Yu and P. Dupree. 2003. Localisation and characterisation of cell wall mannan polysaccharides in *Arabidopsis thaliana*. Planta 218(1): p. 27–36.
Harris, P.J. and B.A. Stone. Chemistry and molecular organization of plant cell walls. pp. 60–93. *In*: M.E. Himmel [ed.]. 2008.Biomass recalcitrance. Blackwell, Oxford.He, X., N. Liu, Z. Zhang, B. Zhang and Y. Ma. 2008. Inducible and constitutive expression of a novel thermostable alkaline β-mannanase from alkalophilic *Bacillus* sp. N16-5 in Pichia pastoris and characterization of the recombinant enzyme. Enzyme Microb. Technol. 43: 13–18.
Hsiao, Y.M., Y.F. Liu, M.C. Fang and Y.H. Tseng. 2010. Transcriptional regulation and molecular characterization of the manA gene encoding the biofilm dispersing enzyme mannan endo-1,4-β-mannosidase in Xanthomonas campestris. J. Agric. Food. Chem. 58: 1653–1663.
Huang, S.P., C.L. Wang, G.M. Zhang and L.X. Ma. 2007. Construction of a double functional recombinant strain of *Pichia pastoris* co-expressing phytase and mannanases and the enzymatic analysis. Wei Sheng Wu Xue Bao 47: 280–4.
Jeon, S.D., K.O. Yu, S.W. Kim and S.O. Han. 2011. A cellulolytic complex from *Clostridiumcellulovorans* consisting of mannanase B and endoglucanase E has synergistic effects on galactomannan degradation. Appl. Microbiol. Biotechnol. 90: 565–572.
Jiang, Z., Y. Wei, D. Li, P. Chai and I. Kusakabe. 2006. High-level production, purification and characterization of a thermostable β-mannanase from the newly isolated *Bacillus subtilis* WY34. Carbohydr. Polym. 66: 88–96.
Jørgensen, H., T. Eriksson, J. Börjesson, F. Tjerneld and L. Olsson. 2003. Purification and characterization of five cellulases and one xylanase from *Penicillium brasilianum* IBT 20888. Enzyme and Microb. Technol. 32: 851–86.

Katrolia, P., P. Zhou, P. Zhang, Q. Yan, Y. Li, Z. Jiang and H. Xu. 2012. High level expression of a novel β-mannanase from *Chaetomium* sp. exhibiting efficient mannan hydrolysis. Carbohy. Pol. 87: 480–490.

Kim, D.Y., S.J. Ham, H.J. Lee, H.Y. Cho, J.H. Kim, Y.J. Kim, D.H. Shin, Y.H. Rhee, K.H. Son and H.Y. Park. 2011. Cloning and characterization of a modular GH5 β-1,4-mannanase with high specific activity from the fibrolytic bacterium *Cellulosimicrobium* sp. strain HY-13. Biores. Technol. 102: 9185–9192. doi: 10.1016/j.biortech.2011.06.073.

Kirk, O., T.V. Borchet and C.C. Fuglsang. 2002. Industrial enzyme applications. Curr. Opin. Biotechnol. 13: 345–351.

Kote, N.V., A.G.G. Patil and V.H. Mulimani. 2009. Optimization of the production of thermostable endo-β-1,4 mannanase from a newly isolated *Aspergillus niger* gr and *Aspergillus flavus* gr. Appl. Biochem. Biotechnol. 152: 213–223.

Kumagai, Y., H. Usuki, Y. Yamamoto, A. Yamasato, J. Arima, T. Mukaihara and T. Hatanaka. 2011. Characterization of calcium ion sensitive region for β-mannanase from *Streptomyces thermolilacinus*. Biochim. Biophys. Acta 1814: 1127–1133.

Li, Y., P. Yang, K. Meng, Y. Wang, H. Luo, N. Wu, Y. Fan and B. Yao. 2008. Gene, cloning, expression and characterization of a novel β-mannanase from *Bacillus circulans* CGMCC 1416. J. Microbiol. Biotechnol. 18: 160–166.

Lin, S.S., W.F. Dou, H. Xu, H.Z. Li, Z.H. Xu and Y. Ma. 2007. Optimization of medium composition for the production of alkaline β-mannanase by alkaliphilic *Bacillus* sp. N16-5 using response surface methodology. Appl. Microbiol. Biotechnol. 75(5): 1015–1022.

Lundqvist, J., A. Teleman, L. Junel, G. Zacchi, O. Dahlman, F. Tjerneld and H. Stalbrand. 2002. Isolation and characterization of galactoglucomannan from spruce (*Picea abies*). Carbohydr. Polym. 48: 29–39.

Luo, H., Y. Wang, H. Wang, J. Yang, Y. Yang and H. Huang. 2009. A novel highly acidic β-mannanase from the acidophilic fungus *Bispora* sp. MEY-1: gene cloning and overexpression in *Pichia pastoris*. Appl. Microbiol. Biotechnol. 82: 453–61.

Lyons, T. P. and K.A. Jacques. 2002. Nutritional Biotechnology in the Feed and Food Industries. T.P. Lyons, K.A. Jacques and J.M. Hower (eds.). Nottingham University Press, Nottingham, UK.

Mabrouk, M.E.M. and A.M.D.E.I. Ahwany. 2008. Production of β-mannanase by *Bacillusamyloliquefaciens* 10A1 cultured on potato peels. Afri. J. Biotechnol. 7: 1123–1128.

Magalhães, P.O. and A.M.F. Milagres. 2009. Biochemical properties of a β-mannanase and a β-xylanase produced by *Ceriporiopsissubvermispora* during biopulping conditions. Int. Biodeterioration & Biodegradation 63: 191–195.

Manjula, S., M. Shinde and J. Lalitha. 2010. Optimization of culture conditions for the production of β-mannanse from an agar utilizing *Paenibacillus* sp. MSL-9. The Bioscan 5(1): 75–79.

Meenakshi, G. Singh, A. Bhalla and G.S. Hoondal. 2010. Solid state fermentation and characterization of partially purified thermostable mannanase from Bacillus sp. MG-33. Bioresources 5(3): 1689–1701.

Mercê, A.L.R., E. Fernandes, A.S. Mangrish and M.R. Sierakowskia and Bruno Szpoganicz. 2001. Fe(III)-Galactomannan Solid and Aqueous Complexes. Potentiometric, EPR Spectroscopy and Thermal Data. J. Braz. Chem. Soc. 12(6): 791–798.

Mohamad, S.N., R.N. Ramanan, R. Mohamad and A.B. Ariff. 2011. Improved mannan degrading enzymes production by *Aspergillus niger* through medium optimization. New Biotechnol 28: 146–152.doi: 10.1016/j.nbt.2010.10.008.

Moreira, L.R. and E.X. Filho. 2008. An overview of mannan structure and mannan-degrading enzyme systems. Appl. Microbiol. Biotechnol. 79(2): 165–78.

Moreira, L.R.S., N.G. Milanezi and E.X.F. Filho. Enzymology of Plant Cell Wall Breakdown. *In*: M.S. Buckeridge and G.H. Goldman [eds.]. 2011. Routes to Cellulosic Ethanol. Springer, New York. doi: 10.1007/978-0-387-92740-4.

Mou, H., F. Zhou, X. Jiang and Z. Liu. 2011. Production, purification and properties of β-mannanase from soil bacterium *Bacillus circulans* M-21. J. Food. Biochem. 35: 1451–1460.

Mudau, M.M. and M.E. Setati. 2008. Partial purification and characterization of endo-β-1,4 mannanases from *Scopularipsis candida* strains isolated from solar salterns. Afri. J. Biotechnol. 7(13): 2279–2285.

Neidleman, S.L. 1992. New applications of biocatalysts. Curr. Opin. Biotechnol. 3(2): 119–23.

Nevalainen, K.M.H., S.J.T. Valentino and P.L. Bergquist. 2005. Heterologous protein expression. in filamentous fungi. Trends Biotechnol. 23: 468–74.

Norita, S., M. Rosfarizan and A.B. Ariff. 2010. Evaluation of the activities of concentrated crude mannan-degrading enzymes produced by *Aspergillus niger*. Mal. J. Microbiol. 6(2): 171–180.

Palackal, N., L. Christopher, S. Zaidi, P. Luginbuhl, P. Dupree and F. Goubet. 2007. A multifunctional hybrid glycosyl hydrolase discovered in an uncultured microbial consortium from ruminant gut. Appl. Microbiol. Biotechnol. 74(1): 113–124.

Pan, X., J. Zhou, A. Tian, K. Le, H. Yuan, Y. Xue, Y. Ma and H. Lu. 2011. High level expression of a truncated β-mannanase from alkalophilic *Bacillus* sp. N16-5 in *Kluyveromycescicerisporus*. Biotechnol. Lett. 33: 565–570.

Parisi, G.C., M. Zilli, M.P. Miani, M. Carrara, E. Bottona, G. Verdianelli, G. Battaglia, S. Desideri, A. Faedo, C. Marzolino, A. Tonon, M. Ermani and G. Leandro . 2002. High-fibre diet supplementation in patients with irritable bowel syndrome (IBS): a multicenter, randomized, open trial comparison between wheat bran diet and partially hydrolyzed guar gum (PHGG) Diges. Disea. Sci. 47(8): 1697–1704.

Petrus, J., V. Zyl, V. Moodely, S.H. Rose, R.L. Roth and W.H.V. Zyl. 2009. Production of the *Aspergillus aculeatus* endo-1,4-β-mannanase in *niger*. J. Ind. Microbiol. Biotechnol. 36: 611–617.

Pham, T.A., J.G. Berrin, E. Record, K.A. To and J.C. Sigoillot. 2010. Hydrolysis of softwood by Aspergillus mannanase: role of a carbohydratebinding module. J. Biotechnol. 148: 163–170.

Politz, O., M. Krah, K.K. Thomsen and R.R. Borriss. 2000. A highly thermostable endo-1,4-beta-mannanase from the marine bacterium *Rhodothermusmarinus*. Appl. Microbiol. Biotechnol. 53(6): 715–721.

Puls, J. and J. Schuseil. Chemistry of hemicellulose: relationship between hemicellulose structure and enzyme required for hydrolysis. pp. 1–27. *In*: M.P. Coughlan and G.P. Hazlewood [eds.]. 1993. Hemicellulose and hemicellulases. Portland, London.

Qiao, J., Z. Rao, B. Dong and Y. Cao. 2010. Expression of *Bacillussubtilis* MA-139 beta-mannanase in *Pichiapastoris* and the enzyme characterization. Appl. Biochem. Biotechnol. 160(5): 1362–1370.

Qiao, Y., X. Chen, H. Ding and M. Yue. 2008. Heterologous expression and characterization of man gene from *Bacillussubtilis* in *Pichiapastoris*. Front. Biol. China 3: 26–31.

Rashid, S.A., I. Darah and I.C. Omar. 2010. Utilization of palm kernel cake for the production of mannanase by an indigenous filamentous fungus, *Aspergillus niger* USM F4 under solid state fermentation. Int. Microbiol. 9: 1.

Rattanasuk, S. and M.K. Cairns. 2009. *Chryseobacteriumindologenes* novel mannanase-producing bacteria. Songklanakarin J. Sci. Technol. 31(4): 395–399.

Roth, R., V. Moodley and P. van Zyl. 2009. Heterologous expression and optimized production of an Aspergillus aculeatus endo-1,4-β-mannanase in *Yarrowialipolytica*. Mol. Biotechnol. 43: 112–20.

Sachslehner, A. and D. Haltrich. 1999. Purification and some properties of a thermostable acidic endo-β-1,4-D-mannanase from *Sclerotium* (Athelia) *rolfsii*. FEMS Microbiol. Lett. (177): 47–55.

Sachslehner, A., B. Nidetzky, K.D. Kulbe and D. Haltrich. 1998. Induction of mannanase, xylanase, and endoglucanase activities in *Sclerotium rolfsii*. Appl. Environ. Microbiol. 64(2): 594–600.

Said, S. and R. Pietro. 2002. Enzimas de interesse industrial e biotecnológico. 1ª edn. Ed. Eventos, Rio de Janeiro, Brazil.
Sakka, M., M. Goto, T. Fujino, E. Fujino, S. Karita, T. Kimura and K. Sakka. 2010. Analysis of a *Clostridium josui* cellulose gene cluster containing the man5A gene and characterization of a recombinant Man5A. Biosci. Biotechnol. Biochem. 74(10): 1–6.
Santos, C.R., F.M. Squina, A.M. Navarro, R. Ruller, R. Prade and M.T. Murakami. 2010. Cloning, expression, purification, crystallization and preliminary X-ray diffraction studies of the catalytic domain of a hyperthermostable endo-1,4-β-D-mannanase from *Thermotoga petrophila* RKU-1. Struc. Bio. Cryst. Communi. F66: 1078–1081.
Sá-Pereira, P., H. Paveia, M. Costa-Ferreira and M.R. Aires-Barros. 2003. A new look at xylanases. Mol. Biotechnol. 24: 257–281.
Scheller, H.V. and P. Ulvskov. 2010. Hemicelluloses.Annu. Rev. Plant Biol. 61: 263–89.
Schroder, R., T.F. Wegrzyn, N.N. Sharma and R.G. Atkinson. 2006. LeMAN4 endo-β- mannanase from ripe tomato fruit can act as a mannan transglycosylase or hydrolase. Plant 224(5): 1091–102.
Schroder, R., R.G. Atkinson and R.J. Redgwell. 2009. Re-interpreting the role of endo-β-mannanases as mannan endotransglycosylase/hydrolases in the plant cell wall. Ann. Bot. 104(2): 197–204.
Setati, M.E., P. Ademark, W.H. Van Zyl, B. Hahn-Hagerdal and H. Stalbrand. 2001. Expression of the *Aspergillusaculeatus* endo-β-1,4-mannanase encoding gene (man1) in *Saccharomyce scerevisiae* and characterization of the recombinant enzyme. Protein Expr.Purif. 21: 105–14.
Shabalin, K.A., H. Brumer and A.A. Kulminskaya. 2009. Transglycosylating and hydrolytic activities of the β-mannosidase from *Trichodermareesei*. Biochimie. 91: 632–638.
Shi, P., T. Yuan, J. Zhao, H. Huang, H. Luo, K. Meng, Y. Wang and B. Yao. 2011. Genetic and biochemical characterization of a protease-resistant mesophilic β-mannanase from *Streptomyces* sp. S27. J. Ind. Microbiol. Biotechnol. 38: 451–458. doi: 10.1007/s10295-010-0789-3.
Stalbrand, H., A. Saloheimo, J. Vehmaanpera, B. Henrissat and M. Penttila. 1995. Cloning and expression in *Saccharomyces cerevisiae* of a *Trichoderma reesei*β-mannanase gene containing a cellulose binding domain. Appl. Environ. Microbiol. 61: 1090–7.
Summpunn, P., S. Chaijan, D. Isarangkul, S. Wiyakrutta and V. Meevootisom. 2011. Characterization, gene cloning and heterologous expression of β-Mannanase from a thermophilic *Bacillus subtilis*. J. Microbiol. 49(1): 86–93.
Sunna, A. 2010. Modular organization and functional analysis of dissected modular β-mannanase CsMan26 from *Caldicellulosiruptor* Rt8B.4. Appl. Microbiol. Biotechnol. 86: 189–200.
Tanaka, M., Y. Umemoto, H. Okamura, D. Nakano, Y. Tamaru and T. Arak. 2009. Cloning and characterization of a β-1,4-mannanase 5C possessing a family 27 carbohydraate-binding module from a marine bacterium, *Vibrio* sp. strain MA-138. Biosci. Biotechnol. Biochem. 73: 109–116.
Tang, C.M., L.D. Waterman, M.H. Smith and C.F. Thurston. 2001. The cel4 gene of *Agaricusbisporus* encodes a β-mannanase. Appl. Environ. Microbiol. 67: 1303–2298.
Takeno, F., H. Yamada, K. Sekiya, B. Fujitani and K. Ohtsu. 1990. Effect of partially decomposed guar gum on high-cholesterol-fed rats and non-dietary fiber-fed rats. J. Jpn. Soc. Nutr. Food. Sci. 43: 421–425.
Titapoka, S., S. Keawsompong, D. Haltric and S. Nitisinprasert. 2008. Selection and characterization of mannanase producing bacteria useful for the formation of pre biotic manno oligosaccharides from copra meal. World J. Microbiol. Biotechnol. 24: 1425–1433.
Van Zyl, W.H., S. Rose, K. Trollope and J.F. Gorgens. 2010. Fungal β-mannanases: Mannan hydrolysis, heterologous production and biotechnological applications. Proc. Biochem. 45: 1203–13.

Vieira, W.B., L.R.S. Moreira, A. Monteiro and E.X. Ferreira. 2007. Production and characterization of an enzyme complex from a new strain of *Clostridiumthermocellum* with emphasis on its xylanase activity. Braz. J. of Microbiol. 38: 237–242.

Wang, J., Z. Shao, Y. Hong, C. Li, X. Fu and Z. Liu. 2010. A novel β- mannanase from *Pantoeaagglomerans* A021: gene cloning, expression, purification and characterization. World J. Microbiol. Biotechnol. 26: 1777–1784. doi: 10.1007/s11274-010-0358-y.

Wang, Y., A. Wong, X. Huang, D. Liu and D. Yao. 2009. Cloning, expression and characterization of mannanase from *Armillariellatabescens* EJLY2098 in *Pichiapastoris*. Sheng Wu Gong Cheng Xue Bao 25: 920–6.

Warren, R.A. 1996. Microbial hydrolysis of polysaccharides. Annu. Rev. Microbiol. 50: 183–212.

Wei, Y.H., A.J. Mao, Y.Z. He, Y. Qiao and D.Y. Dong. 2005. Expression of an endo- β-mannanase gene from *Trichodermareesei* in *Pichiapastoris*. Sheng Wu Gong Cheng Xue Bao 6: 878–83.

Weingartner, V. 2010. Produção, purificação e identificação de mannanase, obtida por fermentação no estado sólido utilizando cascas de soja e *Aspergillusniger*. Dissertação de Mestrado, Universidade Federal do Paraná, Paraná, Brasil.

World Enzymes to 2015, Freedonia Group Incorporated. www.freedoniagroup.com. Access in 01/06/2012.

Ximenes, E.A., H. Chen, I.A. Kataeva, M.A. Cotta, C.R. Felix, L.G. Ljungdahl and X.L. Li. 2005. A mannanase ManA, of the polycentric anaerobic fungus *Orpinomyces* sp. strain PC-2 has carbohydrate binding and docking modules. Can. J. Microbiol. 51: 559–68.

Xu, B., P. Hagglund, H. Stalbrand and J.-C. Janson. 2002. Endo-β-1,4-mannanases from blue mussel, *Mytilusedulis*: Purification, characterisation, and mode of action. J. Biotechnol. 92: 267–77.

Yamabhai, M., S. Emrat, S. Sukasem, P. Pesatcha, N. Jaruseranee and B. Buranabanyat. 2008. Secretion of recombinant Bacillus hydrolytic enzymes using *Escherichiacoli* expression systems. J. Biotechnol. 133: 50–57.

Yang, P., Y. Li, Y. Wang, K. Meng, H. Luo, Y. Yuan, Y. Bai, Z. Zhan and B. Yao. 2009. A novel β-mannanase with high specific activity from *Bacilluscirculans* CGMCC1554: gene cloning, expression and enzymatic characterization. App. Biochem. Biotechnol. 159: 85–94.

Yanhe, MA., Y. Xue, Y. Dou, Z. Xu, W. Tao and P. Zhou. 2004. Characterization and gene cloning of a novel beta-mannanase from alkaliphilic *Bacillus* sp. N16-5. Extremophiles 8: 447–454.

Zakaria, M.M., S. Yamamoto and T. Yagi. 1998. Purification and characterization of an endo-1,4-β-mannanase from *Bacillussubtilis* KU-1. FEMS Microbiol. Letts. 158: 25–3.

Zhang, M., X.L. Chen, Z.H. Zhang, C.Y. Sun, L.L. Chen, H.L. He, B.C. Zhou and Y.Z. Zhang. 2009. Purification and functional characterization of endo-β-mannanase MAN5 and its application in oligosaccharides production from konjac flour. Appl. Microbiol. Biotechnol. 83: 865–873.

Zhang, S.F., J.H. Song, M.C. Wu, J.P. Sheng and J.F. Li. 2008. Mutation breeding of *Aspergillus niger* strain LW-1 for high-yield β-mannanase production. Chin. Agri. Biotechnol. 16(2): 346–350.

Zhao, J., P. Shi, H. Luo, P. Yang, H. Zhao, Y. Bai, H. Huang, H. Wang and B. Yao. 2010. An acidophilic and acid-stable β-mannanase from *Phialophora* sp. P13 with high mannan hydrolysis activity undersimulated gastric conditions. J. Agric. Food. Chem. 58: 3184–3190.

Zhou, H.Y., H.Y. Pan, L.Q. Rao and Y.Y. Wu. 2011. Redesign the α/β fold to enhance the stability of mannanase Man23 from *Bacillussubtilis*. Appl. Biochem. Biotechnol. 163: 186–194. doi: 10.1007/s12010- 010-9027-8.

10

Ligninolytic Enzymes from White-rot Fungi and Application in the Removal of Synthetic Dyes

Gisele Cristina dos Santos Bazanella, Caroline Aparecida Vaz Araújo, Rafael Castoldi, Giselle Maria Maciel, Fabíola Dorneles Inácio, Cristina Giatti Marques de Souza, Adelar Bracht and *Rosane Marina Peralta**

Introduction

Large amounts of chemical dyes are used for different industrial applications such as textile and printing industries. It is estimated that between 10 and 20 percent of about 0.7 million tons of dyestuff that are manufactured each year and used in dyeing processes may be found in wasterwater (Soares et al. 2001, Arantes et al. 2006). Several of these dyes are very conducive to light, temperature and microbial attack, making them recalcitrant compounds (Nyanhongo et al. 2002). Dyes can obstruct the passage of sunlight through water resources, leading to decreased photosynthesis by aquatic

Laboratory of Biochemistry of Microorganisms, State University of Maringá, PR, Brazil.
*Corresponding author: E-mail: rmperalta@uem.br; rosanemperalta@gmail.com

plants coupled to a decreased concentration of dissolved oxygen, and to a diminished biodegradation of organic matters. Currently the removal of dyes from the effluents is brought about by physicochemical means, including adsorption, precipitation, coagulation-flocculation, oxidation, filtration and photo-degradation (Vandevivere et al. 1998, Ahn et al. 1999, Robinson et al. 2001). At the present, the biotechnological approaches have proven to be potentially effective and ecoefficient in the treatment of these pollution sources. Several microorganisms including bacteria, algae, yeast and filamentous fungi have the ability to decolorize and degrade a wide range of dyes (Forgacs et al. 2004, Pandey et al. 2007, El-Sheekh et al. 2009, Kaushik and Malik 2009). The possibility of using filamentous fungi associated with the decay of wood, usually called white-rot fungi, has attracted considerable attention. White-rot fungi possess a group of oxidative enzymes called lignin-modifying enzymes or ligninolytic enzymes involved in lignin degradation, mainly laccases and peroxidases. Due to the fact that ligninolytic enzymes are non-specific, they are able to attack a series of molecules chemically similar to lignin including pesticides, polyclyclic aromatic hydrocabons, synthetic polymers and synthetic dyes (Asgher et al. 2008).

White-rot Fungi

White-rot fungi are Basidiomycetes that participate in the biodegradation of lignin in nature, which is essential for global carbon recycling. Lignin is a polymer found in wood and vascular tissues (ten Have and Teunissen 2001). In the cell wall, lignin is covalently associated to carbohydrate components (cellulose and hemicelluloses) forming a matrix that protects the cell wall against microbial attack (Leonowicz et al. 2001). Unlike most natural polymers, lignin is irregular and non-repeating (Fig. 10.1). Its biosynthesis is the result of oxidative polymerization of several phenyl-propanoid precursors, such as coniferyl alcohol, sinapyl alcohol and *p*-coumaryl alcohol. The polymerization occurs at random by various carbon-carbon and ether bonds resulting in an irregular structure (Ralph et al. 2004).

Lignin resists the attack by most microorganisms; anaerobic processes tend not to attack the aromatic rings at all and the aerobic breakdown of lignin is slow. In nature, only basidiomyceteous white-rot fungi are able to degrade lignin efficiently. Wood decayed by white-rot fungi is pale in color because of oxidative bleaching and loss of lignin and often retains a fibrous texture. There are about 10,000 species of white-rot fungi, with varying capacities to degrade lignin, cellulose, and hemicelluloses. However, not more than three dozen have been properly studied. The majority of the studies have focused on *Phanerochaete chrysosporium*, followed by *Trametes versicolor, Bjerkandera adusta and Pleurotus* sp. (Dashtban et al. 2010).

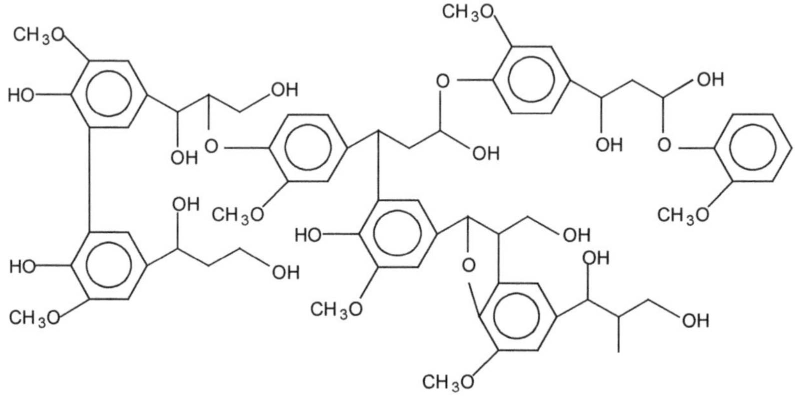

Figure 10.1 Lignin: some possible structures.

Isolation and Laboratory Maintenance of White-rot Fungi

White-rot fungi have great potential for biotechnological applications. For this reason, there has been a growing interest in screening for new white rot fungi species and strains. A classical strategy is to collect basidioma or mycelia of white rot fungi in forests, dead trees, and lignocellulosic crop residues showing signs of attack by fungi. Samples of basidioma or mycelia are aseptically transferred onto potato dextrose agar or malt extract agar and sub-cultured until obtaining pure mycelia. Identification is based on morphological, physiological, biochemical and genetic characteristics of the basidioma, hyphae and spores. More recently, molecular biology techniques are being used in the identification of new isolates. Advancement in molecular methods have permitted a more rational study of the phylogenetic relationships within the various organisms. Non-coding internal transcribed spacer regions (ITS1 and ITS2) of the ribosomal DNA seem to be one of the most frequently used analytical tools (Prewitt et al. 2008).

There are several options for maintenance of white rot fungi in the laboratory. Continuous growth methods of preservation in which white rot fungi are grown on agar (e.g., malt extract agar, potato dextrose agar and yeast extract agar) are typically used for short-term storage. For long-term storage, preservation in distilled water called the method of Castellani (Castellani 1939), mineral oil and anhydrous silica gel are some of the indicated methods. These are low cost methods but none is considered permanent. Lyophilization and liquid nitrogen refrigeration (cryo-preservation) are expensive methods but are considered permanent.

Ligninolytic Enzymes

Lignin is not used as an energy source by white-rot fungi, thus it is not a substrate of primary metabolism. The biodegradation of lignin is an oxidative process that occurs during secondary metabolism in order to gain access to the energy-rich cellulose and hemicelluloses, which are chemically bound to lignin in the plant cell wall. For becoming the most effective wood degraders in nature, the white rot fungi developed non-specific and radical-based mechanisms for degrading lignin. White-rot fungi secrete one or more of four extracellular enzymes that are involved not only in lignin degradation but also in the degradation of several pollutants. The four major ligninolytic enzymes are: lignin peroxidase (LiP, EC 1.11.1.14), manganese peroxidase (MnP, EC 1.11.1.13), versatile peroxidase (VP, EC 1.11.1.16) and laccase (Lac, EC 1.10.3.2). All these enzymes act via generation of free radicals which represent an efficient way to reach their substrates. Laccase was first described over 128-yr ago, becoming one of the oldest known enzymes (Mayer and Staples 2002). LiP and MnP were first discovered in *Phanerochaete chrysosporium* (Tien and Kirk 1984, Glenn and Gold 1985). Versatile peroxidase (VP) is a new addition to the group of ligninolytic enzymes and was discovered in a strain of *Pleurotus eryngii* (Martinez et al. 1996, Martinez 2002). The genomes of some species of white-rot fungi such as *Phanerochaete chrysosporium* (Martinez et al. 2004) and *Schizophyllum commune* (Ohm et al. 2010) have already been sequenced and are useful tools for understanding the enzymatic systems used by different fungal species in lignocelluloses degradation. *P. chysosporium* has genes encoding peroxidases but not laccases, whereas *S. commune* has genes encoding laccases but not peroxidases. Therefore, these fungi probably have different mechanisms to degrade lignin. Due to the redox potential, laccase alone can only oxidize phenolic lignin structures. The addition of a mediator, a small chemical compound, extends the substrate range to non-phenolic lignin structures. It is assumed that the mediator is needed because the large laccase molecule cannot enter the secondary cell wall and oxidize lignin directly. Most of the mediators found contain N-hydroxy-groups, such as 1-hydroxybenzotriazole (HBT), violuric acid (VA) and N-hydroxy-acetanilide (NHA). For an extensive list of native and synthetic mediators in ligninolytic systems the review articles of Wesenberg et al. (2003) and Morozova et al. (2007) can be seen.

White-rot fungi usually secrete one or more of ligninolytic enzymes in different combinations. Distribution of white-rot fungi into groups according to their enzymatic systems have been undertaken (Hatakka 1994, Tuor et al. 1995, Nerud and Misurcova 1996). Generically, white-rot fungi can be distributed into four groups, according to their ability to produce laccases and peroxidases (LiP, MnP and VP):

1) laccase and the peroxidase MnP and LiP such as in *Trametes versicolor* and *Bjerkandera adusta*;
2) laccase and at least one of the peroxidases such as in *Lentinula edodes*, *Pleurotus eryngii* and *Ceriporiopsis subvermispora*;
3) only laccase such as in *Schizophyllum commune*;
4) only peroxidases such as in *Phanerochaete chrysosporium*.

The most frequently observed ligninolytic enzymes among the white-rot fungi species are laccases and MnP and the least are LiP and VP.

The ligninolytic enzymes may function cooperatively or separately from each other, however auxiliary enzymes (unable to degrade lignin on their own) are necessary to complete the process of lignin and/or xenobiotics degradation: aryl alcohol oxidase (AAO, EC 1.1.3.7), aryl alcohol dehydrogenase (AAD, EC 1.1.1.90), glyoxal oxidase (GLOX, EC 1.2.3.5), quinone reductase (QR, EC 1.1.5.1), cellobiose dehydrogenase (CDH, EC 1.1.99.18), superoxide dismutase (SOD, EC 1.15.1.1), glucose 1-oxidase (GOX, EC 1.1.3.4), pyranose 2-oxidase (P2Ox, EC 1.1.3.4) and methanol oxidase (EC 1.1.3.13). These are mostly oxidases generating H_2O_2 and dehydrogenases. Cytochrome P450 monooxygenases are also significant components involved in the degradation of lignin and many xenobiotics. Other important components of the lignin degradative system of white-rot fungi are low molecular mass oxidants such as hydroxyl radicals (•OH) and chelated Mn^{3+} (Mn^{3+} mainly chelated by oxalic acid), produced through the action of the ligninolytic enzymes. These are particularly important during the early stages of wood decay and can also act on the degradation of xenobiotics (ten Have and Teunissen 2001, Hammel et al. 2002, Hofrichter 2002, Watanabe et al. 2002). Recent additions to the enzymatic systems of white-rot fungi include dye-decolorizing peroxidases or DyP involved in the oxidation of synthetic high redox-potential dyes and nonphenolic lignin model compounds (Liers et al. 2010) and aromatic peroxygenases or APOs that catalyze diverse oxygen transfer reactions which can result in the cleavage of ethers (Hofrichter et al. 2010, Liers et al. 2011).

Laboratory Cultivation for Production of Ligninolytic Enzymes

In recent years the traditional screening reagents tannic and gallic acid (Bavendam reaction) have been replaced by synthetic phenolic reagents such as guaiacol and syringaldazine for screening the ligninolytic capability of white-rot fungi. Therefore, guaiacol and syringaldazine are added to the malt extract agar or potato dextrose agar and the ligninolytic activity is observed as a brown halo around the microbial growth (Fig. 10.2A; Erden et al. 2009). Another very common technique for screening the ligninolytic capability of white rot fungi is to evaluate the decolorization halo of

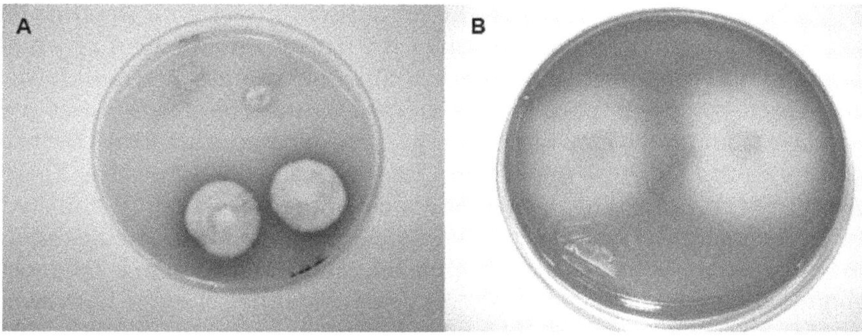

Figure 10.2 Evaluation of the ligninolytic activity on agar plates. In A: Test of oxidation of guaiacol. In B: Test of decolorization of the polymeric dye Poly R478.

Color image of this figure appears in the color plate section at the end of the book.

synthetic dyes on agar plates. The anthraquinone derivative dye remazol brilliant blue R (Machado et al. 2005) and the polymeric dye Poly R478 (Kiiskinen et al. 2004) are frequently used for this purpose (Fig. 10.2B).

Industrially important enzymes have traditionally been obtained from submerged cultures because of the ease of handling and greater control of environmental factors such as temperature, pH and aeration. Additionally, specific factors of the media composition can be relevant in the production of ligninolytic enzymes. For example, the production of ligninolytic enzymes is considerably enhanced by aromatic and phenolic compounds related to lignin or lignin derivatives, such as ferulic acid, 2,5 xylidine and veratryl alcohol (Fig. 10.3), and cations such as manganese and copper (Leonowicz et al. 2001, Lo et al. 2001, Nyanhongo et al. 2002, Gill and Arora 2003, Souza et al. 2004, Lorenzo et al. 2006, Twchanowicz et al. 2006, Saeki et al. 2011). Factors like these are more easily controlled in submerged cultures. Several recent contributions have described the production of laccase and peroxidases, MnP, LiP and VP, in submerged cultures (Linke et al. 2005, Lorenzo et al. 2006, Rogalski et al. 2006, Janusz et al. 2007, Tellez-Tellez et al. 2008, Johnsy and Kaviyarasan 2011, Ding et al. 2012).

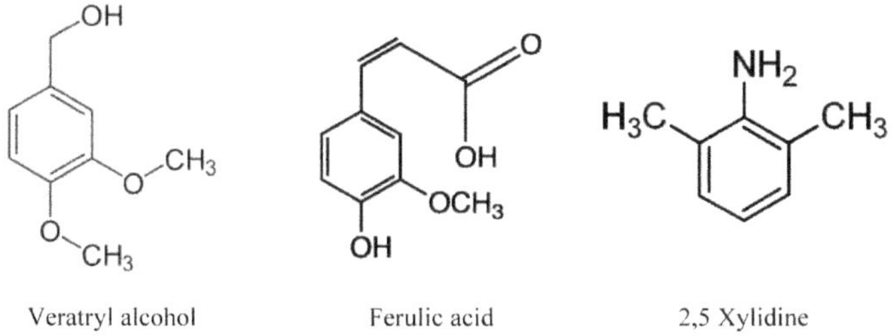

Figure 10.3 Chemical structures of some important inducers of ligninolytic enzymes.

Solid state culture (SSC) is also an attractive option for ligninolytic enzyme production because it mimics the natural environment of white rot fungi and presents higher productivity, simpler operation, and lower cost when compared with the submerged culture. SSC reproduces the conditions under which white rot fungi grow in nature and allows the use of a series of food, agricultural, and forestry wastes as substrates. These wastes are rich in carbohydrates, and most of them are easily assimilated by microorganisms and for this reason can be used for their cultivation in SSC without the necessity of supplementation. Wheat bran is the most commonly used substrate for the cultivation of white-rot fungi. However, the list of possibilities is very large and includes several lignocellulolytic wastes such as cane bagasse, corn cob, wheat straw, oat straw, rice straw, and food processing wastes such as banana, kiwi fruit, and orange wastes (Holker et al. 2004, Tychanowicz et al. 2004, Alexandrino et al. 2007, Osma et al. 2007, Couto 2008). However it is worth searching for new substrates, especially if they are available in large amounts, allowing the growth of white-rot fungi without further supplementations, and obtaining valuable products (Zilly et al. 2012).

Environmental Applications of Ligninolytic Enzymes

Ligninolytic enzymes are highly non-specific and, thus, capable of degrading a wide range of recalcitrant compounds structurally similar to lignin. In recent years, the capability of white rot fungi and their enzymes to biodegrade several xenobiotics and recalcitrant pollutants has generated considerable research interest in the area of industrial/environmental microbiology. As a consequence, a considerable number of reviews detailing the numerous characteristics and applications of ligninolytic enzymes have been published (Conesa et al. 2002, Hofrichler 2002, Mayer and Staples 2002, Wesenberg et al. 2003, Baldrian 2006, Kersten and Cullen 2007, Ruiz-Duenas and Martinez 2009). Among the industrial applications of the ligninolytic enzymes found in WRF are a). paper whitening, b). degradation of industrial dyes from the textile industry and c). degradation of xenobiotic compounds, including polycyclic aromatic hydrocarbons, phenolics, herbicides and other pesticides (Asgher et al. 2008).

Synthetic Dyes

Synthetic dyes are aromatic compounds. According to their dissociation in an aqueous solution, dyes can be classified as acid, direct reactive dyes (anionic), basic dyes (cationic) and disperse dyes (nonionic). They are used in textiles, paper printing, color photography, cosmetics, plastics, pharmaceutical and leather industries. Synthetic dyes have been used increasingly in textile and dyeing industries because of their ease and cost effectiveness in synthesis, firmness and variety in color when compared to that of natural dyes.

About 100,000 commercial dyes are manufactured including several varieties of dyes such as acidic, basic, reactive, azo, diazo and anthraquinone based meta complex dyes. Over 10,000 dyes with an annual production of over 7×10^5 metric tons are commercially available (Campos et al. 2001). Depending on the class of the dye, upto 50 percent of them are released into the industrial effluents. Colored industrial effluents from the dyeing industries represent major environmental problems. Unbound reactive dyes undergo hydrolysis favored by the temperatures and pH values during the dyeing processes. Many dyes are visible in water at concentrations as low as 1 mg/L and have a huge impact on the aquatic environment caused by its turbidity and high pollution strength, in addition to the toxic degradation products that can be formed. Dyes that absorb light with wavelength in the visible range (350–700 nm) are colored. Dyes contain chromophoric electron systems with conjugated double bonds and auxochromes. The latter are electron withdrawing or electron donating substituents that cause the color of the chromophore by altering the overall energy of the electron system. Usual chromophores are –C=C–, –C=N–, –C=O–, –N=N–, $-NO_2$ and quinoid rings, and usual auxochromes are $-NH_3$, –COOH, $-SO_3H$ and –OH.

Dyes can also be classified according to the chemical structure of the chromophoric group. According to this criterion they can be classified, for example, as azo, anthraquinone, triphenylmethane, heterocyclic, and polymeric dyes (Fig. 10.4). Among these, the versatile azo and triphenylmethane dyes account for most textile dyestuffs produced.

Methods for Evaluating the Removal of Synthetic Dyes

The most simple and common methods to evaluate dye decolorization are the spectrophotometric assays. To test the ability of white rot fungi to decolorize synthetic dyes *in vivo*, a dye is added to submerged, liquid or solid state cultures and at different times of cultivation, the decrease of visible light absorbance at the wavelength of maximum absorbance of the dye is measured in a spectrophotometer. In the same way, it is possible to evaluate the capability of a crude or purified ligninolytic enzyme to decolorize a specific dye. As examples, the decolorization of RBBR, Congo red and poly R478 can be estimated at 595, 497 and 520 nm, respectively (Boer et al. 2004). A more accurate method is to obtain UV-VIS or VIS spectra of the material before and after treatment (Fig. 10.5).

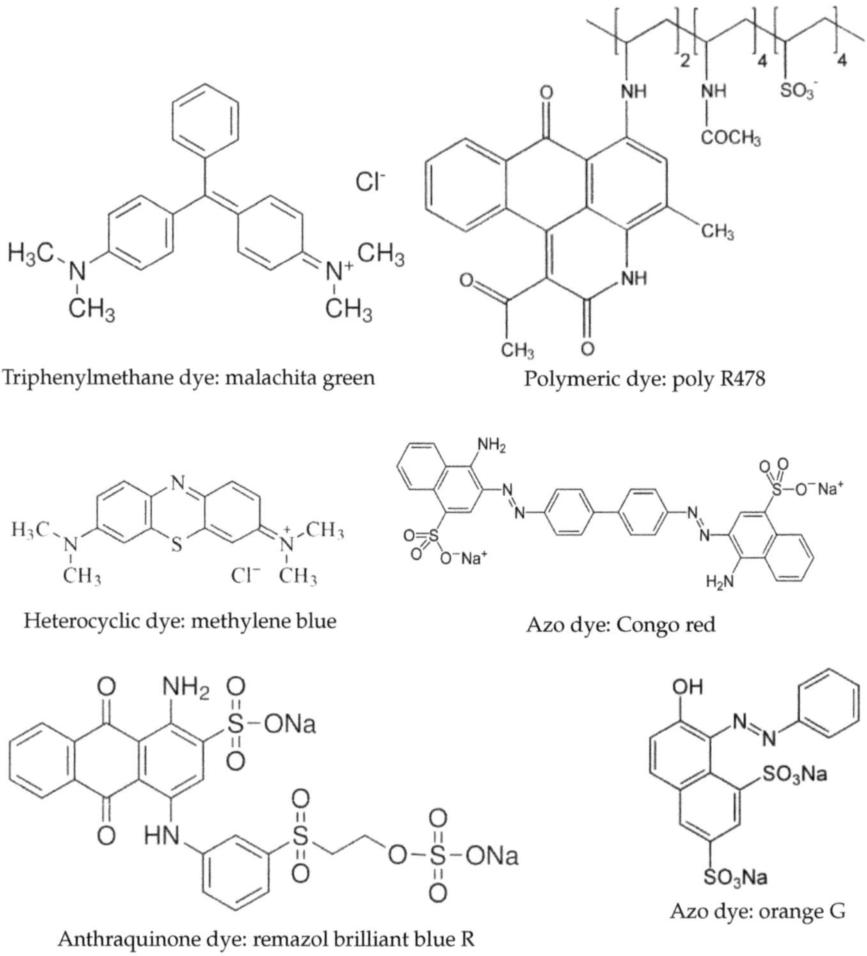

Figure 10.4 Some examples of azo, anthraquinone, triphenylmethane, heterocyclic and polymeric dyes.

Although the majority of studies on dye decolorization are carried out using spectrophotometric analysis, this technique is limited as an analytical tool as it allows only the evaluation of the chemical modifications that occur in the chromophore. New methodologies have been recently introduced with the objective of identifying soluble and insoluble compounds produced by the action of ligninolytic enzymes when acting on the dyes. Some examples are:

1. Zille et al. (2005) used liquid chromatography-mass spectrometry (LC-MS) and ^{13}C-nuclear magnetic resonance (^{13}C-NMR) to identify, soluble and insoluble compounds produced from the decomposition

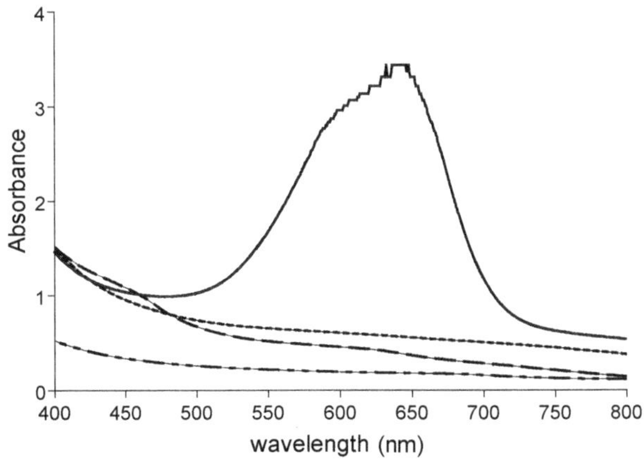

Figure 10.5 Visible spectra of laundry effluent before and after treatment with *Ganoderma lucidum, Pycnoporus* sp. and *Trametes* sp. crude laccases. Crude laccases from white rot fungi were added to an effluent containing the textile effluent and the mixtures maintained under agitation of 120 rpm for 2 hr at 28°C Before treatment (continues line); after treatment with laccase from *G. lucidum, Pycnoporus* sp. *Trametes* sp. (dashed lines) (unpublished data).

of the azo dye 3-(2-hydroxy-1-naphthylazo) by *Trametes villosa* laccase. By using LC-MS it was possible to identify at least two metabolites: hydroxybenzenesulfonic acid and benzenesulfonic acid. ^{13}C-NMR allowed the identification of compounds produced by polymerization reaction during the long-term batch decolorization processes of the azo dye.

2. Michniewicz et al. (2008) used a high performance liquid chromatography system equipped with a diode array detector (HPLC-DAD) to evaluate the modifications caused by a purified *Cerrena unicolor* laccase in the anthraquinone dye acid blue 62. The analysis did not allow the chemical identification of the oxidation products, but revealed the formation of at least two new more polar oxidation products.
3. Murugesan et al. (2009b) used HPLC and liquid chromatography-mass spectrometry-electrospray ionization (LC-ESI-MS) analysis and concluded that *N*-demethylation was the key mechanism of decolorization of the tryphenylmethane dye malachita green by *Ganoderma lucidum* laccase.
4. Zhao et al. (2006) and Zhao and Hardin (2007) used HPLC-DAD to analyze the metabolites produced from the degradation of the azo dyes (in special disperse orange 3) by *Pleurotus ostreatus*. In these studies, no attempt was made to correlate the dye degradation with ligninolytic enzymes, but nitrobenzene, 4-nitroanisole, 4-nitrophenol and 4-nitroaniline were identified as metabolite products of dye degradation.

An aspect normally neglected in dye decolorization studies is the toxicity changes occurring during fungal biodegradation. It is well known that synthetic dyes are toxic, most of them carcinogenic (Wesenberg et al. 2003), but the question is: how much safer are the products formed through chemical transformation of synthetic dyes by white rot fungi and their ligninolytic enzymes? The knowledge of this point has tremendous importance, but only a few works have attempted to present answers. Some examples are:

1. Ramsay and Nguyen (2002) used the white rot fungus *Trametes versicolor* to decolorize eight textile dyes and also evaluated the toxicity of the material before and after the treatment using a Microtox 500 Analyzer (Azur Environmental, Newark, Del.). After decolorization the toxicity of the solutions containing Amaranth, Tropaeolin O and Reactive Black 5 was unchanged; Reactive Blue 15, Remazol Brilliant Blue R and Cibacron Brilliant Red 3G-P decreased to non-toxic levels; and Cibacron Brilliant Yellow 3B-A and Congo Red became more toxic.
2. Abadulla et al. (2000) observed a reduction of toxicity (based on the oxygen consumption rate of *Pseudomonas putida*) of triarylmethane, indigoid, azo, and anthraquinonic dyes after treatment with an immobilized laccase from *Trametes hirsuta*.
3. Apohan and Yesilada (2005) observed a reduction in the toxicity of azo dyes, astrazon blue and astrazon red, after treatment with pellets of the white rot fungus *Funalia trogii*.
4. Anastasi et al. (2010) used Mn-peroxidase and Mn-independent peroxidase from *Bjerkandera adusta* to decolorize 13 industrial dyes. Toxicity of the simulated wastewaters was also evaluated, before and after the fungal treatment, by means of a phytotoxicity test on *Cucumis sativus* L. seeds.
5. Casieri et al. (2008) used *Trametes pubescens* and *Pleurotus ostrealus*, immobilized on polyurethane foam cubes in bioreactors to decolorize two anthraquinone dyes (B49 and RBBR) and one azo dye (R243). To evaluate the toxicity before and after treatments, the authors used the *Lemna minor* (duckweed) ecotoxicity test.
6. Zhuo et al. (2011) used laccase-producing white-rot fungi strain *Ganoderma* sp. En3 to decolorize synthetic dyes and real textile dye effluent. Phytotoxicity study with respect to *Triticum aestivum* and *Oryza sativa* demonstrated that *Ganoderma* sp. En3 was able to detoxify four synthetic dyes, two simulated dye effluents and the real textile dye effluent.
7. Eichlerová et al. (2007) used a bioassay based on the growth inhibition of duckweed *Lemna minor* to evaluate the toxicity of synthetic dyes

orange G ans remazol brilliant blue before and after treatment with *Dichomitus squalens* (static and skaken cultures). The toxicity of both dyes decreased after the decolorization process.

Ligninolytic Enzymes and Dye Decolorization

Representative studies published after the year 2000 where ligninolytic enzymes were used for removing synthetic dyes are presented below. Only those studies are listed in which direct or strong indirect evidence is presented about the participation of enzymes in the dye decolorization processes. For analyses of the articles published before 2000, see previous reviews such as Wesenberg et al. 2003, Forgacs et al. 2004, Husain 2006.

1. Several structurally different synthetic dyes (including azo, triphenylmethane, heterocyclic and polymeric dyes) were decolorized in submerged and solid cultures of *Pleurotus pulmonarius* producing laccase as the main ligninolytic enzyme. Both laccase activity and dye decolorization were related to glucose and ammonium starvation or to induction by ferulic acid. Additionally, decolorization and laccase activity were equally affected by pH and temperature. Laccase can thus be considered to be the major enzyme involved in the ability of *P. pulmonarius* (Zilly et al. 2002).
2. Crude laccase enzyme from *Ganoderma lucidum* obtained in wheat bran solid state cultures demonstrated excellent decolorization activity to anthraquinone dye remazol brilliant blue R (RBBR) without redox mediator, whereas the diazo dye remazol black-5 (RB-5) requires a redox mediator. N-hydroxybenzotriazole at a concentration of 1 mM was the best redox mediator (Murugesan et al. 2007).
3. *Ganoderma lucidum* laccase obtained from wheat bran solid state cultures was able to tolerate mixtures of several metal ions. Treatment of simulates reactive dye effluent (RBBR and RB-5) by laccase showed that the redox mediator system is necessary for effluent decolorization. Syringaldehyde, a natural redox mediator, was more effective than the synthetic mediator 1-hydroxybenzotriazole (HBT) (Murugesan et al. 2009a).
4. *G. lucidum* laccase obtained from wheat bran solid state cultures was able to decolorize the dye malachite green. The dye decolorization was improved when natural phenolic compounds such as acetovanillone, *p*-coumaric acid, ferulic acid, syringaldehyde and vanillin act as redox mediators. Characterization of MG transformation products by HPLC, UV-VIS and liquid chromatography-mass spectrometry-electrospray ionization analysis revealed that N-demethylation was the key mechanism of decolorization of MG by laccase (Murugesan et al. 2009b).

5. *G. lucidum* laccase obtained in passion fruit waste solid state cultures was used to decolorize RBBR. The enzyme was inhibited by NaCl but not by Na_2SO_4. Inhibition by NaCl was of the mixed type with two different inhibition constants. The enzyme was able to completely decolorize RBBR in the presence of 1.0 M Na_2SO_4 and 50 percent decolorization was obtained in the presence of 0.1 M NaCl (Zilly et al. 2011).
6. Direct dye solar golden yellow R was decolorized by *G. lucidum* Mn peroxidase obtained from submerged culture. Various parameters such as optimum pH, incubation time, temperature and amounts of carbon and nitrogen additives were optimized to achieve maximal decolorization of the dye. Addition of 1 percent starch accelerated the decolorization (96 percent), while the addition of nitrogen sources inhibited both enzyme induction and dye decolourisation (Bibi et al. 2009).
7. The dyes orange G and RBBR were treated using ligninolytic enzymes from *Dichomitus squalens*, *Ischnoderma resinosum* and *Pleurotus calyptratus* obtained in liquid cultures. All white rot fungi efficiently decolorized both Orange G and RBBR, but they differed in decolorization capacity depending on cultivation conditions and ligninolytic enzyme production. Two different decolorization patterns were found in these strains: Orange G decolorization in *I. resinosum* and *P. calyptratus* was caused mainly by laccase, while RBBR decolorization was effected by manganese peroxidase (MnP); in *D. squalens* laccase and MnP cooperated in the decolorization processes (Eichlerová et al. 2005, 2007).
8. Anthraquinone and azo dyes were decolorized by *Cerrena unicolor* laccase obtained from submerged cultures. The capability of crude and purified laccase were evaluated for removing different classes of dyes using kinetic studies and HPLC analysis. The decolorization process can be satisfactorily described by the Michaelis–Menten kinetic model. The most suitable pH for decolorization was 3.5. Laccase was able to decolorize recalcitrant azo dyes all without any redox mediators (Michniewicz et al. 2008).
9. Two purified laccases (POXC and POXA3) from *Pleurotus ostreatus* obtained from liquid cultures were able to degrade RBBR *in vitro*, in the absence of any redox mediators. These laccases differ significantly in their efficiency of decolorization of the tested dye, as suggested by comparison of their catalytic efficiency (k_{cat}/K_m values) towards RBBR. Furthermore, using a mixture of both POXC and POXA3 a remarkable improvement in the reaction rate and in the final level of dye decolorization was observed. The extent of RBBR decolorization by laccase mixture also depends on incubation temperature and enzyme

concentration. The dye is decolorized by laccase isoenzymes most efficiently under acidic conditions. Treatment of RBBR with the laccase mixture reduced its toxicity by 95 percent (Palmieri et al. 2005).

10. One laccase isoenzyme (LAC1) from *Pycnoporus sanguineus* obtained from malt extract cultures was able to decolorize several dyes. Regarding the activity profile of the biocatalyst against pH, salts, temperature and target substrates, LAC-1 appears to be a good candidate for application in acid dye bath treatments. Studying the model anthraquinonic dye ABu62 decolorization, it was proved that this dye was a good substrate for LAC-1. LAC-1 showed an unusual kinetic behavior, suggesting that LAC-1 was activated in presence of ABu62 (Trovaslet et al. 2007).
11. *Trametes villosa* laccase (Novo Nordisk, Denmark) was used for direct azo dye degradation, and the reaction products that accumulated after 72 hr of incubation were analyzed. Liquid chromatography-mass spectrometry (LC-MS) analysis showed the formation of phenolic compounds during the dye oxidation process as well as a large amount of polymerized products that retain azo group integrity. The aminophenol reactions were also investigated by ^{13}C-nuclear magnetic resonance and LC-MS analysis, and the polymerization character of laccase was shown. This study highlights the fact that laccases polymerize the reaction products obtained during long-term batch decolorization processes with azo dyes. These polymerized products provide unacceptable color levels in effluents, limiting the application of laccases as bioremediation agents (Zille et al. 2005).
12. Various textile dyes were decolorized with the immobilized *Trametes hirsuta* laccase, and the toxicity of the degradation products was determined. The aquatic toxicities of samples were evaluated based on their inhibitory effects on the oxygen consumption rate of *Pseudomonas putida* (Abadulla et al. 2000).
13. The capability of white-rot fungi *Irpex lacteus* to decolorize reactive Orange 16 and Remazol Brilliant Blue R was mainly related to a laccase-like enzyme activity associated with fungal mycelium. The highest enzymatic activity was obtained with ABTS as substrate. Enzyme activity was fully inhibited with 50 mM NaN_3. Depending on the chemical structure of dyes, redox mediators had a positive effect on the dye decolorization by fungal mycelium. Enzyme isolated from fungal mycelium was able to decolorize synthetic dyes *in vitro* (Svobodová et al. 2008).
14. Native polyacrylamide gel electrophoresis (PAGE) and sodium dodecyl sulfate PAGE confirmed the role of crude laccase from *Rigidoporus lignosus* in the decolorization of remazol brilliant blue and malachite green. Efficient enzymatic decolorization without redox mediator was

achieved. Unusual storage stability at alkaline pH was observed, with the laccase activity being enhanced 1.5–2-fold after 3 hr of incubation (Li et al. 2009).

15. Original textile industry effluents were decolorized using *G. lucidum* Mn peroxidase obtained in submerged cultures. Process optimization could improve color removal efficiency of the fungus to 95 percent within only 2 d, catalyzed by Mn peroxidase (1295 U/L) as the main ligninolytic enzyme at pH 3 and 35°C using 1 percent starch supplemented Kirk's basal medium (Asgher et al. 2010).
16. Direct dye solar golden yellow R was decolorized by *Ganoderma lucidum* Mn peroxidase obtained from submerged culture. Various parameters such as optimum pH, incubation time, temperature and amounts of carbon and nitrogen additives were optimized to achieve maximal decolorization of the dye. Addition of 1 percent starch accelerated the decolorization (96 percent). Addition of nitrogen sources inhibited both enzyme induction and dye decolorization (Bibi et al. 2009).
17. The role of Mn peroxidase in dye decolorization of polymeric dye Poly R478 was assayed with isoenzyme PCH4 from *Phanerochaete chrysosporium* and isoenzyme BOS2 from *Bjerkandera* sp. BOS55. The assays were conducted with semi-continuous addition of H2O2. The two pure MnP have different catalytic properties because MnP from *Bjerkandera* produced a certain decolorization in the absence of Mn+2, whereas MnP from *P. chrysosporium* had very little effect. The enzymatic treatment provoked not only the destruction of the chromophoric groups but also a noticeable breakdown of the chemical structure of the dye. As the experiments were conducted with pure enzymes, MnP proved to be the main factor responsible for the dye decolorization (Moreira et al. 2001a).
18. A novel strain of *Bjerkandera* sp. (B33/3), with particularly high decolorization activities upon Poly R-478 and Remazol Brilliant Blue R (RBBR) dyes, was isolated. The role of the ligninolytic extracellular enzymes produced by this strain on decolorization of RBBR was studied in some depth. The basis of decolorization is an enzyme-mediated process, in which the main enzyme responsible is a recently described peroxidase with capacity for oxidation of manganese, as well as veratryl alcohol and 2,6-dimethoxyphenol in a manganese-independent reaction (Moreira et al. 2001b).
19. From the extracellular fluid of a novel strain of *Bjerkandera* sp., it isolated, purified and identified the main enzyme responsible for Remazol Brilliant Blue R dye decolorization. Such an enzyme is able to oxidize manganese, as well as veratryl alcohol and 2,6-dimethoxyphenol in manganese-independent reactions; hence, it can be included in the new group of versatile peroxidases. The enzyme substrate range for

oxidation of several phenolic and non-phenolic aromatic compounds was determined and the corresponding Michaelis–Menten kinetic constants calculated (Moreira et al. 2006).

20. Manganese-peroxidase and manganese-independent peroxidase from *Bjerkandera adusta* were correlated with capability of fungus to decolorize 13 industrial dyes. In order to evaluate its true bioremediation potential, the fungus was packed in a fixed-bed bioreactor, for treatment of large volumes of real wastewater. The fungus result was effective during 10 cycles of decolorization, remaining active for a very long period, in non-sterile conditions. Toxicity of the simulated wastewaters was also evaluated, before and after the fungal treatment, by means of a phytotoxicity test on *Cucumis sativus* L. seeds (Anastasi et al. 2010).
21. *Phanerochaete chrysosporium* was able to decolorize the azo dye orange in submerged cultures. The dye decolorization was dependent of pH and temperature. The main ligninolytic associated to dye descolorization was Mn peroxidase (Sharma et al. 2009).
22. Manganese peroxidase was the main enzyme involved in dye decolorization of decolorize azo dyes Reactive Orange 16 and Naphthol Blue Black by two white-rot fungi, *Irpex lacteus* and *Phanerochaete chrysosporium*, A significant increase in the Reactive Orange 16 decolorization by the agitated *I. lacteus* cultures was observed after adding 0.1 percent Tween 80, following a higher Mn-dependent peroxidase production. The *in vitro* dye decolorization using the purified enzyme proved its decolorization ability (Svodová et al. 2006).
23. Manganese peroxidase (MnP) from *Phanerochaete chrysosporium* was also responsible by the decolorization of a series of sulfonphthalein dyes, such as bromocresol green, bromophenol blue (Cristian et al. 2003).
24. Manganese peroxidase was the main enzyme responsible by decolorization of azo and anthraquinone dyes by *Phanerochaete sordida*. Decolorization of dye mixtures by *P. sordida* was partially inhibited by polyvinyl alcohol (PVA) that wastewaters from textile industries often contain (Harazono and Nakamura 2005).

Concluding Remarks

The majority of synthetic dyes are chemically stable and resistant to microbiological attack. White rot fungi are excellent candidates for biodegradation of these recalcitrant compounds. The isolation of new strains or the adaptation of existing ones to the decomposition of dyes will probably increase the efficacy of dye bioremediation in the near future.

The use of ligninolytic enzyme preparations can bring considerable benefits over the direct use of white rot fungi, considering that preparations can be more easily standardized, facilitating accurate dosage. Enzyme application is simpler than the use of microorganisms and can be rapidly modified according to the characteristics of the dye or dyes to be removed. In addition to this, analysis of metabolite compounds produced by enzyme preparations is easier than analysis of metabolites produced by using the whole white rot fungus.

There is a need for including new methodologies in elucidating the chemical structures of metabolites produced by the action of ligninolytic enzymes. This should also explain the metabolic routes used by white rot fungi in dye degradation. Furthermore, the verification that the process of decolorization corresponds to detoxification is a crucial step in order to indicate a real application.

Acknowledgements

This work was supported by grants from the Fundação Araucária (PRONEX 276/08) and Conselho Nacional de Desenvolvimento Científico e Tecnológico (CNPq, Edital 047/2010-SISBIOTA, proc. 563260/2010-6). R.M. Peralta and A. Bracht are fellowship holders of CNPq. C.V. Araújo, G.M. Maciel, R. Castoldi, and F.D. Inácio are fellowship holders of the Coordenação de Aperfeiçoamento de Pessoal de Ensino Superior (CAPES).

References

Abadulla, T., T. Tzanov, S. Costa, K.H. Robra, A. Cavaco-Paulo and G.M. Gubitz. 2000. Decolorization and detoxification of textile dyes with a laccase from *Trametes hirsuta*. Appl. Environ. Microbiol. 66: 3357–3362.

Ahn, D., W. Cahng and T. Yoon. 1999. Dyestuff wastewater treatment using chemical oxidation, physical adsorption and fixed bed biofilm process. Process Biochem. 34: 429–439.

Alexandrino, A.M., H.G. Faria, C.G. M.Souza and R.M. Peralta. 2007. Aproveitamento do resíduo de laranja para a produção de enzimas lignocelulolíticas por *Pleurotus ostreatus*. Cien. Tecnol. Alim. 27: 364–368.

Anastasi, A., F. Spina, V. Prigione, V. Tigini, P. Giansanti and G.C. Varese. 2010. Scale-up of a bioprocess for textile wastewater treatment using *Bjerkandera adusta*. Biores. Technol. 101: 3067–3075.

Apohan, E. and O. Yesilada. 2005. Role of white rot fungus *Funalia trogii* in detoxification of textile dyes. J. Bas. Microbiol. 45: 99–105.

Arantes, V., C. Baldocchi and A.M.F. Milagres. 2006. Degradation and decolorization of a biodegradable-resistent polymeric dye by chelator-mediated Fenton reactions. Chemosphere 63: 1764–1772.

Asgher, M., H. Bhatti, M. Ashradf and R. Legge. 2008. Recent developments in biodegradation of industrial pollutants by white rot fungi and their enzyme system. Biodegradation 19: 771–783.

Asgher, M., S. Noreen and H.N. Bhatti. 2010. Decolorization of dye-containing textile industry effluents using *Ganoderma lucidum* IBL-05 in still cultures. Water Environ. Res. 82: 357–361.

Baldrian, P. 2006. Fungal laccases—occurrence and properties. FEMS Microbiol. Rev. 30: 215–242.

Bibi, I., N. Bhatti and M. Asgher. 2009. Decolourisation of direct dyes with manganese peroxidase from white rot basidiomycete *Ganoderma lucidum* IBL5. Can. J. Chem. Eng. 87: 435–440.

Boer, C.G., L. Obici, C.G.M. Souza and R.M. Peralta. 2004. Decolorization of synthetic dyes by solid state cultures of *Lentinula* (*Lentinus*) *edodes* producing manganese peroxidase as the main ligninolytic enzyme. Biores. Technol. 94: 107–112.

Campos, R., A. Kandelbauer, K.H. Robra, C.P. Artur and G.M. Gubitz. 2001. Indigo degradation with purified laccases from *Trametes hirsuta* and *Sclerotim rolfsii*. J. Biotechnol. 8: 131–139.

Casieri, L., G.C. Varese, A. Anastasi, V. Prigione, K. Svobodova and V. Filippelo Marchisio. 2008. Decolorization and detoxication of reactive industrial dyes by immobilized fungi *Trametes pubescens* and *Pleurotus ostreatus*. Folia Microbiol. 53: 44–52.

Castellani, A. 1939. Viability of some pathogenic fungi in distilled water. J. Trop. Med. Hyg. 42: 225–226.

Conesa, A., P.J. Punt and C. van den Hondel. 2002. Fungal peroxidases: molecular aspects and applications. J. Biotechnol. 93: 143–158.

Couto, S.R. 2008. Exploitaion of biological wastes for the production of value-added pruducts under solid state fermentation conditions. Biotechnol. J. 3: 859–870.

Cristian, V.V., R. Shrivastava, C. Novotny and B.R. Vyas. 2003. Decolorization of sulfonphthalein dyes by manganese peroxidase activity of the white-rot fungus *Phanerochaete chrysosporium*. Folia Microbiol. 48: 771–774.

Dashtban, M., H. Schraft, T. Syed and W. Qin. 2010. Fungal biodegradation and enzyme modification of lignin. Int. J. Biochem. Mol. Biol. 1: 36–50.

Ding, Z., L. Peng, Y. Chen, L. Zhang, Z. Gu, G. Shi and K. Zhang. 2012. Production and characterization of thermostable laccase from the mushroom *Ganoderma lucidum* using submerged fermentation. Afri. J. Microbiol. Res. 6: 1147–1157.

Eichlerová, I., L. Homolka, L. Lisá and F. Nerud. 2005. Orange G and Remazol Brilliant Blue R decolorization by white rot fungi *Dichomitus squalens, Ischnoderma resinosum* and *Pleurotus calyptratus*. Chemosphere 60: 398–404.

Eichlerová, I., L. Homolka, O. Benada, O. Kofronová, T. Hubálek and F. Nerud. 2007. Decolorization of Orange G and Remazol Brilliant Blue R by the white rot fungus *Dichomitus squalens*: toxicological evaluation and morphological study. Chemosphere 69: 795–802.

El-Sheekh, M.M., M.M. Gharieb, and G.W. Abou-El-Souod. 2009. Biodegradation of dyes by some green algae and cyanobacteria. Int. Biodeter. Biodegr. 63: 699–704.

Erden, E., M.C. Ucar, T. Gezer and N.K. Pazarlioglu. 2009. Screening for ligninolytic enzymes from autochthonous fungi and applications for decolorization of Remazole Marine Blue. Braz. J. Microbiol. 40: 346–353.

Forgacs, E., T. Cserháti and G. Oros. 2004. Removal of synthetic dyes from wastewaters: a review. Environ. Int. 30: 953–971.

Gill, P.K. and D.S. Arora. 2003. Effect of culture conditions on manganese peroxidase production and activity by some white rot fungi. J. Ind. Microbiol. Biotechnol. 30: 28–33.

Glenn, J.K. and M.H. Gold. 1985. Purification and characterization of an extracellular Mn(II)-dependent peroxidase from the lignin-degrading basidiomycete *Phanerochaete chrysosporium*. Arch. Biochem. Biophys. 242: 329–341.

Hammel, K.E., A.N. Kapich, K.A. Jensen and Z.C. Ryan. 2002. Reactive oxygen species as agents of wood decay by fungi. Enzyme Microb. Tech. 30: 445–453.

Harazono, K. and K. Nakamura. 2005. Decolorization of mixtures of different reactive textile dyes by the white-rot basidiomycete *Phanerochaete sordida* and inhibitory effect of polyvinyl alcohol. Chemosphere 59: 63–68.

Hatakka, A. 1994. Lignin-modifying enzymes from selected white-rot fungi - production and role in lignin degradation. FEMS Microbiol. Rev. 13: 125–135.

Hofrichter, M. 2002. Review: lignin conversion by manganese peroxidase (MnP). Enzyme Microb. Tech. 30: 454–466.

Hofrichter, M., R. Ullrich, M.J. Pecyna, C. Liers and T. Lundell. 2010. New and classic families of secreted fungal heme peroxidases. Appl. Microbiol. Biotechnol. 87: 871–897.

Holker, U., M. Hofer and J. Lenz. 2004. Biotechnological advantages of laboratory-scale solid-state fermentation with fungi. Appl. Microbiol. Biotechnol. 64: 175–186.

Husain, Q. 2006. Potential applications of the oxidoreductive enzymes in the decolorization and detoxification of textile and other Synthetic dyes from polluted water: A Review. Crit. Rev. Biotechnol. 26: 201–221.

Janusz, G., J. Rogalski and J. Szczodrak. 2007. Increased production of laccase by *Cerrena unicolor* in submerged liquid cultures. World J. Microbiol. Biotechnol. 23: 1459–1464.

Johnsy, G. and V. Kaviyarasan. 2011. Nutrient composition on the effect of extracellular peroxidase production by *Lentinus kauffmanii* under submerged culture condition. Int. J. Chem. Technol. Res. 3: 1563–1570.

Kaushik, P. and A. Malik. 2009. Fungal dye decolourization: recent advances and future potential. Int. Biodeter. Biodegr. 35: 127–141.

Kersten, P. and D. Cullen. 2007. Extracellular oxidative systems of the lignin-degrading Basidiomycete *Phanerochaete chrysosporium*. Fungal Gen. Biol. 44: 77–87.

Kiiskinen, L.L., M. Rättö and K. Kruus. 2004. Screening for novel laccase-producing microbes. J. Appl. Microbiol. 97: 640–646.

Leonowicz, A., N.-S. Cho, J. Luterek, A. Wilkolazka, M. Wotjas-Wasilewska, A. Matus-Zewska, M. Hofrichter, D. Wesenberg and J. Rogalski. 2001. Fungal laccase: properties and activity on lignin. J. Bas. Microbiol. 41: 185–227.

Li, L., W. Dai, P. Yu, J. Zhao and Y. Qu. 2009. Decolorisation of synthetic dyes by crude laccase from *Rigidoporus lignosus* W1. J. Chem. Technol. Biotechnol. 84: 399–404.

Liers, C., C. Bobeth, M. Pecyna, R. Ullrich and M. Hofrichter. 2010. DyP-like peroxidases of the jelly fungus *Auricularia auricula-judae* oxidize nonphenolic lignin model compounds and high-redox potential dyes. Appl. Microbiol. Biotechnol. 85: 1869–1879.

Liers, C., T. Arnstadt, R. Ullrich and M. Hofrichter. 2011. Patterns of lignin degradation and oxidative enzyme secretion by different wood- and litter-colonizing basidiomycetes and ascomycetes grown on beech-wood. FEMS Microbiol. Ecol. 78: 91–102.

Linke, D., H. Bouws, T. Peters, M. Nimtz, R.G. Berger and H. Zorn. 2005. Laccases of *Pleurotus sapidus:* characterization and cloning. J. Agric. Food Chem. 53: 9498–9505.

Lo, A.S.C., Y.S. Ho and J.A. Buswell. 2001. Effect of phenolic monomers on the production of laccases by the edible mushroom *Pleurotus sajor-caju* and partial characterization of a major laccase component. Mycologia 93: 413–421.

Lorenzo, M., D. Moldes and M.A.A. Sanromán. 2006. Effect of heavy isoenzymes by *Trametes versicolor* and on their ability to decolourize dyes. Chemosphere 63: 912–917.

Lundell, T.K., M.R. Makela and K. Hilden. 2010. Lignin-modifying enzymes in filamentous basidiomycetes—ecological, functional and phylogenetic review. J. Bas. Microbiol. 50: 5–20.

Lyra E.S., K.A. Moreira, T.S. Porto, M.N. Carneiro da Cunha, F.B. Paz Júnior, B.B. Neto, J.L. Lima-Filho, M.A.Q. Cavalcanti, S. Converti and A.L.P. Porto. 2009. Decolorization of synthetic dyes by basidiomycetes isolated from woods of the Atlantic Forest (PE), Brazil. World J. Microbiol. Biotechnol. 25: 1–5.

Machado, K.M.G., D.R. Matheus and V.L.R. Bononi. 2005. Ligninolytic enzymes production and Remazol brilliant blue R decolorization by tropical brazilian basidiomycetes fungi. Braz. J. Microbiol. 36: 246–252.

Martinez, A.T. 2002. Molecular biology and structure-function of lignin- degrading heme peroxidases. Enzyme Microb. Tech. 30: 425–444.
Martinez, D., L.F. Larrondo, N. Putnam, M.D.S. Gelpke, K. Huang, J. Chapman, K.G. Helfenbein, P. Ramaiva, J.C. Detter, F. Larimer, P.M. Coutinho, B. Henrissat, R. Berka, D. Cullen and D. Rokhsar. 2004. Genome sequence of the lignocellulose degrading fungus *Phanerochaete chrysosporium* strain RP78. Nat. Biotechnol. 22: 695–700.
Martinez, M.J., F.J. Ruiz-Duenas, F. Guillen and A.T. Martinez. 1996. Purification and catalytic properties of two manganese peroxidase isoenzymes from *Pleurotus eryngii*. Eur. J. Biochem. 237: 424–432.
Mayer, A.M. and R.C. Staples. 2002. Laccase: new functions for an old enzyme. Phytochemistry 60: 551–565.
Michniewicz, A., S. Ledakowicz, R. Ullrich and M. Hofrichter. 2008. Kinetics of the enzymatic decolorisation of textile dyes by laccase from *Cerrena unicolor*. Dyes Pigments 77: 295–302.
Moreira, M.T., C. Palma, I. Mielgo, G. Feijoo and J.M. Lema. 2001a. *In vitro* degradation of a polymeric dye (Poly R-478) by manganese peroxidase. Biotechnol. Bioeng. 75: 362–368.
Moreira, P.R., E. Almeida-Vara, G. Sena-Martins, I. Polonia, F.X. Malcata and J.C. Duarte. 2001b. Decolourisation of Remazol Brilliant Blue R via a novel *Bjerkandera* sp. strain. J. Biotechnol. 89: 107–111.
Moreira, P.R., F. Boullenne, E.A. Vara, F.X. Malcata, J.M. Frère and J.C. Duarte. 2006. Purification, kinetics and spectral characterisation of a new versatile peroxidase from a *Bjerkandera* sp. isolate. Enzyme Microb. Tech. 38: 28–33.
Morozova, O.V., G.P. Shumakovich, S.V. Shleev and Y.A. Yaropolov. 2007. Laccase–Mediator Systems and Their Applications: A Review. Applied Biochemistry and Microbiology 43: 523–535.
Murugesan, K., I.H. Nam, Y.-M. Kim and Y.-S. Chang. 2007. Decolorisation of a reactive dyes by a thermostable laccase produced by *Ganoderma lucidum* in solid state culture. Enzyme Microb. Tech. 40: 1662–1672.
Murugesan, K., Y.M. Kim, J.R. Jeon and Y.S. Chang. 2009a. Effect of metal ions on reactive dye decolorization by laccase from *Ganoderma lucidum*. J. Hazard Mater. 168: 523–529.
Murugesan, K., I.-H. Yang, Y.-M.Kim, J.-R. Jeon and Y.-S. Chang. 2009b. Enhanced transformation of malachite green by laccase of *Ganoderma lucidum* in the presence of natural phenolic compounds. Appl. Microbiol. Biotechnol. 82: 341–350.
Nerud, F. and Z. Misurcova. 1996. Distribution of ligninolytic enzymes in selected white-rot fungi. Folia Microbiol. 41: 264–266.
Nyanhongo, G.S., J. Gomes, G.M. Gübitz, R. Zvauya, J. Read and W. Steiner. 2002. Decolorization of textile dyes by laccases from a newly isolated strain of *Trametes modesta*. Water Res. 36: 1449–1456.
Ohm, R. A., J.F. de Jong, L.G. Lugones, A. Aerts, E. Kothe, J.E. Stajich et al. 2010. Genome sequence of the model mushroom *Schizophyllum commune*. Nat. Biotechnol. 28: 957–965.
Osma, J.F., J.L. Toca-Herrera and S.R. Couto. 2007. Banana skin: a novel waste for laccase production by *Trametes pubescens* under solid-state conditions. Application to synthetic dye decolouration. Dyes Pigments 75: 32–37.
Palmieri, G., G. Cennamo and G. Sannia. 2005. Remazol brilliant blue R decoulorisation by the fungus *Pleurotus ostreatus* and its oxidative enzymatic system. Enzyme Microb. Tech. 36: 17–24.
Pandey, A., P. Singh and L. Iyengar. 2007. Review bacterial decolorization and degradation of azo dyes. Int. Biodeter. Biodegr. 59: 73–84.
Prewitt, M.L., S. Diehl, T.C. McElroy and W.J. Diehl. 2008. Comparison of general fungal and basidiomycete-specific ITS primers for identification of wood decay fungi. Forest Prod. J. 58: 68–71.

Ralph, J., K. Lundquist, G. Brunow, F. Lu, H. Kim, P.F. Schatz, J.M. Marita, R.D. Hatfield, S.A. Ralph, J.H. Christensen and W. Boerjan. 2004. Lignins: Natural polymers from oxidative coupling of 4-hydroxyphenyl- propanoids. Phytochem. Rev. 3: 29–60.
Ramsay, J.A. and T. Nguyen. 2002. Decoloration of textile dyes by *Trametes versicolor* and its effect on dye toxicity. Biotechnol. Lett. 24: 1757–1761.
Robinson, G., R. McMullan, R. Marchant and P. Nigam. 2001. Remediation of dyes in textile effluent: a critical review on current treatment technologies with a proposed alternative. Biores. Technol. 77: 247–255.
Rogalski, J., J. Szczodrak and G. Janusk. 2006. Manganese peroxidase production in submerged cultures by free and immobilized mycelia of *Nematoloma frowardii*. Biores. Technol. 97: 469–476.
Ruiz-Duenas, F.J. and A.T. Martinez. 2009. Microbial degradation of lignin: how a bulky recalcitrant polymer is efficiently recycled in nature and how we can take advantage of this. Microbial Biotechnol. 2: 164–177.
Saeki, N., H. Takeda, E. Tanesaka and M. Yoshida. 2011. Induction of manganese peroxidase and laccase by *Lentinula edodes* under liquid culture conditions and their isozyme detection by enzymatic staining on native PAGE. Mycoscience 52: 132–136.
Sharma, P., L. Sing and N. Dilbaghi. 2009. Biodegradation of Orange II dye by *Phanerochaete chrysosporium* in simulated wastewater. J. Sci. Ind. Res. 68: 157–161.
Soares, G.M.B., M.C. Costa-Ferreira and M.T.P. Amorim. 2001. Decolorization of an anthraquinone-type using a laccase formulation. Biores. Technol. 79: 171–177.
Souza, C.G.M., G.K. Tychanowicz, D.F. Souza and R.M. Peralta. 2004. Production of laccase isoforms by *Pleurotus pulmonarius* in response to presence of phenolic and aromatic compounds. J. Bas. Microbiol. 44: 129–136.
Svobodová, K., A. Majcherczyk, C. Novotny and U. Kues. 2008. Implication of mycelium-associated laccase from *Irpex lacteus* in the decolorization of synthetic dyes. Biores. Technol. 99: 463–471.
Téllez-Téllez, M., J.F. Fernandez, A.M. Montiel-Gonzalez, C. Sánchez and G. Diaz-Godinez. 2008. Growth and laccase production by *Pleurotus ostreatus* in submerged and solid state fermentation. App. Microbiol. Biotechnol. 81: 675–679.
ten Have, R. and P.J.M. Teunissen. 2001. Oxidative mechanisms involved in lignin degradation by white-rot fungi. Chem. Rev. 101: 3397–3413.
Tien, M. and T.K. Kirk. 1984. Lignin-degrading enzyme from *Phanerochaete chrysosporium* —purification, characterization and catalytic properties of a unique H_2O_2-requiring oxygenase. P. Natl. Acad. Sci. Biol. 81: 2280–2284.
Tuor, U., K. Winterhalter and A. Fiechter. 1995. Enzymes of white-rot fungi involved in lignin degradation and ecological determinants for wood decay. J. Biotechnol. 41: 1–17.
Trovaslet, M., E. Enaud, Y. Guiavarch, A.-M. Corbisier and S. Vanhulle. 2007. Potential of a *Pycnoporus sanguineus* laccase of an anthraquinone acid dye. Enzyme Microb. Tech. 41: 368–376.
Tychanowicz, G.K., A. Zilly, C.G.M. Souza and R.M. Peralta. 2004. Decolourisation of industrial dyes by solid-state cultures of *Pleurotus pulmonarius*. Process Biochem. 39: 855–859.
Tychanowicz, G.K., D.F. Souza, C.G.M. Souza, M.K. Kadowaki and R.M. Peralta. 2006. Copper improves the production of laccase by the white rot fungus *Pleurotus pulmonarius* in solid state fermentation. Braz. Arch. Biol. Technol. 49: 699–704.
Vandevivere, P.C., R. Bianchi and W.S. Verstraete. 1998. Treatment and reuse of wastewater from the textile wet-processing industry: review of emerging technologies. J. Chem. Technol. Biotechnol. 72: 289–302.
Watanabe, T., H. Teranishi, Y. Honda and M. Kuwahara. 2002. A selective lignin-degrading fungus, *Ceriporiopsis subvermispora*, produces alkylitaconates that inhibit the production of a cellulolytic active oxygen species, hydroxyl radical in the presence of iron and H_2O_2. Biochem. Biophys. Res. Co. 297: 918–923.
Wesenberg, D., I. Kyriakides and S.N. Agathos. 2003. White-rot fungi and their enzymes for the treatment of industrial dye effluents. Biotechnol. Adv. 22: 161–187.

Zhao, X. and I.R. Hardin. 2007. HPLC and spectrophotometric analysis of biodegradation of azo dyes by *Pleurotus ostreatus*. Dyes Pigments 73: 322–325.

Zhao, X., I.R. Hardin and H.-M. Hwang. 2006. Biodegradation of a model azo disperse dye by the white rot fungus *Pleurotus ostreatus*. Int. Biodeter. Biodegr. 57: 1–6.

Zhuo, R., L. Ma, F. Fan, Y. Gong, X. Wan, M. Jiang, X. Zhang and Y. Yang. 2011. Decolorization of different dyes by a newly isolated white-rot fungi strain *Ganoderma* sp.En3 and cloning and functional analysis of its laccase gene. J. Hazard. Mater. 192: 852–873.

Zille, A., B. Górnacka, A. Rehorek and A. Cavaco-Paulo. 2005. Degradation of azo dyes by *Trametes villosa* laccase over long periods of oxidative conditions. Appl. Environ. Microbiol. 71: 6711–6718.

Zilly, A., C.G.M. Souza, I.P. Barbosa-Tessmann and R.M. Peralta. 2002. Decolorisation of industrial dyes by a Brazilian strain of *Pleurotus pulmonarius* producing laccase as the sole phenol-oxidizing enzyme. Folia Microbiol. 47: 315–319.

Zilly, A., J.S. Coelho-Moreira, A. Bracht, C.G.M. Souza, A.E.Carvajal, E.A. Koehnlein and R.M. Peralta. 2011. Influence of NaCl and Na_2SO_4 on the kinetics and dye decolorization ability of crude laccase from *Ganoderma lucidum*. Int. Biodeter. Biodegr. 65: 340–344.

Zilly, A., G.C.S. Bazanella, C.V. Helm, C.A. Araújo, C.G.M. Souza, A. Bracht and R.M. Peralta. 2012. Solid-state bioconversion of passion fruit waste by white-rot fungi for production ox oxidative and hydrolytic enzymes. Food Bioproc. Technol. 5: 1573–1580.

11

The Role of α-Glucuronidases in the Deconstruction of Plant Cell Wall Structure

Antonielle Vieira Monclaro and
*Edivaldo Ximenes Ferreira Filho**

Introduction

The matrix of polysaccharides presents a great structural diversity in the plant cell wall. Each molecule may contain different sugar residues with the same or different bonds, branched to varying degrees. The term holocellulose is used to describe a diverse group of polysaccharides, including cellulose and hemicelluloses that occur in close association with lignin (Andreaus et al. 2008, Khandeparker and Numan 2008). The presence of enzymes that break the various branch points is essential for complete hydrolysis of the plant cell wall holocellulose structure, creating more sites for subsequent enzymatic hydrolysis. They may also have a role in the introduction of specific modifications in the polysaccharide structure, such as changing solubility, polymer configuration and linkages to phenolic compounds and lignin. Within this context, α-glucuronidases are key enzymes in the cleavage

Enzymology Laboratory, Cellular Biology Department, University of Brasília, Brasilia, 70910-900, Federal District, Brazil.
*Corresponding author: eximenes@unb.br

of glucuronic acid side-groups that protect the neighboring β-1,4-linkages of the xylan backbone from being cleaved by xylan-degrading enzymes (Biely et al. 2000, Nagy et al. 2002, de Wet et al. 2006).

Xylan is a polysaccharide that represents one of the major sources of reduced carbon in the environment (Polizeli et al. 2005, Sánchez 2009, Jordan et al. 2012). It contains a backbone of xylose with β-1,4-linked glycosidic and side-chains, showing α-1,2 linked 4-O-methyl-D-glucosyluronic groups, single α-1,2 or α-1,3 linked arabinofuranosyl residues, or more extended side-chains in which the arabinofuranosyl residues contain additional substituents (Aspinall 1959, McCleary and Matheson 1986). The linear framework of xylans from hardwood contains at least 70 1,4-linked β-D-xylopyranose units, every tenth of which has a terminal 4-O-methyl-α-D-glucuronic acid distributed acid residue attached (Timell 1964). The acidic and O-acetyl units are attached to C-2 and C-3 of the xylosyl residues, respectively. Approximately 70 percent of the xylosyl residues of hardwood xylans are acetylated (Wong et al. 1988). The acetylated xylan of hardwoods contains single 4-O-methyl-α-D-glucuronic acid distributed on approximately 10 to 18 percent of its xylosyl residues (Timell 1967). The differences between the various structures of xylan polymer are mainly reflected in the proportions and types of side-chains present and their modes of attachment.

Two types of enzymes are involved in the breaking down of xylan: main-chain enzymes, including β-1,4-D-xylanase (EC 3.2.1.8) and β-1,4-D-xylosidase (EC 3.2.1.37), and side-chain enzymes (Nurizzo et al. 2002, Nagy et al. 2003, Shallom and Shoham 2003). The removal of branched substituent from xylan is eaffected by side-chain enzymes of different specificities such as α-arabinofuranosidase (EC 3.2.1.55), acetyl xylan esterases (EC 3.1.1.72) and α-D-glucuronidases (EC 3.2.1.139) (Shallom and Shoham 2003, Siqueira and Filho 2010, Jordan et al. 2012). Figure 11.1 shows the attack of main-chain (xylanase) and some side-chain enzymes (α-arabinofuranosidase and α-glucuronidase) to 4-*O*- methylglucuronoarabinoxylan structure. Accessory enzymes are very important in the removal of side-chain substituents (Bachmann and McCarthy 1991, Filho 1994). They act by synergistic interaction with main chain-cleaving enzymes, being defined as heterosynergy (Coughlan et al. 1993). The synergism is observed when the action of the main chain enzyme facilitates the release of the substituent by the side chain enzyme or *vice versa* (uniproduct heterosynergy), and when the extent of liberation of the substituent and of hydrolysis of the main chain resulting from the action of the combined enzymes exceed the sum of those observed following the actions of the individual enzymes. This chapter pays particular attention to fungal α-D-glucuronidases with emphasis in their mechanism of action on xylan structure of plant cell wall.

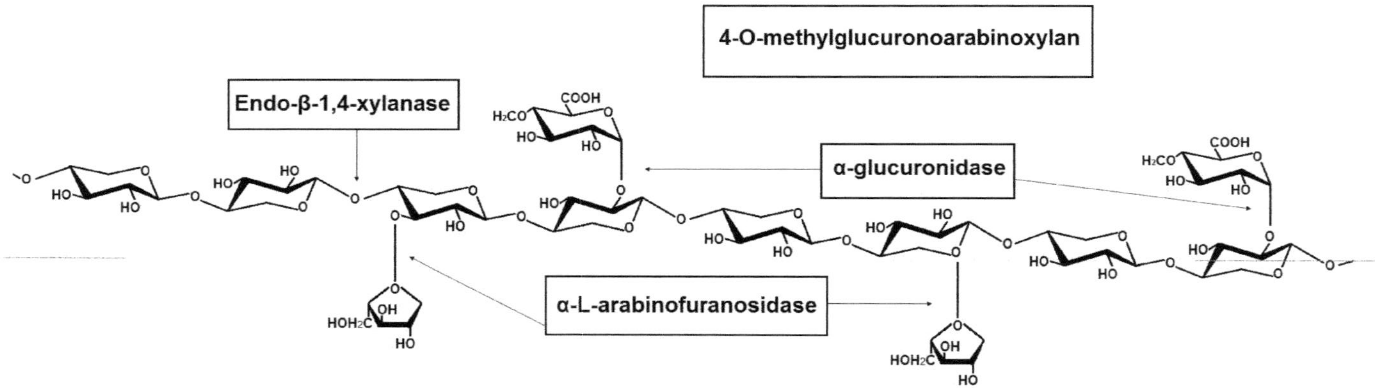

Figure 11.1 Hydrolysis of 4-*O*-methylglucuronoarabinoxylan by xylan-degrading enzymes. Modified from Polizeli et al. (2005).

The Mechanisms of α-glucuronidase Action

The hydrolysis mechanism proposed for α-glucuronidases from fungi involves the cleavage of α-(1→2)-glycosidic linkage between 4-*O*-methyl-α-D-glucopyranosyluronic acid (MeGlcA)/D-glucopyranosyluronic acid (GlcA) and xylopyranosyl residues in xyloolgiosaccharides or glucuronoxylan (Puls et al. 1987, Siika-aho et al. 1994, Ryabova et al. 2009, Chong et al. 2011). According to the CAzY database (http://www.cazy.org/), α-glucuronidases belonging to glycoside hydrolase family GH67 (EC 3.2.1.139) are described to cleave 4-O-methyl-α-D-glucopyranosyluronic acid α-(1→2)-linked to the non-reducing xylopyranosyl residues in short xylooligosaccharides (Cantarel et al. 2009). These enzymes have been only reported in ascomycetes, and are not able to attack glucuronoxylan and aryl glycosides of GlcA or MeGlcA.

Several examples of heterosynergy have been reported between α-glucuronidase from GH67 family and main chain-degrading enzymes (Puls et al. 1987, Coughlan et al. 1993, Várnai et al. 2011, Zhang et al. 2011). Thus, the heterosynergism between α-glucuronidase and main chain-cleaving enzymes (xylanase and β-xylosidase) is required for the release of MeGlcA/GlcA linked to xylan backbone (Puls et al. 1987, Siika-aho et al. 1994, Shao et al. 1995, de Vries et al. 1998, Chong et al. 2011). Puls et al. (1987) reported the synergistic action between α-glucuronidase and β-xylanase after incubation of the substrate (4-*O*-methylglucuronoxylan) with xylanase. The α-glucuronidase released MeGlcA from the products of xylanase action and rendered the xylooligomers for β-xylosidase action. The α-glucuronidase from *Schizophyllum commune* was reported to remove MeGlcA side chains from hardwood arabinoglucuronoxylan (Tenkanen and Siika-aho 2000). However, this enzyme showed limited activity against birchwood acetyl-glucuronoxylan due to the high amount of acetyl substituents. In addition, it needed a xylanase to improve its activity against soluble deacetylated glucuronoxylan. According to Tenkanen and Siika-aho (2000), this enzyme prefers to remove MeGlcA groups mainly from soluble substrates, since it showed a limited action on xylan with low solubility. α-Glucuronidase from *Thermoascus aurantiacus* was described to be highly active towards polymeric xylan, showing a similar initial hydrolysis rate on 4-*O*-methylglucuronic acid side groups in different lengths of xylo-oligosaccharides and xylan (Khandke et al. 1989).

Chong et al. (2011) reported α-glucuronidase with activity against polymeric xylan. This enzyme was classified as member of a novel glycoside hydrolase family (GH115, EC 3.2.1.131). These enzymes are commonly found in basidiomycetes and ascomycete, showing a broader substrate specificity (van den Brink and de Vries 2011). A novel extracellular α-glucuronidase of *Pichia stipitis* was reported to be active against hardwood glucuronoxylan

(Ryabova et al. 2009). This enzyme acts on substrates in which the uronic acid is linked to the non-reducing xylopyranosyl residues of main chain fragments, and was able to remove MeGlcA residues directly from polymeric substrate. Furthermore, it also exhibits typical catalytic activity of α-glucuronidases from GH67, acting on the shortest acidic product of glucuronoxylan hydrolysis by xylanase. This is confirmed by its action on aldotetraouronic acid Xyl(MeGlcA) Xyl–Xyl, the shortest acidic product of glucuronoxylan hydrolysis by GH 10 endoxylanases. However, there is no evidence whether the enzyme recognizes GlcA residues in a similar way as MeGlcA residues. The phylogenetic comparison of this enzyme with GH67 and GH4 α-glucuronidases showed no close relation between these GH families, this enzyme being (PsAgu115) further classified as a member of the GH115 family (Kolenová et al. 2010). According to Ryabova et al. (2009), the heterosynergism involving α–glucuronidases and xylanases of families 10 and 11 will create new sites for the productive binding to the enzymes hydrolyzing the xylan main chain. On the other hand, Vršanská et al. (2007) observed that the debranching activity on glucuronoxylan hindered the action of the appendage-dependent GH5 xylanase that recognizes MeGlcA side chains as its specificity determinants. According to Coughlan et al. (1993), the action of one type of enzyme preventing the action of a second enzyme is known as antisynergy. Therefore, the removal of a substituent such as 4-O-methyl-α-D-glucuronic acid by α-glucuronidase would prevent action by the xylanase.

The inverting mechanism is characteristic of α-glucuronidases from GH67 family (Nurizzo et al. 2002, Golan et al. 2004). The hydrolysis of the glycosidic bond occurs with invertion of anomeric configuration liberating the β-anomer of the MeGlcA as the reaction product. The first report of α-glucuronidase acting by inverting mechanism was made by Biely et al. (2000). In this report, α-glucuronidase from *A. tubingensis*, belonging to GH67 family, was able to release MeGlcA only from those beechwood glucuronoxylan fragments in which the acid is attached to the non-reducing terminal xylopyranosyl residue. In addition, the charged groups in these carbohydrate residues influenced the selection of the hydrolytic mechanism, since other hydrolases attacking α-glycosidic linkages of uronic acids, endo- and exopolygalacturonases also utilized the single displacement mechanism of hydrolysis. An α-glucuronidase of *Aspergillus niger* hydrolyzes *O*-α-D-glucosyluronic acid α-D-glucosyluronic acid to produce α- and β-anomers of D-glucuronic acid in an equimolar ratio (Kiryu et al. 2005), suggesting that inversion of the anomeric configuration of the substrate is involved in the hydrolysis mechanism. In addition, it showed no activity against α-glucuronyl linkages of 4-*O*-methyl-D-glucosyluronic acid-α-(1→2)-xylooligosaccharides derived from xylan. The inverting characteristic was also described for α-glucuronidases of GH115 family of *P. stipitis* (PSAgu115) and *S. commune* (ScAgu115) (Kolenová et al. 2010). Both enzymes released MeGlcA as their β-anomer and were active against a series of aldouronic acids —aldotriouronic acid MeGlcA2Xyl2, aldotetraouronic acids MeGlcA3Xyl3 and MeGlcA2Xyl3, aldopentaouronic acid MeGlcA3Xyl4, aldohexaouronic acid MeGlcA3Xyl5 and beechwood glucuronoxylan. The specific activity of ScAgu115 also increased with the length of xylooligosaccharide chain and the activity against aldotetraouronic acid $MeGlcA_2Xyl_3$, aldopentaouronic acid $MeGlcA_3Xyl_4$ and aldohexaouronic acid $MeGlcA_3Xyl_5$ was similar. The above results suggest the binding of the MeGlcA-substituted xylopyranosyl residue and two adjacent xylopyranosyl residues are the most important factor for the activity of ScAgu115. PSAgu115 lower activity on aldopentaouronic acid $MeGlcA_3Xyl_4$ and aldohexaouronic acid $MeGlcA_3Xl_5$ indicates an inhibitory role of increased chain length. These results indicate that GH115 α-glucuronidases recognize not only the uronic acid substituent but also a part of the xylan main chain. The lowest specific activity of ScAgu115 and PSAgu115 on glucuronoxylan at a concentration corresponding to about 10 mM of MeGlcA residues may be a consequence of the formation of non-productive complexes of the enzymes with unsubstituted parts of the xylan main chain, but it may also be the result of a steric hindrance by substituents in the vicinity of the enzyme active site. Maximal activity of *Aureobasidium pullulans* NRRL Y-2311-1 recombinant α-glucuronidase against aldouronic acids increased with increasing chain length, in spite of a decrease in affinity for the substrate with increasing chain length. Catalytic efficiency was consequently the highest for the shortest possible substrate for α-glucuronidase (aldobiouronic acid), and decreased with aldotriouronic and aldotetraouronic acids as the substrates (de Wet et al. 2006).

Kolenová et al. (2010) suggest that the major difference between the GH115 and GH67 α-glucuronidase families consists in the topography of substrate binding site. Thus, the α-glucuronidases from GH67 family present a pocket type substrate binding site which selects specifically for short acidic xylooligosaccharides, with the uronic acid linked to the non-reducing xylopyranosyl residue. Experiments carried out by Golan et al. (2004) and Nagy et al. (2003) show that these enzyme interactions are not only restricted to the uronic acid residue but also with one or two xylopyranosyl residues, including the residue substituted with the uronic acid. Within this context, the role of the xylose aldobiouronic acid moiety is important for the action of GH67 α-glucuronidases, since these enzymes are not able to act on aryl α-glucuronides as substrates.

The phylogenetic analysis of α-glucuronidases shows that these enzymes can be divided into three subfamilies: groups I and II, composed of bacterial α-glucuronidases, and group III, containing fungal α-glucuronidases (Shallom et al. 2004). The three subfamilies differ in the N- and C-terminal ends and in the lengths of their amino acid sequences. In this case, α-glucuronidases from group II are longer at the C terminus by about 22 residues compared to group I enzymes. On the other hand, the fungal α-glucuronidases (group III) are much longer than the bacterial enzymes, containing about 110 additional residues at the C terminus. Phylogenetic analysis also showed that the fungal GH115 α-glucuronidases are distinctly separate from the prokaryotic clade and distributed over three branches (Chong et al. 2011). The first one consisted exclusively of basidiomycete sequences and included the *S. commune* α-glucuronidase (Agu1). The second branch consisted exclusively of ascomycete sequences and contained the GH115 α-glucuronidase from *P. stipitis* (Ryabova et al. 2009). The third branch presented no biochemically characterized sequences and consisted mostly of sequences from ascomycetes, but it also included a second GH115 protein from *S. commune*. It is noteworthy that this last group was more distant from the first two, suggesting that a distinct enzyme activity or substrate specificity may be found in this group.

The first report of α-glucuronidase (Agu4B) belonging to GH4 family and produced by bacterium was made by Suresh et al. (2003). In this family, α-glucuronidase produced by *Thermotoga maritime* was active against chromophoric synthetic substrate, *p*-nytrophenyl-α-D-glucuronopyranoside (*p*NP-GUA), previously reported by Fontana et al. (1988) and Bronnenmeier et al. (1995a). Agu4B was able to hydrolyze *p*NP-GUA and did not hydrolyze 4-*O*-methyl-D-glucuronoxylan or its fragments oligosaccharides. It showed no sequence similarity to any of the known α-glucuronidases, and therefore it cannot be considered to be a hemicellulolytic glycosidase. The α-glucuronidase of GH4 family requires a cofactor and cysteine for its activity (Hall et al. 2009). Agu4B showed cofactor and reducing agent requirements for increasing its activity, being more active in the presence of NAD^+, Mn^{2+} and thiol compounds. To date, the *p*NP-GUA hydrolyzing α-glucuronidase activity was not reported in fungal species.

Some procedures for the measurement of α-glucuronidase activity are described in the literature (Puls et al. 1987, Fontana et al. 1988, Wood and Wilson 1995, Bronnenmeier et al. 1995a, 1995b), including the liberation of MeGlcA from aldouronic acids followed by determination of the reducing power of the free acid, by the Somogyi and Nelson reagents modified according to Milner and Avigad (1967), the non modified procedure for determining decreased sugars with reduction of aldouronic acids in the presence of $NaBH_4$, and the HPLC determination of free MeGlcA or the equivalent amount of xylooligosaccharide. The alkaline deesterification of 4-nitrophenyl 2-*O*-(methyl 4-*O*-methyl-α-D-glucuronopyranosyl)-β-D-xylopyranoside (Nph-Xyl-MeGlcA) was used by Biely et al. (2000) as an alternative to determine α-glucuronidase activity. In this case, the α-glucuronidase-catalyzed formation of 4-nitrophenyl β-D-xylopyranoside was coupled with hydrolysis by β-xylosidase. Thus, the coordinated action of both enzymes makes the 4-nitrophenyl glycoside of the aldobiuronic acid available as a specific chromogenic substrate of α-glucuronidase (Biely and Puchart 2006).

Purification and Characterization of α-glucuronidases

A number of α-glucuronidases has been purified and characterized from a great variety of fungal species (Khandke et al. 1989, Margolles-Clark et al. 1996, Tenkanen and Siika-aho 2000, Heneghan et al. 2007, Hinz et al. 2009). The relevant enzyme has been detected in filtrates of various microorganisms, such as *Trichoderma reesei, A. bisporus, Talaromyces emersonii, T. aurantiacus* and *Chrysosporium lucknowense*. The purification of α-glucuronidases is based on a number of alternative combinations of procedures, such as ultrafiltration, gel filtration and ion-exchange chromatographic methods (Puls et al. 1987, Kiryu et al. 2005, de Wet et al. 2006). Purification of α-glucuronidases to homogeneity is necessary for detailed characterization studies and for the determination of their structures. An α-glucuronidase from *P. stipitis* was purified to electrophoretic homogeneity in several steps using ultrafiltration, anion-exchange chromatography and hydrophobic interaction chromatography (Ryabova et al. 2009). The enzyme showed a single band of approximately 120 kDa on SDS-PAGE electrophoresis. Uchida et al. (1992) reported the purification of two variants of α-glucuronidases (CM-I and CM-II) from *A. niger* 5–16. CM-I and CM-II were purified about 2200- and 3800-fold, respectively, by ammonium sulfate precipitation, ion-exchange and gel filtration chromatography. An α-glucuronidase was purified from a commercial preparation from *A. niger* by a combination of chromatography procedures, including gel filtration and ion-exchange (Kiryu et al. 2005). The enzyme was purified 586-fold and its molecular weight was estimated to be 120 kDa and 58 kDa by gel filtration and SDS-PAGE, respectively. Two kinds of α-glucuronidase from *Agaricus bisporus* were purified by preparative gel filtration (Puls et al. 1987). The major α-glucuronidase component of *Agaricus bisporus* was identified in the molecular weight range > 450 kDa. The heterologous α-glucuronidase of *Aureobasidium pullulans*, expressed in *Escherichia colli*, was purified to homogeneity by ultrafiltration, Ni-chelation affinity chromatography, and displayed an electrophoretic mobility of 157 kDa on SDS–PAGE. An α-glucuronidase from *S. commune* was purified by anion exchange, hydrophobic interaction and gel filtration chromatography procedures (Tenkanen and Siika-aho 2000). The purification of α-glucuronidase from *Thermoascus aur* antiacus by ammonium sulfate precipitation, desalting on Sephadex G-25, ion-exchange on DEAE-Sephadex A-50 and gel filtration on Ultrogel HA lead to 394-fold purification (Khandke et al. 1989).

Some biochemical properties of α-glucuronidases isolated from fungal sources are listed in Table 11.1. Most of the enzymes show molecular weights varying from 58 to 157 kDa, pI value range from 3.6 to 6.2, and pH and temperature optimum values range from 3.0 to 6.3 and 40 to 70°C, respectively. Most of the enzymes investigated belong to GH67 family and are stable at 40°C. Despite their important role for xylan breakdown, few enzymes from fungal sources have been isolated and characterized. It is also probable that other α-glucuronidases differing in specificity will be discovered.

Conclusions

This chapter provides information on most of the aspects of the mechanism of action of α-glucuronidases. The efficient exploitation of xylans as a source of fuels and industrial chemicals requires an improvement of the understanding of the xylan-degrading enzymes system that effect their conversion (Filho et al. 1996), including main- and side chain-cleaving enzymes. Because of the structural complexity of plant cell walls, a wide variety of enzyme systems have been developed by different sources in nature, including bacteria and fungi, as a strategy to overcome the matrix components. Within this context, α-glucuronidases are key enzymes in the breakdown of xylan structure, which is carried out by ensembling of enzymes which hydrolyze glycosidic bonds in xylooligomers and xylan structure (Siqueira and Filho 2010). Thus, it is clear that one must have a thorough understanding of the enzyme systems required for the hydrolysis of each of the major components of the plant cell wall, as well as the microorganisms that produce these enzymes systems. The discovery and development of novel plant cell wall degrading enzymes is a key step towards more efficient depolymerization of polysaccharides to fermentable sugars for the production of bioethanol and other bioproducts (King et al. 2011). It should be noted that, besides their direct use in the saccharification of xylan, α-glucuronidases have potential applications in several other areas, including the production of chemical feedstocks and fuels and improved products in the pulp and paper industry. Despite extensive research on xylan-degrading enzymes, there are still major gaps in our understanding of how they hydrolyze, act synergistically, and the role of carbohydrate binding modules. The function of each xylan-degrading enzyme in the plant cell wall break down may be defined by enzymes interacting in dynamic and cooperative ways in time and space such as an interactome of enzymes.

Table 11.1 Some properties of α-glucuronidases from fungal source.

Microorganism	Family	Optimum temp (°C)	Thermostability	Optimum pH	Isoeletric point	MW (kDa)	K_m	$V_{m.}$ (μmol/min/mg)	Reference
Agaricus bisporus	NR[d]	52	NR[d]	3.3	NR[d]	>450[b]	NR[d]	NR[d]	Puls et al. 1987
Aspergillus niger	NR[d]	40	NR[d]	3.0–3.5	NR[d]	58[a], 120[b]	4.4[gh]	50[h]	Kiryu et al. 2005
Aspergillus niger 5–16	NR[d]	60	NR[d]	5.0	NR[d]	NR[d]	NR[d]	NR[d]	Uchida et al. 1992
Aspergillus tubingensis	NR[d]	70	1440 hr (60°C)	4.5–6.0	5.18[c]	107[a], 100[b]	0.14 ± 0.03[fi]	NR[d]	De Vries et al. 1998
Aureobasidium pululans	GH67	65	>120 hr (45°C)	5.0–6.0	NR[d]	157[a]	0.25 ± 0.009[fj]	14.1 ± 0.08[j]	de Wet et al. 2006
Aureobasidium pululans							1.54 ± 0.19[fk]	60.0 ± 4.3[k]	de Wet et al. 2006
Aureobasidium pululans							3.7 ± 0.23[fl]	135 ± 6.2[l]	de Wet et al. 2006
Aureobasidium pululans							17.4 ± 2.8f[m]	126 ± 14.1[m]	de Wet et al. 2006
Chrysosporium luckonowense C1	GH67	50	NR[d]	6.0	NR[d]	NR[d]	NR[d]	NR[d]	Hinz et al. 2009
Phanerochaete chrysosporium	NR[d]	50	NR[d]	3.5	4.6[e]	112[a]	NR[d]	NR[d]	Castanares et al. 1995
Pichia stiptis	NR[d]	60	3 hr (40°C)	4.4	4.0–4.64[ce]	120[a]	NR[d]	NR[d]	Ryabova et al. 2009
Piromonas communis	NR[d]	50	NR[d]	5.5	NR[d]	103[b]	NR[d]	NR[d]	Wood and Wilson 1995

Table 11.1 contd....

Table 11.1 contd....

Microorganism	Family	Optimum temp (°C)	Thermostability	Optimum pH	Isoeletric point	MW (kDa)	K_m	$V_{m.}$ (µmol/min/mg)	Reference
Schizophyllum comunne	NR[d]	40	24 hr (40°C)	5.8	3.6[e]	125[a]	NR[d]	NR[d]	Tenkanen and Siika-Aho 2000
Schizophyllum comunne	GH115	NR[d]	NR[d]	NR[d]	4.47[c]	107[c]	NR[d]	NR[d]	Chong et al. 2011
Talaromyces emersonii	GH67	50	NR[d]	5.0	NR[d]	91.6[c]	NR[d]	NR[d]	Heneghan et al. 2007
Thermoascus auranticus	NR[d]	65	6 hr (50°C)	4.5	NR[d]	118[a], 107[b]	0.145[fn]	2.5[n]	Khandke et al. 1989
Trichoderma reesei RUT c30	NR[d]	60	24 hr (40°C)	4.5–6.0	5.0–6.2[e]	91[a]	NR[d]	NR[d]	Siika-aho et al. 1994

[a]Determined by SDS-PAGE
[b]Determined by gel filtration
[c]Predicted by aminoacid sequence
[d]Not reported
[e]Determined by isoelectric focusing
[f]mM
[g]mg/mL
[h]K_m value for trehalose dicarboxylate as the substrate
[i]K_m value for aldotriouronic acid-aldobiuronic acid as the substrate
[j]K_m value for aldobiouronic acid as the substrate
[k]K_m value for aldotriouronic acid as the substrate
[l]K_m value for aldotetrauronic acid as the substrate
[m]K_m value for aldopentauronic acid as the substrate
[n]K_m value for xylotriose as the substrate

Acknowledgments

E.X.F.F. acknowledges a research fellowship from the Brazilian Research Council (CNPq). This work was funded by CNPq (research grants 563260/2010-6 and 563823/2010-0), Foundation for Research Support of Federal District (Brazil, Pronex Program) and National Institute of Science and Technology of Bioethanol.

References

Andreaus, J., E.X.F. Filho and E.P.S. Bon. A review on Biotechnology of Holocellulose-Degrading Enzymes. pp. 1–41. *In*: C.T. Hou and J.F. Shaw [eds.]. 2008. Biocatalysis and Bioenergy. John Wiley & Sons, Inc., Hoboken, New Jersey, USA.

Aspinall, G.O. 1959. Structural chemistry of the hemicelluloses. Adv. Carbohydr. Chem. 14: 429–468.

Bachmann, S.L. and A.J. McCarthy. 1991. Purification and cooperative activity of enzymics constituting the xylan-degrading system of *Thermomonospora fusca*. Appl. Environ. Microbiol. 57: 2121–2130.

Biely, P. and V. Puchart. 2006. Recent progress in the assay of xylanolytic enzymes. J. Sci. Food Agric. 86: 1636–1647.

Biely, P., R.P. de Vries, M. Vranská and J. Visser. 2000a. Inverting characteristic of α-glucuronidase from *Aspergillus tubingensis*. Biochim. Biophys. Acta 1474: 360–364.

Biely, P., J. Hirsch, D.C. la Grange, W.H. van Zyl and B.A. Prior. 2000b. A chromogenic substrate for a β-xylosidase-coupled assay of α-glucuronidase. Anal. Biochem. 286: 289–294.

Bronnenmeier, K., H. Meissner, S. Stocker and W.L. Staudenbauer. 1995a. α-Glucuronidases from the xylanolytic thermophiles *Clostridium stercorarium* and *Thermoanearobacterium saccharolyticum*. Microbiol. 141: 2033–2040.

Bronnenmeier, K., A. Kern, W. Liebl and W.L. Staudenbauer. 1995b. Purification of *Thermotoga maritima* enzymes for the degradation of cellulosic materials. Appl. Environ. Microbiol. 61: 1399–1407.

Cantarel, B.L., P.M. Coutinho, C. Rancurel, T. Bernard, V. Lombard and B. Henrissat. 2009. The carbohydrate-active enymes database (CAZy): an expert resource for glycogenomics. Nucleic Acids Res. 37: D233–D2.

Castanares, A., A.J. Hay, A.H. Gordon, S.I. McCrae and T.M. Wood. 1995. D-Xylan-degrading enzyme system from the fungus *Phanerochaete chrysosporium*: isolation and partial characterisation of an α-(4-O-methyl) -D-glucuronidase. J. Biotechnol. 43: 183–194.

Chong, S.-L., E. Battaglia, P.M. Coutinho, B. Henrissat, M. Tenkanen and R.P. de Vries. 2011. The α-glucuronidase Agu1 from *Schizophyllum commune* is a member of a novel glycoside hydrolase family (GH115). Appl. Microbiol. Biotechnol. 90: 1323–1332.

Coughlan, M.P., M.G. Tuohy, E.X.F. Filho, J. Puls, M. Claeyssens, M. Vrsanská and M.M. Hughes. pp. 53–84. *In:* M.P. Coughlan and G.P. Hazlewwod [eds.]. 1993. Hemicellulose and Hemicellulases. Portland Press Research Monograph IV. Cambridge, Great Britain.

de Vries, R.P., C.H. Poulsen, S. Madrid and J. Visser. 1998. aguA, the gene encoding an extracellular α-glucuronidase from *Aspergillus tubingensis*, is specifically induced on xylose and not on glucuronic acid. J. Bacteriol. 180: 243–249.

de Wet, B.J.M., W.H. van Zyl and B.A. Prior. 2006. Characterization of the *Aureobasidium pullulans* α-glucuronidase expressed in *Saccharomyces cerevisiae*. Enzyme Microb. Technol. 38: 649–656.

Filho, E.X.F. 1994. The xylan-degrading enzyme system. Braz. J. Med. Biol. Res. 27: 1093–1109.

Filho, E.X.F., J. Puls and M.P. Coughlan. 1996. Purification and characterization of two arabinofuranosidases from solid-state cultures of the fungus *Penicillium capsulatum*. Appl. Environ. Microbiol. 62: 168–173.

Fontana, J.D., M. Gebara, M. Blumel, H. Schneider, C.R. McKenzie and K.G. Johnson. 1988. α-4-*O*-Methyl-D-glucuronidase component of xylanolytic complexes. Methods Enzymol. 160: 560–571.

Golan, G., D. Shallom, A. Teplitsky, G. Zaide, S. Shulami, T. Baasov, V. Stojanoff, A. Thompson, Y. Shoham and G. Shoham. 2004. Crystal structures of *Geobacillus stearothermophilus* α-glucuronidase complexed with its substrate and products. J. Biol. Chem. 279: 3014–3024.

Hall, B.G., A. Pikis and J. Thompson. 2009. Evolution and biochemistry of family 4 glycosidases: implications for assigning enzyme function in sequence annotations. Mol. Biol. Evol. 26: 2487–2497.

Heneghan, M.N., L. McLoughlin, P.G. Murray and M. Tuohy. 2007. Cloning, characterisation and expression analysis of α-glucuronidase from the thermophilic fungus *Talaromyces emersonii*. Enzyme Microb. Technol. 61: 677–682.

Hinz, S. W.A., L. Pouvreau, R. Joosten, J. Bartels, M.C. Jonathan, J. Wery and H.A. Schols. 2009. Hemicellulase production in *Chrysosporium lucknowense* C1. J. Cereal Sci. 50: 318–323.

Jordan, D.B., M.J. Bowman, J.D. Braker, B.S. Dien, R.E. Hector, C.C. Lee, J.A. Mertens and K. Wagschal. 2012. Plant cell walls to ethanol. Biochem. J. 442: 241–252.

Khandke, K.M., P.J. Vithayathil and S.K. Murthy. 1989. Purification and characterisation of an α-glucuronidase from a thermophillic fungus, *Thermoascus aurantiacus*. Arch. Biochem. Biophys. 274: 5lll–517.

Khandeparker, R. and M.T. Numan. 2008. Bifunctional xylanases and their potential use in biotechnology. J. Ind. Microbiol. Biotechnol. 35: 635–644.

King, B.C., K.D. Waxman, N.V. Nenni, L.P. Walker, G.C. Bergstrom and D.N. Gibson. 2011. Arsenal of plant cell wall degrading enzymes reflects host preference among plant pathogenic fungi. Biotechnology for Biofuels 4: 4.

Kiryu, T., H. Nakano, T. Kiso and H. Murakami. 2005. Purification and characterization of a novel α-glucuronidase from *Aspergillus niger* specific for *O*-α-D-glucosyluronic acid α-D-glucosiduronic acid. Biosci. Biotechnol. Biochem. 69: 522–529.

Kolenová, K., O. Ryabova, M. Vranská and P. Biely. 2010. Inverting characteristic of family GH115 α-glucuronidases. FEBS Lett. 584: 4063–4068.

Margolles-Clark, E., M. Saloheimo, M. Siika-aho and M. Penttillä. 1996. The α-glucuronidase-encoding gene of *Trichoderma reesei*. Gene 172: 171–172.

McCleary, B.V. and N.K. Matheson. 1986. Polysaccharides having a β-D-xylan backbone. Adv. Carbohydr. Chem. Biochem. 44: 158–164.

Milner, Y. and G. Avigad. 1967. A copper reagent for the determination of hexuronic acids and certain ketohexoses. Carbohydr. Res. 4: 359–361.

Nagy, T., K. Emami, C.M.G.A. Fontes, L.M.A. Ferreira, D.R. Humphry and H.J. Gilbert. 2002. The membrane-bound α-glucuronidase from *Pseudomonas cellulosa* hydrolyzes 4-*O*-methyl-D-glucuronoxylooligosaccharides but not 4-*O*-methyl-D-glucuronoxylan. J. Bacteriol. 184: 4925–4929.

Nagy, T., D. Nurizzo, G.J. Davies, P. Biely, J.H. Lakey, D.N. Bolam and H.J. Gilbert. 2003. The α-glucuronidase, GlcA67A, of *Cellvibrio japonicus* utilizes the carboxylate and methyl groups of aldobiouronic acid as important substrate recognition determinants. J. Biol. Chem. 278: 20286–20292.

Nurizzo, D.T. Nagy, H.J. Gilbert and G.J. Davies. 2002. The structural basis for catalysis and specificity of the *Pseudomonas cellulosa* α-glucuronidase, GlcA67A. Struct. 10: 547–556.

Polizeli, M.L.T.M., A.C.S. Rizzatti, R. Monti, H.F. Terenzi, J.A. Jorge and D.S. Amorim. 2005. Xylanases from fungi: properties and industrial applications. Appl. Microbiol. Biotechnol. 67: 577–591.

Puls, J., O. Schmidt and C. Granzow. 1987. α-Glucuronidase in two microbial xylanolytic systems. Enzyme Microb. Technol. 9: 83–88.

Ryabova, O., M. Vršanská, S. Kaneko, W.H. van Zyl and P. Biely. 2009. A novel family of hemicellulolytic α-glucuronidase. FEBS Lett. 583: 1457–1462.

Sánchez, C. 2009. Lignocellulosic residues: biodegradation and bioconversion by fungi. Biotechnol. Adv. 27: 185–194.

Shallom, D. and Y. Shoham. 2003. Microbial hemicellulases. Curr. Opin. Microbiol. 6: 219–228.

Shallom, D., G. Golan, G. Shoham and Y. Shoham. 2004. Effect of dimer dissociation on activity and thermostability of the α-glucuronidase from *Geobacillus stearothermophilus*: dissecting the different oligomeric forms of family 67 glycoside hydrolases. J. Bacteriol. 186: 6928–6937.

Shao, W., S.K.C. Obi, J. Puls and J. Wiegel. 1995. Purification and characterization of the α-glucuronidase from *Thermoanaerobacterium* sp. strain JW/SL-YS485, an important enzyme for the utilization of substituted xylans. Appl. Environ. Microbiol. 61: 1077–1081.

Siika-aho, M., M. Tenkanen, J. Buchert, J. Puls and L. Viikari. 1994. An α-glucuronidase from *Trichoderma reesei* RUT C-30. Enzyme Microb. Technol. 16: 813–819.

Siqueira, F.G. and E.X.F. Filho. 2010. Plant cell wall as a substrate for the production of enzymes with industrial applications. Mini-Rev. Org. Chem. 7: 54–60.

Suresh, C., M. Kitaoka and K. Hayashi. 2003. A thermostable non-xylanolytic α-glucuronidase *Thermotoga maritima* MSB8. Biosci. Biotechnol. Biochem. 67: 2359–2364.

Tenkanen, M. and M. Siika-aho. 2000. An α-glucuronidase of *Schizophyllum commune* acting on polymeric xylan. J. Biotechnol. 78: 149–161.

Timell, T.E. 1964. Wood hemicelluloses: part I. Carbohydr. Chem. 19: 247–302

Timell, T.E. 1967. Recent progress in the chemistry of wood hemicelluloses. Wood Sci. Technol. 1: 45–70.

Uchida, H., T. Nanri, Y. Kawabata, I. Kusakabe and K. Murakami. 1992. Purification and characterization of intracellular α-glucuronidase from *Aspergillus niger* 5–16. Biosci. Biotechnol. Biochem. 56: 1608–1615.

van den Brink, J. and R.P. de Vries. 2001. Fungal enzyme sets for plant polysaccharide degradation. Appl. Microbiol. Biotechnol. 91: 1477–1492.

Varnái, A., L. Huikko, J. Pere, M. Siika-aho and L. Viikari. 2011. Synergistic action of xylanase and mannanase improves the total hydrolysis of softwood. Biores. Technol. 102: 9096–9104.

Vršanská, M., K. Kolenová, V. Puchart and P. Biely. 2007. Mode of action of glycoside hydrolase family 5 glucuronoxylan xylanohydrolase from *Erwinia chrysanthemi*. FEBS J. 274: 1666–1677.

Wood, T.M. and C.A. Wilson. 1995. α-(4-O Methyl)-D-glucuronidase activity produced by the rumen anaerobic fungus *Piromonas communis*. A study of selected properties. Appl. Microbiol. Biotechnol. 43: 893–900.

Wong, K.K.Y., L.U.L. Tan and J.N. Saddler. 1988. Multiplicity of β-1,4-xylanases in microorganisms: functions and applications. Microbiol. Rev. 52: 305–317.

Zhang, J.P. Tuomainen, M. Siika-aho and L. Viikari. 2011. Comparison of the synergistic action of two thermostable xylanases from GH families 10 and 11 with thermostable cellulases in lignocellulose hydrolysis. Biores. Technology 102: 9090–9095.

12

Biotechnological Applications of Lipases in Biodiesel Production

Antonio Carlos Ferreira Batista,[1,a,*] *Tiago Almeida Silva*,[1,b] *Andressa Tironi Vieira*[2,c] and *Marcelo Firmino de Oliveira*[2,d]

Introduction

The search for alternative fuels to petroleum was stimulated by two main reasons. Firstly, there are predictions that in the near future the main oil reserves would completely run out, which would be a disaster, due to the high-profile global dependence on oil (Leung et al. 2010). Moreover, there are serious environmental and public health concerns related to the burning of this fuel (Leung et al. 2010, Chen et al. 2010), including

[1]Laboratory of Renewable Energy and Environment of Pontal, Faculty of Integrated Sciences of Pontal, Federal University of Uberlândia. Rua Vinte 1600, Tupã. 38304-402, Ituiutaba, MG, Brazil.

[a]E-mail: flash@pontal.ufu.br

[b]E-mail: tiago_almeida@qui.pontal.ufu.br.

[2]Department of Chemistry, Faculty of Philosophy, Sciences and Letters of Ribeirao Preto, University of Sao Paulo. Av. Bandeirantes 3900, Monte Alegre. 14040-901, Ribeirão Preto, SP. Brazil.

[c]E-mail: dessaqtironi@yahoo.com.br

[d]E-mail: marcelex@ffclrp.usp.br.

*Corresponding author

emission of so-called greenhouse gases, among which carbon dioxide (CO_2), methane (CH_4) and nitrous oxide (N_2O) stand out. In addition to these greenhouse gases, the pollutant sulfur dioxide (SO_2) is also released, which has detrimental effects on the environment and human health, such as respiratory disorders, allergies, nervous system and vital organs degeneration, and cancer, just to mention a few (Pinto et al. 2005).

Among the proposals for substitutes of mineral diesel, biodiesel stands out as the main candidate, since it offers several advantages over mineral diesel, such as, reduced emissions of CO_2 and particulate materials, use of renewable sources, and biodegradability (Pinto et al. 2005, Sendzikiene et al. 2006, Conceição et al. 2007).

Biodiesel is produced from renewable sources and may be synthesized from animal fats and various types of vegetable oil. The fact that it comes from a renewable source justifies the use of the prefix "bio" prior to the word fuel. This was supported by many researchers who advocated the establishment of the so-called closed carbon cycle in biodiesel production, where CO_2 released by fuel combustion is absorbed by the plant material during photosynthesis, and so forth, as illustrated in Fig. 12.1 (Oliveira et al. 2008, Harper and Starbuck 2009, Knothe 2010).

An important feature of biodiesel is its adaptability to the diesel cycle engines without any mechanical modification, which has made it an alternative fuel for all the existing fleet (Lianhua et al. 2010, Morshed et al. 2011).

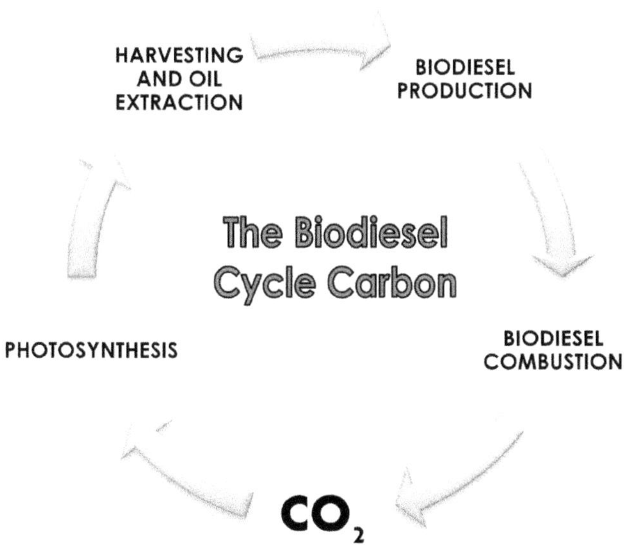

Figure 12.1 Closed carbon cycle for the production of biodiesel.

In chemical terms, biodiesel is defined by the American Society for Testing and Materials (ASTM) as alkyl esters of long-chain carboxylic acids obtained from renewable sources like vegetable oils and animal fats (Zhang et al. 2003, Silva et al. 2006). Biodiesel can be used in engines in its pure form (B100) or blended with the diesel fuel. To represent the percentage of biodiesel existing in the mixture an international nomenclature has been created, namely BX, where X stands for the percentage of biodiesel in the blend (e.g., B2 has 2 percent biodiesel) (Shang et al. 2010).

Biodiesel is a result of the efforts made towards using of vegetable oils as fuels in diesel cycle engines. The direct use of vegetable oils in engines is not feasible, due to its high viscosity and density as well as low volatility, which lead to poor atomization in the combustion chamber and culminate in operational problems, such as the formation of deposits in the engine (Nielsen et al. 2008). The development of processes for the modification of vegetable oils and adjustment of their property has given rise to biodiesel, and currently there are four main methods of biodiesel production, namely: microemulsification, esterification, transesterification, and cracking.

The microemulsification process consists in the formation of emulsions (biodiesel) constituted by vegetable oils and additives, in order to fix different properties. Short-chain alcohols such as methanol, ethanol, propanol and 1-butanol, act as reducers of oil viscosity, while higher alcohols are used as chain surfactants, and alkyl nitrates are employed as adjusters of the cetane number (Ma and Hanna 1999, Ranganathan et al. 2008).

Despite significant improvements, the intensive use of biodiesel produced by microemulsification leads to operational problems in the functioning of engines (Ma and Hanna 1999, Nielsen et al. 2008).

Esterification involves condensation of a fatty acid with a mono-alcohol, to form esters (Solomons and Fryhle 2009). The esterification reactions occur very slowly in the absence of strong acid catalysts, but when a carboxylic acid and an alcohol are refluxed together with a small amount of strong acid, the balance is reached within a few hours (Chongkhong et al. 2009). The most common catalysts are sulfuric acid (Solomons and Fryhle 2009, Chongkhong et al. 2009), p-toluene sulphonic acid, and alkyl sulphonic acid (Chongkhong et al. 2007). This reaction is typically used for the synthesis of biodiesel from vegetable oil with high content of free fatty acids (Lianhua et al. 2010).

Thermal cracking, a process known as pyrolysis, is the term used for the cleavage of bonds in complex molecules, with consequent formation of simpler molecules by action of high temperatures (Ranganathan et al. 2008). Heating is accomplished under drastic temperature conditions, over 450°C, and in the absence of air or oxygen (Ma and Hanna 1999). The catalytic cracking of oils has been studied, and the main catalysts used in this process are aluminum oxide (Al_2O_3) and silicon oxide (SiO_2) (Ma and Hanna 1999, Balat and Balat 2008). Several raw materials can be used in the synthesis of fuel by cracking, including vegetable oils and animal fats (Hanna and Ma 1999). These procedures generate triacylglycerides, alkanes and alkenes to a larger extent, and carboxylic acids to a lesser extent (Ma and Hanna 1999, Limà et al. 2004, Ranganathan et al. 2008).

The cracking technology has not been widely used because the equipment is very expensive (Ma and Hanna 1999). Furthermore, the fuel obtained from cracking is chemically similar to mineral diesel and is deoxygenated, thereby offering fewer environmental benefits as compared to an oxygenated fuel, and thus being less environmentally friendly (Ma and Hanna 1999, Knothe 2005).

Despite the existence of different methods for biodiesel production, as briefly described above, only transesterification has been widely used by the biofuels industry as a synthetic route. A transesterification reaction consists in the reaction between a vegetable oil or animal fat (triacylglycerides) and short-chain alcohols in the presence of a catalyst, with formation of a mixture of esters and glycerin, as illustrated by the equation in Fig. 12.2 (Leung et al. 2010, Oliveira et al. 2011).

Transesterification reduces the viscosity of vegetable oils upon elimination of the glycerol present in the triacylglyceride molecule. The alcohol used in the synthesis should preferably be a short-chain alcohol, and there are articles describing the synthesis of biodiesel using methanol, ethanol, 1- and 2-propanol, and butanol (Ma and Hanna 1999, Nielsen et al. 2008, Leung et al. 2010). Among these alcohols, the most effective ones are methanol and ethanol. Methanol is usually the alcohol of choice, because it is cheaper and very active for reaction (Leung et al. 2010). However, biodiesel production from ethanol has been stimulated, since this alcohol is obtained from renewable sources and is less harmful than methanol (Lenardão et al. 2003, Vieitez et al. 2011).

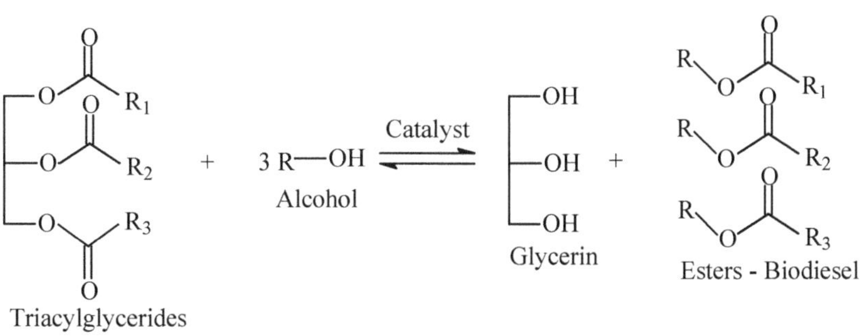

Figure 12.2 General equation for a transesterification reaction.

Different active inorganic catalysts, both homogeneous and heterogeneous, can be utilized in transesterification reactions, but the process led by alkaline catalysts has been the one used most often (Balat and Balat 2008, Leung et al. 2010). A good alternative that has been investigated is the use of lipases as biocatalysts of the transesterification reaction for biodiesel synthesis (Leung et al. 2010).

Transesterification via Chemical Catalysis

The transesterification reaction has been catalyzed since the first trials, when it was carried out in the presence of chemical catalysts, homogeneous inorganic catalysts, acidic or basic.

In the transesterification process conducted by homogeneous acid catalysis, sulfuric acid, phosphoric acid, hydrochloric acid, and organic sulfonic acid have been mainly used as catalysts (Ma and Hanna 1999, Meher et al. 2006).

Acid transesterification is well suited in cases where the vegetable oil to be used in the reaction displays high acid value, such as frying oils waste (Ma and Hanna 1999, Meher et al. 2006). If a high concentration of free fatty acids is present in the oil, the acid transesterification and esterification processes can be conducted simultaneously so as to, to obtain high oil conversion into biodiesel (Lotero et al. 2005, Drapcho et al. 2008). The use of acid homogeneous catalysis on an industrial scale is always presented by the fact that the reaction is considerably slower and demands large oil/alcohol molar ratios, high temperatures (above 100°C) and long reaction time for complete conversion (over 3 hr) (Schuchardta et al. 1998). An additional complication is the corrosion of the reactors and equipment used in the production line by concentrated acids (Dabdoub et al. 2009, Leung et al. 2010).

The main route used for the catalytic transesterification reaction worldwide is the homogeneous basic catalysis. The first patents and articles describing the development of transesterification via alkaline catalysis dates back to the 1940s and 1950s, and since then this has been the method utilized in the study and mass production of biodiesel (Dabdoub et al. 2009).

This catalytic route stands out as compared to acid catalysis due to the increased reaction rate and fewer problems related to corrosive effects (Schuchardta et al. 1998). From an industrial viewpoint, the transesterification reaction is well established for basic catalysts, and alkali metal hydroxides and alkoxides as well as sodium and potassium carbonates excel in this sense (Schuchardta et al. 1998, Dabdoub et al. 2009). The reaction mechanism of transesterification via alkaline catalysis is characterized by three consecutive reaction steps (Ma and Hanna 1999), represented in Fig. 12.3. Initially, the alkoxide is prepared from the reaction of a base with the corresponding alcohol. Then, the first stage of the transesterification reaction itself, there is a nucleophilic attack of the alkoxide at the carbon of the carbonyl groups of the triacylglyceride molecule, leading to the formation of a tetrahedral intermediate. The latter gives rise to diacylglyceride alkyl ester that is neutralized by deprotonation of the water molecule, with consequent regeneration of the basic catalyst. The reaction mechanism of the first stage spans over two steps until the formation of glycerol and a mixture of alkyl esters (biodiesel) (Schuchardta et al. 1998, Ma and Hanna 1999, Meher et al. 2006).

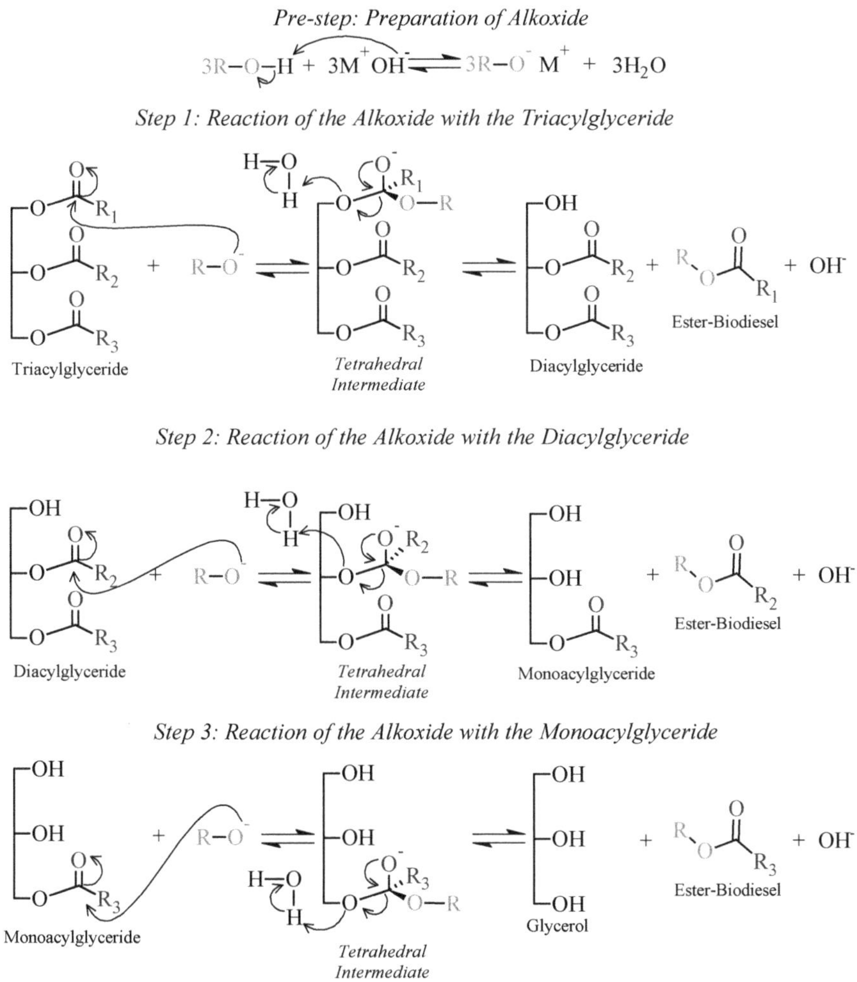

Figure 12.3 Reaction mechanism for the transesterification reaction driven by basic homogeneous catalysis.

This catalytic reaction occurs more rapidly than the acidic pathway, however, as in the case of acid catalysis, the reverse reaction may occur, so the quantity of alcohol used in the process should be larger than the 1:3 oil: alcohol molar ratio (Ma and Hanna 1999, Meher et al. 2006). Generally, 1:6 oil: alcohol molar ratio is sufficient to guarantee satisfactory yields in alkaline catalysis. This value is well below that required by acid catalysis, which often exceeds 1:30 oil: alcohol molar ratio (Schuchardta et al. 1998, Ma and Hanna 1999, Meher et al. 2006).

The flowchart presented in Fig. 12.4 is a simplified scheme of biodiesel production by alkaline transesterification, which occurs in

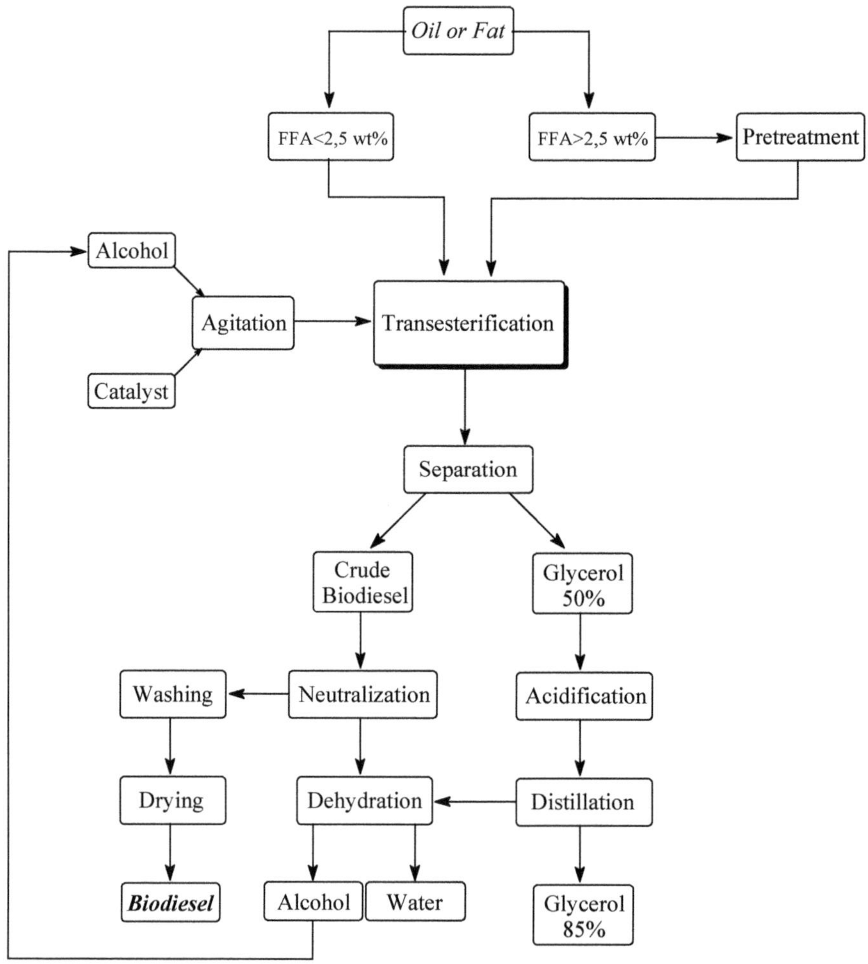

Figure 12.4 Simplified flowchart of the production process of biodiesel via base-catalyzed transesterification reaction.

different steps. Early in the process there might be a major complication, related to the contents of free fatty acids (FFA) and water in the oil sample. This may incur in an additional step involving pretreatment of the raw material for correction of the levels of the interfering substances.

The presence of an elevated amount of free fatty acids (FFA) in the composition of the oily matter culminates in a saponification reaction between the alkaline catalyst and the fatty acids, as shown in Fig. 12.5 (Leung et al. 2010). This reaction is extremely undesirable, because catalyst consumption for soap formation reduces the efficiency of the synthesis. Furthermore, the presence of soap results in the appearance of emulsions, which hinders separation of the biodiesel phase from glycerol (Schuchardta et al. 1998, Leung et al. 2010).

There is a general consensus that an FFA content of 2.5 percent enables accomplishment of the transesterification reaction without significant loss. However, for FFA contents greater than 2.5 percent, saponification becomes a potential hindrance (ISTC 2006, Leung et al. 2010). One way to circumvent this problem is to carry out pretreatment of the oil sample as mentioned in Fig. 12.3, which usually consists of an esterification reaction for conversion of free fatty acids to esters, prior to the transesterification of the triacylglycerides molecules (Canakci and Gerpen 2001).

Due to the presence of water in the system, another reaction can take place in parallel with transesterification: the hydrolysis of triacylglycerides (see Fig. 12.6) (Schuchardta et al. 1998, Leung et al. 2010). Therefore, the moisture content in the oil sample being transesterificated should be

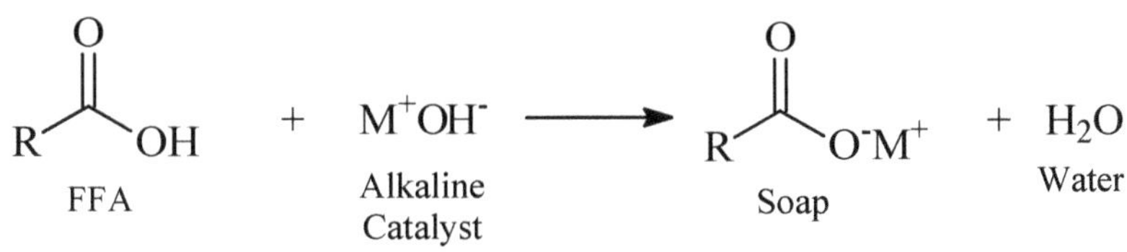

Figure 12.5 Saponification reaction between a free fatty acid and the alkaline catalyst (Leung et al. 2010).

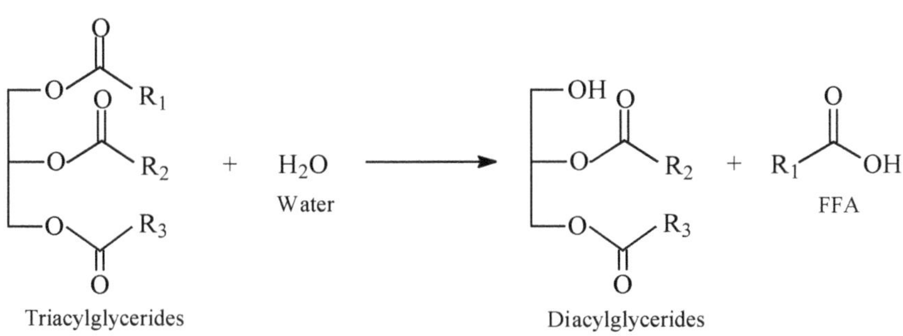

Figure 12.6 Hydrolysis of triglycerides (Leung et al. 2010).

as low as possible, 1 percent at most (Hanna and Ma 1999). Even the water formed during the preparation of the alkoxide may be sufficient to promote hydrolysis.

After the transesterification reaction, the reaction mixture consists mainly of the biodiesel phase and glycerin, both contaminated with excess alcohol and catalyst and the unreacted oil (Leung et al. 2010). As a result, a series of purification steps becomes necessary after the phase separation. A critical and complex process involves catalyst removal, which involves a series of biodiesel washings until the purity required by quality standards is achieved. However, this process does not allow for recovery of catalyst, typically phosphoric acid (H_3PO_4) which is neutralized during the process (Ma and Hanna 1999, Ranganathan et al. 2008, Ribeiro et al. 2011).

Based on the complexity and complications of biodiesel production and purification by transesterification via alkaline catalysis, several investigators have vested in the development of heterogeneous catalysts. The idea is to produce new catalysts, so that the transesterification can be conducted for any oil sample, regardless of their state of conservation, producing biodiesel in high yield and with low content of impurities, not to mention that the heterogeneous catalyst is easily recovered and reused in the process.

The use of heterogeneous catalysts has been encouraged precisely because they are very tolerant to the presence of water in the oil being transesterified, thus dismissing the need for the pre-treatment of oil with high FFA content. Moreover, biodiesel purification is significantly simplified due to the easy separation of the catalyst by filtration (Pinto et al. 2005, Singh Chouhan and Sarma 2011). Reflecting this interest in heterogeneous catalysts, a review article by Pinto et al. (2005) has highlighted the growing attention given to this class of catalysts, illustrated by a graph reproduced in Fig. 12.7, which lists the number of articles published between 2000 and 2004 as a function of the type of catalyst.

Various types of heterogeneous catalysts, including solid catalysts of basic character and inorganic acids, and more recently a new class of biocatalysts, consisting of immobilized enzymes (lipases) (Singh Chouhan and Sarma 2011) have been evaluated. In particular, biocatalysts have attracted the interest of researchers in recent years, also referenced by Pinto et al. (2005) (Fig. 12.8).

The transesterification catalyzed by lipases has a significant number of advantages over traditional methods of biodiesel synthesis. In the following section, we discuss the important biotechnological applications of lipases in biodiesel production, approaching the variables intervening in the process, such as the use of free or immobilized lipase, effect of different types of alcohol, and temperature, among others.

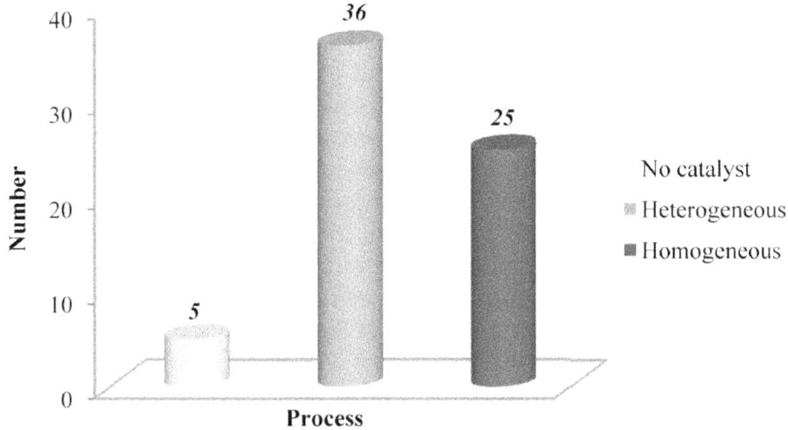

Figure 12.7 Number of articles for each type of catalytic process (Pinto et al. 2005).

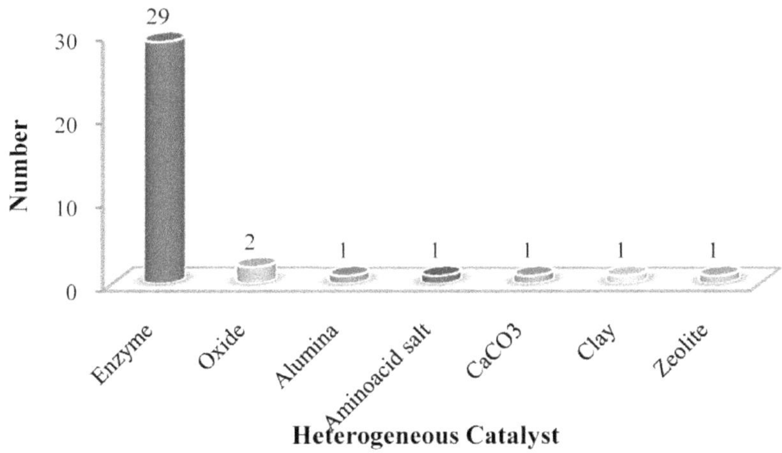

Figure 12.8 Number of articles published for each major type of heterogeneous catalyst (Pinto et al. 2005).

Transesterification via Enzymatic Catalysis

Enzymes, natural catalysts of reactions taking place in biological systems, are characterized by a high catalytic power and specificity and can make a reaction 10^7 times faster than the non-catalyzed process (Berg et al. 2006). The outstanding performance of these biological catalysts has inspired many chemists and engineers to study their application as catalysts in many different conventional chemical processes.

The interest in biodiesel synthesis has been steadily increasing, specifically in regard to the production of biofuels. The enzymes that are active for the present typical transesterification reaction are lipases obtained from microorganisms (Ranganathan et al. 2008). Indeed, a wide range of lipases has been studied for enzymatic biodiesel production (Nielsen et al. 2008). Lipase (triacylglycerol hydrolase, EC 3.1.1.3) acts on an aqueous-organic interface, catalyzing the hydrolysis of carboxylic ester. Figure 12.9 shows the main reactions catalyzed by lipases, namely hydrolysis, esterification, trans- and interesterification (Antczak et al. 2009, Vici 2010, Ribeiro et al. 2011, Rosset 2011).

The reaction mechanism proposed for the enzymatic transesterification is depicted in Fig. 12.10. The reaction begins with the nucleophilic attack of the hydroxyl group of an amino acid residue (serine) on the carbon atom of a carbonyl group, generating the enzyme-substrate complex (step 1). Then, the glycerol oxygen atom is protonated with a proton derived from an amino group of the enzyme, resulting in the cleavage of the CO bond and formation of a diacylglyceride molecule and an intermediate acyl-enzyme (step 2). Next, the acyl enzyme intermediate is attacked by an alcohol, furnishing the alcohol acylated enzyme complex (step 3). In the final step (step 4), the enzyme is reconstituted, and an ester molecule (biodiesel) is produced (Al-Zuhair et al. 2007, Rosset 2011).

In the transesterification reaction, as shown in Fig. 12.9, lipases also catalyze the esterification of carboxylic acids, which means that lipases act synergistically as a catalyst for the two processes on adoption of an appropriate alcohol/water molar ratio, leading to good yields for samples with high oil content (Shah et al. 2003, Antczak et al. 2009, Fjerbaek et al. 2009, Lukovic et al. 2011). This is the first advantage of biocatalysis in relation to chemical catalysis alkaline. There are several

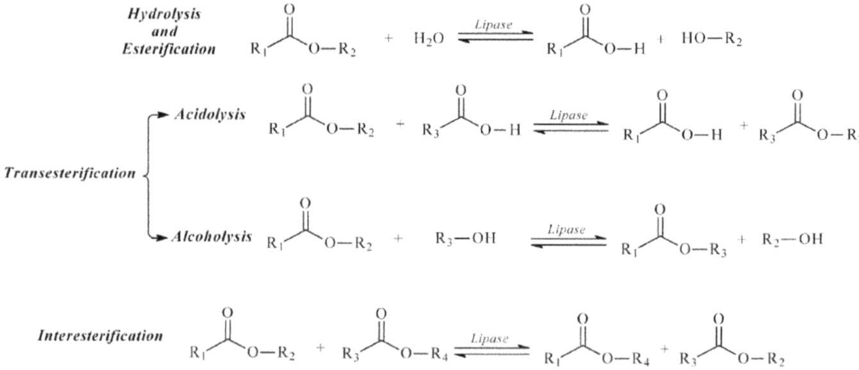

Figure 12.9 Organic reactions catalyzed by lipases (Vici 2010, Rosset 2011).

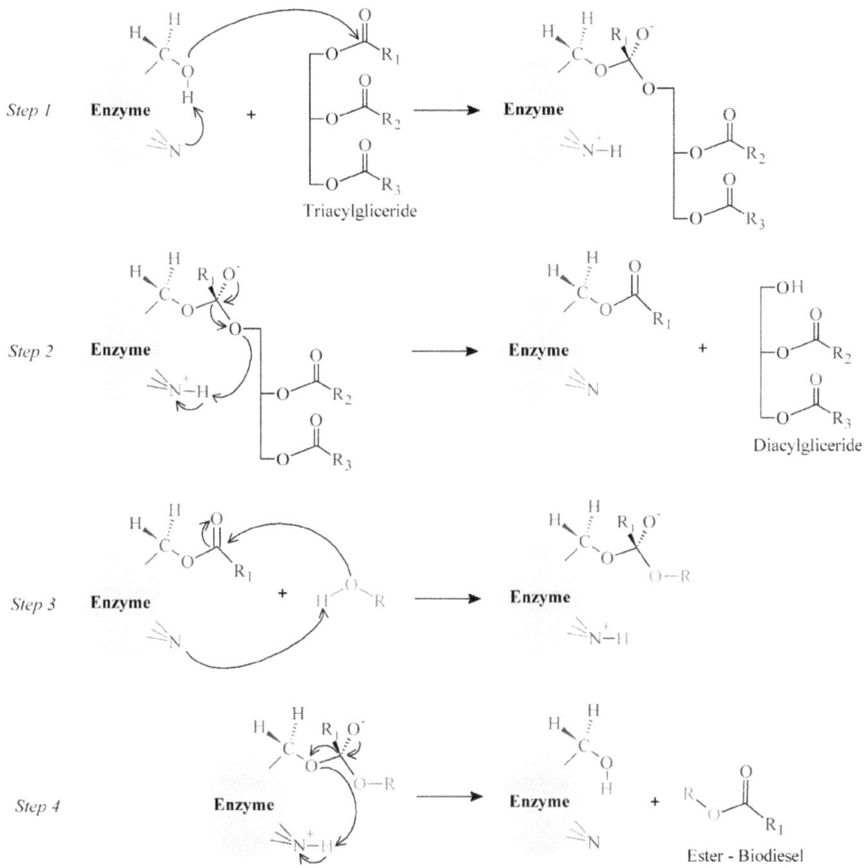

Figure 12.10 Reaction mechanism for the transesterification reaction carried out by enzymatic catalysis (Al-Zuhair et al. 2007, Rosset 2011).

other advantages from both the technical and economic standpoints. A comparison between the two catalytic routes can be made, in order to point out the pros and cons of each type of process (Table 12.1).

Thus the reasons behind the great interest in the application of lipases in biodiesel production can be understood. As already published, lipases catalyze biodiesel formation from both triacylglycerides and fatty acids, allowing the effective application of this synthetic route for oils having a high acid number, such as residual frying oils (Fjerbaek et al. 2009, Lukovic et al. 2011). Concerning the water content, the enzymatic reaction is less affected by moisture and there is no competing hydrolysis reaction or soap formation (Nielsen et al. 2008).

Taking into account the mass yield of the biodiesel reaction, there is no great discrepancy. In several processes, optimized literature results indicate

Table 12.1 Comparison between enzymatic and alkaline catalysis (Shah et al. 2003, Robles-Medina et al. 2009) in transesterification reactions.

Technical issue	Enzymatic catalysis	Alkaline catalysis
Large FFA content in the starting oil	FFA are estherified to biodiesel	FFA are saponified by the catalyst
H_2O content in the starting oil	The catalytic activity of lipases is not affected	Oil hydrolysis with soap formation
Biodiesel Yield[a]	High, around 90%	High, around 96%
Glycerin recovery	Easy, high purity glycerin	Complex, low purity glycerin
Catalyst recovery and reuse	Easy (when immobilized) or waived when operating in a PBR. Degree of reuse is variable	Difficult and impractical, being mostly neutralized with acid and discarded
Energy expenditure	Low, the process occurs under mild temperature (conditions between 20 and 50°C)	Average temperature range of 60 to 80°C
Catalyst cost	High	Low
Environmental impact	Low, the wastewater is discharged	Alkaline and saline effluents are generated
Process productivity[b]	Low	High

[a]Mass of the oil transformed into biodiesel.
[b]Biodiesel mass per reactor volume unit and per time unit.

yields in the range of 90 percent (Robles-Medina et al. 2009). The procedural advantages achieved with enzymatic catalysis offset this income gap. The use of immobilized enzymes as heterogeneous catalysts simplifies catalyst separation, performed by simple filtration. Moreover, when the catalyst is placed in a packed bed reactor (PBR), the separation step is dismissed, enabling catalyst reuse for several cycles (Robles-Medina et al. 2009, Lukovic et al. 2011).

The same is not true of the alkaline process, where the complexity involved in basic catalyst recovery renders the process uneconomical, and neutralizing the base with acid and discarding it is simpler. The amount of effluent produced can reach 0.2 tons per ton of biodiesel produced by chemical catalysis (Suehara et al. 2005) though, in addition to the difficult catalyst separation, the products from alkaline catalysis are very impure, while the products from the enzymatic synthesis are largely pure, dismissing complicated steps of thorough cleaning of the biodiesel phase (Fukuda et al. 2001, Robles-Medina et al. 2009, Lukovic et al. 2011).

A further advantage of the enzymatic process is that less waste and effluents are generated, making it more environment friendly. The amount of energy resources necessary for management of the process also culminate in lower energy expenditure, since enzymatic catalysis is conducted at lower temperatures (Robles-Medina et al. 2009). This set of technical advantages makes lipases ideal catalysts for biodiesel synthesis, leading to a considerably simplified flowchart as verified and presented in Fig. 12.11. The simplification becomes evident on comparison to the flowchart shown in Fig. 12.4 for alkaline transesterification reaction.

Nevertheless, and the consensus among several authors is that the use of enzymes on an industrial scale is prevented due to the high cost of the catalyst (Nielsen et al. 2008, Ranganathan et al. 2008). In addition, issues such as long reaction time, the fact that the enzymatic process is much slower than the alkaline reaction and the inactivation of lipases by methanol and glycerol are drawbacks that still place the alkaline catalysts in a favorable position (Shah et al. 2003, Fjerbaek et al. 2009, Robles-Medina et al. 2009, Lukovic et al. 2011).

On the basis of the numerous advantages of the enzymatic process, researchers have invested heavily in the study of biodiesel enzymatic synthesis, in order to identify the main intervening variables in the process and ways of optimizing them, seeking cheaper and more robust preparations that can replace alkaline catalysts. The following section contains a discussion on these aspects.

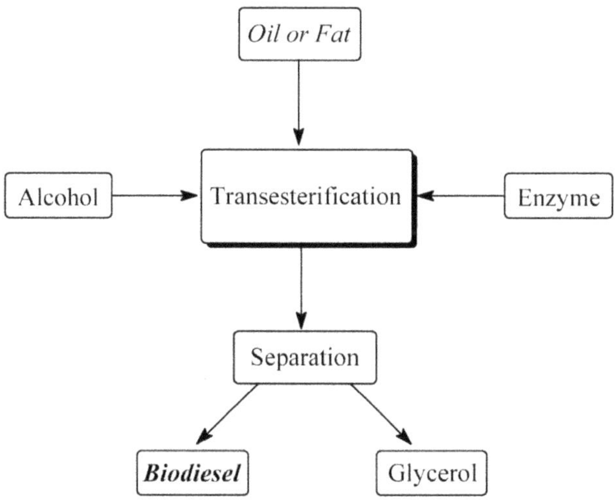

Figure 12.11 Simplified flowchart for biodiesel production via transesterification reaction catalyzed by enzyme.

Technical Aspects of Enzymatic Production of Biodiesel

Use of free vs immobilized lipase. The use of lipases in the free form has been studied and reported in the literature and their lower efficiency when compared to immobilized lipase formulations has been demonstrated (Nielsen et al. 2008). The enzymatic activity of free lipases decays more sharply in the presence of short-chain alcohols (the effects of different types of alcohol will be discussed below), and the glycerine formed in the process prevents the reuse of the enzyme in other production cycles (Nielsen et al. 2008, Ribeiro et al. 2011).

Along with this, the recovery of lipases is difficult, and biodiesel can contain impurities (Ribeiro et al. 2011). The use of free lipases would undoubtedly be ideal in terms of cost (Nielsen et al. 2008). However, in the face of the lack of lipases that can be efficiently used in free form, an alternative has been the use of immobilized biocatalyst. The main advantage of supported enzymes is that they can be easily recovered from the reaction medium by simple filtration (Nielsen et al. 2008, Robles-Medina et al. 2009). Moreover, they are more stable to the biodiesel synthesis conditions and can be reused a large number of times. Currently, a large majority of works have reported on the utilization of immobilized lipases (Nielsen et al. 2008, Fjerbaek et al. 2009, Robles-Medina et al. 2009, Ribeiro et al. 2011).

Most enzymes commercially available today consist of immobilized enzymes indeed, *Candida antarctica* lipase-B (CALB), *Rhizomucor miehei* (RML), and *Thermomyces lanuginosa* (TLL) are marketed in their immobilized form by Novozyme.

Water content. A catalytic transesterification reaction occurs with high yields preferably in the absence of any water (Nielsen et al. 2008, Leung et al. 2010). However, water plays an important role in the catalytic activity of various lipases. The optimal conformation of the lipase is retained by the surrounding water molecules, and the absence of water can cause irreversible alterations in the protein structure (Fjerbaek et al. 2009, Tan et al. 2010, Gog et al. 2012). Therefore, one should attempt to establish a balance in the water content, in order to ensure and maintain good product yields (Nielsen et al. 2008). Indeed, the water content of the enzymatic transesterification reaction depends on the enzyme immobilization on the substrate and the presence of solvents (Fjerbaek et al. 2009). Literature reports on enzymatic processes with high activity in the presence of high levels of water, as well as opposite results. For instance, lipase *Candida antarctica* (Novozym 435) is more active at low water contents (Watanabe et al. 2005, Watanabe et al. 2007). Deng et al. (2005) have also reported this finding for *Candida antarctica*. However, for the lipase *Pseudomonas cepacia*, increasing water content enhances activity while reducing its lifetime. Therefore, it is clear that the water content of the process must be controlled for each type of enzyme, in order to maintain good yields as well as enzyme stability.

Choice of alcohol. Biodiesel production is currently carried out with the short-chain alcohols methanol and ethanol, which are available in bulk and are cheaper. Hence, investigations on the enzymatic synthesis of biodiesel started by making use of these alcohols. However, we have found that short-chain alcohols, especially methanol, exert a negative effect on the stability of lipases, causing enzyme denaturation and inactivation (Fjerbaek et al. 2008, Nielsen et al. 2008, Antczak et al. 2009, Tan et al. 2010). This is so because polar short-chain alcohols have low solubility in oils, which leads to the formation of a new phase in the reaction system in contact with the lipase, thereby causing inactivation (Shimada et al. 2002). This negative effect becomes less pronounced in the case of alcohols with molecular chain > 3, which are better soluble in the oily substrate. Shimada et al. (1999) have demonstrated the inactivation of immobilized lipase *Candida antarctica* using methanol. Some solutions have been proposed in order to overcome this problem and the addiction of alcohol should be highlighted. Shimada et al. (2002) have achieved significant improvement in the product yield on addition of methanol in three successive steps in the case of the immobilized lipase *Candida antarctica*. This methodology enabled lipase reuse for over 100 d. Because of the simplicity of this method, it has been the most widely used as an alternative for prevention of the inactivation effect exerted by the alcohol.

Du et al. 2004 have proposed the replacement of methanol with other acyl acceptors, which determined yields above 90 percent when methyl acetate was utilized as the acyl acceptor and immobilized lipase Novozym 435 as biocatalyst. These same authors demonstrated non-enzymatic inactivation by methyl acetate and showed that the enzymatic activity was retained even after 100 cycles. These results are quite interesting, but the industrial application was frustrated due to the low reaction rate and the high cost of methyl acetate compared to methanol.

Use of solvents. Some authors have proposed the use of solvents to minimize the effect of enzymatic inactivation elicited by methanol, in order to increase the mutual solubility triacylglyceride/methanol and thus protect lipases from denaturation due to high alcohol concentrations (Robles-Medina et al. 2009, Gog et al. 2012). Several types of solvents have been suggested for enzymatic transesterification, such as hexane, petroleum ether, cyclohexane, and heptane (Soumanou and Bornscheuer 2003). The difficulty with using the cited solvents is that the solubility of methanol is not satisfactory, so the problem persists (Royon et al. 2007, Gog et al. 2012). A hydrophilic solvent that has proven to be very effective in improving the stability of the enzyme is the alcohol t-butanol (Li et al. 2006, Wang et al. 2006, Royon et al. 2007). The tertiary alcohol is not covered by lipases due to steric hindrance, and thus it could be used as a solvent for methanol and oil as well as glycerin formed during the process (Royon et al. 2007, Nielsen et al. 2008). In other words, the use of solvents is also recommended as a solution to another problem inherent to the enzymatic synthesis that is denaturation of lipases in the presence of the glycerol formed during the reaction (Robles-Medina et al. 2009), in accordance with literature results (Li et al. 2006, Wang et al. 2006, Royon et al. 2007, Robles-Medina et al. 2009).

Although the use of solvents can overcome the issues related to inactivation caused by both methanol and glycerol, this method still cannot be used in the industry due to difficulties in solvent recovery (Tan et al. 2012).

Effect of Temperature. The temperature at which a particular chemical reaction takes place is a very important parameter for its industrial application . In this sense, reactions occurring at lower temperatures are ideal, due to lower energy consumption. At this point, the energy gain achieved by the enzymes in relation to chemical catalysis can be mentioned. Indeed, the enzymatic process generally happens at lower temperatures, because lipase inactivation takes place at higher temperatures (Gog et al. 2012). Generally, the thermal stability of lipase is increased when it is immobilized as compared to the free enzyme and it remains active for a longer time (Nielsen et al. 2008, Fjerbaek et al. 2009).

The optimum temperature for lipase has been determined to be in the range of 30–70°C. Iso et al. (2001) have found that a temperature rise prompts argumented lipase *Pseudomonas fluorescens* reaction rate. Jeong et al. (2004) have determined an optimum temperature for the immobilized lipase Novozym 435, namely at 40ºC, while for lipase *Rhizopus oryzae* a reaction temperature above 40ºC leads to loss of lipase activity.

In fact, the optimum temperature for the enzymatic process should be studied to ensure being good for both transesterification reaction yields and lipase operational stability (Fjerbaek et al. 2009, Antczak et al. 2009, Gog et al. 2012).

Configuration of the Reactors for Industrial Application. The design of reactors for industrial biodiesel production by enzymatic means has been studied, and among the different reactor configurations that have been presented, batch reactors and packed bed are the ones used most often (Packed Bed Reactor—PBR) (Fjerbaek et al. 2008, Gog et al. 2012).

Batch reactors have been widely used for laboratory-scale studies, due to their simplicity regarding data collection on the enzymatic process (Nielsen et al. 2008, Tan et al. 2010). In this reactor, all the reagents are placed in the reaction tank and kept stirring, which aids dispersion of methanol to the reaction system. It is noteworthy that this alcohol should be preferably added in stages (Nielsen et al. 2008). Nevertheless, the critical problems of this class of reactors arise from the decreased activity of immobilized lipases due to disruption of the interactions between the enzyme and the support owing to immobilizer shear caused by agitation (Tan et al. 2010). Other drawbacks are the long reaction periods and the discontinuous process (Nielsen et al. 2008). Such problems are decisive for the non use of batch reactors in the biodiesel industrial enzyme production.

It is known that for a process to be industrially feasible, it must occur continuously, thereby diminishing costs; this applies to the enzymatic production of biodiesel, and among the proposed reactors that allow for the production of this fuel in continuous flow are the PBRs. A PBR enables biodiesel production by the enzymatic pathway via several denaturation cycles, hence reducing the effects of the enzyme and also the shear, as compared to the batch reactor (Lukovic et al. 2011). Moreover, PBRs allow the reuse of the enzyme without the need to separate the biocatalyst, so that the process can take place continuously (Tan et al. 2010). The main advantage of a PBR is the highest catalyst conversion by weight (Nielsen et al. 2008). However, a marked disadvantage of this reactor refers to the accumulation of the glycerol formed during the reaction at the bottom of the reactor. This is because glycerol can be deposited on the surface of the immobilized lipases, thereby resulting in lower enzymatic activity due to lipase denaturation (Tan et al. 2010, Gog et al. 2012). To solve this problem, it is necessary to remove the glycerol culture broth while it is formed. Shimada et al. (2002) have investigated the technical issues of PBR technology. These authors presented an effective methodology for biodiesel production consisting of the addition of methanol to three reactors PBR packed with lipase *Candida antarctica* arranged in series in three stages, followed by glycerin removal from the reactors (Fig. 12.12). The flow reaction is maintained at 30°C for 100 d without loss of enzymatic activity.

Prospects for the Application of Lipases in Biodiesel Production

In general, the enzymatic synthesis of biodiesel has significant advantages over the route driven by chemical catalysts to begin with, the process is more environmentally friendly, with significant reduction in energy costs and disposal of diminished problems regarding the waste generated during biodiesel purification. However, the main barrier to the effective enforcement of lipases is the high costs of this biocatalyst, which stems from the steps of enzyme immobilization and process adaptation to industrial reactors.

Therefore, research in this field should continue so as to develop efficient immobilization methods that ensure high enzymatic activity and stability. There is no doubt that once the necessary adjustments are achieved lipases will find a promising future as catalysts for biodiesel production.

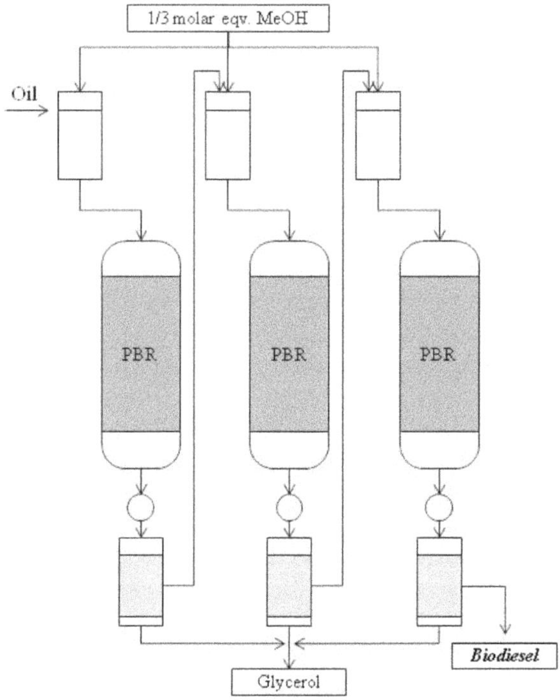

Figure 12.12 Scheme for PBRs operating in series, for the production of materials through the addition of methanol in steps and glycerol separation (Shimada et al. 2002).

Acknowledgements

The authors are grateful to CNPq and FAPEMIG for financial support.

References

Al-Zuhair, S., F.W. Ling and L.S. Jun. 2007. Proposed kinetic mechanism of the production of biodiesel from palm oil using lipase. Process Biochem. 42: 951–960.

Antczak, M.S., A. Kubiak, T. Antczak and S. Bielecki. 2009. Enzymatic biodiesel synthesis—Key factors affecting efficiency of the process. Renewable Energy 34: 1185–1194.

Balat, M. and H. Balat. 2008. A critical review of bio-diesel as a vehicular fuel. Energy Convers. Manage. 49: 2727–2741.

Berg, J.M., J.L. Tymoczko and L. Stryer. 2006. Biochemistry. W.H. Freeman and Company, New York.

Canakci, M. and J.V. Gerpen. 2001. Biodiesel production from oils and fats with high free fatty acid. Trans. ASAE 44: 1429–1436.

Chen, Y.H., J.H. Chen, C.Y. Chang and C.C. Chang. 2010. Biodiesel production from tung (*Vernicia montana*) oil and its blending properties in different fatty acid compositions. Bioresour. Technol. 101: 9521–9526.

Chongkhong, S., C. Tongurai, P. Chetpattananondh and C. Bunyakan. 2007. Biodiesel production by esterification of palm fatty acid distillate. Biomass Bioenergy 31: 563–568.

Chongkhong, S., C. Tongurai and P. Chetpattananondh. 2009. Continuous esterification for biodiesel production from palm fattyacid distillate using economical process. Renewable Energy 34: 1059–1063.

Conceição, M.M., R.A. Candeia, F.C. Silva, A.F. Bezerra, V.J. Fernandes, Jr. and A.G. Souza. 2007. Thermoanalytical characterization of castor oil biodiesel. Renewable Sustainable Energy Rev. 11: 964–975.

Dabdoub, M.J., J.L. Bronzel and M.A. Rampin. 2009. Biodiesel: a critical overview on the current status and perspectives at the academy and industry. Quim. Nova 32: 776–792.

Deng, L., X. Xu, G.G. Haraldsson, T. Tan and F. Wang. 2005. Enzymatic production of alkyl esters through alcoholysis: A critical evaluation of lipases and alcohols. J. Am. Oil Chem. Soc. 82: 341–347.

Drapcho, C.M., N.P. Nhuan and T.H. Walker. 2008. Biofuels Engineering Process Technology. The McGraw-Hill Companies, New York.

Du, W., Y. Xu, D. Liu and J. Zeng. 2004. Comparative study on lipase-catalyzed transformation of soybean oil for biodiesel production with different acyl acceptors. J. Mol. Catal. B Enzym. 30: 125–129.

Fjerbaek, L., K.V. Christensen and B. Norddahl. 2009. A Review of the Current State of Biodiesel Production Using Enzymatic Transesterification. Biotechnol. Bioeng. 102: 1298–1315.

Fukuda, H., A. Kondo and H. Noda. 2001. Biodiesel fuel production by transesterification of oils. J. Biosci. Bioeng. 92: 405–416.

Gog, A., M. Roman, M. Tosa, C. Paizs and F.D. Irimie. 2012. Biodiesel production using enzymatic transesterification—Current state and Perspectives. Renewable Energy 39: 10–16.

Harper, G.D.J. and J. Starbuck. 2009. Run Your Diesel Vehicle on Biofuels. McGraw-Hill Companies, New York.

Iso, M., B. Chen, M. Eguchi, T. Kudo and S. Shrestha. 2001. Production of biodiesel fuel from triglycerides and alcohol using immobilized lipase. J. Mol. Catal. B Enzym. 16: 53–58.

ISTC. 2006. Feasibility report, small scale biodiesel production. Waste Management and Research Center, Illinois, 6–7.

Jeong, G.T., D.H. Park, C.H. Kang, W.T. Lee, C.S. Sunwoo, C.H. Yoon, B.C. Choi, H.S. Kim, S.W. Kim and U.T. Lee. 2004. Production of biodiesel fuel by transesterification of rapeseed oil. Appl. Biochem. Biotechnol. 114: 747–758.

Knothe. G. 2005. Dependence of biodiesel fuel properties on the structure of fatty acid alkyl esters. Fuel Process. Technol. 86: 1059–1070.

Knothe, G. 2010. Biodiesel and renewable diesel: A comparison. Prog. Energy Combust. Sci. 36: 364–373.

Lenardão, E.J., R.A. Freitag, M.J. Dabdoub, A.C.F. Batista and C.C. Silveira. 2003. Green Chemistry—the 12 principles of Green Chemistry and it insertion in the teach and research activities. Quim. Nova 26: 123–129.

Leung, D.Y.C., X. Wu and M.K.H. Leung. 2010. A review on biodiesel production using catalyzed transesterification. Appl. Energy 87: 1083–1095.

Li, L., W. Du, D. Liu, L. Wang and Z. Li. 2006. Lipase-catalyzed transesterification of rapeseed oils for biodiesel production with a novel organic solvent as the reaction medium. J. Mol. Catal. B Enzym. 43: 58–62.

Lianhua, L., L. Pengmei, L. Wen, W. Zhongming and Y. Zhenhong. 2010. Esterification of high FFA tung oil with solid acid catalyst in fixed bed reactor. Biomass Bioenergy 34: 496–499.

Lima, D.G., V.C.D. Soares, E.B. Ribeiro, D.A. Carvalho, E.C.V. Cardoso, F.C. Rassi, K.C. Mundim, J.C. Rubim and P. A. Z. Suarez. 2004. Diesel-like fuel obtained by pyrolysis of vegetable oils, J. Anal. Appl. Pyrolysis. 71: 987–996.

Lotero, E., Y. Liu, D.E. Lopez, K. Suwannakarn, D.A. Bruce and J.G. Goodwin, Jr. 2005. Synthesis of Biodiesel via Acid Catalysis. Ind. Eng. Chem. Res. 44: 5353–5363.

Luković, N., Z. Knežević-Jugović and D. Bezbradica. Biodiesel Fuel Production by Enzymatic Transesterification of Oils: Recent Trends, Challenges and Future Perspectives. pp. 47–72.

In: M. Manzanera [ed.]. 2011. Alternative Fuel. In Tech Open Science, Shanghai. Ma, F. and M.A. Hanna. 1999. Biodiesel production: a review. Bioresour. Technol. 70: 1–15.

Meher, L.C., D.V. Sagar and S.N. Naik. 2006. Technical aspects of biodiesel production by transesterification—a review. Renewable Sustainable Energy Rev. 10: 248–268.

Morshed, M., K. Ferdous, M.R. Khan, M.S.I. Mazumder, M.A. Islam and MD.T. Uddin. 2011. Rubber seed oil as a potential source for biodiesel production in Bangladesh. Fuel 90: 2981–2986.

Nielsen, P.M., J. Brask and L. Fjerbaek. 2008. Enzymatic biodiesel production: Technical and economical considerations. Eur. J. Lipid Sci. Technol. 110: 692–700.

Oliveira, F.C.C., P.A.Z Suarez and W.L.P. Santos. 2008. Biodiesel: Opportunities and Challenges. Quim. Nova Esc. 28: 3–8.

Oliveira, M.F., A.T. Vieira, A.C.F. Batista, H.S. Rodrigues and N.R. Stradiotto. 2011. A simple and fast method for the production and characterization of methylic and ethylic biodiesels from tucum oil via an alkaline route. J. Biomed. Biotechnol. 2011: 1–4.

Pinto, A.C., L.L.N. Guarieiro, M.J.C. Rezende, N.M. Ribeiro, E.A. Torres, W.A. Lopes, P.A.P. Pereira and J.B. de Andrade. 2005. Biodiesel: An Overview. J. Braz. Chem. Soc.16: 1313–1330.

Ranganathan, S.V., S.L. Narasimhan and K. Muthukumar. 2008. An overview of enzymatic production of biodiesel. Bioresour. Technol. 99: 3975–3981.

Ribeiro, B.D., A.M. de Castro, M.A.Z. Coelho and D.M.G. Freire. 2011. Production and use of lipases in bioenergy: a review from the feedstocks to biodiesel production. Enzyme Res. 2011: 1–16.

Robles-Medina, A., P.A. González-Moreno, L. Esteban-Cerdán and E. Molina-Grima. 2009. Biocatalysis: Towards ever greener biodiesel production. Biotechnol. Adv. 27: 398–408.

Rosset, I.G. 2011. Biodiesel production employing biocatalysis by esterification and transesterification reactions. M.S. Thesis, Institute of Chemistry of São Carlos, University of São Paulo, São Carlos, pp. 39–40.

Royon, D., M. Daz, G. Ellenrieder and S. Locatelli. 2007. Enzymatic production of biodiesel from cotton seed oil using *t*-butanol as a solvent. Bioresour. Technol. 98: 648–653.

Schuchardta, U., R. Serchelia and R.M. Vargas. 1998. Transesterification of Vegetable Oils: a Review. J. Braz. Chem. Soc. 9: 199–210.

Schwab, A.W., G.J. Dykstra, E. Selke, S.C. Sorenson and E.H. Pryde. 1988. Diesel fuel from thermal decomposition of soybean oil. J. Am. Oil Chem. Soc. 65: 1781–1786.

Sendzikiene, E., V. Makareviciene and P. Janulis. 2006. Influence of fuel oxygen content on diesel engine exhaust emissions. Renewable Energy 31: 2505–2512.

Shah, S., S. Sharma and M.N. Gupta. 2003. Enzymatic transesterification for biodiesel production. Indian J. Biochem. Biophys. 40: 392–399.

Shang, Q., W. Jiang, H. Lu and B. Liang. 2010. Properties of Tung oil biodiesel and its blends with 0# diesel. Bioresour. Technol. 101: 826–828.

Shimada, Y., Y. Watanabe and T. Samukawa. 1999. Conversion of vegetable oil to biodiesel using immobilized *Candida antarctica* lipase. J. Am. Oil. Chem. Soc. 76: 789–793.

Shimada, Y., Y. Watanabe, A. Sugihara and Y. Tominaga. 2002. Enzymatic alcoholysis for biodiesel fuel production and application of the reaction to oil processing. J. Mol. Catal. B Enzym. 17: 133–142.

Silva, N.L., M.R.W. Maciel, C.B. Batistella and R.M. Filho. 2006. Optimization of Biodiesel Production From Castor Oil. Appl. Biochem. Biotechnol. 129: 405–414.

Singh Chouhan, A.P. and A.K. Sarma. 2011. Modern heterogeneous catalysts for biodiesel production: A comprehensive review. Renewable Sustainable Energy Rev. 15: 4378–4399.

Solomons, T.W.G. and C.B. Fryhle. 2009. Química Orgânica. LTC-Livros Técnicos e Científicos, Rio de Janeiro, pp. 107–109.

Soumanou, M.M. and U.T. Bornscheuer. 2003. Improvement in lipase-catalyzed synthesis of fatty acid methyl esters from sunflower oil. Enzyme Microb. Technol. 33: 97–103.

Suehara, K., Y. Kawamoto, E. Fujii, J. Kohda, Y. Nakano and T. Yano. 2005. Biological treatment of wastewater discharged from biodiesel fuel production plant with alkali-catalyzed transesterification. J. Biosci. Bioeng. 100: 437–442.

Tan, T., J. Lu, K. Nie, L. Deng and F. Wang. 2010. Biodiesel production with immobilized lipase: A review. Biotechnol. Adv. 28: 628–634.

Vici, A.C. 2010. Production, purification and immobilization of lipases produced by *Beauveria brongniartii*. M.S. Thesis, Faculty of Philosophy, Sciences and Letters of Ribeirao Preto, University of Sao Paulo, Ribeirão Preto, pp. 7–9.

Vieitez, I., M.J. Pardo, C. da Silva, C. Bertoldi, F. de Castilhos, J.V. Oliveira, M.A. Grompone and I. Jachmanián. 2011. Continuous synthesis of castor oil ethyl esters under supercritical ethanol. The Journal of Supercritical Fluids 56: 271–276.

Watanabe, Y., P. Pinsirodom, T. Nagao, T. Kobayashi, Y. Nishida, Y. Takagi and Y. Shimada. 2005. Production of FAME from acid oil model using immobilized *Candida antarctica* lipase. J. Am. Oil Chem. Soc. 82: 825–831.

Watanabe, Y., T. Nagao, Y. Nishida, Y. Takagi and Y. Shimada. 2007. Enzymatic production of fatty acid methyl esters by hydrolysis of acid oil followed by esterification. J. Am. Oil Chem. Soc. 84: 1015–1021.

Wang, L., W. Du, D. Liu, L. Li and N. Dai. 2006. Lipase-catalyzed biodiesel production from soybean oil deodorizer distillate with absorbent present in tert-butanol system. J. Mol. Catal. B Enzym. 43: 29–32.

Zhang, Y., M.A. Dubé, D.D. McLean and M. Kates. 2003. Biodiesel production from waste cooking oil: 1. Process design and technological assessment. Bioresour. Technol. 89: 1–16.

13

Pectinases Produced by Microorganisms: Properties and Applications

Maria de Lourdes Teixeira de Moraes Polizeli,[1,*] *André Ricardo de Lima Damásio,*[2,a] *Alexandre Maller,*[2,b] *Hamilton Cabral,*[3] *Aline Moraes Polizeli*[1] and *Mahendra Rai*[4]

Introduction

At present, microbial enzymes are being extensively used in different areas, due to their high efficiency and specificity (Damásio et al. 2010). Microbial enzymes have found applications in areas of genetics, protein engineering and in the development of new industrial processes, including processes

[1]Biology Department, Faculty of Philosophy, Sciences and Letters of Ribeirão Preto, São Paulo University, Brazil.
E-mail: polizeli@ffclrp.usp.br; alinepolizeli@gmail.com
[2]Biochemical and Immunology Department of School of Medicine of Ribeirão Preto, São Paulo University, Brazil.
[a]E-mail: andre.damasio@gmail.com
[b]E-mail: alemaller@yahoo.com.br
[3]Science Pharmaceutical Department of School of Pharmaceutical Science of Ribeirão Preto, São Paulo University, Brazil.
E-mail: hamilton@fcfrp.usp.br
[4]Department of Biotechnology, SGB Amravati University, Maharashtra, India.
E-mail: mkrai123@redifiimail.com
*Corresponding author

related to the environmental pollution control. Pollution is no longer seen by society as an acceptable problem. Among enzymes that show intensive applications in industrial and environmental sectors, it is those from the pectinolytic system that act on pectin or structurally related compounds.

The term "pectin" comes from Greek, and means clotted. It is a natural hydrophilic colloid found in the middle lamella and primary wall of higher plants, where it does not exceed 1 percent of their fresh mass (Rombouts and Pilnik 1980, Hugouvieux-Cotte-Pattat et al. 1994). This term designates a mixture of several compounds, of which the main is pectinic acid. Its active form is located in the cell wall and may be interconnected to other structural polysaccharides and proteins, forming insoluble protopectin (Kashyap et al. 2001). The pectic chain is involved in functions related to physiology, growth and plant development. In green fruits, pectin is found in its insoluble form (protopectin), associated to cellulose microfibrils, giving rigidity to the cell walls. During fruit ripening, enzymes hydrolyze the pectin chain or its lateral chains, making them to become more soluble (Willats et al. 2001).

Pectic substances are degraded by pectinases or pectinolytic enzymes, which are produced by a large number of bacteria, yeasts and fungi, insects, nematodes and plants (Whitaker 1991). These enzymes degrade pectin of the middle lamella resulting in tissue disintegration in order to obtain a source of carbon or a modification during fruit maturation. Pectin degradation of the middle lamella results in the degradation of tissue by cell separation, a process denominated maceration (Bailey and Pessa 1990). Due to the general occurrence of pectic polysaccharides in many vegetables, the enzyme systems capable of degrading them have various mechanisms of action (Rombouts and Pilnik 1980, Kashyap et al. 2001).

Innumerable applications of the enzymes that act on the pectinolytic system stimulate the search for new microorganisms and carbon sources that can act as genetic inducers of the pectinase synthesis in cells. Application in extraction and clarification of fruit juices and vegetables, extraction of olive oil, wine production, animal feed, textile industry, cellulose and paper industries, functional foods and pharmaceuticals, fermentation of tea and coffee, treatment of wastewater pectin and treatment of the banana fiber will be discussed.

This chapter describes recent advances related to the complex structures of pectins, classification of pectinases and their mechanisms of action, production by microorganisms and applications in a number of industrial and environmental control processes.

Pectin Structure

The pectin molecule is included in the family of complex polysaccharides, having a molecular mass between 30 and 300 kDa. It consists of a main chain of 1,4-α-D-galacturonic acid (GalA), partially esterified with methanol or acetyl groups (Villa 1999); its methanolic groups are partly substituted by sodium, potassium or ammonium ions (Fig. 13.1). Some sugars like L-rhamnose, D-galactose, D-xylose and L-arabinose may take part in interrupting the galacturonic chain (Fogarty and Ward 1974, Fogarty and Kelly 1983). The esterification level, the proportion of neutral saccharides and the degree of polymerization are the main heterogeneity elements found in pectic compounds of different origins.

Three great groups of pectic polysaccharides (homogalacturonan, rhamnogalacturonan-I and rhamnogalacturonan-II) have been isolated from plant primary cell walls and structurally characterized (Ridley 2001, Willats et al. 2006) (Fig. 13.1).

Homogalacturonan (HG) consists of a linear chain of GalA, linked by α-1-4 linkages meaning that D-GalA in the HG polymer can be esterified and/ or acetylated. The proportions of methyl esterification and acetylation have a profound impact on functional properties. Pectins are traditionally categorized as much or little esterified, when showing methyl-esterification levels higher or lower than 50 percent, respectively (Vincken et al. 2003). A

Figure 13.1 Representation of the structures of pectin. Dha: Deoxylyxoheptulopyranosylaric acid, KDo: Ketodeoxymannooctulopyranosylonic acid.

Color image of this figure appears in the color plate section at the end of the book.

pectin gel is formed when the portions of HG are interconnected to form three-dimensional crystalline net retaining water and solutes.

Rhamnogalacturonan I (RGI) consists of a repeating chain of alternating GalA and rhamnose, which may have lateral chains of neutral residues like galactose, arabinose and xylose. Some stretches consist of alternating galacturonic acid and rhamnose—"hairy regions", others of lower density of rhamnose—"smooth regions".

Rhamnogalacturonan II (RGII) has a structural support of HG and unlike RGI, complex lateral chains are connected with GalA residues. Until recently, it was accepted that the domains of rhamnogalacturonan and homogalacturonan constitute the structural basis of the pectin polymers. However, an alternative structure has recently been proposed in which a long chain of RGI forms the structure of the molecules (Fig. 13.1). The composition of the RGI sugars may be highly heterogeneous; on the other hand, RGII has a highly preserved structure (Vincken et al. 2003).

Several factors determine the jellifying properties of pectins, including temperature, type of pectin, pH, and presence of sugars, calcium and other solutes. In highly esterified pectins, binding zones are formed by crosslinks of HGs through hydrogen bonds and interactions between methoxyl groups, both resulting from the high concentration of sugar and an acid pH. In pectins with a lower number of esterifications, junctions are formed by crosslinks between calcium ions and free carboxyl groups (Willats et al. 2006).

Based on the type of modifications of the bounds of the main chain, pectic substances may be classified into protopectin, pectic acid, pectinic acid and pectins. Information on each of these terms is provided below.

- **Protopectin** is a term occasionally used to describe water-insoluble pectic substances, connected to cellulose fibers, found in plant tissues and utilized to produce soluble pectic substances. Pectic acids are galacturonan containing small quantities of methoxy groups.
- **Pectic acid** salts are called **pectates.**
- **Pectinic acids** are galacturonans with various amounts of methoxy groups. They possess the property of forming a gel with sugars and acids.

Pectinase Classification

Pectinase classification is based on their action on pectin molecule (Fig. 13.2). As recommended by the **Committee on Pectin Nomenclature** (1944), pectinolytic enzymes are generically called pectinases and basically may be divided into three classes according to the following criteria (Kashyap et al. 2001, Sakai et al. 1993, Hoondal et al. 2002, Uenojo et al. 2007):

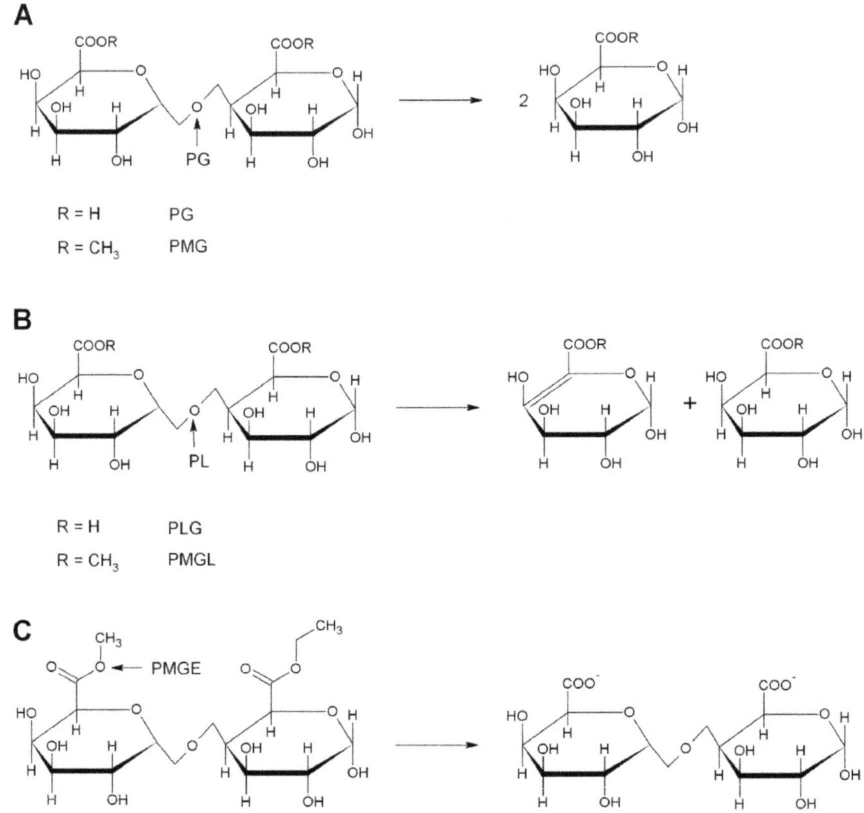

Figure 13.2 Pectinases and action sites on pectin structures. (A) PG: polygalacturonase, PMG: polymethylgalacturonase; (B) PGL: polygalacturonate lyase, PMGL: polymethylgalacturonate lyase; (C) PMGE: polymethylgalacturonate esterase.

(1) **Pectinesterase** (PE), (EC 3.1.1.11; PE): catalyzes the de-esterification of pectin methoxyl groups forming pectic acid and methanol. It acts preferably on the methoxyl ester group of a non-esterified galacturonic unit (Gummadi and Kumar 2005). It can also be classified as (PME) pectin methyl esterase, pectin demethoxylase and pectin methoxylase.

(2) **Depolymerases**: can also be characterized according to the type of cleavage (hydrolysis or trans-elimination) and also to the mode of action on the substrate molecule, that is, by random cleavage of glycosidic bindings in the interior of the molecule (endo-activity) or by cleavage of terminal bindings (exo-activity).

Hydrolases act by splitting glycosidic bonds between galacturonic units by the entrance of a molecule of water and can be classified as:

- **Endopolygalacturonase** (pectin depolymerase, EC 3.2.1.15, endo-PG): randomly cleaves glycosidic α-1,4 bonds of pectate and other galacturonans;
- **Exopolymethylgalacturonase** (poly (galacturonate) hidrolase, EC 3.2.1.67; exo-PG): catalyses sequentially the hydrolysis of the glycosidic α-1,4 bonds of pectate and other galacturonans and acts at the chain terminal.

The lyases act through trans-elimination forming a double bond between C-4 and C-5 of the monomer released from the substrate. Lyases can be classified in:

- **Endo-pectin lyase** (or pectin methyltranseliminase; pectin transeliminase, pectolyase; poly methylgalacturonic transeliminase, EC 4.2.2.10): cleaves randomly pectate and other galacturonas producing unsaturated methyl ester oligogalacturonates;
- **Exopolymethylgalacturonate lyase** (poly (methoxygalacturonide) exolyase; exo-PMGL): causes sequential cleavage of the pectin molecule through a transelimination process;
- **α-1,4-D-Endopolygalacturonic acid lyase or endopectate lyase**—EC 4.2.2.2; endo-PGL): randomly cleaves pectate and little methoxylated pectin, producing dimers and trimers and a series of unsaturated oligogalacturonates;
- **Exopolygalacturonate lyase** (or exopectate lyase—EC 4.2.2.9): cleaves sequential α-1,4 bonds of pectic acid, releasing unsaturated products from the reducing end.

(3) **Protopectinase**: solubilizes protopectin resulting in highly polymerized pectin.

Complete degradation of pectin involves the action of RG rhamnohydrolases (Fig. 13.3), that promote depolymerization of the rhamnogalacturonan chain, releasing rhamnose from the non-reducing ends, **RG galacturonohydrolases** that release monogalacturonic acid from the non-reducing chain ends (Mutter et al. 1998), **RG acetylesterases** that hydrolyze acetylgroups of rhamnogalacturonan (Searle-Van Leeuwen et al. 1992) and **xylogalacturonases** that cleave the bond between galacturonate residues and xylose releasing xylose-galacturonic acid dimers (van der Vlugt-Bergmans et al. 2000).

Polygalacturonases can also be classified into the family 28 of glycoside hydrolases (Table 13.1). In this family, there are also endopolygalacturonases, exopolygalacturonases, RG rhamnopolygalacturonases and endoxylogalacturane

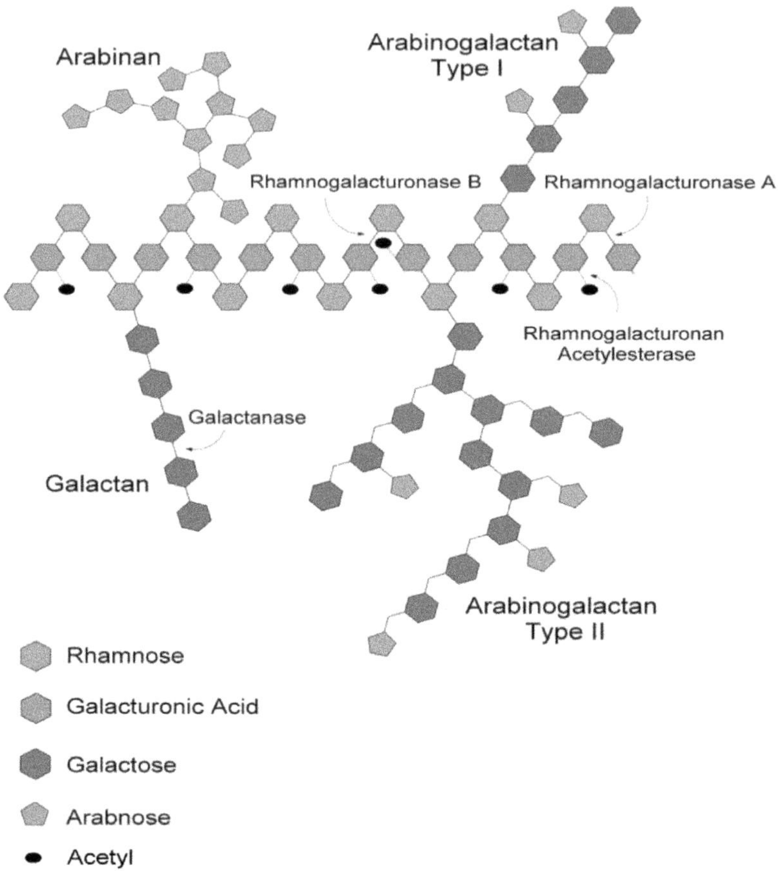

Figure 13.3 Enzymes and action sites on rhamnogalacturonan II structure.
Color image of this figure appears in the color plate section at the end of the book.

hydrolases. These enzymes have a catalytic site of an aspartate residue functioning as a nucleophile and another one that acts as a proton donor. Family 1 of polysaccharide-lyases harbors endopectate lyases, exopectate lyases and pectin lyases. Families 1 and 28 present helical beta parallel structural motives.

Pectinases from Microorganisms

Microbial pectinases have occupied a prominent place in industrial enzymology. Bacteria, yeasts and fungi can produce pectinases. Among reported microorganisms are *Fusarium solani, Aureobasidium pullulans, Selenomonas tuminantium, Aspergillus niger, Penicillium griseoroseum, Candida*

Table 13.1 Classification and structure of pectinases families.

Family	Known activities	3D Structure Status	EC number
GH[1]-28	endopolygalacturonase	Fold: (β)-helix	3.2.1.15
GH-28	exopolygalacturonase	Fold: (β)-helix	3.2.1.67
GH-28	exo-polygalacturonosidase	Fold: (β)-helix	3.2.1.82
GH-28	rhamnogalacturonase	Fold: (β)-helix	3.2.1.-
PL[2]-1	endopectate lyase	Fold: parallel β-helix	4.2.2.2
PL-1	exopectate lyase	Fold: parallel β-helix	4.2.2.9
PL-1	endopectin lyase	Fold: parallel β-helix	4.2.2.10
CE[3]-8	pectin methylesterase	Fold: (β)-helix	3.1.1.11
CE-12	pectin acetylesterase	non-classified	3.1.1.-
CE-12	Rhamnogalacturonan acetylesterase	non-classified	3.1.1.-
CE-13	pectin acetylesterase	non-classified	3.1.1.-

[1]GH: Glicosyl hydrolases
[2]PL: Polysaccharide lyase
[3]CE: Carbohydrate esterase
Source: http://www.cazy.org/

macedoniensis, Fusarium oxysporum, Saccharomyces fragilis, Bacillus pumilus, Bacteroides thetaiomicron and *Rhizoctonia fragarie* (Schuster et al. 2002, Sathyanarayana and Panda 2003).

The genes that encode the pectinases are cloned and expressed in various cell types. Some of the genes of endopolygalacturonases are cloned and expressed in *Saccharomyces cerevisiae* (Jia and Wheals 2000), in *Pichia pastoris* (Blanco et al. 2002), *Escherichia coli* (Wubben et al. 1999) and pectate lyase in *E. coli* (Zhao et al. 2007a,b).

Pectinases present different specificities for substrates and properties (Table 13.2), but basically can be separated into homogalacturonan and rhamnogalacturonan groups. Pectinases produced from *A. niger* are the most extensively used in industry. This fungus produces many enzymes which are active on homogalacturonan, including pectin methyl- and acetyl-esterase, endopolygalacturonase, exopolygalacturonase, pectate lyase and pectin lyase. It possesses a complete family of PG codifying genes and produces isoenzymes with considerable differences in relation to substrate specificity, cleavage patterns and optimal activity of pH (Lang and Dörnenburg 2000).

The biotechnological potential of fungal pectinolytic enzymes has aroused the attention of several researchers. Several agro-industrial residues like wheat bran, sugar-cane bagasse, coffee pulp, corncobs, lemon and apple peels are being explored as inducers of microbial pectinases (Sathyanarayana and Panda 2003, Patil and Dayanand 2006). A polygalacturonase was produced from *Aspergillus niveus* cultured on liquid or solid media supplemented with agro-industrial wastes (Maller et al. 2011). Submerged

Table 13.2 Biochemical properties of some pectinases from microorganisms.

Producer	Enzyme	pH#	Temperature# (°C)	K_m* (mg.ml^{-1})	V_{max}	Reference
Aspergillus niveus	Exo-PG	4.0	55	-	-	Maller et al. 2011
Aspergillus niveus	PL	8.5	55	-	-	Maller et al. 2012
Aspergillus niger NRRL3	PGI	5.0	40	-	-	Fahmy et al. 2008
Bacillus pumilus BK2	PGL	8.5	70	0.24	0.72 (c)	Klug-Santner et al. 2006
Bacillus sp. TS47	PG	8.0	70	-	-	Takao et al. 2000
Fusarium moniliforme	PGI	4.8	45	0.110	111.11 (a)	Niture and Pant 2004
Fusarium moniliforme	PGII	5.3	40	0.166	13.33 (a)	Niture and Pant 2004
Kluyveromyces wickerhamii	crude pectinase	5.0	50	-	-	Moyo et al. 2003
Mrakia frigida	PGL	8.5–9.0	30	-	-	Margesin et al. 2005
Mucor flavus	PGL	3.5–5.5	45	-	-	Gadre et al. 2003
Neurospora crassa	PG	6.0	45	5.0	357 (b)	Polizeli et al. 1991
Neurospora crassa exo-1 mutant	PGI	5.5	40	1.161	177.94 (a)	Crotti et al. 1998a
	PGII	5.5	45	0.023	2.08 (a)	
	PGL	9.0	50	0.50	273.20 (a)	
	PL	10.0	50	0.076	363.40 (a)	
Paecilomyces variotii	Exo PG	4.0	65	1.84	432	Damásio et al. 2010
Penicillium frequentans	PGI	3.9	50	0.68	596.8 (a)	Chellegatti et al. 2002
Penicillium oxalicum	PL	8.0	50	1.1	-	Yadav and Shastri 2007
Rhizopus microsporus var. *rizopodiformis*	PG	3.5	65	-	-	Damásio et al. 2011
Streptomyces lydicus	PG	6.0	50	1.63	677.8 (a)	Jacob et al. 2008
Tetracoccospo-rium sp.	PG	4.3	40	3.23	0.15 (b)	Aminzadeh et al. 2006
Trichoderma harzianum	PGII	5.0	40	3.4	1.28 (a)	Mohamed et al. 2006

#Optimum pH and Temperature

*This values of K_m refer to use of sodium polypectate;

Symbols: (PG) polygalacturonase; (PL) pectin lyase; (PGL) pectate lyase; (a) µmol.min^{-1}.mg^{-1}; (b) µmol.min^{-1}; (c) g.l^{-1}.min^{-1}

fermentation (SbmF) was tested using Czapek media supplemented with 28 different carbon sources. Among these; orange peel was the best PG inducer. On the other hand, for solid state fermentation (SSF), lemon peel was the best inducer. By comparing SbmF with SSF, both supplemented with lemon peel, it was observed that PG levels were 4.4-fold higher under SSF. Pectinase production can be induced by pectins of various sources, sodium polypectate and polygalacturonic acid. These complex polysaccharides can not enter the cells polysaccharides cannot enter the cells to affect gene expression. Thus, the real inducers of pectinolytic enzymes are considered compounds of low molecular mass or monogalacturonic acid itself.

There are reports on the role of the cell surface in the transduction of regulatory signals that control the adaptation of fungal cells of the wild type and mutants with mycelial phenotypes, and cell wall-less mutants (slime) of *Neurospora crassa* to their nutritional medium (Polizeli et al. 1990). These authors observed that the synthesis of cellulases, pectinases and xylanases was induced by polysaccharide substrates. Carbon catabolic repression was observed for all mutants with mycelial phenotypes. On the other hand, for some slime mutants, all enzymes that degrade polysaccharides were constitutively produced and were resistant to catabolic repression.

Polizeli et al. (1991) investigated pectinase production in the exo-1 mutant of *N. crassa*, an hyperproducer strain that produces many exoenzymes and also presents deficiencies in the cell wall composition (Gratzner and Sheehan 1969). This mutant secreted five to six times higher pectinase levels than the wild type. This enzyme was characterized as a monomeric glycoprotein with a high capacity to hydrolyze sodium pectate, characterizing it as an endopolygalacturonase. The exo-1 mutant also induced pectinases with galactose (Crotti et al. 1996, Crotti et al. 1998a, Crotti et al. 1998b). When these authors compared the induction of pectinases produced by the exo-1 of *N. crassa* in the presence of monogalacturonic acid, galactose and pectins, it was shown that polygalacturonase was induced preferably by galactose, but galacturonic acid was the best inducer of lyase activity.

The inducing effect of galactose and galacturonic acid appeared to differ:

(i) The mixture of galacturonic acid and galactose synergistically increased the production of pectic enzymes, when compared to those obtained by the inducers separately;
(ii) The inducing effect of galacturonic acid was partly inhibited by glucose;
(iii) The inducer effect of galactose was increased in the presence of glucose.

Taken together, these results suggest a more complex mechanism of pectinolytic enzymes regulation by pectin-containing molecules.

Heterologous Expression of Fungal Pectinases

Filamentous fungi are widely used for the production of homologous and heterologous proteins, but compared to homologous proteins, production levels of heterologous proteins are usually lower. The production levels of heterologous proteins have been drastically improved (Gouka et al. 1997). Homologous and heterologous protein production by filamentous fungi is often limited by the expression of proteases at high levels. By eliminating specific protease activities, protein production in filamentous fungi can be considerably improved (van den Homberg et al. 1997).

The production of pectinases has been improved by heterologous expression, mainly in filamentous fungi. A rhamno galacturonan acetylesterase (RGAE) was purified to homogeneity from the filamentous fungus *Aspergillus aculeatus*, and the NH_2-terminal amino acid sequence was determined by Kauppinen et al. (1995). Full-length cDNAs encoding the enzyme were isolated from an *A. aculeatus* cDNA library using a polymerase chain reaction-generated product as a probe. The 936-base pair rha1 cDNA encodes a 250-residue precursor protein of 26,350 Da, including a 17-amino acid signal peptide. The rha1 cDNA was overexpressed in *Aspergillus oryzae*, a filamentous fungus that does not possess RGAE activity, and the recombinant enzyme was purified and characterized.

From the onset of gene technology, yeasts have been among the most commonly used host cells for the production of heterologous proteins. At the beginning of this new development, the attention in molecular biology and biotechnology focused on the use of the best characterized species, *Saccharomyces cerevisiae*, leading to an increasing number of production systems for recombinant compounds (Gellissen and Hollenberg 1997).

A polygalacturonase gene of *Aspergillus awamori* IFO 4033 was cloned by genomic Southern hybridization with a probe of a DNA fragment synthesized by PCR. This was done using primers constructed based on the N-terminal amino acid sequence of a polygalacturonase, protopectinase-AS, produced by the strain and consensus internal amino acid sequence of fungal polygalacturonases. The cloned gene was inserted into an expression plasmid for yeast, pMA91, and introduced into *S. cerevisiae* to be expressed (Nagai et al. 2000). A pectinolytic industrial yeast strain of *S. cerevisiae* was generated containing the *S. cerevisiae* endopolygalacturonase gene (PGU1) constitutively expressed under the control of the 3-phosphoglycerate kinase gene (PGK1) promoter. The new strain contains DNA derived exclusively from yeast and expresses a high polygalacturonic acid hydrolyzing activity.

Yeast transformation was carried out by an integrative process targeting a dispensable upstream region of the acetolactate synthase locus (ILV2), which determines sulfometuron methyl resistance (Fernández-González et al. 2005).

Purified proteins are a starting point for biophysical and biochemical characterization methods that can assist in the elucidation of function. The challenge for the production of proteins at the scale and quality required for experimental, therapeutic and commercial applications has led to the development of a diverse set of methods for heterologous protein production.

Bacterial expression systems are commonly used for protein production as these systems provide an economical route for it and require minimal technical expertise to establish a laboratory protein production system (Zerbs et al. 2009).

Pectin lyases from *A. oryzae* and *A. niger* are usually used for the production of traditional fermented foods, but these fungi produce less pectinases under natural conditions. The cDNA coding mature Pell (without signal peptide) was amplified from *A. oryzae* by RT-PCR. Pell cDNA was cloned into pET-28a (+) expression vector, then was transformed into *E. coli* Turner (DE3) plac I cells to express Pell with 6-His tag. For improving the efficiency of Pell expression in *E. coli*, the conditions of expressing the Pell in *E. coli* were optimized. *E. coli* Turner (DE3) plac I cells with pET-28a (+)-pell was first cultivated at 37°C, 220 r/min until OD600 reached about 0.8. Then, cultivation broth was added with 0.05–0.1 mmol/L IPTG and continuously incubated at 15°C, at 170 r/min for 60 hr for expressing of Pell. The recombinant expressed Pell activity could reach 400 u/mL medium, which is 4000-fold of Pell produced naturally by *A. oryzae* and superior than known recombinant amount of pectin lyases expressed in different fungi expression systems (Zhao et al. 2007a,b).

Polygalacturonase Action Mechanism

Through the alignment of primary sequences of endopolygalacturonases, there were four regions conserved Asn-Thr-Asp, Gly-Asp-Asp, Gly and Gly-His-Arg-Ile-Lys (Pickersgill et al. 1998). In addition to these, other clusters were detected by Sony and Ponnuswamy (2007)—Ser-Ile-Gly-Ser. These regions are considered as the functional conservation (Sony and Ponnuswamy 2007).

The catalytic sites of endopolygalacturonases have basically two functions: the binding of substrate and its subsequent catalysis. According to these shares, some catalytic mechanisms were determined. The glycosylated hydrolases present two types of mechanisms, retention or inversion of configuration of the anomeric carbon (Armand et al. 2000).

To understand the role of the catalytic-amino acid residues, site-directed mutation in the fungal polygalacturonase was performed and suggested that three aspartic acid residues Asp-180, Asp-201 and Asp-202 (numbered according to the *A. niger* Polygalacturonase II) are directly involved in the catalysis (Fig. 13.4A). In addition to these residues, Arg-256, Lys-258 and His-223 (Fig. 13.4B) are also involved in the catalysis: the Arg and Lys residues bind the substrate and His is the maintenance of state ownership of ionizing the catalytic aspartic acid (Sony and Ponnuswamy 2007).

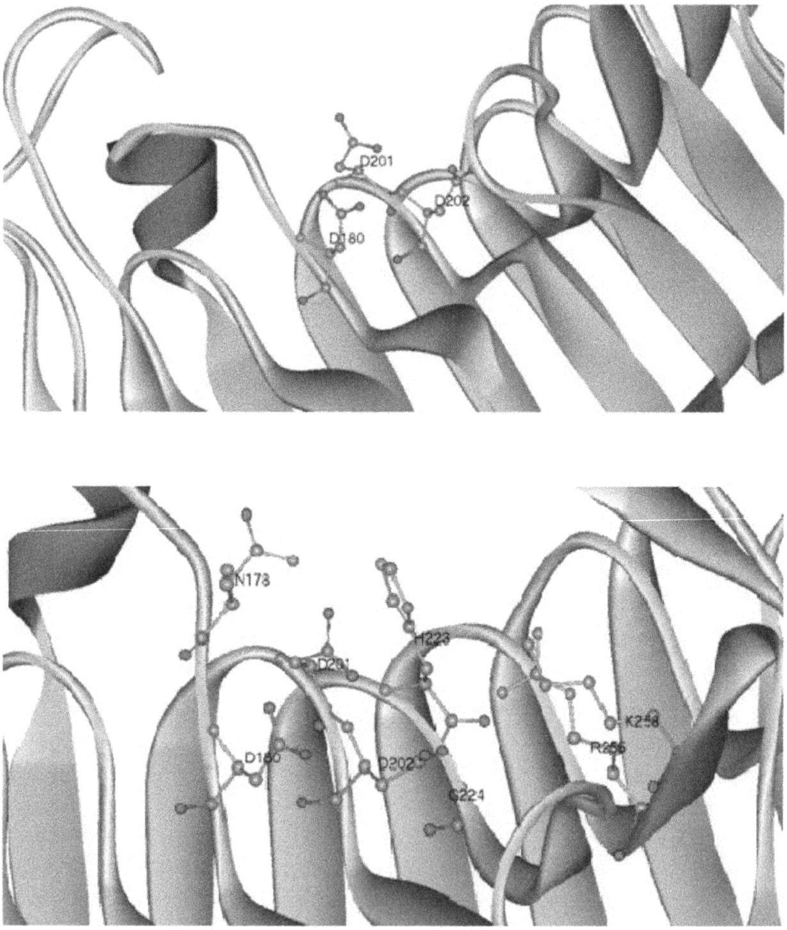

Figure 13.4 Ribbon representation of the conserved amino acid residues of the catalytic site of Polygalacturonase of *Aspergillus niger*: (A) representation of Asp-180, 201 and 202; (B) representation of the catalytic site residues. (The figure was generated using the ViewerLite (Accelrys Inc.) program, based on the authors van Santn, Y. et al. 1999 and Pagès, S. et al. 2000. The sequence using was 1czf PDB.)

Color image of this figure appears in the color plate section at the end of the book.

Polygalacturonases acting by invertion mechanism have a pair of carboxilic acids in the catalytic site, where one of them acts as a proton acid releasing oxygen to the glycosidic linkage and the other acts as a base activated by a water molecule, thus making the nucleophilic attack on the sugar anomeric carbon (Sony and Ponnuswamy 2007).

In the enzymes acting by an inverted mechanism, the glycosidic linkages are hydrolyzed in water by a simple mechanism of direct displacement in the anomeric center. The glycosidases showing the retention mechanism involves the formation of an intermediary (glycosyl-enzyme), in which the sugar is covalently linked to the protein via the carboxylic side chain of aspartic or glutamic acid, and the covalent bond-glycosylated enzyme (glycosyl-enzyme) is formed and hydrolyzed (Withers 1995).

For a better understanding of the Polygalacturonase II catalytic mechanism of the fungus *A. niger* (Armand et al. 2000), we performed site-directed mutations of six amino acid residues conserved between Polygalacturonases (Asp-180, Asp-201, Asp-202, His-223, Arg-256 and Lys-258). All mutated residues are apparently critical for the catalysis, except for the His-223.

The direct involvement of His-223 in the active site of polygalacturonase II, according to Armand et al. (2000) can be a residue to make the donation of proton in the reaction catalyzed by polygalacturonase II.

When the mutation in His-223 was concluded, it was observed that the catalysis was affected, but the frequency at which the cleavage of the connection occurred was not changed dramatically when the Histidine was replaced by Cysteine at position 223. This mutation did not cause disruption in subsites –1 and +1, indicating that His-223 plays a role in the catalysis by maintaining an optimal state of ionization of the carboxylate involved in the catalysis by protons share. This fact is corroborated by the fact that the mutated enzyme in the CYS-223, whose sulfhydryl group is capable of sharing proton, had its activity remaining high (Armand et al. 2000).

The role of aspartic acid in the hydrolysis of glycosidic linkages is confirmed, for example, in the inverted mechanism, where two groups are required for a distance of carboxylic 10Å (± 2) (Armand et al. 2000).

The Asp-201 in general makes the protonation of the products when they are hydrolyzed; there are three assumptions supporting this proposal:

I) When the mutation of this residue occurred, it was possible to see that the enzyme became inactive;
II) The mutation of this residue showed a minor effect on BCFs (Bound Cleavage Frequencies) of the enzyme on olygogalacturonates, suggesting that Asp-201 is not directly interacting with the substrate;

III) The His-223 residue is fundamental for the catalysis. The His-223 shares protons with Asp-201, allowing this last residue to be in the ionization state appropriate to protonate the product (Armand et al. 2000).

After the resolution of the structure from these three aspartic acid residues of polygalacturonase from *Erwinia carotovora*, the amino acids corresponding to Asp-180 and Asp-201 in PGII are directly involved in catalysis, according to Pickersgill et al. (1998).

Applications of Microbial Pectinases

Microbial enzymes are usually applied in many industrial and environmental sectors. The world enzyme market has a commercial annual turnover of US$2 billion dollars and keeps growing every year at a rate of 8–10 percent (data supplied on-line by the Brazilian Ministry of Science and Technology). The food industry has benefited mostly from the use of enzymes, using 15 percent of industrial processes. Furthermore, this sector has been showing an increasing expansion of 9.2 percent in relation to 2004 (Association of Biological Agriculturists of the State of Rio de Janeiro, Brazil). From this total, the industry of natural juices, the largest source of application of pectinases, had a production of 11 million liters between January and March of 2007, a 12.6 percent growth relative to the same period of 2006 (Brazilian Association of Soft Drinks and Non-Alcoholic Drinks).

Pectinases began to be used in the food industry in 1930 (Kertesz 1930). Following technological advances and the need for new sources of fungal pectinases, they became one of the most important industrial enzymes, acquiring high application in textile processes, degumming of plant fibers, treatment of aqueous residues, tea fermentation, food, beer, wine, animal feed, pulp, paper industries and agriculture (Bajpai 1999). A more detailed description of pectinase participation in various industrial sectors is given below.

Extraction and clarification of fruit juices and vegetables

Most preparations of microbial pectic enzymes are used in industrial processes utilizing fruits; even so, these represent only about a quarter of the production of enzymes used in food worldwide. Around 1930, when the food industries started juice production, yields were very low and many difficulties arose for their filtration and clarification. Thereafter, research on pectinases, cellulases and hemicellulases of the fungi *A. niger* and *Trichoderma* sp. together with the increasing knowledge on fruit composition, helped to overcome these difficulties (Grassin and Fauquembergue 1996).

Usually pectinases (pectin lyase, methylesterase, endo and exopolygalacturonases, pectin acetylesterase, rhamnogalacturonase, endo- and exoarabinases), combined with cellulases and hemicellulases —collectively called maceration enzymes—are used for the extraction and clarification of fruits and vegetables (Grassin and Fauquembergue 1996, Galante and Conti 1998, Gummadi and Panda 2003).

During the production of apple and pear juice, all the fruits are ground, forming a mass, which after a mechanical process, yields a limpid juice and a solid phase called pulp (Galante et al. 1998). The use of maceration enzymes increases both yields as well as initial processes without high investments. Maceration enzymes are generally used in two steps:

(1) following grinding, to macerate the fruit pulp until partial or full liquefaction, that not only increases yields, decreases periods for processing, but also increases the extraction of nutritive components from the fruit;
(2) following juice extraction, pectinases are used for clarification, therefore decreasing viscosity prior to concentration, increasing filtration rates and the stability of the final product.

Commercially produced juices by these industries include: sparkling ones (apple, pears and grapes), turbid juices (citric, plums, tomato, and nectars) and unicellular products (resulting from plant tissue transformations in a suspension of intact cells), whose objective is to preserve cell integrity by selective hydrolysis of middle lamella polysaccharides (Kashyap et al. 2001).

Unlike apple pectin that is highly methylated, orange juice naturally contains a large amount of pectinesterases that cleave methoxyl groups from pectin molecules. In the presence of Ca++ ions, insoluble calcium pectinate is formed, resulting in undesirable particle precipitation. To prevent this, two methods are available: the first is to denature the pectinesterases by heating the juices; however, this procedure interferes with the product flavor. An alternative is to freeze the juice concentrate, keeping the enzyme in an active state.

Juices are in general commercialized in this form; however, concentrating and freezing are very expensive for storage and transportation. Pectinases can be used in two ways in the extraction of fruit juice: at the end of pulp extraction (0.5 ± 2.0 g/100 Kg pulp at room temperature), to reduce viscosity or right after finishing its manufacturing at the same concentration, at room temperature, for 30 min (Rebeck 1990). This enables better extraction of sugars and soluble solids, resulting in higher yields and lower viscosity.

The traditional method of clarification of lemon juice uses the natural presence of pectinesterases. Compounds responsible for turbidity are mainly obtained from fruit peels. Peel and pulp are ground into 3–5 mm particles, mixed with water (1:1), heated to 95°C to destroy methylesterases and cooled to 50°C. The mixture contains large amounts of pectin, other polysaccharides, glycoproteins and essential oils; it is then treated with pectinases or pectinases plus cellulases. After 1 hour, until the end of the process; the liquid extract, if necessary, is again depectinized with the enzyme. Centrifugation, pasteurization, concentration, bottling and sales, follow.

Pectinases play an important role in tea and coffee fermentation. The latter uses pectinolytic enzymes to remove the mucilaginous layer present in the grains. Sometimes pectic enzymes are used to remove the mucilaginous coat from the grains. Cellulases and hemicellulases present in the enzyme preparation aid in the digestion of the mucilage. A commercial enzyme preparation is spread on the grains (2–10 g/ton at 15–20°C). The fermentation stage of coffee is accelerated, and reduced from 40 or 80, to 20 hr due to enzyme treatment (Carr 1985, Godfrey 1985). Alkaline pectinases are mainly used to remove fiber and pretreatment of industrial juices, but they are mostly from bacterial sources. Other application examples would be the production of Japanese paper, paper manufacture, and oil extraction (e.g., Canola), as well as, coffee and tea fermentation (Kapoor et al. 2000, Kashyap et al. 2001, Hoondal et al. 2002).

Extraction of olive oil

In the past few years, olive oil has attracted the world market due to its numerous benefits for health. Its extraction involves olive crushing in stone or a knife mill, passage of the resulting paste through a series of horizontal mixers and decanters and centrifugation at high speed for oil recovery. The main advantages of using maceration enzymes during olive oil extraction are: increased extraction (up to 2 Kg oils/100 Kg olives), better fractionation during centrifugation, obtaining of oil with high levels of antioxidants and vitamin E and decreased rancidification (Galante and Conti 1998).

Canola seed, coconut, sunflower seed and olive oils are traditionally produced by the extraction with organic solvent; hexane is the most common and a potential carcinogenic. Cell wall degrading enzymes including pectinases are used in the extraction of vegetable oils in aqueous processes for the liquefaction of the structural components of cell walls of seeds that contain oil (Kashyap et al. 2001).

Wine production

The first enzyme used in the wine industry was a commercial pectinase from *Aspergillus*. Over the last four decades, attempts have been made to improve yeast strains for fermentation of grape juices as well as microbial enzymes during wine production. The three major extracellular enzymes used were pectinases, beta-glucanases and hemicellulases. Their main benefits included clarification and easier filtration, as well as improved quality and stability of the wine. Addition of pectinases during grape crushing improved juice extraction, reduced time for clarification and increased the terpene content of the wine. Furthermore, pectinases with high pectin lyases and low pectin methyl esterase activities are preferred to minimize methanol release from methylpolygalacturonic acid during production (Galante et al. 1998).

Animal feed

Pectinases are used together with other enzymes to reduce viscosity of the feed, increase the absorption and release of nutrients via hydrolysis of non-degradable fibers and of nutrients blocked by fibers (Sharma and Sathyanarayama 2006).

Textile industry

Pectinolytic enzymes can be used in these industries to degrade the pectin layer that covers cellulose fibers, releasing them for posterior processing (Piccoli-Valle et al. 2001), at the treatment of fluid residues and the degumming of natural fibers (Kaur et al. 2004, Klug-Santner et al. 2006). Pectinases are used for the maceration of vegetable fibers, linen, hemp and jute, in the biopreparation of cotton and the enzymatic polishing of mixed jute and cotton tissue (Kaur et al. 2004, Klug-Santner et al. 2006).

Degumming of rami with pectin lyase produces fibers of higher quality than those produced by commercial enzyme complexes or using a chemical process with alkali, and also reduces environmental pollution (Silva et al. 2005). In crude cotton, removal of pectin wax and gumming agents by the use of pectinases plus amylases, lipases and hemicellulases, under adequate conditions, replaces the uses of caustic soda, generating high quality products, for subsequent dyeing and weaving, with lesser energy consumption (Sawada and Ueda 2001).

Cellulose and paper industries

During paper manufacturing, pectinases may depolymerize pectin-containing substances and decrease cationic demand of pectic solutions and result in filtrates from peroxide blanching, solving problems of retention in mechanical blanching of cellulose and treatment of effluents from paper mills (Sharma and Sathyanarayama 2006).

Functional foods and pharmaceuticals

The pectin hydrolysates and polysaccharide pectics are classified as probiotic, since they can be used as promoters of health in human and animal nutrition, through growth stimulus to select and/or the activity of species of resident bacteria in the intestinal colon (Uenojo and Pastore 2007).

Yamada (1995) found two different bioactive pectic polysaccharides, bupleuran 2IIb and 2IIc roots of *Bupleurum falcatum*. Bupleuran 2IIc, consisting mainly of galacturonic partial regions, showed potent anti-ulcer activity. The digestion of bupleuran 2IIc with endo-polygalacturonase provided mainly galacturonic oligosaccharides and small proportions of regions resistant to the enzyme. The results of the PG resistant carbohydrate portion showed a potent activity that is indicative of a role for endo-PG in the production of pectic poly-pectioligosaccharides as pharmaceuticals (Lang and Dörnenburg 2000).

Fermentation of tea and coffee

The pectinases have an important role in the fermentation of tea and coffee. The fermentation of coffee beans using pectinases is used by microorganisms acting on the removal of the mucilage layer of grains. The stage of the fermentation process of coffee is fast and reduces from approximately 80 hr to 20 hr with the enzymatic treatment, thereby improving the final product.

Pectinases from fungi are also used in the manufacturing of tea. The fermentation of tea treated with enzyme accelerates, although the dose may carefully be adjusted to avoid damage to the tea leaves (Kashyap et al. 2001).

Treatment of wastewater pectin

The citrus processing industry uses a certain amount of water that contains little pectinaceus material decomposed by microorganisms in the treatment of activated sludge. Tanabe et al. (1987) attempted to develop a new treatment for wastewater, using processing alkalophillic microorganism that produces an endo-lyase alkaline extracellular pectinase, and the treatment with this lineage caused useful results in the removal of pectin substances from wastewater (Kashyap et al. 2001).

Treatment of the banana fiber

Banana fibers are light weight, soft fibers obtained from the pseudo-stem of the banana. After harvesting the fruit, the stem of the banana is discarded as a waste material. The banana fiber can be used for the production of the Kraft paper (Jacob et al. 2008).

Jocob et al. (2008) performed a study of degradation of the fiber of dried banana and hand stripped with crude extract of pectinase from the *Streptomyces lydicus* lineage and observed that a gradual release of reducing sugar was an effective treatment.

The excessive removal of the non-cellulolytic coating, gummy material part of the fiber of the plant cell is called degumming, and is required before the industrial use of the fiber. The chemical treatment of the fiber is a toxic waste and causes serious threats to the environment as well as biological disturbances. Pectinases play the main role for the processing of these fibers, since 40 percent of dry weight change of the plant cell is comprised by pectin (Jacob et al. 2008).

Conclusions and Perspectives

From the beginning of polysaccharide research, in the first half of the 20th century, phenomenal advances in enzyme production and improvement of their properties for many industrial uses have taken place. Identification of organisms, mainly with thermophilic characteristics, enables the producing processes which work at high temperatures and various pH levels, showing good stability under these conditions. Advances in the field of recombinant DNA technology provides a means for the super expression of enzymes. Yeasts are able to provide ease of growth and genetic manipulation, as well as the capacity to perform specific post-translation modifications for eukaryotic cells. Furthermore, if compared to more complex eukaryote systems, yeast cells are more economical, present better yields and require less time and efforts. Recombinant DNA technology and/or site-directed mutations might lead to the production of pectinases, and genes might be easily introduced in selected microorganisms that use this new genetic material by means of direct expression.

Another great challenge for research is to aggregate value to agroindustrial residues that are being produced in large scale as sugarcane refuse, fruit peels or other agro/animal husbandry products, that when accumulated lead to environmental pollution. The cultivation of microorganisms, mainly filamentous fungi fermented on solid substrate, has become a solid basis for pectinase production and other enzymes with potential industrial application. Thus, the combination of strategies like the cultivation of thermophile fungi with agro-industrial residues, allied to the development of enzyme superexpression by molecular techniques, may benefit various industrial sectors as well as the environment.

Acknowledgements

We thank Fundação de Amparo à Pesquisa do Estado de São Paulo (FAPESP), Conselho de Desenvolvimento Científico e Tecnológico (CNPq) and National System for Research on Biodiversity (Sisbiota-Brazil, CNPq 563260/2010-6/FAPESP n° 2010/52322-3). M.L.T.M.P. and H. Cabral are Research Fellows of CNPq. We also thank Mariana Cereia for English review.

References

Aminzadeh, S., H. Naderi-Manesh, K. Khajeh and M. Naderi-Manesh. 2006. Purification, characterization, kinetic properties, and thermal behavior of extracellular polygalacturonase produced by filamentous fungus *Tetracoccosporium* sp. Appl. Biochem. Biotechnol. 135(3): 193–208.

Armand, S., M.J.M. Wagemaker, P. Sánchez-Torres, H.C.M. Kester, Y. van Santen, B.W. Dijkstra, J. Visser and J.A.E. Benen. 2000. The active site topology of *Aspergillus niger* endopolygalacturonase II as studied by site-directed mutagenesis. J. Biol. Chem. 275(1): 691–696.

Bailey, M.J. and E. Pessa. 1990. Strain and process for production of polygalacturonase. Enzyme Microbial. Technol. 12: 266–271.

Bajpai, P. 1999. Application of enzymes in the pulp and paper industry. Biotechnol. Prog. 15: 147–157.

Blanco, P., G. Thow, C.G. Simpson, T.G. Villa and B. Williamson. 2002. Mutagenesis of key amino acid alters activity of a *Saccharomyces cerevisiae* endo-polygalacturonase expressed in *Pichia pastoris*. FEMS Microbiol. Lett. 210: 187–191.

Carr, J.G. Tea, coffee and cocoa. pp. 133–154. *In*: B.J.B. Wood [ed.]. 1985. Microbiology of Fermented Foods, vol. II. Elsevier Applied Science, London.

CAZy-Carbohydrate-Active Enzymes. Available from: http://www.cazy.org/. Accessed 01/03/2012.

Chellegatti, M.A.S.C., M.J.V. Fonseca and S. Said. 2002. Purification and partial characterization of exopolygalacturonase I from *Penicillium frequentans*. Microbiol. Res. 157: 19–24.

Committee on Pectin Nomenclature of the Agricultural Food Division. 1927. Proc. Am. Chem. Soc. 49: 37.

Crotti, L.B., J.A. Jorge, H.F. Terenzi and M.L.T.M. Polizeli. Purification and characterisation of galactose-induced pectinases from the *exo-1* mutant strain of *Neurospora crassa*. pp. 787–792. *In*: J. Visser and A.G.P. Voragen [eds.]. 1966. Pectins and Pectinases. Elsevier Science BV, Netherlands.

Crotti, L.B., J.A. Jorge, H.F. Terenzi and M.L.T.M. Polizeli. 1998a. Characterization of galactose-induced extracellular and intracellular pectolytic activities from the *exo-1* mutant strain of *Neurospora crassa*. J. Ind. Microbiol. Biotechnol. 20: 238–243.

Crotti, L.B., J.A. Jorge, H.F. Terenzi and M.L.T.M. Polizeli. 1998b. Regulation of the pectic enzymes from the *exo-1* mutant strain of *Neurospora crassa*: effects of glucose, galactose and galacturonic acid. J. Basic. Microbiol. 38(3): 181–188.

Damásio, A.R.L., T.M. Silva, M. Maller, J.A. Jorge, H.F. Terenzi and M.L.T.M. Polizeli. 2010. Purification and partial characterization of an exo-polygalacturonase from *Paecylomyces variotii* liquid cultures. Appl. Biochem. Biotechnol. 160(5): 1496–1507.

Damásio, A.R.L., A. Maller, T.M. Silva, J.A. Jorge, H.F. Terenzi and M.L.T.M. Polizeli. 2011. Biotechnological potential of alternative carbon sources for production of pectinases by *Rhizopus microsporus* var. *rhizopodiformis*. Braz. Arch. Biol. Technol. 54 (1): 141–148.

Fahmy, A.S., F.M. El-beih, S.A. Mohamed, S.S. Abdel-Gany and E.A. Abd-Elbaky. 2008. Characterization of an exopolygalacturonase from *Aspergillus niger*. Appl. Biochem. Biotechnol. 149: 205–217.

Fernández-González, M., J.F. Úbeda, R.R. Cordero-Otero, V.T. Gururajan and A.I. Briones. 2005. Engineering of an oenological *Saccharomyces cerevisiae* strain with pectinolytic activity and its effect on wine. Int. J. Food Microbiol. 102(2): 173–183.

Fogarty, W.M. and O.P. Ward. 1974. Pectinases and pectic polysaccharides. Prog. Ind. Microbiol. 13: 59–119.

Fogarty, M.V. and C.T. Kelly. Pectic enzymes. 1983. *In*: Microbial Enzymes and Biotechnology, Elsevier, Applied Science Publishers Londres.

Gadre, R.V., G.V. Driessche, J.V. Beeumen and M.K. Bhat. 2003. Purification, characterisation and mode of action of an endo-polygalacturonase from the psychrophilic fungus *Mucor flavus*. Enzyme Microb. Technol. 32: 321–330.

Galante, Y.M. and A. Conti. Monteverdi R. Application of *Trichoderma* enzymes in food and feed industries. pp. 327–42. *In*: G.F. Harman and C.P. Kubicek [eds.]. 1998. *Trichoderma* & *Gliocladium*—Enzymes, Biological Control and Commercial Applications, vol 2 Taylor & Francis, London.

Gellissen, G. and C.P. Hollenberg. 1997. Application of yeasts in gene expression studies: a comparison of *Saccharomyces cerevisiae*, *Hansenula polymorpha* and *Kluyveromyces lactis*—a review. Gene. 190(1): 87–97.

Godfrey, A. Production of industrial enzymes and some applications in fermented foods. pp. 345–373. *In*: B.J.B.Woods [ed.]. 1985. Microbiology of Fermented Foods, vol 1 Elsevier, Applied Science London.

Gouka, R.J., P.J. Punt and C.A.M.J.J. van den Hondel. 1997. Efficient production of secreted proteins by *Aspergillus*: progress, limitations and prospects. Appl. Microbiol. Biotechnol. 47: 1–11.

Grassin, C. and P. Fauquembergue. Fruit juices. pp. 226–4. *In*: T. Godfrey and S. West [eds.]. 1996. Industrial Enzymology, 2nd edn. Macmillan UK.

Gratzner, H. and D.N. Sheehan. 1969. *Neurospora* mutant exhibiting hyperproduction of amylase and invertase.J. Bacteriol. 97(2): 544–9.

Gummadi, S.N. and Panda T. 2003. Purification and biochemical properties of microbial pectinases—a review. Process Biochem. 38: 987–996.

Gummadi, N.G. and D.S. Kumar. 2005. Microbial pectic transeliminases. Biotechnol. Lett. 27: 451– 458.

Hoondal, G.S., R.P. Tiwari, R. Tewari and N. Dahiya. 2002. Microbial alkaline pectinases and their industrial applications: a review. Appl. Microbiol. Biotechnol. 59: 409–418.

Hugouvieux-Cotte-Pattat, N., W. Nasser and J. Robert-Baudouy. 1994. Molecular characterization of the *Erwinia chrysanthemi kdgk* gene involved in pectin degradation. J. Bacteriol. 176(8): 2386–2392.

Jacob, N., K.N. Niladevi, G.C. Anisha and P. Prema. 2008. Hydrolysis of pectin: An enzymatic approach and its application in banana fiber processing. Microbiol. Res. 163: 538–544.

Jia, J. and A. Wheals. 2000. Endopolygalacturonase genes and enzymes from *Saccharomyces cerevisiae* and *Kluyveromyces marxianus*. Curr. Genet. 38: 264–270.

Kapoor, M., Q.K. Beg, B. Bhushan, K.S. Dadhich and G.S. Hoondal. 2000. Production and partial purification of a thermo-alkali stable polygalacturonase from *Bacillus* sp. MG-CP-2. Process. Biochem. 36: 467–473.

Kashyap, D.R., P.K. Vohra, S. Chopra and R. Tewari. 2001. Applications of pectinases in the commercial sector: a review. Bioresour. Technol. 77: 215–227.

Kauppinen, S., S. Christgau, L.V. Kofod, T. Halkier, K. Dorreich and H. Dalbøge. 1995. Molecular Cloning and Characterization of a Rhamnogalacturonan Acetylesterase from *Aspergillus aculeatus*. J. Biol. Chem. 270(45): 27172–27178.

Kaur, G., S. Kumar and T. Satyanarayana. 2004. Production, characterization and application of a thermostable polygalacturonase of a thermophilic mould *Sporotrichum thermophile* Apinis. Bioresour. Technol. 94: 239–243.

Kertesz, Z. 1930. A new method for enzymic clarification of unfermented apple juice. US patent # 1.932.833.

Klug-Santner, B.G., W. Schnitzhofer, M. Vrsanská, J. Weber, P.B. Agrawal, V.A. Nierstrasz and G.M. Guebitz. 2006. Purification and characterization of a new bioscouring pectate lyase from *Bacillus pumilus* BK2. J. Biotechnol. 121(3): 390–401.

Lang, C. and H. Dörnenburg. 2000. Perspectives in the biological function and the technological application of polygalacturonases. Appl. Microbiol. Biotechnol. 53(4): 366–75.

Maller, A., A.R.L. Damásio, T.M. Silva, J.A. Jorge, H.F. Terenzi and M.L.T.M. Polizeli. 2011. Biotechnological potential of agro-industrial wastes as a carbon source to thermostable polygalacturonaseproduction in *Aspergillus niveus*. Enz. Res. Article ID 289206, doi:10.4061/2011/289206.

Maller, A., T.M. Silva, A.R.L. Damásio, V.R.A. Reis, J.A. Jorge, H.F. Terenzi and M.L.T.M. Polizeli. 2012. Production of pectin lyase by *Aspergillus niveus* under submerged and solid state fermentations using agro-industrial residues as carbon sources. Int. Res. J. Microbiol. 3(1): 29–35.

Margesin, R., V. Fauster and P.A. Fonteyne. 2005. Characterization of cold-active pectate lyases from psychrophilic *Mrakia frigida*. Lett. Appl. Microbiol. 40(6): 453–459.

Ministério Brasileiro de Ciência e Tecnologia. Available from: http: //www.mct.gov.br/. Accessed 01/08/2011.

Mohamed, S.A., N.M. Farid, E.N. Hossiny and R.I. Bassuiny. 2006. Biochemical characterization of an extracellular polygalacturonase from *Trichoderma harzianum*. J. Biotechnol. 127(1): 54–64.

Moyo, S., B.A. Gashe, E.K. Collison and S. Mpuchane. 2003. Optimising growth conditions for the pectinolytic activity of *Kluyveromyces wickerhamii* by using response surface methodology. Int. J. Food Microbiol. 85: 87–100.

Mutter, M., C.M. Renard, G. Beldman, H.A. Schols and A.G. Voragen. 1998. Mode of action of RG-hydrolase and RG-lyase toward rhamnogalacturonan oligomers. Characterization of degradation products using RG-rhamnohydrolase and RG-galacturonohydrolase. Carbohydr. Res. 311(3): 155–64.

Nagai, M., A. Ozawa, T. Katsuragi, H. Kawasaki and T. Sakai. 2000. Cloning and Heterologous Expression of Gene Encoding a Polygalacturonase from *Aspergillus awamori*. Biosci. Biotechnol. Biochem. 64(8): 1580–1587.

Niture, S.K. and A. Pant. 2004. Purification and biochemical characterization of polygalacturonase II produced in semi-solid medium by a strain of *Fusarium moniliforme*. Microbiol. Res. 159(3): 305–3014.

Pagès, S., W.H.M. Heijne, H.C.M. Kester, J. Visser and J.A.E. Benen. 2000. Subsite Mapping of *Aspergillus niger* Endopolygalacturonase II by Site-directed Mutagenesis. J. Biol. Chem. 275(38): 29348–29353.

Patil, S.R. and A. Dayanand. 2006. Production of pectinase from deseeded sunflower head by *Aspergillus niger* in submerged and solid-state conditions. Bioresour. Technol. 97: 2054–2058.

Piccoli-Valle, R.H., F.M.L. Passos, F.J.V. Passos and D.O. Silva. 2001. Production of pectin lyase by *Penicillium griseoroseum* in bioreactors in the absence of inducer. Braz. J. Microbiol. 32: 135.

Pickersgill, R., D. Smith, K. Worboys and J. Jenkins. 1998. Crystal structure of polygalacturonase from *Erwinia carotovora* ssp. carotovora. J. Biol. Chem. 273(38): 24660–24664.

Polizeli, M.L.T.M., R.C.L.R. Pietro, J.A. Jorge and H.F. Terenzi. 1990. Effects of cell wall deficiency on the synthesis of polysaccharide-degrading exoenzymes, a study on mycelial and wall-less phenotypes of the *fz, sg, os-1* ("slime") triple mutant of *Neurosporacrassa*. J. Gen. Microbiol. 136: 1463–1468.

Polizeli, M.L.T.M., J.A. Jorge and H.F. Terenzi. 1991. Pectinase production by *Neurospora crassa*. Purification and biochemical characterization of polygalacturonase activity. J. Gen. Microbiol. 137: 1815–1823.

Rebeck, H. Processing of citrus juices. *In*: D. Hick [ed.]. 1990. Production and packaging of non-carbohydrate fruit juices and fruit beverages. Van Nosrand Reinhold, New York.
Ridley, B.L., M.A. O'neil and D. Mohnen. 2001. Pectins: structure, biosynthesis, and oligogalacturonide-related signaling. Phytochem. 57: 929–967.
Rombouts, F.M. and W. Pilnik. Pectic enzymes. pp. 227–272. *In*: A.H. Rose [ed.]. 1980. Microbial Enzymes and Bioconversions. Academic Press, London.
Sakai, T., T. Sakamoto, J. Hallaert and E.J. Vandamme. 1993. Pectin, pectinase and protopectinase: production, properties and applications. Adv. Appl. Microbiol. 39: 213–94.
Sathyanarayana, N. and G.T. Panda. 2003. Purification and biochemical properties of microbial pectinases—a review. Process. Biochem. 38: 987–996.
Sawada, K. and M. Ueda. 2001. Enzyme processing of textiles in reverse micellar solution. J. Biotechnol. 89: 263–269.
Schuster, E., N. Dunn-Colemam and J.C. Frisvad. 2002. On the safety of *Aspergillus niger*—a review. Appl. Microbiol. Biotechnol. 59: 425–435.
Searle-Van Leeuwen, M.J.F., L.A.M. Van Der Broek, H.A. Schols, G. Beldman and A.G.J. Voragen. 1992. Rhamnogalacturonan acetylesterase: a novel enzyme from *Aspergillus aculeatus*, specific for the deacetylation of hairy (ramified) regions of pectins. Appl. Microbiol. Biotechnol. 38: 347–349.
Sharma, D.C. and T. Sathyanarayama. 2006. A marked enhancement in the production of a highly alkaline and thermostable pectinase by *Bacillus pumilus* dcsr1 in submerged fermentation by using statistical methods. Bioresour. Technol. 97(5): 727–733.
Silva, D., K. Tokuioshi, E.S. Martins, R. Da Silva and E. Gomes. 2005. Production of pectinase by solid-state fermentation with *Penicillium viridicatum* RFC3. Process. Biochem. 40: 2885–2889.
Sony, S.M.M. and M.N. Ponnuswamy. 2007. Molecular basis behind the substrate specificity of poligalacturonase through computational study. Polymer. 48: 910–916.
Takao, M., T. Nakaniwa, K. Yoshikawa, T. Terashita and T. Sakai. 2000. Purification and characterization of thermostable pectate lyase with protopectinase activity from thermophilic *Bacillus* sp. TS47. Biosci. Biotechnol. Biochem. 64: 2360–2367.
Tanabe, H., K. Yoshihara, K. Tamura, Y. Kobayashi, I. Akamatsu, N. Niyomwan and P. Footrakul. 1987. Pretreatment of pectic wastewater from orange canning process by an alkalophilic *Bacillus* sp. J. Fermentation Technol. 65(2): 243–246.
Uenojo, M. and G.M. Pastore. 2007. Pectinolytic enzymes: Industrial applications and future perspectives. Quím. Nova. 30(2): 388–394.
van den Homberg, J.P.T.W., P.J.I. van de Vondervoort, L. Fraissinet-Tachet and J. Visser. 1997. *Aspergillus* as a host for heterologous protein production: the problem of proteases. Tibtech. 15: 256–263.
van der Vlugt-Bergmans, C.J., P.J. Meeuwsen, A.G. Voragen and A.J. Van Ooyen. 2000. Endo-xylogalacturonan hydrolase, a novel pectinolytic enzyme. Appl. Environ. Microbiol. 66(1): 36–41.
van Santen, Y., J.A.E. Benen, K.H. Schroöter, K.H. Kalk, S. Armand, J. Visser and B.W. Dijkstra. 1.68-Å Crystal Structure of Endopolygalacturonase II from *Aspergillus niger* and Identification of Active Site Residues by Site-directed Mutagenesis. J. Biol. Chem. 274(43): 30474–30480.
Villa, B.S. 1999. Production of pectic enzymes in yeasts. FEMS Microbiol. Lett. 175: 1–9.
Vincken, J.P., H.A. Schols, R.J. Oomen, M.C. Mccann, P. Ulvskov and A.G. Voragen. 2003. If homogalacturonan were side chain of rhamnogalacturonan I. Implications for cell wall architecture. Plant Physiol. 132(4): 1781–1789.
Whitaker, J.R. Microbial pectolytic enzymes. pp. 133–175. *In*: W.M. Fogarty and C.T. Kelly [eds.]. 1991. Microbial Enzymes and Biotechnology. Elsevier, Applied Science London and New York.
Willats, W.G.T., L. Mccartney, W. Mackie and J.P. Knox. 2001. Pectin: cell biology and prospects for functional analysis. Plant. Mol. Biol. 47: 9–27.

Willats, W.G.T., J.P. Knox and J.D. Mikkelsen. 2006. Pectin: new insights into an old polymer are starting to gel. Trends Food Sci. Technol. 17: 97–104.

Withers, S.G. Probing of glycosidase active sites through labeling, mutagenesis and kinetic studies. *In:* Carbohydrate Bioengineering. Proceeding of an International conference Elsinore, Dinmark. S.B. Petersen, B. Svensson and S. Pedersen [eds.]. 1995, chapter 09 Elsevier, Amsterdam, The Netherlands. pp. 97–111.

Wubben, J.P., W. Mulder, A.T. Have, J.A.L.V. Kan and J. Visser. 1999. Cloning and partial characterization of endopolygalacturonase genes from *Botrytis cinerea*. Appl. Environ. Microbiol. 65(4): 1596–1602.

Yadav, S. and N.V. Shastri. 2007. Purification and properties of an extracellular pectin lyase produced by the strain of *Penicillium oxalicum* in solid-state fermentation. Indian J. Biochem. Biophys. 44(4): 247–51.

Yamada, H. 1995. Structure and pharmacological activity of pectic polysaccharides from the roots of *Bupleurum falcatum* L. Nippon Yakurigaku Zasshi 106: 229–237.

Zerbs, S., A.M. Frank and F.R. Collart. 2009. Bacterial systems for production of heterologous proteins. Methods Enzymol. 463: 149–168.

Zhao, Q., S. Yuan, Y. Zhang, H. Zhu, C. Dai, F. Yang and F. Han. 2007a. Expression, purification and characterization of pectate lyase A from *Aspergillus nidulans* in *Escherichia coli*. World J. Microbiol. Biotechnol. 23: 1057–1064.

Zhao, Q.X., S. Yuan and Y.L. Zhang. 2007b. Expression of pectin lyase 1 from *Aspergillus oryzae* in *Escherichia coli*. Chinese J. Biotechnol. 23(5): 873–877.

14

Improving Fungal Enzyme Properties through Protein Engineering

*Lucas Ferreira Ribeiro** and *Liliane Fraga Costa Ribeiro*

Introduction

Protein engineering developed in the 80's with the construction of proteins modified by site directed mutagenesis, studying catalytic mechanisms and the relationship between structure and function (Brannigan and Wilkinson 2002). With the advances in molecular genetics and, in particular, the techniques of gene manipulation, there are now conditions to change the macromolecular structures allowing the manipulation of target functions (Fan et al. 2009). Thus, great interest has been aroused in bioengineers to fit proteins to various medical and industrial applications (Arnold 1996). These studies are meant to understand and subsequently direct the improvement process of the biological activities.

There are a large number of successful works, which include the techniques of protein engineering to modify properties such as: thermostability (Hokanson et al. 2011), catalytic properties (Spadiut et al.

Biochemistry and Immunology Department, Medical School, University of São Paulo. Bandeirantes avenue s/n. 3900 Ribeirão Preto, São Paulo, Brazil.
*Corresponding author: lucasbqi@gmail.com

2009), redox potential (Mate et al. 2010), dependence of pH (Heckmann-Pohl et al. 2006), functional expression of an enzyme (Camarero et al. 2012) or its glycosylation (Taylor et al. 2012).

Protein engineering is a multidisciplinary field, in which different techniques and knowledge are applied. First, it is necessary to know the characteristics and functional properties of the protein of interest, and then modify it in order to improve its performance. Although the approaches contained in these protein engineering studies were made from different perspectives and with different end goals, biochemists and bioengineers have had common issues to mimic the evolution made by nature (Bloom and Arnold 2009). A multidisciplinary approach is used, which is generally represented by the cycle of protein engineering (Fig. 14.1).

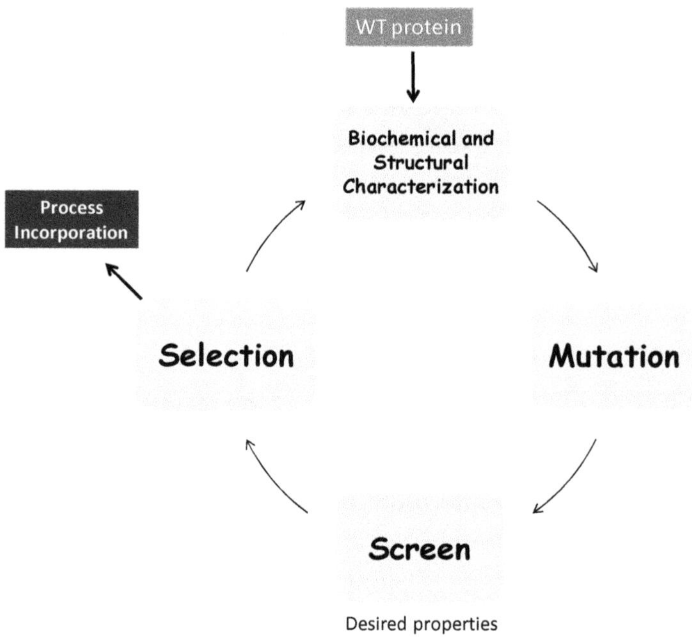

Figure 14.1 Cycle of protein engineering.

How to Engineer New Properties and Functions in Proteins?

The aim of this chapter is to give the reader an overview of some of the most recent techniques of protein engineering applied to fungal enzymes emphasizing the important relationship between basic science and its application. We will divide the engineering strategies in three main points: rational, not rational and semi-rational design.

Rational Enzyme Design

It is necessary to know structural details combined with biochemical information of the enzyme to use the rational design approach. It is therefore a way of testing the limits of the range of experimental knowledge.

One of the most traditional approaches of rational design is Site Directed Mutagenesis, which is a very classical tool used to evaluate the effect of one or more amino acids, predefined on the properties of the protein under study. It is important in the study of the relationship between a protein structure and its function. Site directed mutagenesis, also called rational mutagenesis, implies, by definition, at some prior knowledge of the contribution played by residues or regions on enzyme properties. Sometimes, besides the three-dimensional structure of the protein, it is necessary to know the structure of the complex with its substrate, inhibitors or modulators. When a protein belongs to a large and well characterized family, the alignment of sequences may also provide support for the selection of point mutations. Computational analysis may also help to predict the energetic importance of residues in certain regions and in the definition of mutations. In a structural study of a thermostable endoglucanase from *Humicola grisea* Cel12A, three unusual free cysteines were observed: Cys175, Cys206 and Cys216. Later, the following mutants Cel12A: Cys175Gly, Cys206Pro and Cys216Val were constructed by site directed mutagenesis. Of these, two showed a reduction in Tm (Cys206Pro and Cys216Val) and one (Cys175Gly) showed an increase of 1.3°C in Tm (Sandgren et al. 2005), thus demonstrating the contribution of these cysteines for enzyme stability.

Another rational approach is the creation of multifunctional or chimerical enzymes. This is a potential strategy to reduce the costs of the enzyme in industrial processes. Engineering enzymes to have multi-domains in a single polypeptide chain simplify the operations of production and its purification. In addition, when enzymes are fused to form chimeras, the physical proximity of the catalytic domains can create interfaces between domains and improve the performance, when compared with the activities of separated parental enzymes (Ribeiro et al. 2011). There are many examples of enzymes that presented new properties when fused to another domain, some of them non catalytic, like CBMs (carbohydrate-binding modules) that consist of relatively small domains, generally linked to the N- or C-terminal catalytic domain by a linker, often glycosylated, necessary for the independent folding and flexibility of both domains (Boraston et al. 2004). An example is the fusion of a domain with laccase activity from *Pycnoporus cinnabarinus* with a CBM1 domain of CbhB from *Aspergillus niger* with a natural linker (Ravalason et al. 2009). CBM1 domain allows the connection to cellulose, while laccase domain degrades lignin around the cellulose, resulting in an enzyme with good application for the pulp and paper industry (Ibarra et al. 2006). Another example of fusion to CBMs is a mannanase-CBM that was designed for the biofuels sector. Mannanases can act synergistically with cellulases to improve plant biomass degradation. A chimeric mannanase-CBM1 was produced in *A. niger* with high yield (130 mg/L) by a combination of *Aspergillus aculeatus* mannanase and a CBM from *A. niger* CbhB (Pham et al. 2010).

The construction of a chimeric enzyme containing feruloyl esterase and xylanase activities also showed an overproduction in *Aspergillus niger*, besides a synergistic activity between the feruloyl esterase and xylanase domains on complex substrates degradation (Levasseur et al. 2005).

Directed Evolution

Directed evolution is a powerful approach in protein engineering, which aims to employ the basis of natural selection to create new properties in proteins and RNAs, developing features that nature did not bother to make (by the absence of natural pressure), but that may be important for certain processes (e.g., pH and temperature extremes) (Otten and Quax 2005). The tools employed are generally in character not rational or semi-rational, so it is not necessary to have the knowledge of the structure or the evolution mechanism. Molecular evolution is made by analogy with classical evolution, in which there are three key points: generation of genetic variability, selective pressure and amplification. From these pillars there is a simple algorithm that generates the complexity of life (Fig. 14.2). In comparison with Fig. 14.1, the generation of molecular diversity is accomplished by random mutations and recombinations and the selected

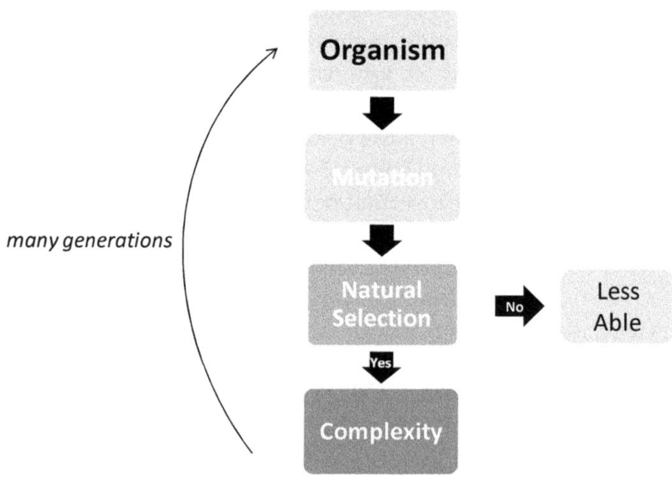

Figure 14.2 Key points for classical evolution.

molecules in the screen are the "survivors" to selective pressure. The cycle can proceed generating an accumulation of mutations (amplification) and greater complexity. Directed evolution is a method with the stochastic nature and the accumulation of beneficial mutations has a probabilistic character. Thus there is the need to generate a high number of mutant strains (library) and to have a feasible experimentally system that is sensitive and fast.

There are four basic requirements for the success of directed evolution:

1. The property to be improved must be physically possible.
2. Creation of genetic diversity.
3. Choice of a suitable host organism for a functional expression.
4. Fast and highly selective screening that reflects the desired property.

Generation of Molecular Diversity

An unimagined diversity can be hidden in a simple code and natural protein sequence. Since an enzyme consist of N residues, and if there are 20 possible amino acids at each position in the chain, then the possible protein sequence space is 20^N (Hughes et al. 2003). The entire imaginary sequence space of an enzyme could be packed with properties so far not detected. This astronomically large number of polypeptides is a potential obstacle to identify and isolate interesting variations. Therefore, a directed evolution experiment should be done carefully providing intelligently designed libraries and improved screening techniques. A large number of different techniques to generate genetic diversity and to perform evolution experiments either within the cells (*in vivo*) or in the laboratory test tube (*in vitro*) has been developed. The most used ways to generate molecular diversity are random point mutagenesis and *in vitro* recombination.

Random mutagenesis methods

An error-prone of the polymerase chain reaction (PCR) is the most used method for random mutagenesis (Cadwell and Joyce 1994). Under certain conditions (e.g., pH and divalent cations), polymerase will mis-incorporate bases at a controllable rate during gene amplification. The most important parameter in the experiments of mutagenesis is random distribution mutations over the entire gene. This can be estimated based on the binomial distribution. For a sequence of length N which is mutated with an error rate ε, the probability of the introduction of k mutations is given by:

$$P\,(k,n,\varepsilon) = (n!/[(n-k)!k!])\ \varepsilon^{k}(1-e)^{n-k} \text{ (Joyce 2004)}$$

However, it is preferable to work with lower error rates, e.g., 2–3 base substitutions or ~1 amino acid substitution per sequence per generation, which causes an accumulation mainly of adaptive mutations, whereas the higher error rates can confound the analysis by generating neutral and deleterious mutations (Arnold et al. 2001). Different fungal enzymes have been improved by error-prone PCR (Xiao et al. 2002, Wang et al. 2005, Lee et al. 2007, Voutilainen et al. 2007, Festa et al. 2008, Mchunu et al. 2009, Nakazawa et al. 2009, Stephens et al. 2009, Miele et al. 2010, Liu et al. 2011, Mate et al. 2011).

Another method equally important and considered the most simple of random mutagenesis is the EP-RCA, which uses Rolling Circle Amplification (RCA) described in 2004 by Fujii et al. (Fujii et al. 2004). It consists in a single RCA step that uses universal random hexamer primers (NNNNNN) and a Phi-29 DNA polymerase using $MnCl_2$ and a fixed incubation temperature, followed by transformation of the host strain (Fujii et al. 2006). EP-RCA can produce a mutation frequency reasonable for directed evolution experiments, not requiring the use of restriction enzymes or ligase. This technique is most often used in cloned genes in a circular array. However an adjustment was made for small DNA (coding peptide), wherein the DNA encoding the signal peptide for glucoamylase in a recombinant *Saccharomyces cerevisiae* was circularized and served as a template for the EP-RCA (Luhe et al. 2010). Using multiple random hexamer primers it created tandem repetitions of the small DNA and simultaneously introduced mutations in the sequence. Beneficial mutations were introduced efficiently with 3.7 mutations per kb, generating a signal peptide capable increasing glucoamilase secretion by 30 percent more than the native peptide.

Several methods based on mutagenesis have emerged in recent years, such as: Random Insertion/Deletion (RID) Mutagenesis (Murakami et al. 2002), Segmental Mutagenesis (Pikkemaat and Janssen 2002), Pentapeptide Scanning Mutagenesis (PSM) (Hayes and Hallet 2000), Triplet Nucleotide Removal (Jones 2005). These are briefly discussed in Table 14.1.

Recombination-based methods

The emergence of punctual mutations is a fine and effective adjustment mechanism, but nature also uses other means to create new molecular diversity that evolution can act, one of them is recombination. Its application in directed evolution is an advance that allows the rearrangement of multiple genes to generate improved enzymes. DNA shuffling method developed by Stemmer (Stemmer 1994a, 1994b) for *in vitro* random recombination presented an initial point on the scientific capacity to mimic the natural evolution mechanism. Recombination allows the rapid accumulation of beneficial mutations in genes of a number of parents and even from different species. In addition, deleterious mutations are removed. It is also a promising approach to combine two or more improved properties that

Table 14.1 Random mutagenesis methods.

Random Mutagenesis			
Method	**Advantages**	**Disadvantages**	**Ref.**
Error-Prone PCR	✓ Well-proven methodology ✓ Easy to perform	✓ Biased library caused by taq polymerase activity ✓ Negative mutations further amplified	(Cadwell and Joyce 1994)
Error-Prone RCA	✓ No specific primers ✓ Isothermal procedure ✓ No restriction enzymes or ligases	✓ Mutations in some regions of plasmid DNA may be lethal ✓ Expensive commercial kit	(Fujii et al. 2006)
RID Mutagenesis	✓ Insertions and deletions are performed concurrently ✓ No codon bias	✓ Complex procedure	(Murakami et al. 2002)
Segmental Mutagenesis	✓ Relatively easy to perform	✓ Majority of fragments are ligated out of frame ✓ It requires a large amount of initial DNA	(Pikkemaat and Janssen 2002)
PSM	✓ KpnI site in fingerprints allows insertion of additional sequences (e.g., epitopes, protein tags)	✓ Needs appropriate purified transposase ✓ Transposon insertions occur along whole vector DNA	(Hayes and Hallet 2000)
Triplet nucleotide removal	✓ Creation of libraries of protein variants with single amino acids removed at random positions. ✓ It allows replacing the previous codon (e.g., as in alanine scanning mutagenesis or for the introduction of unnatural amino acids)	✓ Needs appropriate purified transposase ✓ Transposon insertions occur along whole vector DNA	(Jones 2005)

have evolved simultaneously (Xia and Wang 2009) or separately (Patten et al. 1997). The process of recombination depends especially on the identity level of the parents DNA. The inherent limitation in the techniques of DNA recombination derived from DNA shuffling depends on high identity between the sequences. In general, when the DNA sequences having less than 70 percent of identity are mixed, the parents reconstruction is often observed as the crossover tend to aggregate in regions of high sequence identity due to the annealing-based reassembly (Moore et al. 2001). Recombination techniques that work with parental genes with identity greater than 70 percent are classified as homologous methods. Recent techniques intend to recombine genes less related and with less identity.

Such techniques, without any dependence on sequence or those that recombine sequences with identity less than 70 percent can be classified as non-homologous methods (Zhang et al. 2010).

Homologous recombination. Beneficial mutations from different variants may be grouped together by recombination. One of the most popular methods for recombination is DNA shuffling developed in 1994 by W. Stemmer (Stemmer 1994a, 1994b). In this method DNA from different variants with beneficial mutations are cleaved by the action of DNase I, producing random fragments of 10–50 bp. These fragments are mixed and denatured in a PCR without primers, and since the fragments are complementary the annealing between them will occur. Some of these annealings result in heteroduplexes fragments that come from two copies of the parental gene. During the annealing and extension cycles, mutations are recombined. After 20–30 cycles of assembly, an amplification with external primers is done to amplify the full-length sequence. DNA shuffling is often preceded by error-prone PCR or other random mutagenesis techniques. Some fungal enzymes were also evolved using DNA shuffling method developed by Stemmer. A lignin peroxidase (LiP) is an example that had located combined mutations mainly on the surface, on heme site, and Ca^{2+}-binding site of the enzyme, which resulted in an increased enzyme functionality and stability in the presence of H_2O_2 (Ryu et al. 2008). Another example is a chitinase which presented mutations outside of the catalytic pocket or at the binding site generated by DNA shuffling that resulted in increased catalytic activities (Fan et al. 2007).

A variation of this method is called DNA shuffling family, which allows the creation of chimeric proteins libraries of a homologous genes family (up to 14 parents) with homology > 60% (Crameri et al. 1998, Joern et al. 2002). The low diversity found in shuffling experiments is often attributed to the formation of parental homoduplex coming from the fragments annealing. The number of crossover events in the chimeras is very low (usually 1–4 per gene). ssDNA strategy may increase heteroduplex formation, like the target gene cloning in a phagemid vector and later digestion with DNaseI (Kikuchi et al. 2000); also homologous genes hybridization and subsequent treatment with S1 nuclease. In this case the non-homologous regions of the heteroduplex will form single strands that serve as a site for the S1 nuclease. Undigested fragments produced (double-stranded regions) will serve as internal primers for a reassembly of the gene through PCR increasing the diversity (Hong 2006). Other adaptations as the fragmentation of the parental genes using restriction enzymes instead of DNAase I (Kikuchi et al. 1999) or exploration of λ exonuclease for ssDNA templates generation (Zha et al. 2003) may greatly increase the rate of chimeric genes in the library.

Another method was developed to increase the frequency of recombination of the family shuffling, to include synthetic oligonucleotides that combine sequence elements from different templates. On Degenerate Oligonucleotide Gene Shuffling (DOGS) (Gibbs et al. 2001), genes encoding the desired protein are aligned and divided into segments of conserved regions. DNA segments are amplified by degenerated primers. Each primer has a non-degenerated core flanked by degenerated ends. An extension by overlap of the segments followed by a PCR with nested specific primers remakes the full-length, recombining all segments. This technique was applied to a recombinant barley α-amylase 1 (rAMY1) and 2 (rAMY2) (Fukuda et al. 2005). Even with a sequence identity of 80 percent, native enzymes were produced in very different quantities of 1.1 to <0.05 mg/L respectively, by strain S150-2B of *Saccharomyces cerevisiae*. Low AMY2 production made the mutational analysis of structure-function very difficult. Therefore, different AMY1/AMY2 chimeras were constructed using Degenerate Oligonucleotide Gene Shuffling (DOGS) along with *in vivo* homologous recombination in *S. cerevisiae*. Clones with different levels of secretion were produced in this yeast, contributing to the understanding of the secretion clones levels. It enabled the creation of a AMY2 variant with a higher level of secretion, which was also confirmed in *Pichia pastoris* that secreted 60 mg/L of the mutant in comparison to 3 mg/L of wild-type (Fukuda et al. 2005).

StEP (Staggered Extension Process) is another way of making the recombination, which is based on cross-hybridization of the genes extension in the reaction catalyzed by the polymerase (Fig. 14.3) (Zhao et al. 1998). Short times in sub-optimal extension conditions can generate small fragments that may annular randomly in other variants of the gene. During the short cycles a full-length chimeric gene is produced. A galactose oxidase (GOase) from *Fusarium* was evolved using StEP and expression in functional form in *Escherichia coli*. The evolved enzymes retained the activity and substrate specificity of the native fungal oxidase, but were more thermostable, and presented an increased expression level (Sun et al. 2001). It was also made an improvement of activity of a laccase from *Myceliophthora thermophila* (MtL) and an increase in its expression in *Saccharomyces cerevisiae* via StEP (Bulter et al. 2003).

Homologous recombination can also be performed *in vivo* via *Saccharomyces cerevisiae* or *Hansenula polymorpha* transformation (Adamczak and Krishna 2004). The recombination machinery is highly efficient and the easy handling of these organisms, especially *S. cerevisiae*, have been frequently used to construct libraries containing 10^4–10^{10} variants (Cherry et al. 1999, Abécassis et al. 2000, Swers et al. 2004, Pirakitikulr et al. 2008, Benatuil et al. 2010).

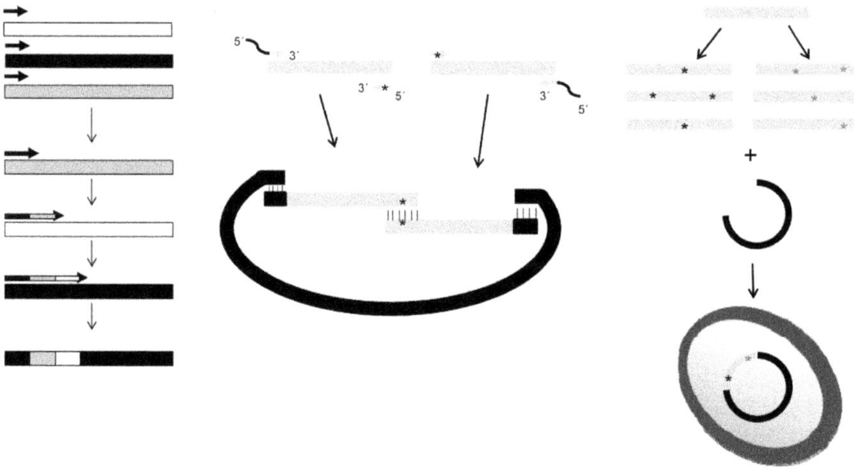

Figure 14.3 Homology-dependent methods for recombining DNA sequences.

When *S. cerevisiae* is co-transformated with linear DNA fragments that have overlap regions, the fragments are subjected to homologous recombination, which restores the circular structure of plasmid (Fig. 14.3) (Orrweaver and Szostak 1985, Mezard and Nicolas 1994). Plasmid repair is an extremely efficient process, and can replace the binding insert/vector step. This repair can be obtained by co-transformation of the opened plasmid with the mutant gene, containing homologous sequences in both ends as short as 30 bp, resulting in more reliable transformation frequency and shortening the protocol for the library expression (Bulter and Alcalde 2003). *In vivo* recombination offers some unique advantages: efficient cloning of the hybrid sequences resulting in an expression vector of yeast and constitution of a second round of DNA shuffling without the use of PCR-based techniques. Episomal vectors are available, thus facilitating the recovery of interesting mutants without integration into the genomic DNA as usually happens in yeasts, such as *Pichia pastoris* (Madzak et al. 2004). Cloning performed in yeast helps to avoid common problems that occur when fungi proteins are expressed in *E. coli*. *In vivo* Assembly of Mutant libraries (IVAM) is another approach that have been successfully used to engineer fungal laccases for a better tolerance in organic co-solvent (Fig. 14.3) (Zumarraga et al. 2008). IVAM allowed it to explore the sequence space of laccase through the *in vivo* DNA shuffling of several different mutant libraries in a single cycle of evolution.

It is also possible to combine the recombinant DNA strategies *in vitro* and *in vivo* to evolve enzymes at the laboratory. An example is the Combinatorial Libraries Enhanced by Recombination in Yeast (CLERY) (Valérie Abécassis et al. 2000). In this method the eukaryotic parental genes are first shuffled by DNA shuffling or family shuffling procedure (based on a method by Stemmer) and then inserted into the yeast expression vector. The last step is a second round of *in vivo* DNA shuffling involving direct cloning, expression, and selection of functional enzymes in yeast. In a similar approach StEP mutagenic was combined with the *in vivo* DNA shuffling to evolve ligninolytic peroxidases (Garcia-Ruiz et al. 2010).

A computational algorithm called SCHEMA was also developed to estimate the disturbances caused in the three dimensional structure of a protein by recombination (Saab-Rincon et al. 2009). SCHEMA uses a protein structure data to define the limits of contiguous blocks of amino acids, which minimize the average number of contacts that are broken between the amino acids side chains in the library when the blocks are exchanged between the different parents. Three cellulases (CBH II) from fungi (*Humicola insolens*: parental 1, *Hypocrea jecorina*: parental 2, and *Chaetomium thermophilum*: parental 3) were used to produce a collection of thermostable chimera. Proteins were divided in eight blocks, and 23 chimeras of 6,561 possible sequences (3^8) were secreted in a heterologous way in *Saccharomyces cerevisiae*. Approximately 22 percent of these chimeras presented half-lives of thermal inactivation greater than the parental, suggesting that chimeras collection contains 100s of cellulases highly thermostable (Heinzelman et al. 2009).

Other important methods based on homologous recombination, such as: Random-Priming *in vitro* Recombination (RPR) (Shao et al. 1998); Random Chimeragenesis on Transient Templates (RACHITT) (Coco et al. 2001) and Multiplex-PCR-based Recombination (MUPREC) (Eggert et al. 2005) are briefly discussed in Table 14.2.

Non-homologous recombination. Recombination processes seen until now require a high sequence identity between the parental genes (greater than 60 percent), because they are related with the annealing and extension of DNA strands. However, in many situations it would be interesting to observe a gene shuffling with low (less than 60 percent) or no sequence identity. Some enzymes with non-homologous sequences can have high structural homology, which can make shuffling between them desirable. Also, the intron-exon organization of the eukaryotes genome allowing the natural recombination of non-homologous genes is an attraction for bioengineers. The "exon shuffling" reflects a natural evolution mechanism, with the exons exchange between genes not related, which results in proteins from different families but that share conserved functional domains (Kolkman and Stemmer 2001). However, homologous recombination methods are inefficient for such purposes, which led to the development of a series of techniques that do not require a template annealing/extension. These methods are not appropriate for the punctual mutations recombination,

Table 14.2 Homologous recombination methods.

Homologous Recombination			
Method	**Advantages**	**Disadvantages**	**Ref.**
DNA Shuffling	✓ Robust ✓ Possible to remove non-adaptive mutations by backcrossing process	✓ High background of unshuffled parental clones ✓ Lower recombination frequency between regions of homology shorter than 15 bp	(Stemmer 1994a, Crameri et al. 1998)
DOGS	✓ Easy implementation via PCR ✓ Adjustable recombination frequency ✓ Reduced self-hybridization of parental genes ✓ Low homology required	✓ Greater number of primers to be prepared ✓ Relatively low crossover rate ✓ Frameshifts may occur	(Gibbs et al. 2001)
StEP	✓ No DNase I digestion (eliminated sequence bias) ✓ Simplicity	✓ It needs fine control of PCR ✓ It needs high homology (over 80%)	(Zhao et al. 1998)
In vivo recombination in yeast	✓ Efficient cloning without the insert/vector ligation step ✓ DNA shuffling without the use of PCR-based techniques ✓ Cloning/expression of proteins of fungi directly in eukaryotic cells	✓ Low recombination frequency between homology regions shorter than 30 bp ✓ Skill to work with eukaryotic expression system ✓ Specialized vectors	(Zumarraga et al. 2008)
CLERY	✓ It combines *in vitro* and *in vivo* DNA shuffling ✓ It avoids biases caused by ligation and propagation in *E. coli*	✓ Proportion of functional clones relatively low ✓ Multiple steps necessary	(Abécassis et al. 2000)
RPR	✓ Independent of the DNA template length ✓ It allows the use of single-stranded DNA or RNA templates ✓ It requires less parental DNA than DNA shuffling	✓ Mispriming is not random, transitions occur more often than transversions	(Shao et al. 1998)

Table 14.2 contd....

Table 14.2 contd.

Homologous Recombination			
Method	**Advantages**	**Disadvantages**	**Ref.**
RACHITT	✓ High recombination frequency (on average 14/gene) ✓ Single hybridization step ✓ It recombines genes of low homology sequence	✓ Quality ssDNA template is crucial and might be diffcult to synthesize ✓ DNase I digestion ✓ Many steps: ssDNA template generation, gap-filling, flaps digestion, etc.	(Coco et al. 2001)
MUPREC	✓ No DNase I digestion ✓ Small size of the library ✓ No background of parental among active clones	✓ It needs fine control of PCR conditions	(Eggert et al. 2005)
SCHEMA	✓ Low homology required ✓ Minimize structure disruption ✓ High functional proteins percentage	✓ Software dependent ✓ 3-D structures of desired proteins and a reliable parental sequence alignment are required	(Saab-Rincon et al. 2009)

the most common use of DNA shuffling, but are useful when it is desired to generate hybrid genes that share little DNA sequence similarity.

In contrast to DNA shuffling and related methods, Incremental Truncation for the Creation of HYbrid enzymes (ITCHY) developed by Ostermeier (Ostermeier et al. 1999) was the first technique designed to generate crossover between non-homologous sequences. In the enhanced procotol (Ostermeier and Lutz 2003), two genes or gene fragments are cloned in tandem using a single plasmid to create an ITCHY library. Plasmid is then digested with restriction enzyme (RE1) in a single point between the two genes. The DNA is subjected to exonuclease III (Exo III), in a truncation time-dependent. Gene A is digested at its 5′ end and gene B at its 3′ end. Exo III digestion rate is controlled by adding NaCl so the reaction rate is limited to 10 bp/min. Blunt ends are prepared by treatment with Klenow fragment and DNA polymerase. The connection between equally truncated fragments in the plasmid is also cut up by digestion with the restriction enzyme RE2. A and B differently truncated genes can be connected and create a chimeric gene with a single crossover site. In this case, crossover is not a result of homologous recombination process, but a fusion point. Another modification of the initial ITCHY version, known as THIO-ITCHY (Fig. 14.4), was designed to avoid the difficult time sampling

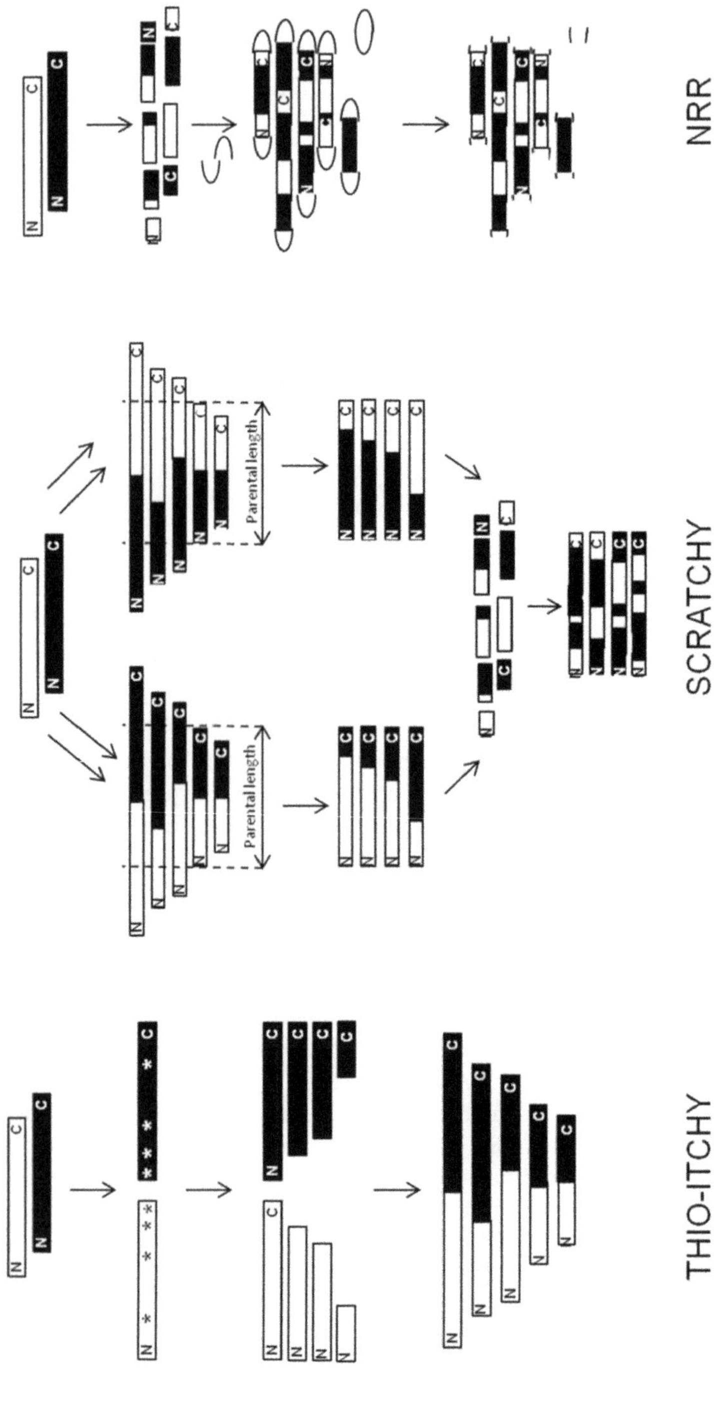

Figure 14.4 Homology-independent methods for recombining DNA sequences.

that was necessary for time dependent truncation (Lutz et al. 2001a). In THIO-ITCHY, the entire plasmid with A and B cloned in tandem genes is amplified by PCR using dNTPs and a small amount of an analogous to α-phosphothioate dNTP (αS-dNTPs) (Fig. 14.4). The digestion with Exo III is stopped by randomly incorporated αS-dNTPs in sequence. This protocol provides incorporation of random mutations using epPCR conditions during the amplification step.

THIO-ITCHY method was used to create a library with different size loops in a lipase B variant from *Candida antarctica* (CALB, CP283, built by circular permutation) that presented a surface loop (Qian et al. 2009). The recovered DNA was linked into pPIC9 and transferred into *Pichia pastoris* for its expression. Library members were identified by the formation of halo in a medium containing tributyrin. Depending on the extent of loop truncation, a change was observed in the enzyme quaternary structure, shifting from a strictly monomeric form to a dimer with a domain-swapped N-terminal segment.

The same authors also proposed a methodology called SCRATCHY which is a combination of the incremental truncation for the creation of hybrid enzymes (ITCHY) with DNA shuffling (Fig. 14.4) (Lutz et al. 2001b). In this first step a library of hybrid enzymes (ITCHY) was created. The linearized hybrid gene is selected by size (parental-size selection) and subcloned into a vector, which provides resistance to ampicillin if the reading frame is correct. Plasmid DNA from these colonies is extracted and can be used for the DNA shuffling step.

Another homology independent method is the Non-homologous Random Recombination (NRR), in which one or more genes are digested with DNase I in a controlled manner by modulating the digestion time. Thus, it is possible to control the sizes of fragments generated. These are treated with T4 DNA polymerase to generate blunt ends. Products from the previous step are purified in gel to separate those which have the desired size range. T4 DNA ligase is added to these products in the presence of poly(ethylene glycol), because this condition strongly favors intermolecular bonds (Bittker et al. 2002). The fragments are linked by blunt ends to each other in a random and non-homologous way. DNA hairpins are added to control the recombinant average size, blocking the connection between the hybrid fragments. Hairpins also create sequences defined in the ends of hybrid products of the library, which may be used to design primers (Fig. 14.4). Recombinants are then amplified, digested with restriction enzymes, cloned into an expression vector and transformed in the model organism (Bittker et al. 2002, Bittker et al. 2004, Doyon and Liu 2007).

NRR application in proteins evolution presents some challenges (Bittker et al. 2004). A single hairpin used at the ends of the sequences does not efficiently clone into expression vectors because both ends are identical.

This problem can be solved using two different hairpins, each of them with different non-palindromic restriction site. This generates a mixture of products, but 50 percent of them have two different hairpins at the ends. Another problem is the presence of internal stop codons that truncate recombined protein products (Bittker et al. 2004). This can be improved by using a vector that has a gene for an antibiotic resistance, which is fused to the C terminus of the hybrid gene, and if the reading frame is correct the resistance gene will be expressed as a bi-functional protein. It is important to observe the ratio between the number of fragments and hairpins added. According to the obtained value the product will present different sizes. With the 2:1 ratio (x hairpin fragment), theoretically, four recombinant products with the same size as the parental genes would be obtained containing an average of three crossovers.

This technique was well applied to identify new upstream activating sequences (UASs) at the promoters KAR2, HAC1 SIL1, using the bases –1 to –1000, relative to the transcription start (Doyon and Liu 2007), with the unfolded protein response (UPR) in yeast. It was then possible to identify known UASs that are upregulated during UPR, as well as new UASs involved that generate transcriptional response in the presence of a UPR inductor. According to the authors this method has many advantages for the discovery of UAS, such as the speed with which it is done, the sensitivity and the lack of false positive and negative clones generated.

There are some other new recombinant techniques for recombinantion of sequence without homology, such as: Sequence Homology-Indepenedent Protein Recombination (SHIPREC) (Sieber et al. 2001) and Exon Shuffling (Kolkman and Stemmer 2001), that are briefly described in Table 14.3.

Random circular permutation

Circular permutation represents a specific approach that does not involve random mutagenesis or recombination. Conceptually it can be defined as the intra molecular changing of a protein N and C termini. In general, the N and C terminals of the original protein should be close enough (~ 5–10 A) to allow their union (Meister et al. 2009). Relocation of the internal N and C-terminus of a protein might relax the structural constraints and result in an increased chain flexibility and accessibility of the active site (Qian and Lutz 2005). The first circular permutation example involving a protein genetic rearrangement was of the yeast phosphoribosyl anthranilate isomerase (yPRAI) (Luger et al. 1989), in which the gene 3′end was removed by restriction enzyme and reinserted into the 5′ end, creating a circularly permuted gene.

Table 14.3 Non-Homologous recombination methods.

Non-Homologous Recombination			
Method	**Advantages**	**Disadvantages**	**Ref.**
ITCHY	✓ Potential to create novel enzymes by domain and domain swapping ✓ It eliminates recombination bias	✓ Limited to two parents ✓ Labor-intensive ✓ Large numbers of progeny out-of-frame ✓ One crossover per iteration length of chimeric gene is not conserved	(Ostermeier and Lutz 2003)
THIO-ITCHY	✓ Same advantages as ITCHY ✓ Simpler protocol than ITCHY ✓ It can be accomplished by random mutagenesis	✓ Same disadvantages as ITCHY	(Lutz et al. 2001a)
SCRATCHY	✓ Multiple crossover libraries ✓ It enables application of DNA-shuffling on sequences with low homology	✓ Limited to two parents ✓ Labor-intensive ✓ Large numbers of progeny out-of-frame	(Lutz et al. 2001b)
NRR	✓ Crossover rate can be controlled by stoichiometric reaction ✓ It creates multiple crossovers in chimeric gene ✓ Great exploration of nucleic acid and protein diversity	✓ Significant number of frameshift mutants encoding inactive proteins: preselection of library necessary ✓ DNaseI digestion bias	(Bittker et al. 2002)
SHIPREC	✓ Crossovers occur mainly at structurally related sites ✓ Larger fraction of active recombinants ✓ It maintains proper sequence alignment ✓ Recombination at homologous positions within overall structure	✓ Limited to two parents ✓ Labor-intensive ✓ One crossover per iteration ✓ DNaseI digestion bias ✓ Significant number of frameshift mutants encodes inactive proteins: preselection of library necessary	(Sieber et al. 2001)
Exon Shuffling	✓ It preserves exon function ✓ Large fraction of active recombinants	✓ It needs known intron-exon organization ✓ Limited diversity	(Kolkman and Stemmer 2001)

Although circular permutation technique is relatively simple, evaluating all possible opening points is not. One of the ways to evaluate this was developed by Graf and Schachman (Graf and Schachman 1996). In this, the gene to be circularly permuted is removed from the plasmid using a restriction enzyme that cuts both sides. The gene must be first designed to not have a stop codon, and to have a linker between the N and C original termini. The removed gene is then circularized in the presence of high DNA ligase concentration and very low DNA concentrations, to favor intra molecular instead of inter molecular linking. Circularized DNA is isolated (linear DNA can be removed by the action of exonuclease) (Ostermeier and Benkovic 2001), and digested by diluted DNase I to obtain a single cut at the double strand per molecule. DNA is repaired to produce blunt ends and subsequently linked to a vector that has an ATG start codon on one side and on the other side it has a series of three stop codons in all the three frames. A lipase from *Candida antarctica* suffered random circular permutation using this approach (Qian and Lutz 2005), and a high frequency of variants presented an improvement at the kinetic parameters. Two of the recombinants that presented a new termini located near the active site, showed a catalytic efficiency 100 times greater than the wild-type enzyme, crediting this increase to the greater accessibility and flexibility of the active site (Qian and Lutz 2005).

Semi-rational Design

Semi-rational mutagenesis is an intersection between a directed and random mutagenesis in which the "hot spots" are defined by structural and/or functional data and that are then individually or combinatorially randomized for all amino acids. It is known that certain types of mutations in specific regions of the enzymes are more likely to produce beneficial effects than in other sites. For example, to increase the activity or to alter the substrate specificity or enantioselectivity mutations near the active site specifically close to residues that are related in binding or catalysis may be more advantageous than those on the surface of an enzyme (Bornscheuer and Pohl 2001). These sites can be identified from three-dimensional structures, from homology modeling, or from directed evolution experiments. The identification of these functionally significant regions can avoid the necessity of screening large libraries. A semi-rational approach may be advantageous in these cases, since an intelligent library can be constructed with the same representation than a random library (Chica et al. 2005, Lutz 2010).

In order to study the effect of the residue 342 in optimum pH of endoglucanase II from *Trichoderma reesei*, previously identified by site directed mutagenesis, a library was constructed via saturation mutagenesis by Strand Overlap Extension (SOE), in which most variants exhibited an up-shift in optimum pH (Qin et al. 2008). In this method, two separated PCR reactions are made to amplify two fragments which overlap in the codon/target sequence region. Each primers pair is synthesized with a random nucleotide sequence (such as-NNN-) encoding for the 20 amino acids, flanked on both sides by nucleotides that specifically anneal in target regions. After an initial PCR amplification, another PCR reaction is used to mount the sequence with the target codon randomly mutated (Georgescu et al. 2003).

Synthesis of the 64 possible codons (NNN) for all 20 amino acids may lead to codon bias, since the genetic code is degenerate and there is, for example, six times more codons for Ser than for Trp or Met. One technique that increases its efficiency is MAX randomization, which removes the genetic redundancy by using a single codon per amino acid and thus requires a smaller library (Neylon 2004).

Another robust and reliable method is the *in vivo* Overlap Extension (IVOE). This strategy mimics the classic SOE, but greatly decreases the number of steps including cloning *in vitro* because the two fragments generated are directly mixed with the vector and recombined *in vivo* in *Saccharomyces cerevisae.* This technique has been successfully explored in semi-rational studies of ascomycete and basidiomycete (Mate et al. 2011, Camarero et al. 2012). Total laccase activity from *Pycnoporus cinnabarinus* was increased 8000 times in a work that describes a platform for directed evolution (Camarero et al. 2012). In this case, the authors used error-prone PCR techniques associated with *in vivo* shuffling and site directed mutation/recombination by IVOE. After random mutations and appropriate screenings, modified bases were introduced to generate proteins containing the following modifications R(α2)S, A(α9)D, A240H e P394H, that were chosen after a previous study of another laccase also evolved in the same laboratory that showed beneficial results for the enzymatic activity. A(α9)D mutation presented the best result, leading to an increase in secretion by 13 times. Thus, the association between random mutations generated by site directed modifications proved to be a strong approach for laccase evolution, which besides being more secreted, its activity also increased.

A fungal laccase variant T2 from *Myceliophthora thermophila* was evolved in a same-rational way, via Combinatorial Saturation Mutagenesis (CSM) by IVOE (Alcalde et al. 2006). After exploring more than 1,700 clones, one of the mutants (7E1) presented higher kinetic parameters when compared to parental enzyme, having a synonymous mutation (L513L, CGT/TTG) and a beneficial mutation S510G (TCG/GGG), which cannot be achieved by conventional techniques of error-prone PCR (Alcalde et al. 2006).

Another example was directed evolution of a Pyranose 2-oxidase (P2Ox) from *Trametes multicolor*, in which, based on the crystal structure, the authors tested a saturated mutagenesis of the catalytic residue V546 (Spadiut et al. 2010). The best mutant generated, V546C, presented the highest turnover value for both D-galactose and D-glucose as for the alternative electron acceptors investigated (1,4-benzoquinone, BQ e ferricenium íon). Furthermore, additional mutations T169G, E542K and L537W, which previously had positive effects on the enzyme catalytic properties (Spadiut et al. 2010), were introduced in the V546C mutant, leading to interesting results in which the catalytic constants have changed pronouncedly. Therefore, combining specific mutations, some of them located in the active site, it was possible to obtain mutants that showed better catalytic properties when compared to the parental enzyme. The results from this semi-rational approach have shown that the combination of rational and random proteins modification techniques is efficient and beneficial.

Recently a method was developed for rapid and simple screen that optimizes combinatorial mutations of selected amino acids called Look-Through Mutagenesis (LTM™) (Rajpal et al. 2005). In this method protein variants are created by direct and systemic introduction of an amino acid subset representing the main chemical properties of the side chain, such as size, shape, charge, hydrophobicity, etc. Considering this chemical property, instead of replacing the residue to be mutated for all 20 possible amino acids, only nine would be necessary (G, S, H, L, P, Y, D, Q, K). LTM™ method rapidly generates information about the effect of the chemical group in a specific position. Then the generated library is relatively small, however diverse enough to allow the discovery of rare mutants. After the identification of positive variants and side chains with favorable effect, another platform is applied, called Combinatorial Beneficial Mutagenesis (CBM) (Rajpal et al. 2005, Hokanson et al. 2011). This platform aims to combine the beneficial mutations and to identify synergism between them. The results of the CBMTM library identifies immutable positions structurally required by the protein, and dramatically reduces the library sizes. The LTMTM/CBMTM technique demonstration was made with a wild-type GH11 xylanase 2 from *Hypocrea jecorina* (*Trichoderma reesei*), an enzyme that has been engineered for 20 yr. LTMTM/CBMTM library generated a diverse set of new mutations, including N71D, Y73G, T95G e Y96Q. When these mutations were combined in a single construction (Hjx-81), the purified protein was active, even after heating at 100°C for 20 min (Hokanson et al. 2011).

What have we Learned?

Engineering proteins through rational design and directed evolution has generated a large number of results, which presents a promising way to adapt enzymes to various bioprocesses. Here we briefly show different techniques and concepts applied to fungal enzymes that enable the development of new properties, and allow the creation of genetic diversity. A large set of molecular biology tools has emerged in recent years, in order to create equilibrated libraries and appropriate screening methods. Directed evolution can be used as a complement to rational design in a semi-rational approach, giving rise to new biocatalysts adapted to processes (resistant to extreme pH, high temperatures, different solvents, etc.). Furthermore, it can also help to solve challenging problems in protein and metabolic engineering. The perspective presented here is still very recent when it comes to fungal proteins, however, these methodologies will prepare a way for innovative and exciting areas that will surely emerge in the coming years.

References

Abécassis, V., D. Pompon and G. Truan. 2000. High efficiency family shuffling based on multi-step PCR and *in vivo* DNA recombination in yeast: statistical and functional analysis of a combinatorial library between human cytochrome P450 and 1A2. Nucleic Acids Research 28(e88).

Adamczak, M. and S.H. Krishna. 2004. Strategies for improving enzymes for efficient biocatalysis. Food Technology and Biotechnology 42(4): 251–264.

Alcalde, M., M. Zumarraga, J. Polaina, A. Ballesteros and F.J. Plou. 2006. Combinatorial saturation mutagenesis by *in vivo* overlap extension for the engineering of fungal laccases. Combinatorial Chemistry & High Throughput Screening 9(10): 719–727.

Arnold, F.H. 1996. Directed evolution: Creating biocatalysts for the future. Chemical Engineering Science 51(23): 5091–5102.

Arnold, F.H., P.L. Wintrode, K. Miyazaki and A. Gershenson. 2001. How enzymes adapt: lessons from directed evolution. Trends in Biochemical Sciences 26(2): 100–106.

Benatuil, L., J.M. Perez, J. Belk and C.M. Hsieh. 2010. An improved yeast transformation method for the generation of very large human antibody libraries. Protein Engineering Design & Selection 23(4): 155–159.

Bittker, J.A., B.V. Le and D.R. Liu. 2002. Nucleic acid evolution and minimization by nonhomologous random recombination. Nature Biotechnology 20(10): 1024–1029.

Bittker, J.A., B.V. Le, J.M. Liu and D.R. Liu. 2004. Directed evolution of protein enzymes using nonhomologous random recombination. Proceedings of the National Academy of Sciences of the United States of America 101(18): 7011–7016.

Bloom, J.D. and F.H. Arnold. 2009. In the light of directed evolution: Pathways of adaptive protein evolution. Proceedings of the National Academy of Sciences of the United States of America 106: 9995–10000.

Boraston, A.B., D.N. Bolam, H.J. Gilbert and G.J. Davies. 2004. Carbohydrate-binding modules: fine-tuning polysaccharide recognition. Biochemical Journal 382: 769–781.

Bornscheuer, U.T. and M. Pohl. 2001. Improved biocatalysts by directed evolution and rational protein design. Current Opinion in Chemical Biology 5(2): 137–143.

Brannigan, J.A. and A.J. Wilkinson. 2002. Protein engineering 20 years on. Nature Reviews Molecular Cell Biology 3(12): 964–970.

Bulter, T. and M. Alcalde. Preparing Libraries in Saccharomyces cerevisiae. *In:* F.H. Arnold and G. Georgiou [eds.]. 2003. Directed Evolution Library Creation. Humana Press, Totowa, New Jersey.

Bulter, T., M. Alcalde, V. Sieber, P. Meinhold, C. Schlachtbauer and F.H. Arnold. 2003. Functional expression of a fungal laccase in Saccharomyces cerevisiae by directed evolution. Applied and Environmental Microbiology 69(2): 987–995.

Cadwell, R.C. and G.F. Joyce. 1994. Mutagenic Pcr. Pcr-Methods and Applications 3(6): S136–S140.

Camarero, S., I. Pardo, A.I. Canas, P. Molina, E. Record, A.T. Martinez, M.J. Martinez and M. Alcalde. 2012. Engineering Platforms for Directed Evolution of Laccase from Pycnoporus cinnabarinus. Applied and Environmental Microbiology 78(5): 1370–1384.

Cherry, J.R., M.H. Lamsa, P. Schneider, J. Vind, A. Svendsen, A. Jones and A.H. Pedersen. 1999. Directed evolution of a fungal peroxidase. Nature Biotechnology 17(4): 379–384.

Chica, R.A., N. Doucet and J.N. Pelletier. 2005. Semi-rational approaches to engineering enzyme activity: combining the benefits of directed evolution and rational design. Current Opinion in Biotechnology 16(4): 378–384.

Coco, W.M., W.E. Levinson, M.J. Crist, H.J. Hektor, A. Darzins, P.T. Pienkos, C.H. Squires and D.J. Monticello. 2001. DNA shuffling method for generating highly recombined genes and evolved enzymes. Nature Biotechnology 19(4): 354–359.

Crameri, A., S.A. Raillard, E. Bermudez and W.P.C. Stemmer. 1998. DNA shuffling of a family of genes from diverse species accelerates directed evolution. Nature 391(6664): 288–291.

Doyon, J.B. and D.R. Liu. 2007. Identification of eukaryotic promoter regulatory elements using nonhomologous random recombination. Nucleic Acids Research 35(17): 5851–5860.

Eggert, T., S.A. Funke, N.M. Rao, P. Acharya, H. Krumm, M.T. Reetz and K.E. Jaeger. 2005. Multiplex-PCR-based recombination as a novel high-fidelity method for directed evolution. Chembiochem 6(6): 1062–1067.

Fan, Y.H., W.G. Fang, Y.H. Xiao, X.Y. Yang, Y.J. Zhang, M.J. Bidochka and Y. Pei. 2007. Directed evolution for increased chitinase activity. Applied Microbiology and Biotechnology 76(1): 135–139.

Fan, Z.M., J.R. Werkman and L. Yuan. 2009. Engineering of a multifunctional hemicellulase. Biotechnology Letters 31(5): 751–757.

Festa, G., F. Autore, F. Fraternali, P. Giardina and G. Sannia. 2008. Development of new laccases by directed evolution: Functional and computational analyses. Proteins-Structure Function and Bioinformatics 72(1): 25–34.

Fujii, R., M. Kitaoka and K. Hayashi. 2004. One-step random mutagenesis by error-prone rolling circle amplification. Nucleic Acids Research 32(19).

Fujii, R., M. Kitaoka and K. Hayashi. 2006. Error-prone rolling circle amplification: the simplest random mutagenesis protocol. Nature Protocols 1(5): 2493–2497.

Fukuda, K., M.H. Jensen, R. Haser, N. Aghajari and B. Svensson. 2005. Biased mutagenesis in the N-terminal region by degenerate oligonucleotide gene shuffling enhances secretory expression of barley alpha-amylase 2 in yeast. Protein Engineering Design & Selection 18(11): 515–526.

Garcia-Ruiz, E., D. Mate, A. Ballesteros, A.T. Martinez and M. Alcalde. 2010. Evolving thermostability in mutant libraries of ligninolytic oxidoreductases expressed in yeast. Microbial Cell Factories 9.

Georgescu, R., G. Bandara and L. Sun. Saturation Mutagenesis. *In:* F.H. Arnold and G. Georgiou [eds.]. 2003. Directed Evolution Library Creation: Methods and Protocols, Humana Press Inc., Totowa, NJ.

Gibbs, M.D., K.M.H. Nevalainen and P.L. Bergquist. 2001. Degenerate oligonucleotide gene shuffling (DOGS): a method for enhancing the frequency of recombination with family shuffling. Gene 271(1): 13–20.

Graf, R. and H.K. Schachman. 1996. Random circular permutation of genes and expressed polypeptide chains: Application of the method to the catalytic chains of aspartate transcarbamoylase. Proceedings of the National Academy of Sciences of the United States of America 93(21): 11591–11596.

Hayes, F. and B. Hallet. 2000. Pentapeptide scanning mutagenesis: encouraging old proteins to execute unusual tricks. Trends in Microbiology 8(12): 571–577.

Heckmann-Pohl, D.M., S. Bastian, S. Altmeier and I. Antes. 2006. Improvement of the fungal enzyme pyranose 2-oxidase using protein engineering. Journal of Biotechnology 124(1): 26–40.

Heinzelman, P., C.D. Snow, I. Wu, C. Nguyen, A. Villalobos, S. Govindarajan, J. Minshull and F.H. Arnold. 2009. A family of thermostable fungal cellulases created by structure-guided recombination. Proceedings of the National Academy of Sciences of the United States of America 106(14): 5610–5615.

Hokanson, C.A., G. Cappuccilli, T. Odineca, M. Bozic, C.A. Behnke, M. Mendez, W.J. Coleman and R. Crea. 2011. Engineering highly thermostable xylanase variants using an enhanced combinatorial library method. Protein Engineering Design & Selection 24(8): 597–605.

Hong, S.G. 2006. Effective family shuffling method using complementary DNA fragments produced by S1 nuclease. Journal of Microbiology and Biotechnology 16(12): 2004–2007.

Hughes, M.D., D.A. Nagel, A.F. Santos, A.J. Sutherland and A.V. Hine. 2003. Removing the redundancy from randomised gene libraries. Journal of Molecular Biology 331(5): 973–979.

Ibarra, D., J. Romero, M.J. Martinez, A.T. Martinez and S. Camarero. 2006. Exploring the enzymatic parameters for optimal delignification of eucalypt pulp by laccase-mediator. Enzyme and Microbial Technology 39(6): 1319–1327.

Joern, J.M., P. Meinhold and F.H. Arnold. 2002. Analysis of shuffled gene libraries. Journal of Molecular Biology 316(3): 643–656.

Jones, D.D. 2005. Triplet nucleotide removal at random positions in a target gene: the tolerance of TEM-1 beta-lactamase to an amino acid deletion. Nucleic Acids Research 33(9).

Joyce, G.F. 2004. Directed evolution of nucleic acid enzymes. Annual Review of Biochemistry 73: 791–836.

Kikuchi, M., K. Ohnishi and S. Harayama. 1999. Novel family shuffling methods for the *in vitro* evolution of enzymes. Gene 236(1): 159–167.

Kikuchi, M., K. Ohnishi and S. Harayama. 2000. An effective family shuffling method using single-stranded DNA. Gene 243(1-2): 133–137.

Kolkman, J.A. and W.P.C. Stemmer. 2001. Directed evolution of proteins by exon shuffling. Nature Biotechnology 19(5): 423–428.

Lee, H.S., J.S. Jang, S.K. Choi, D.W. Lee, E.J. Kim, H.C. Jung and J.G. Pan. 2007. Identification and expression of GH-8 family chitosanases from several Bacillus thuringiensis subspecies. Fems Microbiology Letters 277(2): 133–141.

Levasseur, A., D. Navarro, P.J. Punt, J.P. Belaich, M. Asther and E. Record. 2005. Construction of engineered bifunctional enzymes and their overproduction in Aspergillus niger for improved enzymatic tools to degrade agricultural by-products. Applied and Environmental Microbiology 71(12): 8132–8140.

Liu, Y.H., M. Ye, Y. Lu, X. Zhang and G. Li. 2011. Improving the decolorization for textile dyes of a metagenome-derived alkaline laccase by directed evolution. Applied Microbiology and Biotechnology 91(3): 667–675.

Luger, K., U. Hommel, M. Herold, J. Hofsteenge and K. Kirschner. 1989. Correct Folding of Circularly Permuted Variants of a Beta-Alpha-Barrel Enzyme *in vivo*. Science 243(4888): 206–210.

Luhe, A.L., E.N.Y. Ting, L. Tan, J.C. Wu and H. Zhao. 2010. Engineering of small sized DNAs by error-prone multiply-primed rolling circle amplification for introduction of random point mutations. Journal of Molecular Catalysis B-Enzymatic 67(1-2): 92–97.

Lutz, S. 2010. Beyond directed evolution-semi-rational protein engineering and design. Current Opinion in Biotechnology 21(6): 734–743.

Lutz, S., M. Ostermeier and S.J. Benkovic. 2001a. Rapid generation of incremental truncation libraries for protein engineering using alpha-phosphothioate nucleotides. Nucleic Acids Research 29(4).

Lutz, S., M. Ostermeier, G.L. Moore, C.D. Maranas and S.J. Benkovic. 2001b. Creating multiple-crossover DNA libraries independent of sequence identity. Proceedings of the National Academy of Sciences of the United States of America 98(20): 11248–11253.

Madzak, C., C. Gaillardin and J.M. Beckerich. 2004. Heterologous protein expression and secretion in the non-conventional yeast Yarrowia lipolytica: a review. Journal of Biotechnology 109(1-2): 63–81.

Mate, D., C. Garcia-Burgos, E. Garcia-Ruiz, A.O. Ballesteros, S. Camarero and M. Alcalde. 2010. Laboratory Evolution of High-Redox Potential Laccases. Chemistry & Biology 17(9): 1030–1041.

Mate, D., E. Garcia-Ruiz, S. Camarero and M. Alcalde. 2011. Directed Evolution of Fungal Laccases. Current Genomics 12(2): 113–122.

Mchunu, N.P., S. Singh and K. Permaul. 2009. Expression of an alkalo-tolerant fungal xylanase enhanced by directed evolution in Pichia pastoris and Escherichia coli. Journal of Biotechnology 141(1-2): 26–30.

Meister, G.E., M. Kanwar and M. Ostermeier. Circular Permutation of Proteins. *In:* S. Lutz and U.T. Bornscheuer [eds.]. 2009.Protein Engineering Handbook WILEY-VCH Verlag GmbH & Co. KGaA, Weinheim.

Mezard, C. and A. Nicolas. 1994. Homologous, Homeologous, and Illegitimate Repair of Double-Strand Breaks during Transformation of a Wild-Type Strain and a Rad52 Mutant Strain of Saccharomyces-Cerevisiae. Molecular and Cellular Biology 14(2): 1278–1292.

Miele, A., P. Giardina, G. Sannia and V. Faraco. 2010. Random mutants of a Pleurotus ostreatus laccase as new biocatalysts for industrial effluents bioremediation. Journal of Applied Microbiology 108(3): 998–1006.

Moore, G.L., C.D. Maranas, S. Lutz and S.J. Benkovic. 2001. Predicting crossover generation in DNA shuffling. Proceedings of the National Academy of Sciences of the United States of America 98(6): 3226–3231.

Murakami, H., T. Hohsaka and M. Sisido. 2002. Random insertion and deletion of arbitrary number of bases for codon-based random mutation of DNAs. Nature Biotechnology 20(1): 76–81.

Nakazawa, H., K. Okada, T. Onodera, W. Ogasawara, H. Okada and Y. Morikawa. 2009. Directed evolution of endoglucanase III (Cel12A) from Trichoderma reesei. Applied Microbiology and Biotechnology 83(4): 649–657.

Neylon, C. 2004. Chemical and biochemical strategies for the randomization of protein encoding DNA sequences: library construction methods for directed evolution. Nucleic Acids Research 32(4): 1448–1459.

Orrweaver, T.L. and J.W. Szostak. 1985. Fungal Recombination. Microbiological Reviews 49(1): 33–58.

Ostermeier, M. and S.J. Benkovic. 2001. Construction of hybrid gene libraries involving the circular permutation of DNA. Biotechnology Letters 23(4): 303–310.

Ostermeier, M. and S. Lutz. The creation of ITCHY hybrid protein libraries. *In:* F.H. Arnold and G. Georgiou [eds.]. 2003. Directed Evolution Library Creation: Methods and Protocols, Humana Press Inc., Totowa, NJ.

Ostermeier, M., A.E. Nixon and S.J. Benkovic. 1999. Incremental truncation as a strategy in the engineering of novel biocatalysts. Bioorganic & Medicinal Chemistry 7(10): 2139–2144.

Otten, L.G. and W.J. Quax. 2005. Directed evolution: selecting today's biocatalysts. Biomolecular Engineering 22(1-3): 1–9.

Patten, P.A., R.J. Howard and W.P.C. Stemmer. 1997. Applications of DNA shuffling to pharmaceuticals and vaccines. Current Opinion in Biotechnology 8(6): 724–733.

Pham, T.A., J.G. Berrin, E. Record, K.A. To and J.C. Sigoillot. 2010. Hydrolysis of softwood by Aspergillus mannanase: Role of a carbohydrate-binding module. Journal of Biotechnology 148(4): 163–170.

Pikkemaat, M.G. and D.B. Janssen. 2002. Generating segmental mutations in haloalkane dehalogenase: a novel part in the directed evolution toolbox. Nucleic Acids Research 30(8).

Pirakitikulr, N., P. Peralta-Yahya and V.W. Cornish. 2008. CHED 790-Cassette mutagenesis via yeast homologous recombination. Abstracts of Papers of the American Chemical Society 235.

Qian, Z. and S. Lutz. 2005. Improving the catalytic activity of Candida antarctica lipase B by circular permutation. Journal of the American Chemical Society 127(39): 13466–13467.

Qian, Z., J.R. Horton, X.D. Cheng and S. Lutz. 2009. Structural Redesign of Lipase B from Candida antarctica by Circular Permutation and Incremental Truncation. Journal of Molecular Biology 393(1): 191–201.

Qin, Y.Q., X.M. Wei, X. Song and Y.B. Qu. 2008. The role of the site 342 in catalytic efficiency and pH optima of endoglucanase II from Trichoderma reesei as probed by saturation mutagenesis. Biocatalysis and Biotransformation 26(5): 378–382.

Rajpal, A., N. Beyaz, L. Haber, G. Cappuccilli, H. Yee, R.R. Bhatt, T. Takeuchi, R.A. Lerner and R. Crea. 2005. A general method for greatly improving the affinity of antibodies by using combinatorial libraries. Proceedings of the National Academy of Sciences of the United States of America 102(24): 8466–8471.

Ravalason, H., I. Herpoel-Gimbert, E. Record, F. Bertaud, S. Grisel, S. de Weert, C.A.M.J.J. van den Hondel, M. Asther, M. Petit-Conil and J.C. Sigoillot. 2009. Fusion of a family 1 carbohydrate binding module of Aspergillus niger to the Pycnoporus cinnabarinus laccase for efficient softwood kraft pulp biobleaching. Journal of Biotechnology 142(3-4): 220–226.

Ribeiro, L.F., G.P. Furtado, M.R. Lourenzoni, A.J. Costa, C.R. Santos, S.C.P. Nogueira, J.A. Betini, M.D.T.M. Polizeli, M.T. Murakami and R.J. Ward. 2011. Engineering Bifunctional Laccase-Xylanase Chimeras for Improved Catalytic Performance. Journal of Biological Chemistry 286(50): 43026–43038.

Ryu, K., S.Y. Hwang, K.H. Kim, J.H. Kang and E.K. Lee. 2008. Functionality improvement of fungal lignin peroxidase by DNA shuffling for 2,4-dichlorophenol degradability and H2O2 stability. Journal of Biotechnology 133(1): 110–115.

Saab-Rincon, G., Y. Li, M. Meyer, M. Carbone, M. Landwehr and F.H. Arnold. Protein Engineering by Structure-Guided SCHEMA Recombination. *In:* S. Lutz and U.T. Bornscheuer [eds.]. 2009. Protein Engineering Handbook. WILEY-VCH Verlag GmbH & Co KGaA, Weinheim.

Sandgren, M., J. Stahlberg and C. Mitchinson. 2005. Structural and biochemical studies of GH family 12 cellulases: improved thermal stability, and ligand complexes. Progress in Biophysics & Molecular Biology 89(3): 246–291.

Shao, Z.X., H.M. Zhao, L. Giver and F.H. Arnold. 1998. Random-priming *in vitro* recombination: an effective tool for directed evolution. Nucleic Acids Research 26(2): 681–683.

Sieber, V., C.A. Martinez and F.H. Arnold. 2001. Libraries of hybrid proteins from distantly related sequences. Nature Biotechnology 19(5): 456–460.

Spadiut, O., C. Leitner, C. Salaheddin, B. Varga, B.G. Vertessy, T.C. Tan, C. Divne and D. Haltrich. 2009. Improving thermostability and catalytic activity of pyranose 2-oxidase from Trametes multicolor by rational and semi-rational design. Febs Journal 276(3): 776–792.

Spadiut, O., T.T. Nguyen and D. Haltrich. 2010. Thermostable Variants of Pyranose 2-Oxidase Showing Altered Substrate Selectivity for Glucose and Galactose. Journal of Agricultural and Food Chemistry 58(6): 3465–3471.

Stemmer, W.P.C. 1994a. DNA Shuffling by Random Fragmentation and Reassembly *in vitro* Recombination for Molecular Evolution. Proceedings of the National Academy of Sciences of the United States of America 91(22): 10747–10751.

Stemmer, W.P.C. 1994b. Rapid Evolution of a Protein *in vitro* by DNA Shuffling. Nature 370(6488): 389–391.

Stephens, D.E., S. Singh and K. Permaul. 2009. Error-prone PCR of a fungal xylanase for improvement of its alkaline and thermal stability. Fems Microbiology Letters 293(1): 42–47.

Sun, L.H., I.P. Petrounia, M. Yagasaki, G. Bandara and F.H. Arnold. 2001. Expression and stabilization of galactose oxidase in Escherichia coli by directed evolution. Protein Engineering 14(9): 699–704.

Swers, J.S., B.A. Kellogg and K.D. Wittrup. 2004. Shuffled antibody libraries created by *in vivo* homologous recombination and yeast surface display. Nucleic Acids Research 32(3).

Taylor, C.B., M.F. Talib, C. McCabe, L.T. Bu, W.S. Adney, M.E. Himmel, M.F. Crowley and G.T. Beckham. 2012. Computational Investigation of Glycosylation Effects on a Family 1 Carbohydrate-binding Module. Journal of Biological Chemistry 287(5): 3147–3155.

Valérie, A., D. Pompon and A.G. Truan. 2000. High efficiency family shuffling based on multi-step PCR and *in vivo* DNA recombination in yeast: statistical and functional analysis of a combinatorial library between human cytochrome P450 1A1 and 1A2. Nucleic Acids Research 28: e88.

Voutilainen, S.P., H. Boer, M.B. Linder, T. Puranen, J. Rouvinen, J. Vehmaanpera and A. Koivula. 2007. Heterologous expression of Melanocarpus albomyces cellobiohydrolase Cel7B, and random mutagenesis to improve its thermostability. Enzyme and Microbial Technology 41(3): 234–243.

Wang, T., X. Liu, Q. Yu, X. Zhang, Y. Qu, P. Gao and T. Wang. 2005. Directed evolution for engineering pH profile of endoglucanase III from Trichoderma reesei. Biomolecular Engineering 22(1-3): 89–94.

Xia, T. and Q. Wang. 2009. Directed evolution of Streptomyces lividans xylanase B toward enhanced thermal and alkaline pH stability. World Journal of Microbiology & Biotechnology 25(1): 93–100.

Xiao, Z.Z., P. Wang, Y.B. Qu, P.J. Gao and T.H. Wang. 2002. Cold adaptation of a mesophilic cellulase, EG III from Trichoderma reesei, by directed evolution. Science in China Series C-Life Sciences 45(4): 337–343.

Zha, D.X., A. Eipper and M.T. Reetz. 2003. Assembly of designed oligonucleotides as an efficient method for gene recombination: A new tool in directed evolution. Chembiochem. 4(1): 34–39.

Zhang, T., Z.F. Wu, H. Chen, Q. Wu, Z.Z. Tang, J.B. Gou, L.H. Wang, W.W. Hao, C.M. Wang and C.M. Li. 2010. Progress in strategies for sequence diversity library creation for directed evolution. African Journal of Biotechnology 9(54): 9277–9285.

Zhao, H.M., L. Giver, Z.X. Shao, J.A. Affholter and F.H. Arnold. 1998. Molecular evolution by staggered extension process (StEP) *in vitro* recombination. Nature Biotechnology 16(3): 258–261.

Zumarraga, M., S. Camarero, S. Shleev, A. Martinez-Arias, A. Ballesteros, F.J. Plou and M. Alcalde. 2008. Altering the laccase functionality by *in vivo* assembly of mutant libraries with different mutational spectra. Proteins-Structure Function and Bioinformatics 71(1): 250–260.

15

Genetic Tools for Production of Proteins in Fungi

Estela Y. Valencia[1] and *Felipe S. Chambergo*[2,*]

Introduction

The central dogma of molecular biology describes genes as DNA sequences that are transcribed to messenger RNA (mRNA), which in turn, is translated into a functional protein. In 1953, Watson and Crick reported the linear arrangement of genes and how the DNA replication process takes place. In 1966, Khorana and Niremberg's groups independently determined the genetic code, allowing the prediction of protein sequences from DNA sequences.

In 1995, the complete genomes of two microorganisms were determined for the first time. Fleischmann et al. (1995) and Fraser et al. (1995) have published, respectively, the entire DNA sequence of *Haemophilus influenzae* (1.83 Mb) and *Mycoplasma genitalium* (0.58 Mb). The following year, Goffeau et al. (1996) published the complete genome sequence of *Saccharomyces cerevisiae* (12 Mb), making this the first eukaryotic organism to have its DNA sequence completely unraveled. On February 2001, a draft of the human

[1]Instituto de Ciências Biomédicas, Av. Prof. Lineu Prestes 1374, Butantã.
E-mail: ynes2002@usp.br.
[2]Escola de Artes, Ciências e Humanidades, Av. Arlindo Bettio 1000. Ermelino Matarazzo, Universidade de São Paulo. São Paulo, Brazil.
E-mail: fscha@usp.br.
*Corresponding author

genome was announced (Lander et al. 2001, Venter et al. 2001). Since then, 173 eukaryotic genomes have been sequenced, including the genomes of several fungi, such as *Phanerochaete chrysosporium*, *Aspergillus nidulans*, *A. oryzae*, *A. niger*, *Neurospora crassa*, *Magnaporthe grisea*, *Trichoderma reesei* and others (GOLD 2012).

Computational analysis of the data obtained from genome projects allows us to identify potential coding sequences or ORFs (open reading frames) that determine the metabolic ability of organisms, which could be explored in biotechnological processes. The advent of recombinant DNA techniques has enabled the functional analysis of putative ORFs encoding proteins with medical, scientific and commercial value; in order to take advantage of this, genetic tools for analysis and expression of recombinant proteins have been developed.

Several systems for production of recombinant proteins in bacteria, yeast, baculoviruses, cell lineages for culture etc. are commercially available (Life Technologies, USA; Novagen, USA; Clontech, USA; Qiagen, Germany; and others). However, expression systems utilizing filamentous fungi have not been developed on a commercial scale.

The ability to produce and secrete large quantities of metabolites to the extracellular medium has enabled the use of filamentous fungi for the industrial production of antibiotics (penicillin, cephalosporin), organic acids (citric acid, fumaric acid) and native proteins or homologous/heterologous recombinant proteins (lipases and proteases) (Fowler and Berka 1991, Wiebe 2003). Some filamentous fungi display many advantages, justifying their use in the production of homologous or heterologous proteins, such as easy cultivation, the presence of a good secretory machinery with post-transcriptional modifications typical of eukaryotic proteins, and the fact that they are safe, non-pathogenic organisms for humans (GRAS, Generally Regarded as Safe) (Palamarczyk et al. 1998, Nevalainen et al. 1994, Miettinen-Oinonen et al. 2005, Nevalainen et al. 2005, Bouws et al. 2008).

In comparison to other eukaryotic cells, such as yeast, algae, insect cells or human cells, filamentous fungi possess a more efficient secretion capacity and this exceptional advantage can be used to produce proteins of interest, which has resulted in the development of expression systems for production of recombinant proteins. According to the Association of Manufacturers and Formulators of Enzyme Products (AMFEP 2009), the filamentous fungi are the most represented organisms (55 percent) in the list of producers of protein.

This chapter presents a summary of the genetic tools for production of proteins used in filamentous fungi, bottlenecks and future perspectives.

The Protein Expression System

Plasmids for cloning, transformation and expression purposes in filamentous fungi have been developed (http://www.fgsc.net). However, only a limited number of expression systems are currently available for efficient overproduction of proteins. In order to express a recombinant protein in a fungal host, it is necessary to construct an expression system in which the DNA sequence coding for the protein of interest is inserted.

The basic components of an expression system are (i) a strong promoter with an efficient transcription start region, (ii) a signal peptide sequence for an efficient secretion, (iii) the gene sequence coding for the protein of interest, (iv) transcription, termination and polyadenylation sequences and (v) a selection marker of fungal transformants (Berka and Barnett 1989, Fowler and Berka 1991). The expression unit is usually assembled in a plasmid vector which can replicate in bacteria for easy preparation of DNA (Fig. 15.1), and that will be inserted into the host cell by transformation.

DNA can be introduced by transformation into fungal cells without a fungal replication origin, in which case the maintenance of transformation depends on integration of the DNA into the chromosomes. Integration occurs either randomly in the genome by non-homologous/illegitimate recombination or by homologous recombination. Multiple insertions (2–10 copies) are common and are frequently found to improve protein

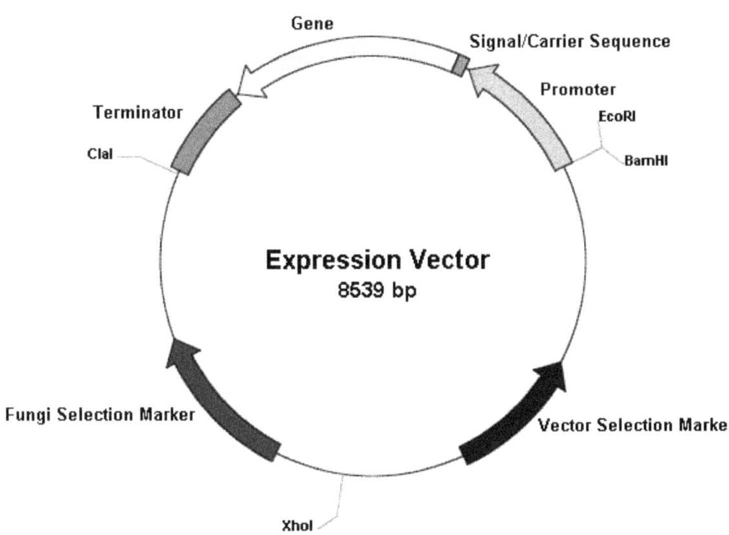

Figure 15.1 Circle map showing typical features of a transformation vector for a secreted gene product in filamentous fungi. A vector for expression of a gene product, comprising a promoter operatively connected to a gene encoding the desired gene product and terminator sequence.

production; however, increasing the copy number does not necessarily raise expression (Fleissner and Dersch 2010).

Expression vectors for gene replacement, protein expression and protein purification in the filamentous fungi *Neurospora crassa* (Honda and Selker 2009), *Trichoderma reesei* (Lv et al. 2012, Zou et al. 2012) and *Aspergillus niger* (Storms et al. 2005, Roth and Dersch 2010) have been developed.

Homologous proteins are usually well expressed in these fungi, but many heterologous proteins are expressed very poorly. Protein expression could be limited by (i) a weak promoter, (ii) copy number (iii) insertion location, (iv) inefficient processing of the protein, or (v) bottlenecks in protein traffic and translocation through the membrane systems of the protein secretory pathway.

Promoter

An expression system should be optimized through the use of a DNA region containing the regulatory elements required for an efficient transcription. DNA regions controlling the expression of several fungal genes have been studied in detail, and several elements (TATA box, initiation transcription site, C/TAAG and CAAT motifs) and specific regulatory motifs have been identified (MacKenzie et al. 1993).

An ideal promoter should be tightly regulated, induced by low-cost means and should yield a high level of expression. Therefore, the selection of a strong promoter which drives an efficient transcription of the target DNA is a key requirement in every expression system. Promoters employed in the production of recombinant proteins can be divided into two categories: constitutive or induced. Table 15.1 shows some of the promoters utilized in the construction of expression systems; it also shows that the majority of these promoters are regulated by carbon or nitrogen sources.

A difficulty encountered by the use of promoters of genes involved in carbon source catabolism is the existence of a regulatory circuit determining carbon catabolite repression, by which the expression of genes encoding enzymes involved in catabolism of other carbon sources are inhibited (Portnoy et al. 2011). The promoter of the gene encoding cellobiohydrolase I (*cbh1*) from *Trichoderma reesei* displays a 1000-fold induction, at least, in the presence of cellulose (Carle-Urioste et al. 1997) and it is strongly repressed by the degradation product of cellulose, that is, glucose (Abrahão-Neto et al. 1995, Ilmén et al. 1996). The strong *cbh1* promoter of *T. reesei* is frequently used in expression systems; however, the carbon catabolite repressor CRE1 may reduce its strength by binding to the *cbh1* promoter at several binding sites (Zou et al. 2012). A major disadvantage of using the *cbh1* promoter is its repression by glucose.

Table 15.1 Selected promoter used in fungal expression system.

Promoter	Gene Function	Condition / Inducer	Source DNA	Reference
trpC	Tryptophan biosynthesis	Constitutive	*A. nidulans*	Hamer and Timberlake 1987
TAKA-A amylase	Amylose hydrolysis	Starch, maltose, maltodextrin	*A. oryzae*	Tada et al. 1991
alcA/AlcR	Alcohol dehydrogenase	Ethanol	*A. nidulans*	Waring et al. 1989
amyA, amyB	Amylase	Starch, maltose, maltodextrin	*A. oryzae*	Huge-Jensen et al. 1989
glaA	Glucoamylase	Starch, 5-thio-glucose, 2-deoxyglucose	*A. niger*	Smith et al. 1990
gpdA	Glyceraldehide-3-phosphate dehydrogenase	Constitutive	*A. nidulans*	Punt et al. 1991
cbhI	Cellobiohydrolase I	Cellulose, sophorose	*T. reesei*	Harkki et al. 1991
pgkA	Phosphoglycerate kinase	Quinate or acetate	*A. nidulans*	Streatfield et al. 1992
aphA	Phosphatase	Inorganic phosphate	*A. niger*	MacRae et al. 1993
oliC/acuD hybrid	ATP synthase + acetate	Acetate	*A. nidulans*	De Lucas et al. 1994
exlA	1,4,-β-Endoxylanase A	Xylose	*A. awamori*	Gouka et al. 1996
pcbC	isopenicillin N synthase	Lactose	*Penicillium chrysogenum*	Gutierrez et al. 1997
tef1	Tanslation-elongation factor 1 α	Constitutive, glucose	*A. oryzae*	Kitamoto et al. 1998
gdhA	Glutamate dehydrogenase	Sucrose, ammonium sulphate	*A. awamori*	Moralejo et al. 1999
sodM	Manganese superoxide dismutase	H_2O_2	*A. oryzae*	Ishida et al. 2004
pn2/tpi hybrid	Neutral amylase + triose phosphate isomerase	Constitutive	*A. nidulans, A. niger*	Olempska-Beer et al. 2006
cbh1m (modified)	The three CREI binding sites in cbh1 promoter were replaced by the binding sites of positive regulator ACEII or HAP2/3/5 complex.	cellulose	*T. reesei*	Liu et al. 2008, Zou et al. 2012
sucA	β-fructofuranosidase	Sucrose, inulin	*A. niger*	Roth and Dersch 2010
hlyA	hemolysin-like protein	Glucose, maltose	*A. oryzae*	Bando et al. 2011
citA	citrate synthase	acetate, glycerol	*A. niger*	Dave and Punekar 2011
catR	catalase	H_2O_2, $CaCO_3$	*A. niger*	Sharma et al. 2011

Glucose-mediated repression in filamentous fungi is made by the repressor protein CreA/CRE1, which binds to the sequence 5′– SYYGGRG – 3′ of the target promoter and regulates the expression of genes involved in degradation of polysaccharides and the use of ethanol and amino acids as a carbon source (Cubero and Scazzocchio 1994, Portnoy et al. 2011). This zinc-finger-type regulator has been identified in *Aspergillus* (Dowzer and Kelly 1991), *Trichoderma* (Strauss et al. 1995), *Humicola* (Takashima et al. 1998), *Sclerotinia* (Vautard et al. 1999), *Trametes* (Stapleton and Dobson 2003) and *Neurospora* species (Ziv et al. 2008), and it regulates the expression of a great number of genes (Portnoy et al. 2011, Sun and Glass 2011). More recently, the CRE1 binding sites have been modified within the *cbh1* promoter through their replacement by a CCAAT box and the ACEII binding site (Liu et al. 2008) or the binding sites of transcription activator ACEII and the HAP2/3/5 complex (Zou et al. 2012), to improve the promoter efficiency. The modified promoter resulted in an increased expression level of the reporter protein.

The promoter of the *A. niger* glucoamylase gene *glaA* and the *A. nidulans* alcohol dehydrogenase gene *alcA* are the most frequently used inducible promoters in *Aspergillus* expression systems. Introduction of a series of CCAAT repeats into the most upstream located CCAAT box has gradually increased transcriptional activity of P*glaA*. Other inducible promoters exploited for high-level gene expression in *Aspergillus* include the transcriptional control region of the *A. nidulans* alcohol dehydrogenase gene *alcC*, the xylose-inducible *A. awamori* endoxylanase (*exlA*) promoter, the thiamine-dependent *A. oryzae thiA* promoter, the TAKA amylase promoter from *A. oryzae* and the sucrase Agene (*sucA*) promoter from *A. niger* (Fleissner and Dersch 2010). Pachlinger et al. (2005) developed a strong but metabolically independent, tightly regulated expression system, which uses the protein expression of human estrogen receptor (hER) and a chimeric promoter containing three copies of the human estrogen-responsive elements (EREs). This system uses the transcriptional activation of the chimeric promoter by estrogenic substances, showed that *Aspergillus* species are highly sensitive to estrogenic compounds.

Understanding the detailed mechanisms of transcriptional control and the interaction of regulatory proteins with fungal promoter regions will help in designing modified promoter elements or regulatory factors optimized for a given set of growth conditions (Mackenzie et al. 1993). The conservation of the mechanism for regulation of translation and transcription between fungal species (Fowler and Berka 1991) allows the use of promoters in different species (Punt et al. 1991), and the construction of hybrid promoters or transcription factors could be useful in the design of expression systems with the ability to respond to specific growth conditions.

Signal Peptide or Carrier/Tags/Terminator and Polyadenylation Sequence

Expressed proteins may be categorized into two classes on the basis of their relationship with respect to the host: a-Intracellular (extraction of the desired protein requires the host to be broken open or lysed in order to free the product for eventual purification), or b-Extracellular (proteins are secreted out of the cell; in this case, purification is performed in the extracellular medium).

Based on the secretory capacity of the filamentous fungi, the homologous/heterologous protein expression system is used for expression and secretion of recombinant proteins. Secretion of proteins in eukaryotes is driven by a secretion signal peptide, which is usually located at the amino terminus of the protein (Kapp et al. 2009).

Signal sequences are N-terminal extensions of newly synthesized secretory proteins and are usually cleaved off by signal peptidase, and the resulting signal peptides are presumably rapidly degraded. They are usually 16 to 30 amino acid residues in length and comprised of a hydrophilic, usually positively charged N-terminal region, a central hydrophobic domain and a C-terminal region with the cleavage site for signal peptidase. Besides these common characteristics, signal sequences do not share sequence similarity and some are more than 50 amino acid residues long (Kapp et al. 2009).

The 5′ end of the DNA coding sequence may (i) inherently contain a signal peptide coding region naturally linked in translation reading frame with the region encoding the secreted polypeptide (ii) contain a signal peptide coding region which is foreign to the coding sequence. The foreign signal peptide coding region may be required where the coding sequence does not naturally contain a signal peptide coding region, or (iii) alternatively, the foreign signal peptide coding region may simply replace the natural signal peptide coding region in order to enhance secretion of the polypeptide.

The identification of ORFs that encode proteins with secretion potential has been possible due to the development of computational methods for prediction of N-terminal signal peptides (Zhang and Henzel 2004), which has allowed the establishment, in filamentous fungi, of a "secretome", a term which refers to the collection of proteins that contain a signal peptide and are processed via the endoplasmic reticulum and Golgi apparatus before secretion (Choi et al. 2010).

Signal sequences from fungal extracellular hydrolases are usually used in expression systems. The signal sequences from the gene encoding cellobiohydrolase I (*cbh1*) (Schuster and Schmoll 2010) and glucoamylase (*glaA*) from *T. reesei* and *A. niger*, respectively, are the most commonly used.

Expression systems presenting N- or C-terminal fusion proteins with FLAG, HA, 6x histidine, Myc, GFP, HAT–FLAG TAP tags and poly-glycine flexible linkers (Honda and Selker 2009), hydrophobin fusion (Mustalahti et al. 2011) or protein N-terminal carrier (α-amylase *glaA*, which enhances protein translocation, folding and prevents degradation) (Conesa et al. 2000), have been developed.

Downstream of the C-terminal fusion regions, the expression vectors contain a transcription terminator and a signal to add polyadenosine residues to transcribed mRNA. In higher eukaryotes it is known that the 3′ non-translated regions not only control transcription termination, but also have important functions with regard to mRNA stability (Radzio and Kuck 1997). The terminators from *cbh1* (T*cbh1*) and the tryptophan biosynthesis *trpC* gene (T*trpC*) from *T. reesei* and *A. nidulans* are most commonly used.

Target Gene

Table 15.2 shows the diversity of genes from diverse organisms that have been used by recombinant homologous/heterologous protein production in filamentous fungi. Many enzymes such as endoglucanases, beta-glucanases, cellobiohydrolases, beta-glucosidases, hemicellulases, xylanases, xylosidases, mannanases, galactanases, galactosidases, pectin methyl esterases, pectin lyases, pectate lyases, endopolygalacturonases, exopolygalacturonases, rhamnogalacturonases, arabanases, arabinofuranosidases, arabinoxylan hydrolases, galacturonases, lyases, amylases, phytases, lipases, proteases, peptidases, and others, have been produced in several fungi.

The worldwide industrial enzyme market is traditionally divided into three segments: (i) technical enzymes (65 percent of sales), including enzymes used in the detergent, starch, textile, leather, pulp and paper, and personal care industries; (ii) food enzymes (25 percent of sales), including enzymes employed in the dairy, brewing, wine and juice, fats and oils, baking industries and (iii) feed enzymes (10 percent of the market), comprising enzymes used in animal feeds (Cherry and Fidantsef 2003). In this context, hydrolases represent 75 percent of the industrial enzymes and carbohydrases are the second largest group of industrial enzymes. Cellulases contribute to 8 percent of the worldwide industrial enzyme demands, their major application is in animal feed (Bon and Ferrara 2007).

Table 15.2 Selected expression system by recombinant proteins production.

Vector/Transformation/ Host strain	Promoter/Sequence sinal or carrier/Terminator (Source)	Marker	Gene (DNA source)	Product (Production)	Reference
pLipA and pMn1/ Protoplast /*A. niger*	*PglaA (A. niger)/ssglaA or glaA cDNA (A. niger)/TtrpC (A.nidulans)*	*pyrG*	*lipA* and *mnP (Phanerochaete chrysosporium)*	Lignin peroxidase (not active) and manganese peroxidase (5–10 mg/L)	Conesa et al. 2000
pHEN44-HXYN /Biolistic /*T. reesei*	*Pcbh1/sscbh1/Tcbh1 (T. reesei)*	*hph*	*xyn2 (Humicola grisea)*	Xylanase (0.5–1 g/L)	Bergquist et al. 2002
pLLK13/Protoplast /*T. reesei*	*Pcbh1/Tcbh1 (T. reesei)*	*hph*	*lac1 (Melanocarpus albomyces)*	Laccase (230 mg/L)	Kiiskinen et al. 2004
pMS190 /Protoplast /*T. reesei*	*Pcbh1/Tcbh1 (T. reesei)*	*hph*	*tyr2 (T. reesei)*	Tyrosinase (0.3–1 g/L)	Selinheimo et al. 2006
pALK1502 /Protoplast /*T. reesei*	*Pcbh1/sscbh1/Tcbh1 (T. reesei)*	*amdS*	*xyn11A (Nonomuraea flexuosa)*	Xylanase (1.85 mg/mL)	Paloheimo et al. 2007
pTRIL-efe /Protoplast /*T. viride*	*Pcbh1/Tcbh1 (T. reesei)*	*hph*	*efe (Pseudomonas syringae)*	Ethylene-forming enzyme (1,059.7 nl h^{-1} g^{-1} dry wt)	Tao et al. 2008
pEstA /Protoplast *T. reesei*	*Pcbh1/Tcbh1 (T. reesei)*	*hph*	*EstA (Piromyces equi)*	Cinnamoyl esterase (33 mg/L)	Poidevin et al. 2009
pRsSymEG1/Protoplast /*A.oryzae*	*PamyB /cDNA-amyB/Tamy-B (A. oryzae)*	*niaD*	*EG1 (Reticulitermes speratus)*	Endoglucanase (603 U/mg)	Todaka et al. 2010
pANJil /Electroporation /*A. niger*	*PgpdA/TtrpC (A. nidulans)*	*pyrG*	*xlnD (A. niger)*	β-xylosidase (2,094 U/mg protein)	Amaro-Reyes et al. 2011

Table 15.2 contd....

Table 15.2 contd.

Vector/Transformation/ Host strain	Promoter/Sequence sinal or carrier/Terminator (Source)	Marker	Gene (DNA source)	Product (Production)	Reference
pWEF31 pWEF32 /*Agrobacterium* /*T. reesei*	*Pcbh1)/sscbh1/Tcbh1* *(T. reesei)*	*hph*	*DsRed2* (*Discosoma* sp.)	Red fluorescent protein (active)	Lv et al. 2011
pXBthg-pbgl1 /*Agrobacterium* /*T. reesei*	*Pcbh1(T. ressei)/TtrpC* *(A. nidulans)*	*hph*	*bgl1* (*Penicilliun decumbens*)	β-glucosidase I (101 IU/mg)	Ma et al. 2011
pGFP and pGFP-HFBI / Protoplast /*T. reesei*	*Pcbh1/sscbh1/HFBI tag-Tcbh1* *(T. reesei)*	*amdS*	*GFP* (*Aequorea victoria*)	Green fluorescent protein (52.2 mg/L)	Mustalahti et al. 2011
Pcat/xylanase-pAN56-1 /Electroporation /*A. niger*	*PcatR/TtrpC* *(A. niger)*	*hph*	*alx* (*A. niger*)	Alkaline xylanase (5.85–20.72 U/mg)	Sharma et al. 2011
pUSAFEI /Protoplast /*A. oryzae*	*PamyB/(A. niger)*		AFUEI (*A. fumigatus*)	Elastase inhibitor (56.1 μg/mL)	Yamashita et al. 2011
pTrCBH-EPO /Protoplast /*T. reesei*	*Pcbh1m/sscbh1/Tcbh1* *(T. reesei)*	*hph*	*HuEPO* (*Homo sapiens*)	Erythropoietin (46 to 97 mg/L)	Zhong et al. 2011
pSK-Lip/Protoplast/ *T. reesei*	*Pcbh1/ sscbh1/ Tcbh1* *(T. reesei)*	*pyr4*	*Lip* (*A. niger*)	Lipase (310 mg/L)	Qin et al. 2012

Selection Marker

Transfected cells are usually plated on agar medium that is selective for the desired type of transformant. A number of selectable markers are available. Table 15.3 lists selectable markers used to complement auxotrophic deficiencies of the host, to provide drug resistance or the ability to use a substrate.

A wide range of genes have been found to be suitable as selectable markers for fungi. The *hph* gene (hygromycin B resistance) from *E. coli* inactivates the antibiotic hygromicin B, and it is the most commonly used selection system because of its effectiveness in most, but not all, systems. Other selective agents such as phleomycin, sulfonylurea, nourseothricin, bialophos, carboxin, blasticidin S and benomyl have also been used. An alternative to drug resistance genes for transformation of fungi is to use auxotrophic markers such as *pyrG* (a homologue of the *Saccharomyces cerevisiae ura3* gene). Mutants that lack *pyrG* are auxotrophic for uracil, so vectors containing *pyrG* allow selection on uracil-deficient media (Berka and Barnett 1989, Punt et al. 2002, Nevalainen et al. 2005).

Transformation System

Several transformation methods have been described and modified in order to make them easier and more precise for each fungal strain. Transformants obtained results from the integration of DNA sequence from the vector into the host cell genome. The integration site is dependent on the homology of the plasmid with the chromosomal DNA, or integration of the vector may occur at a random genomic location. The number of integrated vector copies can be highly variable and it is necessary to test the stability of the transformants by successive platings on selective media (Berka and Barnett 1989). A commonly applied strategy is introducing multiple copies of the gene product of interest, expressed under a strong homologous gene promoter into a high-protein-secreting mutant strain (Nevalainen et al. 2005).

Most commonly used methods for transformation:

1. Using protoplasts (spheroplasts) (Fincham 1989), prepared from fungal hyphae or germinating conidiospores through treatment with commercial cell wall degrading enzymes and stabilization of the resulting protoplasts with 1 M sorbitol, before the transfection with DNA in the presence of calcium chloride. For preparation of protoplasts, the choice of enzyme for digesting the cell walls is crucial. Filamentous fungi often offer a choice of cell types from which protoplasts may be prepared; the most commonly used cell type is young mycelia, and

Table 15.3 Selected markers used for transformation of filamentous fungi.

Gene marker	Name	Resistance/Nutrition/Deficient	Source Organism	Reference
argB	Ornithine carbamoyltransferase	Arginine synthesis	*A. nidulans*	Berse et al. 1983
β-tub	β-tubulin	Benomyl	*Neurospora crassa*	Orbach et al. 1986
amdS	Acetamidase	Acetamide	*A. nidulans*	Rodriguez and Yoder 1987
bar	Bar	Bialaphos (phosphinothricin)	*Streptomyces hygroscopicus*	Thompson et al. 1987
hph	Hygromycin B phosphotransferase	Hygromycin	*Escherichia coli*	Punt et al. 1987
pyrG/pyr4	Pyr	Orotidine-5′-phosphate decarboxylase	*A. niger*	De Ruiter-Jacobs et al. 1989
Ble/Bm	Bleomycin/phleomycin	Bleomycin/phleomycin	*Streptoalloteichus hindustanus*	Austin et al. 1990
bsd	Blasticidin S deaminase	Deamination of blasticidin S	*A. terreus*	Kimura et al. 1994
cbx	Carboxin	Carboxin-resistant succinate dehydrogenase	*Mycosphaerella graminicola*	Skinner et al. 1998
ptrA	Ptra	Pyrithiamine	*A. oryzae*	Kubodera et al. 2000
nat1	Nourseothricin	Nourseothricin acetyltransferase	*Streptomyces noursei*	Alshahni et al. 2010

protoplasts released from the hyphae by enzyme treatment can be easily separated from the hyphal debris. In *Aspergillus* and *Penicillium* species, both germinating conidia and mycelia are used; one procedure is to cultivate the mycelia for protoplast formation on a cellophane membrane covering the surface of an agar plate.

All protoplast preparations have to be protected by the presence of an osmotic stabilizer in the suspending medium. Sorbitol (0.8–1.2 M), mannitol (0.8 M), NaCl (0.6 or 0.7 M), sucrose (0.6 M) or magnesium sulfate (1.2 M) have been used during protoplast preparation (Fincham 1989). The typical components of a transformation mixture are protoplasts at a density of about 10^8 to 10^9/ml, DNA (5 ug/mL), 10 or 50 mM calcium chloride, and a buffer which may be 10 mM Tris-Cl (pH 7.5 or 8.0) or 10 mM MOPS pH 6. Double-stranded DNA, either linear or circular, has usually been used. An incubation time of 15 to 30 min at room temperature has generally been found sufficient for DNA uptake. After transformation, regeneration of protoplasts and selection of transformants is necessary to obtain stable transformants.

2. Electroporation is also usually used to transform protoplasts, though sometimes germinating conidia are used. During electroporation, protoplasts or conidia are exposed briefly to a high amplitude electric field, which permeabilizes the membrane, permitting uptake of foreign DNA. In this procedure, 1 µg of DNA (circular or linearized) is added to 100 µL of ice cold conidial suspension (germinated conidia at 10^6–10^8/mL) in electroporation buffer (50 mM mannitol, 1 mM HEPES or 10 mM Tris-HCl pH 7.5, 270 mM sucrose, 1 mM Li-acetate), and the mixture is incubated on ice for 15 min and then transferred to a 0.2-cm cuvette. Electroporation is performed using 1,000 V, 25 uF capacitance and 400 Ω resistance (pulse length varying between 5.1 and 8.5 ms) (Dobrowolska and Staczek 2009, Gangavaram et al. 2009). After transformation, selection of transformants is necessary to obtain stable transformants.
3. The biolistic system or particle bombardment is used for the delivery of genetic material directly into intact cells and tissues. For particle bombardment, gold or tungsten beads are coated with the transforming DNA and used to bombard the intact fungal tissue. In this procedure, beads are coated with circular or linear DNA by precipitation with 0.1 M spermidine and 2.5 M $CaCl_2$. Following precipitation, the mixture of particles coated with DNA is deposited on a disk. Target cells are spread on potato dextrose agar (PDA) plates and bombarded with the DNA-coated beads in a vacuum chamber with high gas pressure, incubated for 6–12 hr at 28°C and then overlaid with agar containing antibiotic or selection marker. Transformants are visible after 4–5 d of incubation (Te'o et al. 2002).

4. *Agrobacterium* mediated transformation (AMT) system (Michielse et al. 2005). The gram-negative bacterium *Agrobacterium tumefaciens* is a plant pathogen, which causes crown gall tumors on plants. *A. tumefaciens* induces this tumorous growth by transferring a part of its DNA (T-DNA), located on its 200-kbp tumor-inducing (Ti) plasmid, to the host. The expression vector consists of modified versions of the Ti plasmid, where the DNA expression system replaces the region contained between the T-DNA borders. The use of protoplasts, spores, mycelia, and fruiting body tissue as target cells for AMT all have resulted in successful transformation. The co-cultivation conditions are very important and include the ratio between *A. tumefaciens* and the fungal recipient, the length of the co-cultivation period, temperature (range from 20°C to 37°C), pH (between 5.0 and 5.3) and choice of filters (nitrocellulose, hybond, filter paper, cellophane sheets, and polyvinylidene difluoride). Several studies have shown that each fungus has an optimal combination of co-cultivation period and temperature to obtain a maximum number of transformants. Under the right conditions, *A. tumefaciens* is able to transfer T-DNA to a very wide range of fungi and fungal tissues.

Ma et al. (2011) introduced the genome of *T. reesei* strain Rut-C30, by *Agrobacterium*-mediated transformation, the beta-glucosidase I gene from *Penicillium decumbens* ligated with the *cbh1* promoter of *T. reesei,* and obtained transformants with a 6- to 8-fold increase in β-glucosidase activity. In this study, *A. tumefaciens* cells were mixed with an equal volume of conidial suspension of *T. reesei* Rut-C30 and spread on induction medium plate containing acetosyringone. After incubation at 25°C for 2 d, the cultures were scraped with H_2O containing 0.02 percent Tween 80 and plated onto potato dextrose agar plates containing hygromycin B as a selection agent for transformants and cefotaxime to kill *A. tumefaciens* cells.

Hosts

The choice of an expression host strain cannot be made solely on the basis of production yields. Other aspects, such as regulatory issues, have a very important role in this choice.

Works showing filamentous fungi being excellent hosts for the expression and production of homologous/heterologous proteins have been described (Radzio and Kuck 1997, Wiebe 2003, Nevalainen et al. 2005, Ward 2011). Host strains of *Aspergillus* (*A. awamori, A. niger A. sojae, A. japonicus* and *A. oryzae*), *Trichoderma* (*T. reesei*), *Penicillium, Rhizopus, Chrysosporium lucknowense* and *Fusarium graminearum* (reviewed in Nevalainen et al. 2005, Lubertozzi and Keasling 2009, Ward 2011) are now widely used in biotechnology. Fungi of the genera *Penicillium, Acremonium* and *Chrysosporium* might be alternatives to *T. reesei* and *Aspergillus* (Gusakov 2011). Diverse expression systems using induced or constitutive promoters have been developed and the potential of the filamentous fungi *Trichoderma* and *Aspergillus* for producing recombinant proteins has been demonstrated (Table 15.2).

Bottlenecks and Perspectives

The potential of expression systems and of the filamentous fungus for producing recombinant proteins has been extensively exploited; however, the ability to produce the recombinant protein may still be lost for a variety of reasons:

1. Loss of the recombinant DNA from the host cell or loss of expression of this DNA leads to reduced productivity. One theory to account for this observation is that, in abortive colonies the DNA has not been stabilized by integration or by autonomous replication and is only transiently expressed. Thus, the transformant can grow initially, but as the gene product that relieves the selective pressure is diluted out, growth ceases (Rambosek and Leach 1987).
   The development of autonomously replicating vectors constructed using AMA1 (autonomous maintenance in *Aspergillus*) sequences derived from *Aspergillus nidulans* or others organisms, which have increased transformation efficiency and are maintained as plasmid in the transformants (Fierro et al. 2004, Shimizu et al. 2012), can be an interesting approach to avoid the abortive transformation or loss of DNA.
2. The yield of heterologous enzymes from fungi may be low, especially when the donor organism is not a fungus. Factors that can limit protein production are low transcriptional levels, mRNA instability, inefficient translation, post-transcriptional bottlenecks, and finally degradation of the protein of interest. For some heterologous proteins, glycosylation is essential to obtain the proper conformation and high secretion levels, and in such cases low yield of heterologous protein produced could be due to ineffective glycosylation (Maras et al. 1999). To improve heterologous protein expression in filamentous fungi, several strategies have been suggested, both genetic (gene fusion with a gene encoding all or part of a well-expressed and secreted protein; the introduction of multiple copies of the gene; engineering of the glycosylation pathway; induction of the unfolded protein response; construction and use of protease-deficient host strains; use of strong promoters and introducing multiple copies of protein-binding sequence—CCAAT—to

the promoter) as bioprocessing (conditions of culture or fermentation) (Sharma et al. 2009).

3. Filamentous fungi secrete significant amounts of proteolytic enzymes, which efficiently degrade heterologous proteins (Van den Hombergh et al. 1997). Proteases may either be localized in the cell wall or secreted, and may act intracellularly in lytic compartments such as the vacuole or in the extracellular medium (Gouka et al. 1996). Many attempts have been undertaken to construct protease-deficient production strains. For example, a double disruption of two proteinase genes (*tppA* and *pepE*) resulted in a 63 percent increase in human lysozyme production (Fleissner and Dersch 2010) and disruption of five protease genes (*tppA, pepE, nptB, dppIV* and *dppV*) increased the level of bovine chymosin (CHY) in 34 percent, both in *A. oryzae* (Yoon et al. 2009). For recombinant protein production, protease-deficient fungal host strains must be developed with the specific aim of reducing their proteolytic activities.
4. Secretory pathway: Considering that only properly folded, modified, and assembled functional proteins are exported to the Golgi apparatus, a high flux of recombinant proteins through the endoplasmic reticulum (ER) results in their incorrect or delayed folding and misassembly, being retained and degraded in the ER (ER quality control), a situation termed ER stress or protein secretion stress (Fleissner and Dersch 2010). The design of a fine-tuning system for regulated expression in fungi is important for the production of toxic proteins that may have a negative effect on cell performance. Only a few components of the fungal secretory pathway have been described; a better understanding of the fungal secretory pathway, protein trafficking in fungal hyphae and the elucidation of the interactions between proteins in the secretory pathway and proteins in resident cell organelles could lead to the identification of components important for the process of cellular secretion.

   The use of protein N-terminal carrier which enhances protein translocation, folding and prevents degradation (Conesa et al. 2000) can cause its temporary accumulation in the endoplasmic reticulum and a significant induction of the unfolded protein response (UPR) (Ohno et al. 2011).

   An interesting line would be to connect the research of secretion more closely with the protein production bioprocess. Studies on the transcriptional regulation of the secretory pathway have revealed similarities, but also interesting differences, to other organisms, such as a different induction mechanism of the unfolded protein response and the repression of genes encoding secreted proteins under secretion stress conditions (Saloheimo and Pakula 2012).

Kosalkova et al. (2012) showed that casein-derived phosphopeptides drastically stimulate the secretion of recombinant bovine chymosin in *Aspergillus awamori*, triggering the traffic of chymosin through the secretory pathway, which leads to the export of this and other extracellular proteins. Qin et al. (2012) determined that the main problem for heterologous expression in *T. reesei* is that production of large amounts of endogenous CBHI might cause saturation of the secretion mechanism *in vivo*. An approach for an efficient down-regulation of *cbh1* gene expression using RNAi-mediated gene silencing showed that the yield of heterologous lipase from *A. niger* was significantly enhanced in a *cbh1* knockout *T. reesei* strain.

5. Many proteins secreted by fungi are glycosylated, and the glycan component can affect the stability and activity of some, but not all, proteins. Studies with the main cellobiohydrolase (Cel7A) of *T. reesei* have confirmed the fact that the composition of the cultivation medium has an effect on protein glycosylation. Full N- and O-glycosylation of the enzyme was obtained on minimal medium, whereas varying forms and degrees of glycosylation were observed on complex growth media, owing to postulated post-secretion modification by extracellular glycan-modifying enzymes (Nevalainen et al. 2005). Correct glycosylation is especially important for proteins of pharmaceutical potential (Archer 2000). When administered to patients, non-mammalian glycans might provoke an immunological response, and incorrectly glycosylated proteins can have a very short half-life in the bloodstream, making them impractical for targeted therapeutics.
6. Influence of fungal morphology: productivity also strongly depends on the overall morphology of the fungus in submerged culture, ranging from freely dispersed mycelium to growth as compact hyphal aggregates or pellets (Grimm et al. 2005). Many different attempts have been made in order to influence and optimize the morphogenic development, including mutagenesis and variation of culture conditions (pH, temperature, osmolarity) (Fleissner and Dersch 2010). Addition of talc, titanate or aluminium microparticles into the culture medium strongly influences the morphological development of filamentous fungi, resulting in higher biomass formation, prevention of pellet formation and thus leading to higher specific productivities of the biomass (Kaup et al. 2008, Driouch et al. 2012). Addition of surfactants (rhamnolipid or Triton X-100) promotes pellet growth by easy separation and enhances cellulase production (Callow and Ju 2012).
7. Insufficient accumulation and lack of efficient purification methods. One strategy commonly used to enhance yields of secreted heterologous proteins is to express the gene as a protein fusion with a kexin

endoproteolytic cleavage site (Lys–Arg is the preferred recognition and cleavage motif) at the junction of the carrier and target protein. The carrier protein is one protein which is naturally well-secreted by the fungus, for example, glucoamylase or cellobiohydrolase (Archer 2000). Mustalahti et al. (2011) showed that hydrophobin fusion induces formation of large protein bodies in the filamentous fungus *T. reesei*. The fusion protein remained soluble in the ER-derived protein bodies and a simple and scalable aqueous two-phase system was used to recover the fusion protein efficiently from intracellular extract.

8. New host: research demonstrates that fungi of the genera *Penicillium, Acremonium, Fusarium, Neurospora, Mortierella* and *Chrysosporium* might represent alternative hosts for protein production (Punt et al. 2002, Gusakov et al. 2011).
9. Environmental conditions affect the rate at which protein-producing strains are displaced by non-producing or low-producing strains. The amount of inoculum, pH and medium composition have been shown to enhance or reduce stable production of recombinant proteins in filamentous fungi, although the reasons for this are yet unknown (Domingues et al. 2000, Wiebe 2003). The cultivation of filamentous fungi is often accompanied by technical problems, such as high broth viscosity, clumpy growth, insufficient oxygen and nutrient transfer which results in reduced productivity or even inactivation, predominantly of cells in the center of the pellets. The formation of pellets with different sizes or the adhesion of growing mycelia to the wall of shaking flasks or parts of bioreactors accompanied with the decrease of metabolic activity was frequently observed (Kaup et al. 2008).

Concluding Remarks

The tools for genetic manipulation of filamentous fungi have been greatly expanded and further developed to improve high-level expression and export of recombinant proteins. In nature, seven phyla of fungi are recognized: Chytridiomycota, Blastocladiomycota, Neocallimastigomycota, Zygomycota, Glomerulomycota, Ascomycota, and Basidiomycota. However, 70 percent of the fungi whose genomes have been sequenced to date are from a single phylum, Ascomycota (Baker et al. 2008). In the next few years, the study and identification of new species will be very important, along with an increase in the number of the fungal genome sequencing programs to determine their metabolic potential for biotechnology (Erjavec et al. 2012).

The obtention of "omics" data can increment the knowledge for the development of a more efficient strategy for the construction of a particular expression strain. To accomplish this, the gene(s) responsible for the synthesis of the desired product need(s) to be identified, and information about the control mechanism and functional properties (e.g., post-transcriptional modifications, toxicity) of the product is necessary for choosing the appropriate expression (e.g., inducible/constitutive promoters) and export system (e.g., with signal sequence/carrier protein). An appropriate transformation system has to be chosen depending on whether random insertions of multiple copies or a single site-specific integration is desired; whether the expression profile of the gene product needs to be analyzed and optimized, by means of adjusting the codon usage; and to determine the need for overproduction of rate-limiting cofactors, metabolites or secretion components, and/or silencing of undesired byproducts. In addition, modulating the interactions of these proteins with their target sites by creating 'hybrid' transcription factors or promoter elements will be especially useful in the design of novel expression cassettes capable of responding to particular growth conditions. The fact that several control sites from filamentous fungi are interchangeable amongst different fungal species increases the feasibility of creating modified promoters with elements derived from various fungal sources.

Development of genetically modified enzymes, chimeric and bi-functional proteins, using genetic engineering tools, has provided the physicochemical properties needed to improve the activity and decrease the cost of the production. The metabolic diversity and secretory capacity exhibited by filamentous fungi will continue to supply the biosynthetic tools for synthesis of a myriad of novel products. The new data from metabolic flux analysis, system biology, synthetic biology, "omics" and the identification of homologous genes from other organisms paves the way for resolving the bottlenecks and addressing the recombinant protein production in fungal cell factories.

References

Abrahão-Neto, J., C.H. Rossini, S. El-Gogary, F. Henrique-Silva, O. Crivellaro and H. El-Dorry. 1995. Mitochondrial functions mediate cellulase gene expression in *Trichoderma reesei*. Biochemistry 34(33): 10456–10462.

Alshahni, M.M., K. Makimura, T. Yamada, K. Takatori and T. Sawada. 2010. Nourseothricin acetyltransferase: a new dominant selectable marker for the dermatophyte *Trichophyton mentagrophytes*. Med. Mycol. 48(4): 665–668.

Amaro-Reyes, A., B.E. Garcia-Almendarez, D.G. Vazquez-Mandujano, S. Amaya-Llano, E. Castano-Tostado, R.G. Guevara-Gonzalez, O. Loera and C. Regalado. 2011. Homologue expression of a beta-xylosidase from native *Aspergillus niger*. J. Ind. Microbiol. Biotechnol. 38(9): 1311–1319.

AMFEP. 2009. List of Enzymes. *In:* Association of Manufacturers and Formulators of Enzyme Products (http://www.amfep.org).

Archer, D.B. 2000. Filamentous fungi as microbial cell factories for food use. Curr. Opin. Biotechnol. 11(5): 478–483.

Austin, B., R.M. Hall and B.M. Tyler. 1990. Optimized vectors and selection for transformation of *Neurospora crassa* and *Aspergillus nidulans* to bleomycin and phleomycin resistance. Gene 93(1): 157–162.

Baker, S.E., J. Thykaer, W.S. Adney, T.S. Brettin, F.J. Brockman, P. D'haeseleer, A. Martinez, R. Miller, D. Rokhsar C. Schadt et al. 2008. Fungal genome sequencing and bioenergy. Fungal Biology Reviews I(5): 5.

Bando, H., H. Hisada, H. Ishida, Y. Hata, Y. Katakura and A. Kondo. 2011. Isolation of a novel promoter for efficient protein expression by *Aspergillus oryzae* in solid-state culture. Appl. Microbiol. Biotechnol. 92(3): 561–569.

Bergquist, P., V. Te'o, M. Gibbs, A. Cziferszky, F.P. de Faria, M. Azevedo and H. Nevalainen. 2002. Expression of xylanase enzymes from thermophilic microorganisms in fungal hosts. Extremophiles 6(3): 177–184.

Berka, R.M. and C.C. Barnett. 1989. The development of gene expression systems for filamentous fungi. Biotechnol. Adv. 7(2): 127–154.

Berse, B., A. Dmochowska, M. Skrzypek, P. Weglenski, M.A. Bates and R.L. Weiss. 1983. Cloning and characterization of the ornithine carbamoyltransferase gene from *Aspergillus nidulans*. Gene 25(1): 109–117.

Bon, E. and M. Ferrara. 2007. Bioethanol Production via Enzymatic Hydrolysis of Cellulosic Biomass. The Role of Agricultural Biotechnologies for Production of Bioenergy in Developing Countries. Rome, Italy: FAO (http://www.fao.org/biotech/seminaroct 2007.htm).

Bouws, H., A. Wattenberg and H. Zorn. 2008. Fungal secretomes—nature's toolbox for white biotechnology. Appl Microbiol Biotechnol. 80(3): 381–388.

Callow, N.V. and L.K. Ju. 2012. Promoting pellet growth of *Trichoderma reesei* Rut C30 by surfactants for easy separation and enhanced cellulase production. Enzyme Microb. Technol. 50(6-7): 311–317.

Carle-Urioste, J.C., J. Escobar-Vera, S. El-Gogary, F. Henrique-Silva, E. Torigoi, O. Crivellaro, A. Herrera-Estrella, and H. El-Dorry. 1997. Cellulase induction in *Trichoderma reesei* by cellulose requires its own basal expression. J. Biol. Chem. 272(15): 10169–10174.

Cherry, J.R. and A.L. Fidantsef. 2003. Directed evolution of industrial enzymes: an update. Curr. Opin. Biotechnol. 14(4): 438–443.

Choi, J., J. Park, D. Kim, K. Jung, S. Kang and Y.H. Lee. 2010. Fungal secretome database: integrated platform for annotation of fungal secretomes. BMC Genomics 11: 105.

Conesa, A., C.A. van den Hondel and P.J. Punt. 2000. Studies on the production of fungal peroxidases in *Aspergillus niger*. Appl. Environ. Microbiol. 66(7): 3016–3023.

Cubero, B. and C. Scazzocchio. 1994. Two different, adjacent and divergent zinc finger binding sites are necessary for CREA-mediated carbon catabolite repression in the proline gene cluster of *Aspergillus nidulans*. EMBO J. 13(2): 407–415.

Dave, K. and N.S. Punekar. 2011. Utility of *Aspergillus niger* citrate synthase promoter for heterologous expression. J. Biotechnol. 155(2): 173–177.

De Lucas, J.R., S. Gregory and G. Turner. 1994. Analysis of the regulation of the *Aspergillus nidulans* acuD gene, encoding isocitrate lyase, by construction of a hybrid promoter. Mol. Gen. Genet. 243(6): 654–659.

De Ruiter-Jacobs, Y.M., M. Broekhuijsen, S.E. Unkles, E.I. Campbell, J.R. Kinghorn, R. Contreras, P.H. Pouwels and C.A. van den Hondel. 1989. A gene transfer system based on the homologous pyrG gene and efficient expression of bacterial genes in *Aspergillus oryzae*. Curr. Genet. 16(3): 159–163.

Dobrowolska, A. and P. Staczek. 2009. Development of transformation system for *Trichophyton rubrum* by electroporation of germinated conidia. Curr. Genet. 55(5): 537–542.

Domingues, F.C., J.A. Queiroz, J.M. Cabral and L.P. Fonseca. 2000. The influence of culture conditions on mycelial structure and cellulase production by *Trichoderma reesei* Rut C-30. Enzyme Microb. Technol. 26(5-6): 394–401.
Dowzer, C.E. and J.M. Kelly. 1991. Analysis of the creA gene, a regulator of carbon catabolite repression in *Aspergillus nidulans*. Mol. Cell Biol. 11(11): 5701–5709.
Driouch, H., R. Hansch, T. Wucherpfennig, R. Krull and C. Wittmann. 2012. Improved enzyme production by bio-pellets of *Aspergillus niger*: targeted morphology engineering using titanate microparticles. Biotechnol. Bioeng. 109(2): 462–471.
Erjavec, J., J. Kos, M. Ravnikar, T. Dreo and J. Sabotic. 2012. Proteins of higher fungi—from forest to application. Trends Biotechnol. 30(5): 259–273.
Fierro, F., F. Laich, R.O. Garcia-Rico and J.F. Martin. 2004. High efficiency transformation of *Penicillium nalgiovense* with integrative and autonomously replicating plasmids. Int. J. Food Microbiol. 90(2): 237–248.
Fincham, J.R. 1989. Transformation in fungi. Microbiol. Rev. 53(1): 148–170.
Fleischmann, R.D., M.D. Adams, O. White, R.A. Clayton, E.F. Kirkness, A.R. Kerlavage, C.J. Bult, J.F. Tomb, B.A. Dougherty, J.M. Merrick et al. 1995. Whole-genome random sequencing and assembly of *Haemophilus influenzae* Rd. Science 269(5223): 496–512.
Fleissner, A. and P. Dersch. 2010. Expression and export: recombinant protein production systems for *Aspergillus*. Appl. Microbiol. Biotechnol. 87(4): 1255–1270.
Fowler, T. and R.M. Berka. 1991. Gene expression systems for filamentous fungi. Curr. Opin. Biotechnol. 2(5): 691–697.
Fraser, C.M., J.D. Gocayne, O. White, M.D. Adams, R.A. Clayton, R.D. Fleischmann, C.J. Bult, A.R. Kerlavage, G. Sutton, J.M. Kelley et al. 1995. The minimal gene complement of *Mycoplasma genitalium*. Science 270(5235): 397–403.
Gangavaram, L.P., N. McHunu, P. Ramakrishnan, S. Singh and K. Permaul. 2009. Improved electroporation-mediated non-integrative transformation of *Thermomyces lanuginosus*. J. Microbiol. Methods 77(2): 159–164.
Goffeau, A., B.G. Barrell, H. Bussey, R.W. Davis, B. Dujon, H. Feldmann, F. Galibert, J.D. Hoheisel, C. Jacq, M. Johnston et al. 1996. Life with 6000 genes. Science 274(5287): 546: 563–547.
GOLD. Genomes Online Database. 2012. (http://www.genomesonline.org/ access 03.30.2012).
Gouka, R.J., J.G. Hessing, P.J. Punt, H. Stam, W. Musters and C.A. Van den Hondel. 1996. An expression system based on the promoter region of the *Aspergillus awamori* 1,4-beta-endoxylanase A gene. Appl. Microbiol. Biotechnol. 46(1): 28–35.
Grimm, L.H., S. Kelly, R. Krull and D.C. Hempel. 2005. Morphology and productivity of filamentous fungi. Appl Microbiol Biotechnol. 69(4): 375–384.
Gusakov, A.V. 2011. Alternatives to *Trichoderma reesei* in biofuel production. Trends Biotechnol 29(9): 419–425.
Gutierrez, S., J. Velasco, A.T. Marcos, F.J. Fernandez, F. Fierro, J.L. Barredo, B. Diez and J.F. Martin. 1997. Expression of the cefG gene is limiting for cephalosporin biosynthesis in *Acremonium chrysogenum*. Appl. Microbiol. Biotechnol. 48(5): 606–614.
Hamer, J.E. and W.E. Timberlake. 1987. Functional organization of the *Aspergillus nidulans* trpC promoter. Mol. Cell Biol. 7(7): 2352–2359.
Harkki, A., A. Mantyla, M. Penttila, S. Muttilainen, R. Buhler, P. Suominen, J. Knowles and H. Nevalainen. 1991. Genetic engineering of *Trichoderma* to produce strains with novel cellulase profiles. Enzyme Microb. Technol. 13(3): 227–233.
Honda, S. and E.U. Selker. 2009. Tools for fungal proteomics: multifunctional neurospora vectors for gene replacement, protein expression and protein purification. Genetics 182(1): 11–23.
Huge-Jensen, B., F. Andreasen, T. Christensen, M. Christensen, L. Thim and E. Boel. 1989. *Rhizomucor miehei* triglyceride lipase is processed and secreted from transformed *Aspergillus oryzae*. Lipids 24(9): 781–785.

Ilmen, M., M.L. Onnela, S. Klemsdal, S. Keranen and M. Penttila. 1996. Functional analysis of the cellobiohydrolase I promoter of the filamentous fungus *Trichoderma reesei*. Mol. Gen. Genet. 253(3): 303–314.

Ishida, H., Y. Hata, A. Kawato, Y. Abe and Y. Kashiwagi. 2004. Isolation of a novel promoter for efficient protein production in *Aspergillus oryzae*. Biosci. Biotechnol. Biochem. 68(9): 1849–1857.

Kapp, K., S. Schrempf, M.K. Lemberg and B. Dobberstein. 2009. Post-Targeting Functions of Signal Peptides. *In:* Protein Transport into the Endoplasmic Reticulum: Madame Curie Bioscience Database. (http://www.landesbioscience.com).

Kaup, B.A., K. Ehrich, M. Pescheck and J. Schrader. 2008. Microparticle-enhanced cultivation of filamentous microorganisms: increased chloroperoxidase formation by *Caldariomyces fumago* as an example. Biotechnol. Bioeng. 99(3): 491–498.

Kiiskinen, L.L., K. Kruus, M. Bailey, E. Ylosmaki, M. Siika-Aho and M. Saloheimo. 2004. Expression of *Melanocarpus albomyces* laccase in *Trichoderma reesei* and characterization of the purified enzyme. Microbiology 150(Pt 9): 3065–3074.

Kimura, M., T. Kamakura, Q.Z. Tao, I. Kaneko and I. Yamaguchi. 1994. Cloning of the blasticidin S deaminase gene (BSD) from *Aspergillus terreus* and its use as a selectable marker for *Schizosaccharomyces pombe* and *Pyricularia oryzae*. Mol. Gen. Genet. 242(2): 121–129.

Kitamoto, N., J. Matsui, Y. Kawai, A. Kato, S. Yoshino, K. Ohmiya and N. Tsukagoshi. 1998. Utilization of the TEF1-alpha gene (TEF1) promoter for expression of polygalacturonase genes, pgaA and pgaB, in *Aspergillus oryzae*. Appl. Microbiol. Biotechnol. 50(1): 85–92.

Kosalkova, K., C. Garcia-Estrada, C. Barreiro, M.G. Florez, M.S. Jami, M.A. Paniagua and J.F. Martin. 2012. Casein phosphopeptides drastically increase the secretion of extracellular proteins in *Aspergillus awamori*. Proteomics studies reveal changes in the secretory pathway. Microb. Cell Fact. 11: 5.

Kubodera, T., N. Yamashita and A. Nishimura. 2000. Pyrithiamine resistance gene (ptrA) of *Aspergillus oryzae*: cloning, characterization and application as a dominant selectable marker for transformation. Biosci. Biotechnol. Biochem. 64(7): 1416–1421.

Lander, E.S., L.M. Linton, B. Birren, C. Nusbaum, M.C. Zody, J. Baldwin, K. Devon, K. Dewar, M. Doyle, W. FitzHugh et al. 2001. Initial sequencing and analysis of the human genome. Nature 409(6822): 860–921.

Liu, T., T. Wang, X. Li and X. Liu. 2008. Improved heterologous gene expression in *Trichoderma reesei* by cellobiohydrolase I gene (cbh1) promoter optimization. Acta Biochim. Biophys. Sin (Shanghai) 40(2): 158–165.

Lubertozzi, D. and J.D. Keasling. 2009. Developing *Aspergillus* as a host for heterologous expression. Biotechnol. Adv. 27(1): 53–75.

Lv, D., W. Wang and D. Wei. 2012. Construction of two vectors for gene expression in *Trichoderma reesei*. Plasmid 67(1): 67–71.

Ma, L., J. Zhang, G. Zou, C. Wang and Z. Zhou. 2011. Improvement of cellulase activity in *Trichoderma reesei* by heterologous expression of a beta-glucosidase gene from *Penicillium decumbens*. Enzyme Microb. Technol. 49(4): 366–371.

MacKenzie, D.A., D.J. Jeenes, N.J. Belshaw and D.B. Archer. 1993. Regulation of secreted protein production by filamentous fungi: recent developments and perspectives. J. Gen. Microbiol. 139(10): 2295–2307.

MacRae, W.D., F.P. Buxton, D.I. Gwynne and R.W. Davies. 1993. Heterologous protein secretion directed by a repressible acid phosphatase system of *Aspergillus niger*. Gene 132(2): 193–198.

Maras, M., I. van Die, R. Contreras and C.A. van den Hondel. 1999. Filamentous fungi as production organisms for glycoproteins of bio-medical interest. Glycoconj J. 16(2): 99–107.

Michielse, C.B., P.J. Hooykaas, C.A. van den Hondel and A.F. Ram. 2005. Agrobacterium-mediated transformation as a tool for functional genomics in fungi. Curr. Genet. 48(1): 1–17.

Miettinen-Oinonen, A., M. Paloheimo, R. Lantto and P. Suominen. 2005. Enhanced production of cellobiohydrolases in *Trichoderma reesei* and evaluation of the new preparations in biofinishing of cotton. J. Biotechnol. 116(3): 305–317.

Moralejo, F.J., R.E. Cardoza, S. Gutierrez and J.F. Martin. 1999. Thaumatin production in *Aspergillus awamori* by use of expression cassettes with strong fungal promoters and high gene dosage. Appl. Environ Microbiol. 65(3): 1168–1174.

Mustalahti, E., M. Saloheimo and J.J. Joensuu. 2011. Intracellular protein production in *Trichoderma reesei* (*Hypocrea jecorina*) with hydrophobin fusion technology. N. Biotechnol.

Nevalainen, H., P. Suominen and K. Taimisto. 1994. On the safety of *Trichoderma reesei*. J. Biotechnol. 37(3): 193–200.

Nevalainen, K.M., V.S. Te'o and P.L. Berquist. 2005. Heterologous protein expression in filamentous fungi. Trends Biotechnol 23 (9): 468–474.

Ohno, A., J. Maruyama, T. Nemoto, M. Arioka and K. Kitamoto. 2011. A carrier fusion significantly induces unfolded protein response in heterologous protein production by *Aspergillus oryzae*. Appl. Microbiol. Biotechnol. 92(6): 1197–1206.

Olempska-Beer, Z.S., R.I. Merker, M.D. Ditto and M.J. DiNovi. 2006. Food-processing enzymes from recombinant microorganisms—a review. Regul Toxicol. Pharmacol. 45(2): 144–158.

Orbach, M.J., E.B. Porro and C. Yanofsky. 1986. Cloning and characterization of the gene for beta-tubulin from a benomyl-resistant mutant of *Neurospora crassa* and its use as a dominant selectable marker. Mol. Cell Biol. 6(7): 2452–2461.

Pachlinger, R., R. Mitterbauer, G. Adam and J. Strauss. 2005. Metabolically independent and accurately adjustable *Aspergillus* sp. expression system. Appl. Environ. Microbiol. 71(2): 672–678.

Palamarczyk, G., M. Maras, R. Contreras and J. Kruszewska. Protein secretion and glycosylation in *Trichoderma*. *In*: C. Kubicek and G. Harman [eds.]. 1998. Trichoderma & Gliocladium, Basic biology, taxonomy and genetics. Taylor & Francis Inc., 1900 Frost Road, Suite 101, Bristol, PA 19007, USA.

Paloheimo, M., A. Mantyla, J. Kallio, T. Puranen and P. Suominen. 2007. Increased production of xylanase by expression of a truncated version of the xyn11A gene from *Nonomuraea flexuosa* in *Trichoderma reesei*. Appl. Environ. Microbiol. 73(10): 3215–3224.

Poidevin, L., A. Levasseur, G. Paes, D. Navarro, S. Heiss-Blanquet, M. Asther and E. Record. 2009. Heterologous production of the Piromyces equi cinnamoyl esterase in *Trichoderma reesei* for biotechnological applications. Lett Appl. Microbiol. 49(6): 673–678.

Portnoy, T., A. Margeot, R. Linke, L. Atanasova, E. Fekete, E. Sandor, L. Hartl, L. Karaffa, I.S. Druzhinina, B. Seiboth et al. 2011. The CRE1 carbon catabolite repressor of the fungus *Trichoderma reesei*: a master regulator of carbon assimilation. BMC Genomics 12: 269.

Punt, P.J., R.P. Oliver, M.A. Dingemanse, P.H. Pouwels and C.A. van den Hondel. 1987. Transformation of *Aspergillus* based on the hygromycin B resistance marker from *Escherichia coli*. Gene 56(1): 117–124.

Punt, P.J., N.D. Zegers, M. Busscher, P.H. Pouwels and C.A. van den Hondel. 1991. Intracellular and extracellular production of proteins in *Aspergillus* under the control of expression signals of the highly expressed *Aspergillus nidulans* gpdA gene. J. Biotechnol. 17(1): 19–33.

Punt, P.J., N. van Biezen, A. Conesa, A. Albers, J. Mangnus and C. van den Hondel. 2002. Filamentous fungi as cell factories for heterologous protein production. Trends Biotechnol. 20(5): 200–206.

Qin, L.N., F.R. Cai, X.R. Dong, Z.B. Huang, Y. Tao, J.Z. Huang and Z.Y. Dong. 2012. Improved production of heterologous lipase in *Trichoderma reesei* by RNAi mediated gene silencing of an endogenic highly expressed gene. Bioresour Technol. 109: 116–122.

Radzio, R. and U. Kuck. 1997. Synthesis of biotechnologically relevant heterologous proteins in filamentous fungi. Process Biochemistry 32(6): 11.

Rambosek, J. and J. Leach. 1987. Recombinant DNA in filamentous fungi: progress and prospects. Crit. Rev. Biotechnol. 6(4): 357–393.

Rodriguez, R.J. and O.C. Yoder. 1987. Selectable genes for transformation of the fungal plant pathogen *Glomerella cingulata f. sp. phaseoli* (*Colletotrichum lindemuthianum*). Gene 54(1): 73–81.

Roth, A.H. and P. Dersch. 2010. A novel expression system for intracellular production and purification of recombinant affinity-tagged proteins in *Aspergillus niger*. Appl. Microbiol. Biotechnol. 86(2): 659–670.

Saloheimo, M. and T.M. Pakula. 2012. The cargo and the transport system: secreted proteins and protein secretion in *Trichoderma reesei* (*Hypocrea jecorina*). Microbiology 158(Pt 1): 46–57.

Schuster, A. and M. Schmoll. 2010. Biology and biotechnology of *Trichoderma*. Appl. Microbiol. Biotechnol. 87(3): 787–799.

Selinheimo, E., M. Saloheimo, E. Ahola, A. Westerholm-Parvinen, N. Kalkkinen, J. Buchert and K. Kruus. 2006. Production and characterization of a secreted, C-terminally processed tyrosinase from the filamentous fungus *Trichoderma reesei*. FEBS J. 273(18): 4322–4335.

Sharma, R., M. Katoch, P. Srivastava and G. Qazi. 2009. Approaches for refining heterologous protein production in filamentous fungi. World J. Microbiol. Biotechnol. 25: 12.

Sharma, R., M. Katoch, N. Govindappa, P.S. Srivastava, K.N. Sastry and G.N. Qazi. 2012. Evaluation of the catalase promoter for expressing the alkaline xylanase gene (alx) in *Aspergillus niger*. FEMS Microbiol. Lett. 327(1): 33–40.

Shimizu, T., T. Ito and S. Kanematsu. 2012. Transient and multivariate system for transformation of a fungal plant pathogen, *Rosellinia necatrix*, using autonomously replicating vectors. Curr. Genet.

Skinner, W., A. Bailey, A. Renwick, J. Keon, S. Gurr and J. Hargreaves. 1998. A single amino-acid substitution in the iron-sulphur protein subunit of succinate dehydrogenase determines resistance to carboxin in *Mycosphaerella graminicola*. Curr. Genet. 34(5): 393–398.

Smith, T.L., J. Gaskell, R.M. Berka, M. Yang, D.J. Henner and D. Cullen. 1990. The promoter of the glucoamylase-encoding gene of *Aspergillus niger* functions in *Ustilago maydis*. Gene 88(2): 259–262.

Stapleton, P.C. and A.D. Dobson. 2003. Carbon repression of cellobiose dehydrogenase production in the white rot fungus *Trametes versicolor* is mediated at the level of gene transcription. FEMS Microbiol. Lett. 221(2): 167–172.

Storms, R., Y. Zheng, H. Li, S. Sillaots, A. Martinez-Perez and A. Tsang. 2005. Plasmid vectors for protein production, gene expression and molecular manipulations in *Aspergillus niger*. Plasmid. 53(3): 191–204.

Strauss, J., R.L. Mach, S. Zeilinger, G. Hartler, G. Stoffler, M. Wolschek and C.P. Kubicek. 1995. Cre1, the carbon catabolite repressor protein from *Trichoderma reesei*. FEBS Lett 376(1-2): 103–107.

Streatfield, S.J., S. Toews and C.F. Roberts. 1992. Functional analysis of the expression of the 3′-phosphoglycerate kinase pgk gene in *Aspergillus nidulans*. Mol. Gen. Genet. 233(1-2): 231–240.

Sun, J. and N.L. Glass. 2011. Identification of the CRE-1 cellulolytic regulon in *Neurospora crassa*. PLoS One 6(9): e25654.

Tada, S., K. Gomi, K. Kitamoto, C. Kumagai, G. Tamura and S. Hara. 1991. Identification of the promoter region of the Taka-amylase A gene required for starch induction. Agric. Biol. Chem. 55(7): 1939–1941.

Takashima, S., A. Nakamura, M. Hidaka, H. Masaki and T. Uozumi. 1998. Isolation of the creA gene from the cellulolytic fungus *Humicola grisea* and analysis of CreA binding sites upstream from the cellulase genes. Biosci. Biotechnol. Biochem. 62(12): 2364–2370.

Tao, L., H.J. Dong, X. Chen, S.F. Chen and T.H. Wang. 2008. Expression of ethylene-forming enzyme (EFE) of *Pseudomonas syringae pv. glycinea* in *Trichoderma viride*. Appl. Microbiol. Biotechnol. 80(4): 573–578.

Te'o, V.S., P.L. Bergquist and K.M. Nevalainen. 2002. Biolistic transformation of *Trichoderma reesei* using the Bio-Rad seven barrels Hepta Adaptor system. J. Microbiol. Methods 51(3): 393–399.

Thompson, C.J., N.R. Movva, R. Tizard, R. Crameri, J.E. Davies, M. Lauwereys and J. Botterman. 1987. Characterization of the herbicide-resistance gene bar from *Streptomyces hygroscopicus*. EMBO J. 6(9): 2519–2523.

Todaka, N., C.M. Lopez, T. Inoue, K. Saita, J. Maruyama, M. Arioka, K. Kitamoto, T. Kudo and S. Moriya. 2010. Heterologous expression and characterization of an endoglucanase from a symbiotic protist of the lower termite, *Reticulitermes speratus*. Appl Biochem Biotechnol 160(4): 1168–1178.

Van den Hombergh, J.P., P.J. van de Vondervoort, L. Fraissinet-Tachet and J. Visser. 1997. *Aspergillus* as a host for heterologous protein production: the problem of proteases. Trends Biotechnol. 15(7): 256–263.

Vautard, G., P. Cotton and M. Fevre. 1999. The glucose repressor CRE1 from *Sclerotinia sclerotiorum* is functionally related to CREA from *Aspergillus nidulans* but not to the Mig proteins from *Saccharomyces cerevisiae*. FEBS Lett. 453(1-2): 54–58.

Venter, J.C., M.D. Adams, E.W. Myers, P.W. Li, R.J. Mural, G.G. Sutton, H.O. Smith, M. Yandell, C.A. Evans, R.A. Holt et al. 2001. The sequence of the human genome. Science 291(5507): 1304–1351.

Ward, O.P. 2011. Production of recombinant proteins by filamentous fungi. Biotechnol Adv.

Waring, R.B., G.S. May and N.R. Morris. 1989. Characterization of an inducible expression system in *Aspergillus nidulans* using alcA and tubulin-coding genes. Gene 79(1): 119–130.

Wiebe, M.G. 2003. Stable production of recombinant proteins in filamentous fungi—problems and improvements. Mycologist 17(3): 5.

Yamashita, N., Y. Komori, Y. Okumura, K. Uchiya, T. Matsui, A. Nishimura, K. Ogawa and T. Nikai. 2011. High-yields heterologous production of the novel *Aspergillus fumigatus* elastase inhibitor AFUEI in *Aspergillus oryzae*. J. Biosci. Bioeng. 112(2): 114–117.

Yoon, J., S. Kimura, J. Maruyama and K. Kitamoto. 2009. Construction of quintuple protease gene disruptant for heterologous protein production in *Aspergillus oryzae*. Appl. Microbiol. Biotechnol. 82(4): 691–701.

Zhang, Z. and W.J. Henzel. 2004. Signal peptide prediction based on analysis of experimentally verified cleavage sites. Protein Sci. 13(10): 2819–2824.

Zhong, Y., X. Liu, P. Xiao, S. Wei and T. Wang. 2011. Expression and secretion of the human erythropoietin using an optimized cbh1 promoter and the native CBH I signal sequence in the industrial fungus *Trichoderma reesei*. Appl. Biochem. Biotechnol. 165(5-6): 1169–1177.

Ziv, C., R. Gorovits and O. Yarden. 2008. Carbon source affects PKA-dependent polarity of *Neurospora crassa* in a CRE-1-dependent and independent manner. Fungal Genet Biol. 45(2): 103–116.

Zou, G., S. Shi, Y. Jiang, J. van den Brink, R.P. de Vries, L. Chen, J. Zhang, L. Ma, C. Wang and Z. Zhou. 2012. Construction of a cellulase hyper-expression system in *Trichoderma reesei* by promoter and enzyme engineering. Microb. Cell Fact. 11(1): 21.

16

Lipases: Imperative Fat-degrading Enzymes

Fernanda Dell Antonio Facchini,[1,a,*]
Marita Gimenez Pereira[2] and *Ana Claudia Vici*[1,b]

Introduction

All lipases are members of the α/β hydrolase fold super-family and are defined as glycerol ester hydrolases that act on acylglycerols to release fatty acids and glycerol. Lipases can be divided into two main groups: (i) Carboxylesterases (EC 3.1.1.1) and (ii) "true" lipases or triacilglicerol lipases (EC 3.1.1.3), which differ in several biochemical features (Messaoudi et al. 2010).

The distinction between lipases and esterases has been studied by several authors and is still controversial, as esterase—defined as carboxylic ester hydrolases which catalyze the esters hydrolysis—is often difficult to differentiate from lipase. The difference between lipase and esterase is that the first catalyzes reaction of water-insoluble substrates, while an esterase

[1]Departamento de Bioquímica e Imunologia, Faculdade de Medicina de Ribeirão Preto, Universidade de São Paulo.
[a]E-mail: fer_facchini@yahoo.com.br
[b]E-mail: acvici@yahoo.com.br
[2]Departamento de Biologia, Faculdade de Filosofia, Ciências e Letras de Ribeirão Preto, Universidade de São Paulo. Av. Bandeirantes, 3900 Monte Alegre, Ribeirão Preto, Brasil.
E-mail: maritagimenez@hotmail.com
*Corresponding author

acts on soluble substrates. According to Sarda and Desnuelle (1958) and Alvarez-Macarie (1999), lipases are activated in the presence of emulsified esters (in presence of a lipid/water interface), although esterases do not show this activation and exercises it hydrolytic capacity on water-soluble substrates following the kinetic model of Michaelis-Menten.

Moreover, the difference between lipases and esterases can also be proved through the specific substrate preference of both enzymes. The natural substrates for lipase are oils and fats containing triacylglycerols, generally constituted by a long chain of fatty acids, while the esterase substrates act on ester bonds releasing fatty acids with low molecular weight (Bier 1955, Brockman 1984) and, although esterases have structures very similar to lipases, they exhibit a motif sequence Gly-X-X-Leu which is not detected in lipases (Bornschewer 2002, Reis 2009). However, the majority of lipases are able to hydrolyze esterases substrates, but the opposite is not true (Jaeger et al. 1999).

Since 1990, after determining the three dimensional lipase structure (Brady et al. 1990, Winkler et al. 1990, Schrag et al. 1991), two factors became the basis of classification: one is that it should contain a kind of "lid", which is known as a protein surface loop that covers the enzyme active site and is directly related with the "interfacial activation", that is the second criteria. This interfacial activation promotes a conformational change in enzyme structure causing the lid movement on contact with this interface, exposing the active site and activating the "true" lipases. Its activity should sharply increase as soon as the triglyceride substrate forms an emulsion (Sarda and Desnuelle 1958, Jaeger et al. 1999). This criteria is worth considering but is inappropriate, mainly because in a large number of cases lipases have a lid but do not exhibit interfacial activation (Verger 1997).

Consequently, lipases constitute a special class of carboxylesterases which hydrolyzes a long chain of acylglicerols. The standard substrates to lipase are glycerolesters with an acyl chain length higher than 10 carbon atoms, while esterase standard substrates are glycerolesters with an acyl chain length lower than 10 carbon atoms (Jaeger et al. 1994, 1999, Egloff et al. 1995).

Furthermore, another enzyme which could be confused with lipase is cutinase. Cutinase is a serine esterase, which hydrolyzes cutin, an insoluble polyester which covers the surface of plants; however cutinase is also able to hydrolyze fatty acid esters and emulsified triacylglycerol as efficiently as lipases. The main difference between both enzymes is that cutinase does not exhibit interfacial activation. According to Martinez et al. (1992), for that reason, cutinases can be considered as a bridge between "true esterases" and "true lipases", since they have the capabilities of both families: it is an efficient catalyst both in solution and at the lipidic interface. These enzymes have been applied as a lipolytic enzyme in a wide variety of industrial processes such as laundry and dishwashing detergents to remove fats (Melo et al. 2003).

Lipase Structure

Three-dimensional structures of several lipases showed that this enzyme shares a structural motif, the α/β hydrolase fold. This motif has been found in both eukaryotes and prokaryotes lipases and is common in others enzymes as many esterases and peptidases. The canonical α/β hydrolase fold consists of eight β-strands surrounded by six α-helices (Fig. 16.1) (Ollis et al. 1992, Schrag and Cygler 1997, Holmquist 2000, Pouderoyen et al. 2001). Some lipases have variations in this basic structure such as *Bacillus subtilis* lipase. This hydrolase has a minimal α/β hydrolase fold with six β-strands in a parallel β-sheet and five α-helices (Pouderoyen et al. 2001).

Lipases and esterases possess a consensus pentapeptide GXSXG with S as the catalytic nucleophile serine in active site. This serine is positioned in C-terminal portion of β5 strand (Jaeger et al. 1994, Schrag and Cygler 1997, Jaeger et al. 1999, Akoh et al. 2004). The α/β hydrolases superfamily has a catalytic triad Ser, His and Asp and lipases, in general, have Ser, His and Asp/Glu residues in its active site, which forms a catalytic triad comparable in arrangement to those of serine proteases (Jaeger et al. 1999, Pouderoyen et al. 2001). Evidence of this catalytic mechanism similar to serine proteases

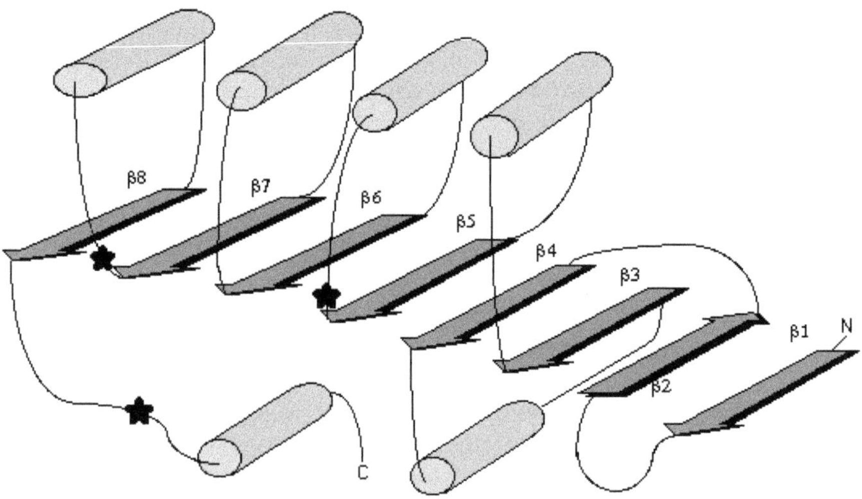

Figure 16.1 Canonical α/β hydrolase fold.α helices and β strands are represented by light gray cylinders and dark gray arrows, respectively. The location of the catalytic triad is indicated by black stars (modified from Ollis et al. 1992).

was confirmed by lipase inhibitors studies and structural analysis of these proteins (Jaeger et al. 1999).

In the catalytic site the triad Ser, His and Asp/Glu is positioned such that the carboxylated residues (Asp or Glu) makes hydrogen bound with a nitrogen in the His ring. Consequently, another nitrogen in the His ring is positioned in such a way to make hydrogen bonds with the hydroxyl group of Ser, which keeps the Ser hydrogen relatively far from its covalent oxygen (Schrag and Cygler 1997, Jaeger and Reetz 1998).

In triacylglycerol hydrolysis the reaction starts with a nucleophilic attack by Ser oxygen on the carbonyl carbon atom of the lipid ester bond. A tetrahedral intermediate is formed where –O is stabilized by interaction with two –NH group; the imidazole His ring is protonated—positively charged —and Asp/Glu negative charge stabilizes it. The oxygen of the ester linkage is protonated and a fatty acid is formed and becomes free. A diacylglycerol remains covalently bonded to Ser oxygen and must be removed (Jaeger et al. 1994, Bornscheuer and Kazlauskas 1999, Bornscheuer 2002).

The second step involves the removal of enzyme diacylglycerol. For that, a water molecule is activated by His residue and a result in a hydroxyl ion performs a nucleophilic attack at the carbonyl carbon atom of the acyl-lipase complex. His residue provides a proton to the Ser oxygen atom breaking the ester bond between Ser and substrate, releasing diacylglycerol and regenerating the lipase. The next hydrolysis reaction occurs to catalyze the newly formed diglyceride into a fatty acid and a monoglyceride and, after, another fatty acid and glycerol (Jaeger et al. 1994, Bornscheuer and Kazlauskas 1999, Bornscheuer 2002).

Studies with lipases three-dimensional structures, including bacteria, fungi and humans lipases, showed the presence of an α-helix loop which forms a "lid". This "lid" is not presented in all lipases, but when it is present the chain with His catalytic triad extends over the catalytic site as a covering lid and making it inaccessible to substrate (Jaeger et al. 1994, Jaeger et al. 1999, Schrag and Cygler 1997, Brzozowski et al. 2000). The mobility degree of the lid varies a lot. In some cases there are apparently two states, open and closed, with energy levels significantly lower than transition states; and the position of the lid is very mobile, with many intermediate states with an energy level comparable to the closed and opened state (Louwrier et al. 1996, Gonzαlez-Navarro et al. 2001). The lipase surface is essentially hydrophilic when the lid is closed. But when the lid is opened conformational change exposes a large hydrophobic surface which allows the site active access for the substrate (Fig. 16.2) (Cygler and Schrag 1997, González-Navarro et al. 2001, Miled et al. 2003).

The lid is present in different lipases family always upon the active site, so it is believed that the opening and closing mechanism is involved in the catalytic lipase mechanism (Cygler and Schrag 1997). The low lipase activity

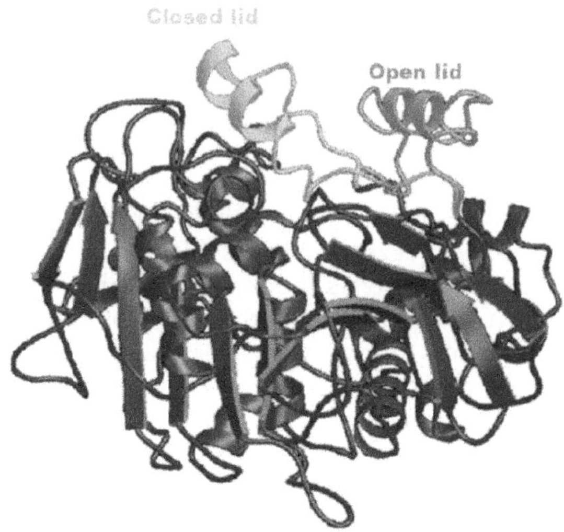

Figure 16.2 Three-dimensional structure of the *Candida rugosa* lipase showing overlapping closed and open lid forms (Grochulski et al. 1994).

in aqueous medium without interfaces suggests that closed conformation is predominant in this condition—there are open structures or at least transient forms. In contrast, in water-oil interface the lid could interact with this interface undergoing a conformational change, opening up and exposing the active. This characteristic is called interfacial activation and is necessary for maximum activation of most true lipases (Jaeger et al. 1994, Schrag and Cygler 1997, Jaeger and Reetz 1998, González-Navarro et al. 2001, Freire and Castilho 2008).

In addition, some lipases may be in dynamic equilibrium between an open and closed form when in solution, and organic solvents facilitate the lid opening probably by reducing the medium dielectric constant. However, the lipid-protein interaction is complex and can not be explained by simple kinetic models (Brzozowski et al. 2000).

The interfacial activation phenomenon is a lipase characteristic that differ from esterases. In water-lipid interface the lid opening and, consequently, the lipase activation would be favored by hydrophobic interaction between the "lid" with the hydrophobic substrates which form aggregates or micelles. Therefore, the "lid" side is the face of the lipase surface which is directed toward the lipid layer. Some biophysical data such as lipid monolayers experiments indicate that the adsorbed lipase to the interface and probably penetrates partially in the lipid layer (González-Navarro et al. 2001, Freire and Castilho 2008). The lipase catalysis, when interfacial activation is necessary, depends strongly on the organization of the substrate present in interfacial structures such as membrane bilayers, monolayers, micelles, vesicles, and water-lipid emulsions (Aloulou et al. 2006).

However Ferrato et al. (1997) showed that neither the lid nor the interfacial activation are characteristics of all lipases. Nevertheless, literature data shows that most lipases undergoes profound conformational changes in interfacial activation, and the lid moves leaving the enzyme in active form exposing the catalytic site and providing a large hydrophobic surface for protein-lipid interaction (Brzozowski et al. 2000, Aloulou et al. 2006).

Reaction Catalyzed by Lipases

Lipases are enzymes which catalyze a large variety of reactions. The most commonly known is the triacylglycerols hydrolysis (Fig. 16.3), providing diacylglycerols, monoacylglycerols, glycerol and free fatty acids (Jaeger et al. 1999). As a consequence of the lipase ability to hydrolyze other esters besides glycerides, has made it valuable in organic synthesis. Additionally, some lipases from different origins can also accept thioesters and hydrolyses esters of chiral alcohols more willingly than esters of chiral acids. The lipase hydrolytic activity can be directly related to its synthesis activity, but is independent of its interesterification activity. It is worth mentioning that lipases from different sources are able to catalyze the same reaction, but

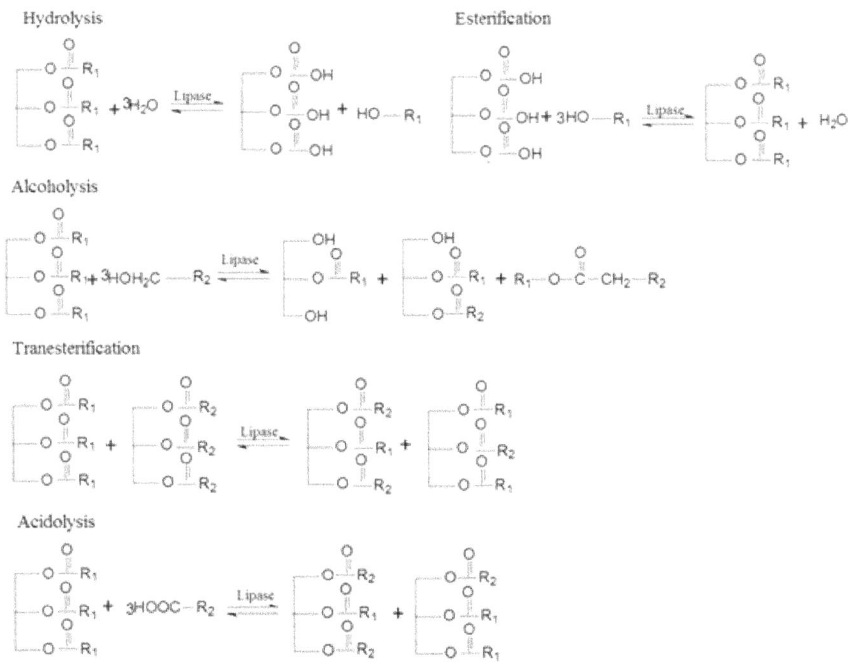

Figure 16.3 Reactions catalyzed by lipases.

they differ in the performance under the same reaction conditions (Faber 1997, Yahya et al. 1998).

According to Tsujisaka et al. (1977) the triacylglycerols hydrolysis reaction can be reversed in the direction of the esterification reaction by controlling medium water content, which could be a complex procedure (Fig. 16.3). Consequently, it also becomes possible to use lipases as interesterification biocatalysts of oils and greases (Quinlan and Moore 1993). Water is used as a solvent in biotransformations and affects the productivity of esterification reactions through the chemical equilibrium, consequently the use of organic solvent overcomes these problems while improving the enzyme specificity. The esterification reaction changes the composition and distribution of fatty acids in the triacylglycerol molecule. Among the processes catalyzed by lipases in organic medium, the synthesis of fatty acid esters appears as a very promising aspect due to their biotechnological importance in several esters in daily and in the industry (Jaeger et al. 1994).

In general, the interesterification reaction refers to the exchange of the acyl radical between an ester and an acid (acidolysis), an ester and an alcohol (alcoholysis) or between two esters (Fig. 16.3). The last reaction is known as transesterification, as described by some authors. The potential of lipase catalyzes esterification and transesterification reactions in organic media is industrially important and therefore the production of esters by a biocatalytic route presents itself as an attractive alternative (Fernandes et al. 2007, Bancerz and Ginalska 2007, Sulaiman et al. 2007, Bisen et al. 2010).

Lately, transesterification reactions have appeared as an attractive option, mainly due to the search for clean and renewable energy, such as biodiesel. In this reaction, three alcohol molecules, preferably with short carbon chain (ethanol or methanol), react sequentially in the presence of lipases with a triglyceride molecule, providing firstly a diglyceride, subsequently a monoglyceride and finally, a glycerol and three monoester molecules (biodiesel). All steps are reversible and the total equilibrium promotes the products formation rate. Lipases catalyzed procedures do not produce side reactions, as the majority of enzymes reactions do. However they are applied to a limited extent due to cost-intensive nature of pure enzymes (Freedman et al. 1984, Fukuda et al. 2001, Rottig et al. 2010).

Other solvents besides water and alcohol may be used as nucleophiles in reactions catalyzed by lipases. Consequently, they can take part in lactonization reactions (synthesis of lactones), amides synthesis, oxymolisis and tio-tranesterifications in organic solvents with high selectivity (Faber 1997).

Biocatalysis reactions in anhydrous organic solvents have been used to obtain numerous lactones due to its easy purification and its stereoselectivity applied to many cases. The lactone synthesizes the hidroxyacids or hidroxiesters are catalyzed by these enzymes and via cyclization by intramolecular esterification or acyl transfer reactions the lactones are obtained. The product depends not singly on the length of the hydroxyl acid, but also on the type of lipase, the solvent, dilution and even on the temperature. Nevertheless, the most important synthetic aspect and enzymatic lactone formation is the possibility of directing the condensation reaction towards the formation of macrocyclic lactones and dilactones. In amide synthesis, when N-nucleophiles such as ammonia, amine or hydrazine are exposed to acyl-transfer reactions, the corresponding N-acyl detivatives are formed via interception of the acyl enzyme intermediate by N-nucleophile (Faber 1997). The use of lipases in esters aminolysis in anhydrous medium has been successful in the peptide synthesis and fatty acid amides synthesis (Klazlaukas and Bornsheuer 1998).

Lipases Classification

As indicated by literature, the lipase specificity is controlled by the enzyme molecular properties, the substrate structure and by factors which affect the enzyme-substrate binding (Jensen 1983).

According to the enzyme classification, lipases can be divided as follows (Macrae and Hammond 1985, Faber 1997).

1) **Regioselectivity:** (i) Specific sn-1,3- lipase: catalyze the release of fatty acids of specific sn-1 and sn-3 glycerides positions. The total hydrolysis can occur if the reaction is prolonged, considering the formation of 2-monoacylglicerols or 1.2;(2.3)-diacylglycerols, chemically unstable, can be isomerized spontaneously to 1 or 3 positions. (ii) Non-specific lipase: catalyze the hydrolysis of triacylglycerols to free fatty acid and glycerol randomly. Do not show specificity to the nature of the acyl group or to the position in which it is esterified to glycerol. (iii) Fatty acid specific lipases: catalyze the hydrolysis of specific types of acyl groups in triacylglycerol molecules. This capacity of regioselectively is particularly important in the synthesis of biologically and industrially interesting carbohydrates, where selective protection and unprotection of hydroxyl groups is the central problem.
2) **Substrate selectivity:** (i) to the different hydrolysis rates under triacylglycerols, diacylglycerols or monoacylglycerols (ii) to the fatty acids making up the substrate relative to the carbon chain size and/ or to the number of unsaturation.
3) **Stereoespecificity:** discriminates between enantiomers in the case of racemic substrates. In these reactions, both the enantiomeric substrates or the two forms of mirror-image orientation of a pro-chiral substrate involving its enantropic groups compete for the enzymatic atalytic site.

   The combination of all these types is also observed.

Lipase Kinetics and Interfacial Activation

Lipases are very interesting enzymes and exhibit a low hydrolytic activity when their substrates are dissolved in an aqueous medium and are monomers. The physicochemical interaction with different substrates is an important characteristic of lipases. In contrast to the esterases and some others enzymes, lipases are more active in heterogeneous systems and do not show a normal Michaelis-Menten activity, considering that this is valid for homogeneous systems. It displays almost no activity as long as the substrate is in a dissolved monomeric state (Fig. 16.4). However when the substrate concentration is progressively enhanced beyond its solubility limit, a sharp increase in lipase-activity takes place. An increase in activity is pronounced when the substrate concentration reaches the solubility limit (the "critical micelle concentration" or CMC), making aggregates or micelles. The fact that lipases display a high activity beyond CMC has been called interfacial activation. This characteristic of lipases makes the determination of kinetic parameters difficults, due to the complexity of these reactions that occur in heterogeneous systems, such as emulsions (Brockman 1984, Jaeger et al. 1994, Faber 1997, Holmquist 2000).

The molecular justification for this phenomenon has been shown to be a rearrangement process within the lipase. The "inactive" state resides in a freely dissolved lipase in the absence of an aqueous-lipid interface [E] in which the enzyme is covered by the "lid". However, when the lipase is exposed to this aqueous-lipid interface, the lid is folded back and opens the lipase active site constituting the active state [E*]. As a result, the hydrolysis catalyzed by lipases should be performed in a biphasic medium. It is sufficient to use the substrate alone at elevated concentrations,

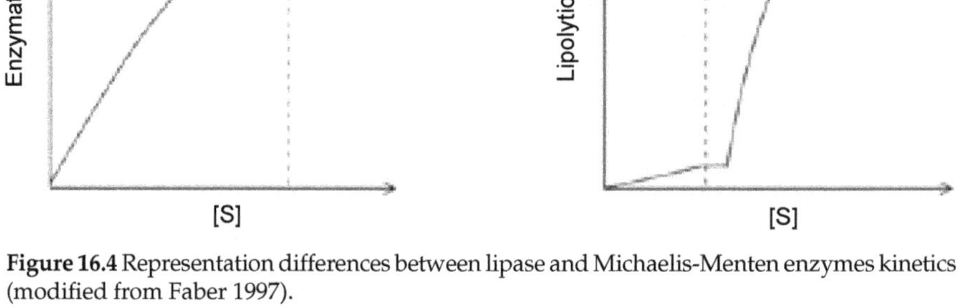

Figure 16.4 Representation differences between lipase and Michaelis-Menten enzymes kinetics (modified from Faber 1997).

constituting the second organic phase, or, the substrate may be dissolved in a water-immiscible organic solvent such as hexane, isopropanol or ether (Faber 1997).

To explain how the lipase/substrate interaction occurs, a model was suggested by Verger et al. (1973). There are two balances in the reaction equation, and the first is between the inactive soluble enzyme [E] and the active enzyme in lipid-water interface [E*]. Second there is another balance between the amount of activated enzyme [E*] in the interface and the amount of complex enzyme/substrate [ES*]. The catalysis occurs then, regenerating the enzyme in the active form [E*] and releasing the product:

$$[E] \overset{\text{Interface}}{\leftrightarrow} [E^*] \overset{+S}{\leftrightarrow} [E^*S] \rightarrow [E^*] + P$$

In another case, Sarda and Desnuelle (1958) revealed that lipase kinetics in emulsions depends on the interface, and this kinetic follows the Michaelis-Menten model, only if the term "substrate concentration" was replaced by "interfacial area". Since then, many methods have been developed based on diverse techniques such as microscopic or lipid monolayer to study the lipase kinetics (Verger and Haas 1976, Wang et al. 1988).

According to the literature, some lipases showed kinetics which follow a similar pattern to the Michaelis-Menten, with the rate reaction proportional to the increase in substrate concentration in the system. To distinguish the classical Michaelis-Menten kinetic, constants are called "apparent", whereas the substrate is not soluble in the system and the enzyme acts at the interface (Guit et al. 1991, Lee et al. 2004).

Lipase Sources

Lipases are ubiquitous in nature and are produced by several plants, microorganisms and mammals. Microbial lipases, such as bacteria, fungi and yeast, are the most widely used enzymes in the biotechnological applications in organic chemistry. This is because microbial enzymes are usually more stable than those derived from animals and plants, making the lipase production more convenient, safer, easier, inexpensive and generally stable in organic solvents. Moreover, this source of lipase has versatility on its properties with regard to the enzymatic activity, to the substrate specificity and to their ease of mass production (Hasan et al. 2006, Singh and Mukhopadhyay 2012).

Most fungal lipases have commercial importance, including lipases from: *Rhizopus* sp., *Penicillium* sp., *Aspergillus* sp., *Mucor* sp., *Ashbya* sp., *Geotrichum* sp., *Beauveria* spp., *Humicola* sp., *Rhizomucor* sp., *Fusarium* sp., *Acremonium* sp., *Alternaria* sp., *Eurotrium* sp., *Ophiostoma* sp., etc. (Cihangir and Sarikava 2004).

Many fungi are able to secrete large quantities of enzymes in the extracellular environment. In addition, a large amount of agro-industrial waste is abundant in Brazil and at a low cost has been successfully tested as culture media for lipase production by fungi, making it economically viable in industrial applications. Therefore microbial lipase enzymes are considered a potential choice for biotechnological applications and fungi have widely used to produce lipase in the last decades (Saxena et al. 2003, Cammarota and Freire 2006, Hasan et al. 2006, Metzger and Bornscheuer 2006, Nagy et al. 2006).

The fungi can usually be separated in mesophilic and termophilic. Mesophilic fungi present optimum reaction temperature between 30 and 40°C and for the majority, low thermal stability. Enzymes produced by thermophilic microorganisms have a higher thermal stability that prevents thermal inactivation. In particular, the advantages of using lipases in processes involving high temperatures include a solubility increase of lipids and other hydrophobic substrates in water and reduce the viscosity of the medium (Hasan et al. 2006).

Although there are several commercially available lipases, the potential application of new enzymes, with particular characteristics in different branches of industrial activity is enormous. In this context, intense lipase application in the future depends on the identification of novel enzymes with desired characteristics in relation to the substrate selectivity, thermostability, pH and temperature optimum. The successful use of these enzymes also depends on the optimization of its production, resulting directly in reduced costs.

Lipase Production

Currently, the most useful fermentation for lipase production is submerged fermentation (SmF). However, while used less, the solid-state fermentation (SSF) is also a growing process. The SmF requires a strict control on composition of the medium, since the lipase production is influenced by several parameters such as: type and concentration of carbon and nitrogen source, presence of metal ions, pH, growth temperature, dissolved oxygen concentration (Elibol and Ozer 2001, Sharma et al. 2001), agitation and aeration, type of inductor and reactor geometry (Martins et al. 2008). In SSF these factors are more flexible.

Specifically, lipase production depends on the influence of buffers, carbohydrates, vitamins, nutrient sources, fats and many nutrient combinations (Aulakh and Prakash 2010, Sharma et al. 2001, Saxena et al. 2003). Lipids as carbon sources are generally essential for a high yield in lipases production; however, a few authors have produced lipases with good yields in the absence of fats and oils (Sharma et al. 2001).

The nitrogen source depends largely on the microorganism in study; in general, the use of organic nitrogen favors high yields (Trevisan 2004). Among the nitrogen sources, peptone were further tested (Salleh et al. 1993, Hiol et al. 2000), soybean meal (Ohnishi et al. 1994, Okeke and Okolo 1990), yeast extract (Ohnishi et al. 1994) and corn steep water (Hiol et al. 2000). Nahas (1988) related that some factors affect growth and production of extracellular lipase by the fungus *Rhizopus oligosporus*. The highest yield was obtained when Tween was used as a carbon source. However, carbohydrate, vegetable oils, proteins and amino acids did not stimulate the lipase production and the fungus grew better using carbohydrate or protein in the culture medium, instead of oils.

The SSF is a suitable technology for enzymes production and agroindustrial waste can be exploited as a source of nutrients to the fermentation process, which favor the costs and benefits to direct marketing large-scale processes catalyzed by enzymes. Common carbon sources used in SSF are agro-industrial wastes such as, coffee pulp, bran, rice and wheat straw, bagasse, husks processed fruits, potatoes, cereal flour, cassava, etc. Moreover, research on SSF has shown higher lipase production, as compared to SmF. This advantage is attributed to the high rate of biomass growth and low proteolytic activity (Makkar and Cameotra 2002, Cammarota and Freire 2006, Diaz et al. 2006, Martins et al. 2008).

Mala et al. (2007) developed a new substrate mixture for lipase production by *Aspergillus niger* MTCC 2594 using wheat bran (WB) and gingerly oil cake (GOC). The result shows that the addition of GOC to WB (WB:GOC, 3:1, w/w) increases the lipase activity around 36 percent. Pie babassu and castor bean were used as carbon sources for *Penicillium brevicompactum* (Silva et al. 2011). Sugar cane bagasse was used to produce lipase by *Rhizomucor pusillus* and *Rhizopus rhizopodiformi* (Cordova et al. 1998). Dominguez et al. (2003) studied a lipase produced by *Yarrowia lipolytica* CECT 1240 using nylon sponges as support brackets, and foods and agro-industrial residues (barley bran and triturated nut) as carbon sources. They obtained a high activity with chopped nuts followed by sunflower meal and barley soaked in oil.

SmF is defined as a fermentation whose substrate is dissolved or suspended in liquid, usually water, which constitutes about 90–99 percent of the total weight of the material to be fermented. SmF presents some advantages as easy monitoring of product formation and substrate consumptionas well as some disadvantages as the large quantity of waste generated and the difficulty of separating the product from the substrate(Mitchell et al. 2000).

According to Nahas (1988) some extracellular lipases are able to grow with Tween and not any oil as a carbon source. On the other hand, most lipase described in literature used SmF for producing different kind of oils.

Messias et al. (2009) observed a lipase production for nine isolates of *Botryosphaeria* spp. using soybean oil and glycerol. Among these nine strains, the *Botryosphaeria ribis* EC-01 was the best lipase producer and, this same strain were also able to produce lipase in some different carbon sources as fatty acids, surfactants, carbohydrate and triglyceride. Moreover the production was increased adding Triton X-100 in the culture medium, which can emulsify stearic, palmitic and oleic acids, soybean oil and castor oil.

A comparison study between both kinds of fermentation was performed with *Penicillium simplicissimun* by Gutarra et al. (2009). SSF was performed using cake babassu as a carbon source while the SmF was performed in a semi-synthetic medium. The biomass production yield, specific activity and conidial production were higher in SSF. Thus, the high production of thermostable lipase from *P. simplicissimum*, using a non-supplement, agro-industrial residues with low cost, demonstrated a biotechnological potential of SSF for the production of industrial enzymes.

Isolation and Screening of Lipolytic Fungi

Currently, the search for new and cheap lipases with variables catalytic characteristics that encourage the selection and isolation of new strains has been constant. Lipase-producing fungi are found in several habitats such as soil contaminated with oil, oilseeds, factories, oil processing plants, decaying food, dairy, industrial wastes, soil with oils from oilseeds, compost piles, coal tips, hot spring water (Sharma et al. 2001) etc. It is possible to perform a preliminary screening of lipolytic filamentous fungi in Petri dishes containing agar supplemented with olive oil or tributyr with gum arabic in order to isolate in a first screening (Singh and Mukhopadhyay 2012). In these dishes it is possible to observe clear zones around the colonies, indicating lipase production. So the fungus can be sorted and selected for lipase production in a liquid or solid medium (Cardenas et al. 2001). Another simple method of screening for lipase production by fungal species was described by Kumar et al. (2011) using agar plates containing Bromophenol blue dye (pH 4.6) and olive oil as a carbon source. After growth clear halo are formed around the colonies, indicating lipase production.

Chromatographic Purification

Many lipases have been extensively purified and characterized according to their properties such as: pH, temperature, stability, metal ions effects and chelating agents. Purification methods used are generally dependent on non-specific techniques such as precipitation, hydrophobic chromatography, gel filtration and ion exchange chromatography (Sharma et al. 2001).

Enzymes purification allows to determine the sequence of amino acids and 3-D structure (Taipa et al. 1992, Aires-Barros et al. 1994, Saxena et al. 2003). The main limitations in purification strategies include low yield and a long period. Protocols used to purify lipases have been constantly reviewed (Saxena et al. 2003). Table 16.1 shows a summary of lipases purification from fungi.

However, in most cases, the lipase purification in a single chromatographic step is not sufficient to reach the desired purity level and often a combination of them is necessary. The gel filtration is the most used method (70 percent) and the columns used present lower load capacity and can be used in initial or final stages, in the latter case, refining the purification protocol. The ion exchange chromatography is the second most used method (63 percent), and more frequently uses the group diethylaminoethyl (DEAE) as anion exchangers and carboxymethyl (CM) as cation exchangers. Nevertheless, ion exchangers based on triethylaminoethyl groups are also used (Veeraragavan et al. 1990) and Q-Sepharose (Menge et al. 1990) is becoming widespread in lipase purification.

Affinity chromatography is used around 21 percent and although in the early stages, and are preferred after the initiation of purification due to its high cost. Hydrophobic interaction chromatography is applied in 45 percent of schemes, the octyl or phenyl group being the most popular. Concanavalin A (ConA) and heparin are used in the purification of fungal and mammals lipases due to the glycoprotein nature of these lipases (Trombs and Blake 1982, Aires-Barros and Cabral 1991).

The choice of the chromatographic purification method depends on the crude extract, and the same initial preparation may have different purification protocols. For instance, to purify a protein to homogeneity with 30 percent yield and purification factor of 320, four or five steps could be necessary (Aires-Barros et al. 1994). These values are the average of all yields and purification factors obtained in each purification scheme (Aires-Barros et al. 1994). The usual procedures applied to the lipase purification are sometimes challenging, time consuming and have low final yield. To remedy these problems, some new technologies have been developed such as membrane processes, immunopurification polymers, poly-vinyl alcohol as a stationary phase in chromatography, aqueous two-phase systems, and hydrophobic interaction chromatography, using spacers activated with epoxy groups and binders such as polyethylene glycol-Sepharose gel (Saxena et al. 2003).

Pre-purification Steps

Most microbial lipases are extracellular and the fermentation process removes cells from the culture supernatant by centrifugation and/or

Table 16.1 Purification procedures for lipases.

Fungal lipases	Purification steps	Recovery (%)	Purification fold	Molecular weight (kDa)	Reference
Aspergillus awamori (BTMFW032)	Ammonium sulfate fractionation (40–90%) and DEAE-Cellulose	33.7%	30.2	90	Basheer et al. 2011
Aspergillus carneus	Ammonium sulfate precipitation and octyl-Sepharose			27	Saxena et al. 2003
Aspergillus fumigatus	Supernatant of cell lysate and Ni-NTA agarose	86.1	8.47		Shangguan et al. 2011
Aspergillus nidulans WG 312	Phenyl-Sepharose and affinity binding on linolenic acid-agarose	10	1186	29	Mayordomo et al. 2000
Aspergillus niger NCIM 1207	Trimellitic anhydride (TMA)-crosslinked deacetylated chitin and Sephacryl-100	48%	9		Trimukhe et al. 2008
Aspergillus niger	Precipitation ammonium sulfate, butyl Toyopearl 650 M, Sephadex G-75, DEAE-Sepharose CL-6B and adsorption on hydroxyapatite	34%	600	35	Sugihara et al. 1988
Aspergillus niger (acid-resistant)	Alcohol precipitation, acid treatment, Biogel P-100 and Mono Q	36%	40		Torossian and Bell 1991
Aspergillus niger F044	Ammonium sulfate precipitation, dialysis, DEAE-Sepharose Fast Flow and Sephadex G-75	34%	73,71	35–40	Shu et al. 2007
Aspergillus niger (NCIM 1207)	Ammomium sulfate precipitation, phenyl sepharose CL4B, Sephacryl S-100	54%	150	32.2	Mhetras et al. 2009
Aspergillus oryzae	Ammonium sulfate fractionation, Econo-Pac Q, phenyl Superose, Mono Q and again Mono Q	5.5%	1242	41	Toida et al. 1995

Aspergillus oryzae	Ammonium sulfate fractionation, acetone precipitation, Econo-Pec Q and Superose 12	11%	792	25 (29)	Toida et al. 1998
Aspergillus terreus	Ammonium sulfate and two consecutive acetone precipitation, Sephadex G-100 and Q-Sepharose	18.04%	11.90	41	Yadav et al. 1998
Botrytis cinerea	Ammonium sulfate fractionation, octyl-Sepharose CL-4B, Con A-Sepharose and Sephacryl S-200	21%	31	60	Commenil et al. 1995
Fussarium sp. YM-30	Ultrafiltration, ammonium sulfate precipitation and DEAE-Toyopearl, CM-Toyopearl and butyl-Toyopearl	37%	2,770	12	Mase et al. 1995
Fussarium heterosporium	SP-Sephadex and gel filtration	38%		31	Shimada et al. 1993
Fussarium globulosum	Ultrafiltration, acetone precipitation, octyl-Sepharose CL-4B and Con A-Sepharose	31%	29		Saxena et al. 2003
Geotrichum candidum ATCC 66592	Sephadex G-100 high-performance liquid chromatography			61/57	Jacobsen and Poulsen 1992
Geotrichum candidum	Ethanol precipitation, Sephacryl 200 HR, Mono Q and Polybuffer exchanger 94	I-62%/ II-18%	I -35 II - 94	56	Veeraragavan et al. 1990
Kurtzmanomyces sp. I-11	Ammonium sulfate precipitation, DEAE-Sephadex A-25, SP-Sephadex C-50 and Sephadex G-100	19.6%	3140	49	Kakugawa et al. 2002
Mortierella alliacea	Acetone precipitation, DEAE-Sepharose and Superdex 200	4%	6.2	11	Jermsuntiea et al. 2011
Mucor miehei	DE-Cellulose 52, DEAE-Sepharose, Phenyl-Sepharose, concavalina A (ConA)-Sepharose and Sepharose 4B	I-57% II-32%	I-15 II-8		Huge-Jensen et al. 1987
Neurospora crassa	Charcoal treatment and dialysis and Sephadex G-100	2%	25.88	27/54	Kundu et al. 1987

Table 16.1 contd....

Table 16.1 contd.

Fungal lipases	**Purification steps**	**Recovery (%)**	**Purification fold**	**Molecular weight (kDa)**	**Reference**
Neurospora spp. TT-241	Ammonium sulfate precipitation, G-100 gel filtration,Toyopearl phenyl-650 M and Ultrogel-HA	16%	370	55	Lin et al. 1996
Penicillium camembertii U-150	Alcohol precipitation, ammonium sulfate fractionation, DEAE-Sepharose, aminooctyl-Sepharose and hydroxyapatite Con A-Sepharose	32%	213	37–39	Isobe et al. 1992
Penicillium candidum	Octyl-Sepharose CL-4B and DEAE-Sephadex		37	29	Ruiz et al. 2001
Penicillium citrinum	Extraction and back-extraction using AOT reversed micelles in isooctane and phenyl-Superose	68%	810	33	Krieger et al. 1997
Penicillium citrinum	Ammonium sulfate precipitation, Superose 6 and phenyl Superose	15.2%	379	63 (31–33)	Krieger et al. 1999
Penicillium chrysogenum	Ultrafiltration, phenyl-Sepharose, Mono Q HR 5/5 and PD-10 column on Sephadex G-25	44%	30	40	Ferrer et al. 2000
Penicillium cyclopium-M1	Ammonium sulfate precipitation, DEAE-Cellulose, DEAE-Sepharose, and hydroxyapatite chromatografies and gel filtration on Cellulofine GC-700	27%	1380	11	Isobe et al. 1988
Penicillium cyclopium	Centrifugation, ammonium sulfate precipitation, phenyl-Sepharose, DEAE-Sepharose and Sephadex G-75	33.2%	16.5	27.5	Wu et al. 1999
Penicillium cyclopium	Ammonium sulfate fractionation, Sephadex G-75, DEAE-Sephadex and again Sephadex G-75	30%	590	40-43	Chahinian et al. 2000
Penicillium expansum	Ultrafiltration-concentrate, ammonium sulfate precipitation and phenyl-Sepharose	5%	219	25	Stocklein et al. 1993

Penicillium simplicissimum	Ammonium sulfate precipitation, phenyl-Sepharose CL-4B, Ultrogel AcA-54 and hydroxyapatite	20%	788	56	Sztajer et al. 1992
Pythium ultimum 144	Ammonium sulfate precipitation, diethylaminoethyl-Sepharose CL-6B and Sephacryl S-200	22%		68 (270)	Mozaffar and Weete 1993
Rhizomucor hiemalis f. hiemalis	Ultrafiltration, ammonium sulfate fractionation, Sephadex G-75, Q-Sepharose and Sephacryl S-200		2200	49	Hiol et al. 1999
Rhizomucor miehei	Ammonium sulfate precipitation, phenyl-Sepharose 6 Fast-Row and DEAE-Sepharose Fast Flow	32%	42	31.6	Wu et al. 1996
Rhizomucor miehei (A)	DE-Cellulose 52 and Con A-Sepharose	57%	15		Huge-Jensen et al. 1987
Rhizomucor miehei (B)	DEAE-Cellulose and phenyl-Sepharose,	32%	120		Huge-Jensen et al. 1987
Rhizopus sp.	Fractionation ammonium sulfate, DEAE-Sephadex (I) and DEAE-Sephadex (II)	(I) 0.034 (II) 0.044	(I) 5.14 (II) 1.9		Pastore et al. 2003
Rhizopus arrhizus	Ammonium sulfate fractionation and Sephadex G-100 gel filtration	42%	720	67	Chattopadhyay et al. 1999
Rhizopus chinensis	Ammonium sulfate precipitation, Butyl-Sepharose and Superdex 75	0.7%	138.3	33	Sun et al. 2009
Rhizopus chinensis	CM-Cellulofine C-500, Toyopearl 650 M, Super Q Toyopearl and CM-Cellulofine C-500	27.6%		28.4	Yasuda et al. 1999
Rhizopus delemar	Oleic acid affinity chromatography, CM-Sephadex	30%	10.3	30.3	Haas et al. 1992
Rhizopusdelemar	A lipase foi concentrada em colunas Ultrogel e Sephadex G-150.	I-11% II-3% III-4%	I-106 II-71 III-47	I-76 II-60 III-45	Tahoun and Ali 1986
Rhizopus japonicus NR 400	Hydroxyapatite, octyl-Sepharose and Sephacryl S-200	31%	93	30	Suzuki et al. 1986

Table 16.1 contd....

Table 16.1 contd.

Fungal lipases	Purification steps	Recovery (%)	Purification fold	Molecular weight (kDa)	Reference
Rhizopus japonicus	Ammomium sulfate, Sephacryl S-200 and Mono S-Sepharose	12.6%	17	29	Salah et al. 2006
Rhizopus homothallicus	SmF—Butyl-Sepharose and Superdex G-200	40%	42	32 (29)	Diaz et al. 2006
	SSF—Butyl-Sepharose and Superdex G-200	15%	77	32 (29)	
Rhizopus niveus (I)	DEAE-Toyopearl, two consecutive runs on CM-Toyopearl	31.1%	54.4	34	Kohno et al. 1994
Rhizopus niveus (II)	DEAE-Toyopearl, two consecutives runs on CM-Toyopearl	3.1%	67.9	30	Kohno et al. 1994
Rhizopus oryzae	Acetone precipitation (80%), Sephadex G-100	64%	160		Razak et al. 1997
Rhizopus oryzae	Ammonium sulfate fractionation, sulfopropyl-Sepharose, Sephadex G-75 and sulfopropyl-Sepharose	22%	1260	32	Hiol et al. 2000
Talaromyces thermophilus	Ammomiun sulfate precipitation, chromatography S-200 and Mono Q-FFLC	29.13%	105.75	39	Romdhane et al. 2010
Thermomyces lanuginosus (Humicola lanuginose)	Acetone precipitation, Sephadex G-75, DEAE-Sepharose CL-6B and hydroxyapatite	15%	150	39	Omar et al. 1987
Thermomyces lanuginosus	adsorption to acrylic microsporus membranes in a pleated capsule filter				Taylor 1989
Trichoderma viride	Fractionation ammonium sulfate, DEAE-Cellulose and Sephadex G-100	46%	134		Kashmiri et al. 2006
Ustilago maydis ATCC 14826 (lipase I)	Q-Sepharose and butyl-Sepharose	12%			Lang et al. 1991
Ustilago maydis ATCC 14826 (lipase II)	Q-Sepharose, DEAE-Sepharose and butyl-Sepharose	13.5%			Lang et al. 1991

filtration. The cell free culture supernatant is generally concentrated by ultrafiltration, ammonium sulfate precipitation or by extraction with organic solvents. About 81 percent of the purification schemes involve precipitation steps, using 71 percent of ammonium sulfate and 26 percent use ethanol, acetone or an acid (typically hydrochloric) followed by a combination of many chromatography methods. The success of the precipitation depends on the concentration of ammonium sulfate used in the solution (Pabai et al. 1995) or the organic solvents concentration used in the precipitation method; moreover it is possible to use large amounts of materials without being affected by the interference from non-protein materials, as observed in the chromatographic methods. In comparison with other techniques, precipitation usually performs better than other chromatographic methods (Saxena et al. 2003).

Biotechnological Application of Lipases

Detergent industry

Lipases have been widely used in detergent formulas due to its potential ability to hydrolyze fats. Consequently, the following requirements in formulating should be considered: (i) the high variation in triglycerides containing fatty acids, requiring lipases with low substrate specificity; (ii) relative resistance to extreme conditions (pH 10–11 and 30–60°C) requiring stable enzymes; (iii) stability in the presence of surfactants and other enzymes, especially proteases and the effect of chemical denaturants and/or degradation caused by proteolytic detergent additives such as surfactants linear alkyl benzene sulfonate (LAS) and proteases. To solve these problems the combination of scanning to develop lipases and attempting to improve the lipase properties on the basis of protein engineering could be an alternative (Jaeger and Reetz 1998). In 1994 Novo Nordisk first introduced recombinant lipase (Lipolase™), originating from fungi and *Thermomyces lanuginosus* expressed in *Aspergillus oryzae* (Trevisan 2004).

Food industry

The lipase used in food processing industry aims at the modification of biomaterials and breaks. The position, chain length and degree of unsaturation not only potentially influence the physical properties, but also the sensory and nutritional value data of triglycerides. Lipases allow changing the fatty acids position in triglycerides or replacing one or more of them. By this mechanism, a lipid can be exchanged for another, making it more attractive for commercial purposes, e.g., for the preparation of cocoa butter (Pabai et al. 1995). A large number of lipases are produced on an industrial scale. Most of the commercial lipases produced are used to improve the taste of dairy products and processing of other foods such as meat, vegetables, fruit, smoked carp, milk products, cooked foods, and beer (Singh and Mukhopadhyay 2012).

Polyunsaturated fatty acids (PUFAs) have a role as an agent's nutriceuticals (Gill and Valivety 1997). Many of them have become essential fatty acids, which are required for membrane-lipid and prostaglandin synthesis. Microbial lipases are used to enrich PUFAs from plant and animal lipids such as Menhaden, tuna or borage oil. PUFAs free and their mono-and diglycerideos are subsequently used to produce a variety of pharmaceuticals including anticholesteremic, thrombolytic and anti-inflammatory (Gill and Valivety 1997, Jaeger and Reetz 1998).

Tea processing

The quality of black tea depends on the dehydration, mechanism of breakdown and the fermentation of the tea bud. During manufacture of black tea the enzymatic breakdown the membrane lipids and initiates the formation of volatile products with taste properties, improving the flavor during tea processing. Lipase from *R. miehei* improves levels of polyunsaturated fatty acids by reducing the total lipid content, improving the flavor (Ramarethinam et al. 2002, Singh and Mukhopadhyay 2012).

Citric acid production in food industry

Citric acid (CA) is an intermediate of the tricarboxylic acid cycle (TCA) containing a key position in the flow of the cell central metabolism. Due to its acidulant and flavoring agents and antioxidant properties, CA is used as a preservative food and beverage industry. *Yarrowia lipolytica* is known to produce a large amount of organic acids, including intermediates of the TCA cycle, such as CA or isocitric acids (ICA). *Y. lipolytica* strains used in this process are characterized by its great resistance in a high substrate and metal ions concentrations, thereby using less refined substrates (Rane and Sims 1993). The production of citric acid was done in a bioreactor using different strains from *Y. lipolytica* with tallow or olive oil (Kamzolova et al. 2005, 2007).

Pulp and paper industry

Another important application is the use of lipases to remove the wax present in the pulp produced in the paper industry, mainly lipases from *Aspergillus* sp., *C. rugosa* and *C. cylindracea*. "Pitch" is the term used to describe collectively the hydrophobic wood components, named triglycerides and waxes, which cause a lot of problems in the manufacture of pulp and paper (Farrell et al. 1997). A control system of a single enzymatic step has been developed to degrade triglycerides present in the suspension mechanisms of the Pitch (Reetz and Jaeger 1998). This system has been successful in two paper mills in Japan the Jujo Paper Co., and the Ishinomaki and Yatusushiro factories. In the Yatusushiro factory (160 tons/day of wood pulp), the goal was to demonstrate the use of enzymes with logs of immature wood.

Textile industry

Lipases are an auxiliary used to remove lubricants providing greater ink absorbency of the tissue to help in dyeing. They reduce the frequency of cracks and grooves in denim due to abrasion system. Commercial lipase preparations are used for desizing denim and cotton (Hasan et al. 2006).

Diagnostic tools and medical applications

Lipases are used as diagnostic tools indicating increased levels of certain diseases. Lipase levels in blood can be used as tools to diagnose and detect conditions such as acute pancreatitis and pancreatic lesion (Lott and Lu 1991). The determination of the lipase activity/level are also important in the diagnosis of heart diseases (McNeill et al. 1991). Moreover, lipase is an activator of the tumor necrosis factor and is used in the treatment of malignant tumors. Others therapeutic applications are found in the treatment of dyspepsia, gastrointestinal disorders, cutaneous manifestations of allergies, intestinal etc. (Singh and Mukhopadhyay 2012).

Biosensors

A biosensor based on enzymatic catalysis by dissolving the biodegradable polymer films has been revealed recently. Three enzyme-polymer systems have been studied: a dextran hydrogel, which is degraded by dextranase; a poly (ester amide), which is degraded by a proteolytic enzyme α-chymotrypsin; and a poly (trimethylene) succinate, degraded by a lipase. Potential application fields such as sensor system are the detection of enzyme concentration and the construction of devices imunosensores in which polymer degrading is used as a marker enzyme. The basic concept of the use of lipases is to produce biosensors as glycerol from the triglycerol analytical sample and to measure and free glycerol by an enzymatic method. This principle enables the physician to diagnose patients with cardiovascular problems, lipase from *C. rugosa* functioning as a biosensor which was developed as a DNA probe (Hasan et al. 2006, Singh and Mukhopadhyay 2012).

Leather industry

Degreasing is an essential step in clothing and leather glove manufacture. In this process, the natural fats excess is removed from greasy skin. Lipases can degrade the layer of grease on the wool, which comes from the sheep skin (Singh and Mukhopadhyay 2012).

Environmental applications

In water treatment: Lipases are used to remove fat from the surface of aerobic tanks allowing the transport of oxygen. The breakdown of fats, cleaning and the prevention of fat blocking film in water systems are important in many industrial operations. Examples include: wastewater treatment, tank cleaning, septic tanks, grease traps, degradation of organic debris—a commercial mixture of lipase, cellulase, inorganic nutrients, wheat bran, etc. (Gandhi 1997).

Degradation of polymers: Polymer degradation is a serious environmental hazard. The lipase from *Thermomyces lanuginosus* (TLL) has been applied as catalysts of vinyl acetate, poly (-caprolactone) and poly (bisphenol-A carbonate). Moreover lipases from TLL and cutinases from *Thermobifida fusca* and *Fussarium solani* were able to hydrolyze poly (ethylene terephthalate) films, fabrics and a (benzoyloxyethyl) terephthalate) endo-wise (Fernandes-Lafuente 2010) and the linear aromatic polyester poly (trimethylene terephthalate) was also treated with TLL and enzymes from *Penicillium citrinum, Thermobifida fusca* and *Fussarium solani pisi* (Eberl et al. 2008).

Wastewater treatment of meat industry: The beef industry wastewater usually has a high content of oils and fats, and low biodegradability. Enzymatic hydrolysis may contribute to the fat degradation in wastewater accelerating the treatment process. Rigo et al. (2008) studied the treatment of meat and pig industry wastewaters in anaerobic reactors with and without enzymatic pretreatment and observed that the high amount of enzyme maximized hydrolysis of oils and grease which increased the degradation.

Pre-treatment of wool: The presence of lipase in wool pretreatment was on the ability to dye woollen fabrics with reactive dyes using TLL (Kantouch et al. 2005). The treatment improved when wool was dyed with reactive dyes. The use of enzymes in the pretreatment enabled the dyeing of wool with mild conditions of temperature, increase of consumption of the dye and increasing the rate of dyeing (Fernandes-Lafuente 2010).

Sensors of fat quality: Due to the specificity of lipase a method was developed using a TLL microtiter plate to determine the fat quality used in large scale processing in which the fats were interesterified with *p*-nitrophenol laurate using TLL (Bertram et al. 2007).

Cleaning application: Lipases have been used in some cleaning material, mixed with or without detergents. In the presence of lipases, the dirt was removed by three consecutive mechanisms: (i) removal of dirt by basic flow bath (ii) dirt emulsion in a washing medium, and (iii) hydrolysis of dirt dispersed (Jurado et al. 2007).

References

Aires-Barros, M.R. and J.M.S. Cabral. 1991. Selective separation and purification of two lipases from *Chromobacterium viscosum* using AOT reversed micelles. Biotechnol. Bioeng. 38: 1302–1307.

Aires-Barros, M.R., M.A. Taipa and J.M.S. Cabral. Isolation and purification of lipase. pp. 243–270. *In:* P. Woolley, S.B. Petersen [eds.]. 1994. Lipases: Their Structure, Biochemistry and Application. Cambridge Univ. Press. Cambridge.

Akoh, C.C., G.C. Lee and J.F. Shaw. 2004. Protein engineering and applications of *Candida rugosa* lipase isoforms. Lipids, Champaign. 39: 513–526.

Aloulou, A., J.A. Rodriguez, S. Fernandez, D. van Oosterhout, D. Puccinelli and F. Carrière. 2006. Exploring the specific features of interfacial enzymology based on lipase studies Biochim. Biophys. Acta. 1761: 995–1013.

Alvarez-Macarie, E., V. Augier-Magro, J. Guzzo and J. Baratti. 1999. Molecular characterisation of the gene encoding an esterase from *Bacillus licheniformis* sharing significant similarities with lipases. Biotechnol. Lett. 21: 313.

Aulakh, S.S. and R. Prakash. 2010. Optimization of medium and process parameters for the production of lipase from an oil-tolerant *Aspergillus* sp. (RBD-01). J Basic Microbiol. 50: 37–42.

Bancerz, R. and G. Ginalska. 2007. A novel thermostable lipase from Basidiomycete *Bjerkandera adusta* R59: characterization and esterification studies. J. Ind. Microbiol. Biotechnol. 34: 553–560.

Basheer, S.M., S. Chellappan, P.S. Beena, R.K. Sukumaran, K.K. Elyas and M. Chandrasekaran. 2011. Lipase from marine *Aspergillus awamori* BTMFW032: Production, partial purification and application in oil effluent treatment. N. Biotechnol. 28(6): 627–38.

Bertram, M., C. Manschot-Lawrence, E. Floter and U.T. Bornscheuer. 2007. A microtiter plate-based assay method to determine fat quality. Eur. J. Lipid Sci. Technol. 109(2): 180–185.

Bier, M. 1955. Lipases. Methods of Enzymology, New York. 1: 627–642.

Bisen, P.S., B.S. Sanodiya, G.S. Thakur, R.K. Baghel and G.B.K.S. Prasad. 2010. Biodiesel production with special emphasis on lipase-catalyzed transesterification. Biotechnol. Lett. 32: 1019–1030.

Bornscheuer, U.T. 2002. Microbial carboxyl esterases: classification, properties and application in biocatalysts. FEMS Microbiol. Rev. 26: 73–81.

Bornscheuer, U.T. and R.J. Kazlauskas. 1999. Hydrolases in organic synthesis regio and stereoselective biotransformations. Wiley-VCH, Weinheim.

Brady, L., A.M. Brzozowski, Z.S. Derewenda, E. Dodson, G. Dodson, S. Tolley, J.P. Turkenburg, L. Christiansen, B. Huge-Jensen, L. Norskov, L. Thim and U. Menge. 1990. A serine protease triad forms the catalytic center of a triacylglycerol lipase. Nature 343: 767–770.

Brockman, H.L. General features of lipolysis: reaction scheme, interfacial structure and experimental approaches. pp. 1–46. *In:* B. Borgstrom and H.L. Brockman [eds.]. 1984. Lipases. Elsevier Science Publishers, Amsterdam.

Brzozowski, A.M., H. Savage, C.S. Verna, J.P. Turkenburg, D.M. Lawson, A. Svendsen and S. Patkar. 2000. Structural origins of the interfacial activation in *Thermomyces (Humicola) lanuginose* lipase. Biochem. 39: 15071–15082.

Cammarota, M.C. and D.M.G. Freire. 2006. A review on hydrolytic enzymes in the treatment of wastewater with high oil and grease content. Bioresour. Technol. 97: 2195–2210.

Cardenas, F., E. Alvarez, M.S. Castro-Alvarez, J.M. Sanchez-Montero, M. Valmaseda, S.W. Elson and J.V. Sinesterra. 2001. Screening and catalytic activity in organic synthesis of novel fungal and yeast lipases. J. Mol. Catal B Enzym. 14: 111–123.

Chahinian, H., G. Vanot, A. Ibrik, N. Rugani, L. Sarda and L.C. Comeau. 2000. Production of extracellular lipases by *Penicillium cyclopium*. Purification and characterization of a Partial acylglycerol lipase. Biosci. Biotechnol. Biochem. 64 (2): 215–22.

Chattopadhyay, M., A.K. Banik and S. Raychaudhuri. 1999. Production and purification of lipase by a mutant strain of *Rhizopus arrhizus*. Folia Microbiol. 44: 37–40.

Cihangir, N. and E. Sarikava. 2004. Investigation of lipase production by a new isolated of *Aspergillus* sp. World J. Microbiol. Biotechnol. 20: 193–197.

Commenil, P., L. Belingheri, M. Sancholle and B. Dehorter. 1995. Purification and properties of an extracellular lipase from the fungus *Botrytis cinerea*. Lipids. 30: 356–357.

Cordova, J., M. Nemmaoui, M. Ismaili-Alaoui, A. Morin, S. Roussos, M. Raimbault and B. Benjilali. 1998. Lipase production by solid state fermentation of olive cake and sugar cane bagasse. J. Mol. Catal B Enzym. 5: 75–78.

Cygler, M. and J.D. Schrag. 1997. Structure as basis for understanding interfacial properties of lipases. Meth.Enzymol. Part. A 284: 3–27.

Diaz, J.C.M., J.A. Rodríguez, S. Roussos, J. Cordova, A. Abousalham, F. Carriere and J. Baratti. 2006. Lipase from the thermotolerant fungus *Rhizopus homothallicus* is more thermostable when produced using solid state fermentation than liquid fermentation procedures. Enzyme Microb. Technol. 39: 1042–50.

Dominguez, A., M. Costas, M.A. Longo and A. Sanromán. 2003. A novel application of solid state culture: production of lipase by *Yarrowia lipolytica*. Biotechnol. Lett. 25(15): 1225–29.

Eberl, A., S. Heumann, R. Kotek, F. Kaufmann, S. Mitsche, A. Cavaco-Paulo and G.M. Gubitz. 2008. Enzymatic hydrolysis of PTT polymers and oligomers. J. Biotechnol. 135(1): 45–51.

Egloff, M.P., S. Ransac, F. Marguet, E. Rogalska, H. Van Tilbeurgh, G. Buono, C. Cambillau and R. Verger. Les lipases: cinétiques, spécificites et aspects structuraux. *In:* F.X. Malcata [ed.]. 1995. Engineering of/with lipases. Kluwer Academic Publishers, Dordrecht.

Elibol, M. and D. Ozer. 2000. Influence of oxygen transfer on lipase production by *Rhizopusarrhizus*. Process Biochem. 34(4): 325–29.

Essamri, M., D. Valerie and C. Louis. 1998. Optimization of lipase production by *Rhizopus oryzae* and study on the stability of lipase activity in organic solvents. Biotechnol. J. 60: 97–103.

Faber K. 1997. Biotransformations in organic chemistry. 3rd edn. Springer Produktions-Gesselschaft, Berlin, Germany.

Farrell, R.L., K. Hata and M.B. Wall. 1997. Solving pitch problems in pulp and paper processes by the use of enzymes or fungi. Adv. Biochem. Eng. Biotechnol. 57: 197–212.

Fernandes, M.L.M., E.B. Saad, J.A. Meira, L.P. Ramos, D.A. Mitchell and N. Krieger. 2007. Esterification and transesterification reactions catalysed by addition of fermented solids to organic reaction media. J. Molec. Catal. B: Enzymatic. 44: 8–13.

Fernandes-Lafuente, R. 2010. Lipase from *Thermomyces lanuginosus*: Uses and prospects as an industrial biocatalyst. J. Mol. Catal B Enzym. 62: 197–212.

Ferrato, F., F. Carriere, L. Sarda and R. Verger. 1997. A critical reevaluation of the phenomenon of interfacial activation. Methods Enzymol. Part B 284: 327–347.

Ferrer, M., F.J. Plon, O.M. Nuero, F. Reyes and A. Ballesteros. 2000. Purification and properties of a lipase from *Penicillium chrysogenum* isolated from industrial wastes. J. Chem. Technol. Biotechnol. 75: 569–576.

Freedman, B., E.H. Pryde and T.L. Mounts. 1984. Variable affecting the yield of fatty esters from transesterified vegetable oil. J. Am. Oil Chem. Soc. 63: 1638–1643.

Freire, D.M.G. and L.R. Castilho. Lipases em biocatálise. *In*: E.P.S. Bon, M.A. Ferrara, M.L. Corvo, A.B. Vermelho, C.L.A. Paiva, R.B. Alencastro and R.R.R. Coelho [eds.]. 2008. Enzimas em biotecnologia: Produção, Aplicações e Mercado. Editora Interciência, Rio de Janeiro.

Fukuda, H., A. Kondo and H. Noda. 2001. Biodiesel fuel production by transesterification of oils. J. Biosci. Bioeng. 92: 405–416.

Gandhi, N.N. 1997. Applications of lipase. J. Am. Oil Chem. Soc. 74: 621–634.

Gill, I. and R. Valivety. 1997. Polyunsaturated fatty acids: Part 1. Occurrence, biological activities and applications. Trends Biotchnol. 15: 401–9.

González-Navarro, H., M.C. Bañó and C. Abad. 2001. The closed/open model for lipase activation. Addressing intermediate active forms of fungal enzymes by trapping of conformers in water-restricted environments. Biochem. 40 (10): 3174–3183.

Grochulski, P., Y. Li, J.D. Schrag and M. Cygler. 1994. Two conformational states of *Candida rugosa* lipase. Protein Science 3: 82–91.

Guit, R.P.M., M. Kloosterman, G.W. Meindersma, M. Mayer and E.M. Meijer. 1991. Lipase Kinetics: Hydrolysis of Triacetin by Lipase from *Candida cylindracea* in a Hollow-Fiber Membrane Reactor. Biotechnol. Bioeng. 38: 727–732.

Gutarra, M.L.E., M.G. Godoy, L.R. Castilho and D.M.G. Freire. 2009. Lipase production and *Penicillium simplicissimum* morphology in solid-state and submerged fermentations. Biotechnol. J. 4: 1450–59.

Haas, M.J., D.J. Cichowicz and D.G. Bailey. 1992. Purification and characterization of an extracellular lipase from the fungus *Rhizopus delemar*. Lipids 27: 571–76.

Hasan, F., A.A. Shah and A. Hameed. 2006. Industrial applications of microbial lipases. Enzyme Microb. Technol. 39: 235–51.

Hiol, A., M.D. Jonzo, N. Rugani, D. Druet, L. Sarda and L.C. Comeau. 2000. Purification and characterization of an extracellular lipase from a thermophilic *Rhizopus oryzae* strain isolated from palm fruit. Enzyme Microb. Technol. 26(5-6): 421–30.

Hiol, A., M.D. Jonzoa, D. Drueta and L. Comeaua. 1999. Production, purification and characterization of an extracellular lipase from *Mucor hiemalis f. Hiemalis*. Enzyme Microb. Technol. 25(1–2): 80–87.

Holmquist, M. 2000. Alpha/Beta-hidrolases fold enzymes: structures, functions and mechanisms. Curr. Prot. Pep. Sci. 1(2): 209–235.

Huge-Jensen, B., D.R. Galluzzo and R.G. Jensen. 1987. Partial purification and characterization of free and immobilized lipases from *Mucor miehei*. Lipids 22(8): 559–565.

Iizumi T., K. Nakamura and T. Fukase. 1990. Purification and characterization of a thermostable lipase from newly isolated *Pseudomonas* sp. KWI-56. Agric. Biol. Chem. 54: 1253–58.

Isobe, K., T. Akiba and S. Yamaguchi. 1988. Crystallization and characterization of lipase from *Penicillium cyclopium*. Agric. Biol. Chem. 52: 41–47.

Jacobsen, T. and O.M. Poulsen. 1992. Separation and characterization of 61- and 57 kDa lipases from *Geotrichum candidum* ATCC 66592. Can J. Microbiol. 38: 75–80.

Jaeger, K.E. and M.T. Reetz. 1998. Microbial lipases form versatile tools for biotechnology. Trends Biotechnol. 16: 396–403.

Jaeger, K.E., B.W. Dijkstra and M.T. Reetz. 1999. Bacterial Biocatalist: molecular biology, three dimensional structures and biotechnological applications of lipases. Annu. Rev. Microbiol. 53: 315–351.

Jaeger, K.E., S. Rasanc, B.W. Dijstra, C. Colson, M. Van Heuvel and O. Misset. 1994. Bacterial lipases. FEMS Microbiol. Rev. 15: 29–63.

Jaeger, K.E., S. Rasanc, B.W. Dijstra, C. Colson, M. Van Heuvel and O. Misset. 1994. Bacterial lipases. FEMS Microbiol. Rev. 15: 29–63.

Jensen, R.G. 1983. Detection and determination of lipase (acylglycerol hydrolase) activity from various sources. Lipids 18: 650–657.

Jermsuntiea, W., T. Aki, R. Toyoura, K. Iwashita, S. Kawamoto and K. Ono. 2011. Purification and characterization of intracellular lipase from the polyunsaturated fatty acid-producing fungus *Mortierella alliacea*. N. Biotechnol. 28(2): 158–64.

JIsobe, K., K. Nokihara, S. Yamaguchi, T. Mase and R.D. Schmid. 1992. Crystallization and characterization of monoacylglycerol and diacylglycerol lipase from *Penicillium camembertii*. Eur. J. Biochem. 203: 233–237.

Jurado, E., V. Bravo, G. Luzón, M. Fernández-Serrano, M. García-Román, D. Altmajer-Vaz and J.M. Vicaria. 2007. Hard-Surface Cleaning Using lipases: Enzyme-Surfactant Interactions and Washing Tests. J. Surfact. Deterg. 10: 61–70.

Kakugawa, K., M. Shobayashi, O. Suzuki and T. Miykawa. 2002. Purification and characterization of a lipase from the glycolipid-producing yeast *Kurtzmanomyces* sp. I-11. Biosci. Biochnol. Biochem. 66(5): 978–85.

Kamzolova, S.V., I.G. Morgunov, A. Aurich, O.A. Perevoznikova, N.V. Shishkanova, U. Stottmeister and T.V. Finogenova. 2005. Lipase Secretion and citric acid production in *Yarrowia lipolytica* yeast grown on animal and vegetable fat. Food Technol. Biotechnol. 43(2): 113–22.

Kamzolova, S.V., T.V. Finogenova, Yu. N. Lunina, O.A. Perevoznikova, L.N. Minachova and I.G. Morgunov. 2007. Characteristics of the growth on rapeseed oil and synthesis of citric and isocitric acids by *Yarrowia lipolytica* yeasts. Microbiology 76(1): 20–24.

Kantouch, A., W.M. Raslan and H. El-Sayed. 2005. Effect of Lipase Pretreatment on the Dyeability of Wool Fabric. J. Natural Fibers. 2: 35–48.

Kashmiri, M.K., A. Adnan and B.W. Butt. 2006. Production, purification and partial characterization of lipase from *Trichoderma viride*. Afr J. Biotechnol. 5: 878–82.

Klazlaukas, R.J. and U.T. Bornsheuer. *In:* H.J. Rehm and P. Starder [eds.]. 1998. A Multi-volume Comprehensive Treatise Biotechnology. 8A: 38, Wiley-VCH, Weinheim, Germany.

Kohno, M., W. Kugimiya, Y. Hashimoto and Y. Morita. 1994. Purification, characterization and crystallization of two types of lipases from *Rhizopus niveus*. Biosci. Biotechnol. Biochem. 58: 1007–12.

Krieger, N., M.A. Taipa, E.H.M. Melo, J.L. Lima-Filho, M.R. Aires-Barros and J.M.S. Cabral. 1997. Kinetic characterization of *penicillium citrinum* lipase in AOT/Isooctane-reversed micelles. Appl. Biochem. Biotechnol. 67(1-2): 87–95.

Krieger, N., M.A. Taipa, E.H.M. Melo, J.L. Lima-Filho, M.R. Aires-Barros and J.M.S. Cabral. 1999. Purification of the *Penicillium citrinum* lipase by chromatographic processes. Bioprocess Eng. 20: 59–65.

Kumar, S.N., Katiyar, P. Ingle and S. Negi. 2011. Use of evolutionary operation (EVOP) factorial design to develop a bioprocess using grease waste as a substrate for lipase production. Bioresour Technol. 102: 4909–12.

Kundu, M., J. Basu, M. Guchhait and P. Chakrabarti. 1987. Isolation and Characterization of an Extracellular Lipase from the Conidia of *Neurospora crassa*. J. Gen. Microbiol. 133(1): 149–153.

Lang, S., E. Katsiwela, F. Kleppe and F. Wagner. *Ustilago maydis* lipolytic enzymes: characterization and partial purification. pp. 361–64. *In:* L. Alberghina, R.D. Schmid and R. Verger [eds.]. 1991. Lipases: Structure, Mechanism and Genetic Engineering. GBF Monographs. VCH, Weinheim.

Lee, S.H., J.I. Choi, S.J. Park, S.Y. Lee and B.C. Park. 2004. Display of bacterial lipase on the *Escherichia coli* cell surface by using fadl as an anchoring motif and use of the enzyme in enantioselective biocatalysis. Appl. Environ. Microbiol. 70: 5074–80.

Lin, S., J.C. Lee and C.M. Chiou. 1996. Purification and characterization of a lipase from *Neurospora* sp. TT-241. J. Am. Oil Chem. Soc. 73(6): 739–745.

Lin, S.F., J.C. Lee and C.M. Chiou. 1996. Purification and characterization of a lipase from *Neurospora* sp. TT-241. J. Am. Oil Chem. Soc. 73: 739–45.

Lott, J.A. and C.J. Lu. 1991. Lipase isoforms and amylase isoenzymes—assays and applications in the diagnosis of acute pancreatitis. Clin. Chem. 37: 361–68.

Louwrier, A., G.J. Drtina and A.M. Klibanov. 1996. On the issue of interfacial activation of lipase in nonaqueous media. Biotechnol. Bioeng. 50(1): 1–5.

Macrae, A.R. and R.C. Hammond. 1985. Present and future applications of lipases. Biotechnol. Gen. Eng. Rev. 3: 193–217.

Makkar, R.S. and S.S. Cameotra. 2002. An update on the use of unconventional substrates for biosurfactant production and their new applications. Appl. Microbiol. Biotechnol. 58(4): 428–34.

Mala, J.G., N.G. Edwinoliver, N.R. Kamini and R. Puvanakrishnam. 2007. Mixed substrate solid state fermentation for production and extraction of lipase from *Aspergillus niger* MTCC 2594. J. Gen. Appl. Microbiol. 53(4): 247–53.

Martinez, C., P. De Geus, M. Lauwereys, G. Matthyssens and C. Cambillau. 1992. *Fusarium solani* cutinase is a lipolytic enzyme with a catalytic serine accessible to solvent. Nature 356: 615–618.

Martins, V.G., S.J. Kalil and J.A.V. Costa. 2008. Co-produção de lipase e biosurfactante em estado sólido para utilização em biorremediação de óleos vegetais e hidrocarbonetos. Quim Nova. 31(8): 1942–47.

Mase, T., Y. Matsumiya and T. Akiba. 1995. Purification and characterization of a new lipase from *Fusarium* sp. YM-30. Biosci. Biotech. Biochem. 59(9): 1771–72.

Mayordomo, I., F. Randez-Gil and J.A. Prieto. 2000. Isolation, purification and characterization of a cold-adaptive lipase from *Aspergillus nidulans*. J. Agric. Food Chem. 48: 105–09.

McNeill, G.P., S. Schimizu and T. Yamane. 1991. High-yield enzymatic glycerolysis of fats and oils. J. Am. oil Chem. Soc. 68(1): 1–5.

Melo, E.P., R.P. Baptista and J.M.S. Cabral. 2003. Improving cutinase stability in aqueous solution and in reverse micelles by media engineering. J. Molec. Catal. B: Enzymatic 22: 299–306.

Menge, U., H.J. Hecht, D. Schomburg and R.D. Schmid. Crystallization and preliminary X-ray studies of lipase from *Geotrichum candidum*. pp. 59–62. *In:* L. Alberghina, R.D. Schmid and R. Verger [eds.]. 1990. Lipases: Structure, Mechanism and Genetic Engineering. GBF Monographs. VCH, Weinheim. Messaoudi, A., H. Belguith, I. Gram and J.B. Hamida. 2010. Classification of EC 3.1.1.3 bacterial true lipases using phylogenetic analysis. Afr. J. Biotechnol. 9(48): 8243–8247.

Messias, J.M., B.Z. Costa, V.M.G. Lima, R.F.H. Dekker, M.I. Rezende, N. Krieger and A.M. Barbosa. 2009. Screening *Botryosphaeria species* for lipase: Production of lipase by *Botryosphaeria ribis* EC-01 grown on soybean oil and other carbon sources. Enzyme Microb. Technol. 45: 426–31.

Metzger, J.O. and U. Bornscheuer. 2006. Lipids as renewable resources: current state of chemical and biotechnological conversion and diversification. Appl. Microbiol. Biotechnol. 71: 13–22.

Mhetras, N.C., K.B. Bastawde and D.V. Gokhale. 2009. Purification and characterization of acidic lipase from *Aspergillus niger*NCIM 1207. Bioresour Technol. 100: 1486–90.

Miled, N., C. Bussetta, A. De Caro, M. Riviere, L. Berti and S. Canaan. 2003. Importance of the lid and cap domains for the catalytic activity of gastric lipases Comp. Biochem. Physiol. Part B 136: 131–38.

Mitchell, D.A., N. Krieger, D.M. Stuart and A. Pandey. 2000. New developments in solid state fermentation. II. Rational approaches to the design, operation and scale-up of bireactors. Process Biochem. 35: 1211–1225.

Mozaffar, Z. and J.D. Weete. 1993. Purification and properties of an extracellular lipase from *Pythium ultimum*. Lipids 28(5): 377–382.

Nagy, V., E.R. Toke, L.C. Keong, G. Szatzker, D. Tbrahim, I.C. Omar, G. Szakacs and L. Poppe. 2006. Kinetic resolutions with novel, highly enantioselective fungal lipases produced by solid state fermentation. J. Mol. Catal B Enzym. 39: 141–8.

Nahas, E. 1988. Control of lipase production by *Rhizopus oligosporus* under various growth conditions. J. Gen. Microbiol. 134: 227–33.

Ohnishi, K., Y. Yoshida, J. Toita and J. Sekiguchi. 1994. Purification and characterization of a novel lipolytic enzyme from *Aspergillus oryzae*. J. Ferment Bioeng. 78: 413–19.

Okeke, C.N. and B.N. Okolo. 1990. The effect of cultural conditions on the production of lipase by *Acremonium strictum*. Biotechnol. Lett. 12: 747–50.

Ollis, D.L., E. Cheah, M. Cygler, B. Dijkstra, F. Frolow, S.M. Franken, M. Harel, S.J. Remington, I. Silman, J. Schrag, J.L. Sussman, K.H.G. Verchueren and A. Goldman. 1992. The α/β hydrolase fold. Protein Eng. 5: 1997–211.

Omar, I.C., M. Hayashi and S. Nagai. 1987. Purification and some properties of a thermostable lípase from *Humicola lanuginosa*. Agric. Biol. Chem. 51: 37–45.

Pabai, F., S. Kermasha and A. Morin. 1995. Lipase from Pseudomonas fragi CRDA 323: partial purification, characterization and interesterification of butter fat. Appl. Microbiol. Biotechnol. 43: 42–51.

Pastore, G.M., V.S.R. Costa and M.G.B. Koblitz. 2003. Purificação parcial e caracterização bioquímica de lipase extracelular produzida por nova linhagem de *Rhizopus* sp. Ciênc Tecnol. Aliment. 23(2): 135–140.

Pimentel, M.C., N. Krieger, L.C. Coelho, J.O. Fontana, E.H. Melo, W.M. Ledingham and J.L. Lima Filho. 1994. Lipase from a Brazilian strain of *Penicillium citrinum*. Appl. Biochem. Biotechnol. 49: 59–74.

Pouderoyen, G., T. Eggert, K.E. Jaeger and B.W. Dijkstra. 2001. The crystal structure of *Bacillus subtilis* lipase: a minimal α/β hydrolase fold enzyme. J. Mol. Biol. 309: 215–226.

Quinlan, P. and S. Moore. 1993. Modification of triglycerides by lipases: process technology and its application to the production of nutritionally improved fats Inform. Inform. 4: 582–585.

Ramarethinam, S., K. Latha and N. Rajalakshmi. 2002. Use of a fungal lipase for enhancement of aroma in black tea. Food Sci. Technol. Res. 8(4): 328–32.

Rane, K.D. and K.A. Sims. 1993. Production of citric acid by *Candida lipolytica* Y1095: Effect of glucose concentration on yield and productivity. Enzyme Microb. Technol. 15(88): 646–51.

Razak, C.N.A., A.B. Salleh, R. Musani, M.Y. Samad and M. Basri. 1997. Some characteristics of lipase from palm oil mill effluent. J. Mol. Catal B Enzym. 3: 153–59.

Reetz, M.T. and K.E. Jaeger. 1998. Overexpression, immobilization and biotechnological application of *Pseudomonas* lipases. Chem. Phys. Lipids 93(1–2): 3–14.

Reis, P. 2009. Lipases at interface: A review Advances in Colloid and Interfaces Science. 147(18): 237–250.

Rigo, E., R.E. Rigoni, P. Lodea, D. Oliveira, D.M.G. Freire, H. Treichel and M. Lucio. 2008. Comparison of two lipases in the hydrolysis of oil and grease in wastewater of the swine meat industry. Ind. Eng. Chem. Res. 47: 233–42.

Romdhane, I.B.B., A. Fendri, Y. Gargouri, A. Gargouri and H. Belghith. 2010. A novel thermoactive and alkaline lipase from *Talaromyces thermophilus* fungus for use in laundry detergents. Biochem. Eng. J. 53: 112–20.

Rottig, A., L. Wenning, D. Broker and A. Steinbuchel. 2010. Fatty acid alkyl ester: perspectives for production of alternative biofuels. Appl. Microbiol. Biotechnol. 85: 1717–1733.

Ruiz, B., A. Farres, F. Masso and S. Sanchez. 2001. Purification and characterization of an extracellular lipase from *Penicillium candidum*. Lipids. 36: 283–89.

Salah, R.B., H. Mosbah, A. Fendri, A. Gargouri, Y. Gargouri and H. Mejdoub. 2006. Biochemical and molecular characterization of a lipase produced by *Rhizopus oryzae*. FEMS Microbiol. Lett. 260: 241–48.

Salleh, A.B., R. Musani, M. Basri, K. Ampon, W.M.Z. Yunus and C.N.A. Razak. 1993. Extra- and intra-cellular lipases from a thermophilic *Rhizopus oryzae* and factors affecting their production. Can J. Microbiol. 39(10): 978–81.

Sarda, L. and P. Desnuelle. 1958. Action de la lipase pancreatique sur les em imulsion Biochim. Biophys. Acta. 30: 513–521.

Saxena, R.K., A. Sheoran, B. Giri and W.S. Davidson. 2003. Purification strategies for microbial lipases—Review. J. Microbiol. Methods 52: 1–18.
Schrag, J.D. and M. Cygler. 1997. Lipases and alpha/beta hydrolase fold. Methods. Enzymol. 284: 85–107.
Schrag, J.D., Y.G. Li, S. Wu and M. Cygler. 1991. Ser-His-Glu triad forms the catalytic site of the lipase from *Geotrichum candidum*. Nature 27; 351(6329): 761–4.
Shangguan, J.J, Y.Q. Liu, F.J. Wang, J. Zhao, L.Q. Fan, S.X. Li and J.H. Xu. 2011. Expression and characterization of a novel lipase from *Aspergillus fumigatus* with high specific activity. Appl. Biochem. Biotechnol. 165: 949–62.
Sharma, R., Y. Chisti and U.C. Banerjee. 2001. Production, purification, characterization, and applications of lipases. Biotechnol. Adv. 19(8): 627–62.
Shimada, Y., C. Koga, A. Sugihara, T. Nagao, N. Takada, S. Tsunasawa and Y. Tominaga. 1993. Purification and characterization of a novel solvent-tolerant lipase from *Fusarium heterosporum*. J. Ferment Bioeng. 75: 349–52.
Shu, Z.Y., J.K. Yang and Y.J. Yan. 2007. Purification and characterization of lipase from *Aspergillus niger* F044. Sheng Wu Gong Cheng Xue Bao. 23(1): 96–100.
Silva, M.F., D.M.G. Freire, A.M. Castro, M. di Luccio, M.A. Mazutti, J.V. Oliveira, H. Treichel and D. Oliveira. 2011. Production of multifunctional lipases by *Penicillium verrucosum* and *Penicillium brevicompactum* under solid state fermentation of babassu cake and castor meal. Bioprocess Biosyst. Eng. 34(2): 145–52.
Singh, A.K. and M. Mukhopadhyay. 2012. Overview of fungal lipase: A Review. Appl. Biochem. Biotechnol. 166: 486–520.
Stocklein, W., H. Sztajer, U. Menge and R.D. Schmid. 1993. Purification and properties of a lipase from *Penicillium expansum*. Biochim. Biophys. Acta 1168: 181–89.
Sugihara, A., Y. Shimada and Y. Tominaga. 1988. Purification and characterization of *Aspergillus niger* lipase. Agric. Biol. Chem. 52(6): 1591–1592.
Sulaiman, A., F.L. Wei and S.J. Lim. 2007. Proposed kinetic mechanism of the production of biodiesel from palmoil using lipase. Process. Biochem. 42: 951–960.
Sun, S.Y., Y. Xu and D. Wang. 2009. Novel minor lipase from *Rhizopus chinensis* during solid-state fermentation: Biochemical characterization and its esterification potential for ester synthesis. Bioresour Technol. 100(9): 2607–2612.
Suzuki, M., H. Yamamoto and M. Mizugaki. 1986. Purification and General Properties of a Metal-Insensitive Lipase from *Rhizopus japonicus* NR 4001. J. Biochem. 100(5): 1207–1213.
Sztajer H. and I. Maliszewska. 1989. The effect of culture conditions on lipolytic of *Penicillium citrinum*. Biotechnol. Lett. 11: 895–98.
Sztajer, H., H. Lunsdorf, H. Erdman, U. Menge and R. Schimid. 1992. Purification and properties of lipase from *Penicillium simplicissimum*. Biochim. Biophys Acta. 1124: 253–61.
Tahoun, M.K. and H.A. Ali. 1986. Specificity and glyceride synthesis by mycelial lipases of *Rhizopus delemar*. Enzyme Microb. Technol. 8(7): 429–432.
Taipa, M.A., M.R. Aires-Barros and J.MS. Cabral. 1992. Purification of lipases. J. Biotechnol. 26: 111–42.
Taylor, F. 1989. Stabilization of lipase from *Thermomyces lanuginosus* with *p*-chloromercuribenzoic acid. J. Ferment. Bioeng. 68: 141–143.
Toida, J., K. Kondou, M. Fukuzawa, K. Ohnishi and J. Sekiguchi. 1995. Purification and characterization of a lipase from *Aspergillus oryzae*. Biosci. Biotecnol. Biochem. 59: 1199–203.
Toida, J., Y. Arikawa, K. Kondou, M. Fukuzawa and J. Sekiguchi. 1998. Purification and characterization of triacylglycerol lipase from *Aspergillus oryzae*. Biosci. Biotechnol. Biochem. 62(4): 759–63.
Torossian, K. and A.W. Bell. 1991. Purification and characterization of an acid-resistent triacylglycerol lipase from *Aspergillus niger*. Biotechnol. Appl. Biochem. 13: 205–11.
Trevisan, H.C. Lipases. pp. 115–135. *In*: S. Said and R.C.L.R. Pietro [eds.]. 2004. Enzimas como agente Biotecnológico. Legis Summa, Ribeirão Preto, Brazil.

Trimukhe, K.D., N.D. Mahadik, D.V. Gokhale and A.J. Varma. 2008. Environment friendly crosslinked chitosan as a matrix for selective adsorption and purification of lipase of *Aspergillus niger*. Int. J. Biol. Macromol. 43: 422–25.

Trombs, M.P. and G.G. Blake. 1982. Stability and inhibition of *Aspergillus* and *Rhizopus* lipases. Biochim. Biophys. Acta. 700: 81–89.

Tsujisaka, Y., S. Okumura and M. Iwai. 1977. Glyceride synthesis by four kinds of microbial lipase. Biochim. Biophys. Acta. 489: 415–422.

Veeraragavan, K., T. Colpitts and B.F. Gibbs. 1990. Purification and characterization of two distinct lipases from *Geotrichum candidum*. Biochim. Biophys. Acta. 1044: 26–33.

Verger, R. 1997. Interfacial activation of lipases: facts and artifacts. Trends Biotechnol. 15: 32–38.

Verger, R. and G.H. De Haas. 1976. Interfacial enzyme kinetics of lipolysis. Ann. Rev. Biophys. Bioeng. 5: 77–117.

Verger, R., M.C.E. Mieras and G.H. Haas. 1973. Action of phospholipase A at interfaces. J. Biol. Chem. 248: 4023–4034.

Wang, I.F., J.J. Lalonde, M. Momogan, D.E. Bergbreiter and C.H. Wong. 1988. Lipase catalyzed irreversible transesterifications using enol esters as acylating reagents: preparative enantio- and regioselective synthesis of alcohols, glycerol derivatives, sugars and organometallics. J. Am. Chem. Soc. 110: 7200–7205.

Winkler, F.K., A. D'Arcy and W. Hunziker. 1990. Structure of human pancreatic lipase. Nature 343: 771–774.

Wu, M.C., S. Wang, W.D. Huang and C.R. Sun. 1999. Purification and characterization of an alkaline lipase from *Penicillium cyclopium* PG37. Acta Biochim. Biophys. Sin. 31(6): 664–68.

Wu, X.Y., S. Jaaskelainen and Y.Y. Linko. 1996. Purification and partial characterization of *Rhizomucor miehei* lipase for ester synthesis. Appl. Biochem. Biotechnol. 59: 145–50.

Yadav, R.P., R.K. Saxena, R. Gupta and W.S. Davidson. 1998. Purification and characterization of a regiospecific lipase from *Aspergillus terreus*. Biotechnol. Appl. Biochem. 28(3): 243–249.

Yahya, A.R.M., W.A. Anderson and M. Moo-Young. 1998. Ester synthesis in lipase-catalyzed reactions. Enz. Microbiol. Technol. 23: 438–450.

Yasuda, M., H. Ogino, T. Kiguchi, T. Kotani, S. Takakura, T. Ishibashi, T. Nakashima, H. Fukuda and H. Ishikawa. 1999. Purification and characterization of lipase from *Rhizopus chinensis* cells. J. Biosci. Bioeng. 88(5): 571–573.

17

Fungal Amylase: Production System

*Heloiza Ferreira Alves do Prado,**
Aline Aparecida dos Reis, Erica Aparecida de Oliveira Santos and *Raisa Deli de Oliveira Sanches*

Introduction

Filamentous fungi are extraordinary microorganisms, which widely impact on many aspects of our lives (Ward 2011). They are eukaryotic microorganisms and characterized branched filamentous hyphae and include mainly *Aspergillus, Penicillium, Trichoderma, Fusarium, Rhizopus* and *Mucor* species (Ward 2011).

With the fast growth of the biotechnological industry, filamentous fungi have been widely used in the fermentation industry. They become principal sources of enzymes and metabolites and therefore have been widely investigated. In the fermentation process, filamentous fungi are dominant producers of a range of primary metabolites, including organic acids, such as citric, gluconic, fumaric, kojic, itaconic acid and fatty acids. In enzymes production, there are many fungal enzymes, for example,

Departamento de Fitotecnia, Tecnologia de Alimentos e Sócio Economia, Faculdade de Engenharia-Campus Ilha Solteira, Univ. Estadual Paulista—UNESP. Av. Brasil n° 56, 15385-000, Ilha Solteira, Brazil.
*Corresponding author: heloiza@agr.feis.unesp.br.

amylases, amyloglucosidases, cellulases, pectinases, laccases/ligninases, phytase, proteases, lipases and others (Wang et al. 2005).

Filamentous fungi are known to produce a broad range of enzymes in large quantities. Knowledge of fungal genetics and fungal biochemical pathways has accumulated during many years of research and industrial applications (Ward 2011). Many species have Generally Regarded As Safe (GRAS) status in the food and food processing industry. Filamentous fungi can secrete larger amounts of proteins, perform post-translational modifications such as glycosylation and disulfidation correctly, and grow on a less expensive medium (Wang et al. 2005).

The amylases are among the most important industrial enzymes showing great biotechnological importance, mainly in the food industry on liquefaction and saccharification of starch in syrup production, beverages, bakery and baby food. Other potential applications are related to textile, feed, chemistry and pharmaceutical industries (Gupta et al. 2003). These enzymes can be derived from various sources, including plants, animals and microorganisms. The microbial amylases are the main industrial demand. Currently, large quantities of microbial amylases are commercially available and have complete hydrolysis of starch to application in the starch processing industry (Gupta et al. 2003, Pandey et al. 2000a).

This chapter is about the traditional processes of solid state fermentation and submerged fermentation using new substrate sources, as agro industrial waste, on systems of amylase production. On the other hand, new methods for increasing the production systems are adopted such as the recombinant protein technology that has advances on use of host yeast with specific properties for amylases production.

Fungi Amylase Production

There are two basic processes for enzyme production: solid state fermentation (SSF) and submerged fermentation (SmF). The main characteristic of these two systems is the free water quantity in the culture medium. During the SmF process, there is more free water than in SSF. So, the SSF process is defined as the cultivation of microorganisms on moist solid supports, either on inert carriers or on insoluble substrates that can additionally be used as carbon and energy source. The fermentation takes place in the absence or near absence of free water, thus, being close to the natural environment in which microorganisms are adapted (Pandey et al. 2000b, Hölker et al. 2004). When the SmF process is used for enzyme production, extracellular enzymes and metabolites stay dissolved in water of the culture medium.

Another characteristic for these systems is related to the microbial adaptation. The aim of the SSF process is the tight contact of the cultivated filamentous fungi or yeast and bacteria with the insoluble substrate, achieving the highest substrate concentrations for fermentation. This technology results, although so far only on a small scale, in several processing advantages of significant potential economic and ecological importance as compared with the SmF process. However, there are also several disadvantages of the SSF process, which have discouraged the use of this technique for industrial production. The main obstructions are due mainly to the build-up of gradients, such as temperature, pH, moisture, substrate concentration during cultivation, which are difficult to control under limited water availability (Hölker et al. 2004).

The amylase production from filamentous fungi is improved using solid state fermentation. Table 17.1 shows some fungi amylase producers using the SSF process. Here it is possible to observe that wheat bran is a carbon source largely used in amylase production. Many agro industrial wastes had been used for amylases production such as corncob, rice, waste food, cassava peels and flour or cakes of coconut, cotton or coffee (Table 17.1). However, amylase production on the SmF process is very significant, because there are some fungal strains producing amylase in this system. During the SmF process starch of different botanic sources or some agro industrial waste as carbon source is used for amylase production (Table 17.2).

The great quantity of biomass available in the environment can be a problem for the governmental sector. But, on the other hand, this biomass available in form of agro industrial, industrial or food waste can be used to produce important metabolites such as microbial enzymes. The importance and potential of enzyme production can be observed in many studies using agro industrial waste from different sources. Therefore some residues discarded in the environment, can be of better use when utilized on systems for enzyme production.

Recombinant Amylase

Recently, new attractive host organisms for the production of foreign proteins obtained from developments in recombinant DNA technology have been reported (Li et al. 2011a). Extensively compared to *Escherichia coli* expression system, the yeast expression system has important advantages for the expression of proteins from eukaryotic microorganisms. These advantages are related to effective posttranslational modification system and ability to produce foreign proteins in soluble and correctly folded form (van Ooyen et al. 2006).

Genomic sequence and cDNA library from *Phanerochaete chrysosporium* BKM-F-1767 was screened and cDNA sequence encoding an α-amylase gene was selected. The amplified cDNA sequence of *pcamy1* was inserted into the prokaryotic expression vector pET-32a(+) and the result was designated as pETPCAmy1. Then, this recombinant plasmid was transformed into

Table 17.1 Microorganisms and carbon source for amylase production using solid state fermentation.

Microorganism	Carbon source	Fermentation system	Amylase type	Reference
Aspergillus awamori MTCC 6652	Soyaflour and 1% wheat bran	Solid state fermentation	α-amylase	Negi and Benerjee 2010
	Wheat bran, 1% casein and 1% starch	Solid state fermentation	α-amylase and protease	
Aspergillus awamori IOC-3914	Babassu cake	Solid state fermentation	Endoamylase and exoamylase	Castro et al. 2010
Aspergillus flavus	cassava peels	Static fermentation Solid state fermentation	β-amylase amyloglucosidase	Adeniran and Abiose 2009
	brewery spent grain (BSG)	Static fermentation	β-amylase	
	plantain peels	Solid state fermentation	β-amylase	
	yam peels	Static fermentation Solid state fermentation	β-amylase amyloglucosidase	
Aspergillus fumigatus	cassava peels	Static fermentation Solid state fermentation	β-amylase Amyloglucosidase	Adeniran and Abiose 2009
	banana peels	Solid state fermentation	amyloglucosidase	
	brewery spent grain (BSG)	Solid state fermentation	amyloglucosidase	
	plantain peels	Solid state fermentation	β-amylase	
	yam peels	Solid state fermentation	amyloglucosidase	
Aspergillus niger	brewery spent grain (BSG)	Static fermentation Solid state fermentation	β-amylase β-amylase and amyloglucosidase	Adeniran and Abiose 2009
	plantain peels	Static fermentation Solid state fermentation	β-amylase β-amylase	
	yam peels	Static fermentation Solid state fermentation	β-amylase β-amylase and amyloglucosidase	
Aspergillus niger AF-1	soyabean powder	Solid state fermentation	α-amylase	Su et al. 2009
Aspergillus niger LPB 28	Cassava starch:sugar cane bagasse (2:1)	Solid state fermentation	α-amylase amyloglucosidase	Spier et al. 2006

Aspergillus niger MTCC-104	rice bran and salt solution	Solid state fermentation	α-amylase	Kumar and Duhan 2011
Aspergillus niger MTCC 281	wheat bran, rye straw, corncob leaf, coconut oil cake and wheat straw	Solid state fermentation	α-amylase	Upadhyay and Pandey 2009
Aspergillus niger S14,	Wheat bran	Solid state fermentation	α-amylase and acid protease	Bhatnagar et al. 2010
Aspergillus niger UO-01	Sugarcane bagasse	Solid state fermentation	α-amylase	Perés Rosés and Guerra 2009
Aspergillus niger UV-60	Smashed ktichen waste : corn stover (3,75:1)	Solid state fermentation	glucoamylase	Wang et al. 2010
Aspergillus niger	Food waste	Solid state fermentation	glucoamylase and lactic acid	Wang et al. 2009
Aspergillus oryzae As 3951	spent brewing grains and salt solution	Solid state fermentation	α-amylase	Xu et al. 2008
Aspergillus oryzae	banana fruit stalk	Solid state fermentation	α-amylase	Ragunathan and Swaminathan 2005
Aspergillus oryzae	Wheat bran	Solid state fermentation	glucoamylase	Zambare 2010
Aspergillus sp. MK07	Wheat bran, starch and peptone	Solid state fermentation	α-amylase	Chamita et al. 2010
Aspergillus sp. HA-2.	wheat bran : coarse rice flakes (1:1)	Solid state fermentation	glucoamylase	Hema et al. 2006
Clostridium thermosulfurogenes SV2	16.5% potato starch and 2.5% corn steep liquor	Solid state fermentation	β-amylase	Reddy et al. 2003
Fusarium solani	Wheat bran	Solid state fermentation	glucoamylase	Bhatti et al. 2007
Helminthosporium oxysporum	brewery spent grain (BSG)	Solid state fermentation	β-amylase and amyloglucosidase	Adeniran and Abiose 2009
	cassava peels	Static fermentation Solid state fermentation	β-amylase Amyloglucosidase	
	plantain peels	Static fermentation	β-amylase	
	yam peels	Solid state fermentation	amyloglucosidase	
Neurospora crassa CFR 308	coffee pulp and mixture of coffee wastes	Solid state fermentation	α-amylase	Murthy et al. 2009

Table 17.1 contd....

Table 17.1 contd.

Microorganism	Carbon source	Fermentation system	Amylase type	Reference
Penicillium brevicompactum	Wheat bran, rice husks, and sunflower oil meal	Solid state fermentation	α-amylase	Balkan and Ertan 2010
Penicillium chrysogenum	Wheat bran	Solid state fermentation	α-amylase	Balkan and Ertan 2007
Penicillium frequestans	brewery spent grain (BSG)	Solid state fermentation	amyloglucosidase	Adeniran and Abiose 2009
	cassava peels	Static fermentation Solid state fermentation	β-amylase amyloglucosidase	
	plantain peels	Static fermentation Solid state fermentation	β-amylase amyloglucosidase	
	yam peels	Solid state fermentation	amyloglucosidase	
Rhizopus microsporus	tapioca peel, 30% fungal inoculum, 1% xylose and 1% urea	Solid state fermentation	α-amylase	Vijayaraghavan et al. 2011
Rhizopus microsporus var. *rhizopodiformis*	Wheat bran : corncob : SR solution (2:4:2)	Solid state fermentation	α-amylase	Peixoto-Nogueira et al. 2008
Saccharomycopsis fibuligera A11	wheat bran, rice husk and cassava starch	Solid state fermentation	α-amylase	Chen et al. 2010
Thermomucor indicae-seudaticae	Wheat bran	Solid state fermentation	glucoamylase	Kumar and Satyanarayana 2004
Thermomyces lanuginosus ATCC 58157	Wheat bran or Wheat bran, soluble starch, peptone and mineral solution	Solid state fermentation	α-amylase	Kunamneni et al. 2005
Trichothecium roseum	wheat bran, rye straw, corncob leaf, sunflower oil meal and rice husk	Solid state fermentation	α-amylase	Balkan et al. 2011

Static fermentation: medium containing 2% agricultural dry waste, 0.2% peptone, 0.3% ammonium sulfate, 0.1% potassium dihydrogen phosphate, 0.03% magnesium sulfate heptahydrate, and 0.03% w/v calcium chloride in distilled water (dH_2O). Inoculated flasks are incubated on non-stirred cultivate.

Table 17.2 Microorganisms and carbon sources for amylase production using submerged fermentation.

Microorganisms	Carbon source	Amylase type	Reference
Aspergillus awamori (gene expression *Saccharomyces cerevisiae*)	Soluble starch	glucoamylase	Pavezzi et al. 2008
Aspergillus awamori (gene expression *Saccharomyces cerevisiae*)	Potato starch	glucoamylase	Pavezzi et al. 2011
Aspergillus flavus	Soluble starch and coconut meal (coconut oil cake)	α-amylase	Sasi et al. 2010
Aspergillus niger UV-60	Smashed food waste 2.50% and raw food waste 3.75%	glucoamylase	Wang et al. 2008
Aspergillus niger NIAB 280	Wheat bran and agro industrial waste	glucoamylase	Rajoka and Yasmeen 2005
Aspergillus niveus	cassava peel : soluble starch (1:1)	α-amylase and glucoamylase	Silva et al. 2009
Aspergillus oryzae CBS 819.72	Gruel (semolina factory processing durum wheat)	α-amylase	Kammoun et al. 2008
Candida famata	Soluble starch	glucoamylase	Mohamed et al. 2007
Clostridium thermosulfurogenes SV2	Potato flour	β-amylase and pullulanase	Ramesh et al. 2001
Chrysosporium asperatum	Potato starch	α-amylase	Sanghvi et al. 2011
Penicillium purpurogenum	Khanna medium and 1% soluble starch	α-amylase	Silva et al. 2011
Rhizopus microsporus TISTR 3518	Liquefied cassava starch	glucoamylase	Arnthong et al. 2010
Saccharomycopsis fibuligera DSM-70554	Cassava starch	α-amylase and glucoamylase	González et al. 2008
Thermomucor indicae-seudaticae	sucrose	glucoamylase	Kumar et al. 2007
Xanthophyllomyces dendrorhous CECT1690	Soluble starch and maltose	β-amylase	Dias et al. 2003

Escherichia coli strain BL2l(DE3). The transformant was incubated at 37°C in Luria-Bertani (LB) medium. The expression of α-amylase was induced until 1mM of isopropyl β-D-thiogalactopyranoside (IPTG). The expressed α-amylase shows about 60 kDa on active form and some inclusion bodies (Wu et al. 2007).

The α-amylase encoding gene was amplified by polymerase chain reaction from *Saccharomycopsis fibuligera* and inserted into a shuttle vector YEp352, together with the yeast phosphoglycerate kinase 1 promoter and α-factor signal gene of the expression vector pMETaB of *Pichia methanolica*. The recombinant expression plasmid pLA8α was transformed into an industrial strain of *Saccharomyces cerevisiae* Sc-11 (Liu et al. 2005). Therefore these authors constructed and introduced the recombinant *amy* expression plasmid into an industrial strain of *Saccharomyces cerevisiae*. The cloned *amy* gene expressed active α-amylase and the transformant Sc-ll-pLA8a could use 42 percent starch in YPS medium under shake flask conditions. The yeast α-factor signal gene was able to guide the α-amylase to secrete efficiently into the culture medium and enhance the amylase expression in relation to control strain (Liu et al. 2005).

Li et al. (2011a) have used a non-conventional yeast species, *Kluyveromyces lactis* genetically modified, to express high level of α-amylase from *Rhizopus oryzae*. The α-amylase gene from *Rhizopus oryzae* was inserted on recombinant plasmid pMD-RoAmy that was constructed to express recombinant α-amylase in *Escherichia coli* and *Saccharomyces cerevisiae*. α-amylase from *Rhizopus oryzae* showed high-levels of expression and secretion in the culture medium when transformed on the recombinant *Kluyveromyces lactis*. Recombinant *Kluyveromyces lactis* has grown efficiently using soluble starch (Li et al. 2011a). The recombinant α-amylase from *Rhizopus oryzae*, RoAmy, was also expressed in *Pichia pastoris* yeast (Li et al. 2011b). These authors constructed a *Pichia pastoris* strain with the capability to express recombinant α-amylase under the control of constitutive glyceraldehyde-3-phosphate dehydrogenase (GAP) and methanolinducible alcohol oxidase 1 promoters. The levels of inducibly expressed recombinant RoAmy (rRoAmy) were higher than those of originally expressed. Low levels of intracellular amylase activities, which had been detected after shake-flask fermentation, have indicated that both signal sequences could effectively direct the secretion of rRoAmy under all studied conditions (Li et al. 2011b).

An acid α-amylase from *Aspergillus niger* was expressed in *Pichia pastoris* SMD1168 (Zeng et al. 2011). The gene of acid-stable α-amylase was amplified by PCR using *Aspergillus niger* genomic DNA as a template. Then, the gene was cloned into the vector of pPIC9K and the recombinant vector pPIC9K-asAA was transformed into *Pichia pastoris*. The expression of the recombinant amylase was strictly induced by methanol, in shaking culture condition, after 168 hr induction with 2 percent of methanol. The amylase has reached maximal activity of 2838 U mL^{-1}. The recombinant amylase, has exhibited maximal activity at pH 4.0 and 70°C. This amylase was stable at the pH ranging from 3.0 to 6.0, and was stable for a long time and with a high level of activity at common industrial temperature 50°C. The molecular mass was about 58 kDa (Zeng et al. 2011).

The α-amylase from *Aspergillus oryzae* was expressed in yeast. The fragment containing the full length cDNA of α-amylase was amplified and designed to favor by modifying its upstream untranslated region. It was subcloned in the expression vector pYEXα1, placed under the control of the yeast CYC1-GAL10 promoter and used to transform *Saccharomyces cerevisiae*. So, cells were expressed and secreted active α-amylase to the medium in a regulated fashion. The recombinant enzyme had similar electrophoretic mobility and catalytic properties to the original *Aspergillus oryzae* α-amylase (Randez-Gil and Sanz 1993).

Yamada et al. (2009) constructed a recombinant *Saccharomyces cerevisiae* for ethanol production from raw starch. The authors have studied three recombinant yeast strains, two haploid strains and one diploid strain. The haploid recombinant strains were constructed by integrating the glucoamylase gene from *Rhizopus oryzae* fused with the 3'-half of the α-agglutinin gene as the anchor protein (MT8-1SS), and the α-amylase gene from *Streptococcus bovis* (NBRC1440SS). The diploid recombinant strain MN8140SS was constructed by mating these opposite types of integrant haploid strains in order to enhance the expression of integrated amylase genes. The diploid strain had the highest ethanol productivity and reusability during fermentation from raw starch. The authors have suggested that these results clearly show that a diploid strain developed by mating two integrant haploid strains is useful for the establishment of an efficient ethanol production process (Yamada et al. 2009).

A novel glucoamylase PCR fragment from *Aureobasidium pullulans* NRRL 12974 (APGA1) was cloned into pMD19-T Simple vector. The PCR product was digested and ligated into digested pPIC9k. This recombinant plasmid linearized with *Sac*I was transformed into *Pichia pastoris* KM71 by electroporation. Thus the authors have reported that this is the highest raw starch hydrolysis efficiency on recombinant fungal glucoamylase (Li et al. 2011c).

Glucoamylase from thermophilic fungus *Thermomyces lanuginosus* was sequenced and cloned. The sequence of glucoamylase (GA) was amplified and inserted into the pPIC9 expression vector, which enables secretion of GA. The recombinant plasmid was linearized with *Dra*I or *Stu*I and it was transformed in *Pichia pastoris* GS115 by electroporation. This study resulted in four clones able to secrete recombinant glucoamylase with characteristics similar to initial glucoamylase from *Thermomyces lanuginosus* (Thorsen et al. 2006).

Yang et al. (2008) constructed a *Rhizopus arrhizus* glucoamylase gene (RaGA)—introns artificially spliced by PCR—suitable for expression in *Saccharomyces cerevisiae* host and tried expressing in *Pichia pastoris*. The RaGA gene, that is artificially constructed, was transferred into *Pichia pastoris* integrative expression vectors pPIC9. The glucoamylase expression was induced by 0.75 percent methanol for 72 hr and led to the synthesis of secreted glucoamylase. It was demonstrated that the glucoamylase gene has been expressed in and secreted from *Pichia pastoris*.

Pichia pastoris was used to transform a thermostable glucoamylase gene from *Chaetomium thermophilum* CT2. In this case, *Pichia pastoris* GS115 was used as a host for expression of glucoamylase cDNA. Recombinant plasmids were constructed from pMD18-T for subcloning and sequencing from pPIC9K, for expression of glucoamylase cDNA. *Pichia pastoris* GS115 was transformed with the SacI—linearized recombinant plasmid pPIC9K /gla by electroporation. The *Chaetomium thermophilum* glucoamylase gene was expressed in *Pichia pastoris*, and the glucoamylase was secreted into the culture medium by the yeast in a functionally active form. The purified recombinant glucoamylase was a glycoprotein with a size of about 66 kDa, with the optimum catalytic activity at pH 4.5–5.0 and 65°C (Chen et al. 2007).

Recently, the heterologous expression and efficient ethanol production from RaGA gene (glucoamylase from *Rhizopus arrhizus*) was studied using *Saccharomyces cerevisiae* as a host (Yang et al. 2011). For the experiment on fermentation flask, the RaGA was transformed into *Saccharomyces cerevisiae* strain. This strain was grown on raw starch and it was possible to use the starch as an energy source to produce ethanol upto a final concentration of 5 percent.

Paciello et al. (2010) have studied a system to recombinant glucoamylase production from Erlenmeyer flasks and Biostat® B bioreactor systems. The recombinant strain used in this experiment was a *Kluyveromyces lactis* strain JA6-GAA, transformed with a multicopy plasmid vector pTS32x carrying glucoamylase gene from *Arxula adeninivorans* and expressed under GAP promoter of *Saccharomyces cerevisiae*. These authors observed that lactose is a good carbon source for growing of the recombinant strain *Kluyveromyces lactis* JA6-GAA in batch and fed batch reactor system cultivation.

Conclusion and Future Perspectives

Agro industrial waste is used as a carbon source for many fungal strains on processes for amylases production (Table 17.1 and Table 17.2). These alternative sources contribute to reducing the final costs of the production system. Wheat bran is used more as a substrate on different process of fermentation and it presents the best level of amylases on the production system, as mentioned in literature. So, the use of alternative carbon sources is a very important strategy for future processes of enzyme production.

Recombinant amylase is another technology applied to increase the amylase level production. Recent studies have been conducted for improving the conditions to express fungal proteins. These studies have demonstrated that the best expression of recombinant amylase genes from fungal strains, occur when eukaryotic microorganisms were used as a host. *Pichia pastoris* is the main yeast in experiments for expression of glucoamylase and α-amylase. Some studies have searched for other eukaryotic hosts and good results have been obtained with *Saccharomyces cerevisiae* and *Kluyveromyces* ssp. *Saccharomyces cerevisiae* is an important strain because it was used in many industrial processes, mainly on bakery products and ethanol production. Therefore the recombinant protein or heterologous expression is an important tool for the development of recombinant microorganisms with the potential for enzyme production and application on specific industrial sectors.

References

Adeniran, A.H. and S.H. Abiose. 2009. Amylolytic potentiality of fungi isolated from some Nigerian agricultural wastes. African Journal of Biotechnology 8(4): 667–672.

Arnthong, J., B. Wanitchaploy, K. Sakai, J.J. Sanglier and V. Kitpreechavanich. 2010. Statistical screening of factors affecting glucoamylase production by a thermotolerant *Rhizopus microsporus* TISTR 3518 using Plackett-Burman design. African Journal of Biotechnology 9(43): 7312–7316.

Balkan, B. and F. Ertan. 2007. Production of α-amylase from *Penicillium chrysogenum* under solid-state fermentation by using some agricultural by-products. Food Technology and Biotechnology 45(4): 439–442.

Balkan, B. and F. Ertan. 2010. The production of a new fungal α-amylase degraded the raw starch by means of solid-state fermentation. Preparative Biochemistry & Biotechnology 40(3): 213–228.

Balkan, B., S. Balkan and F. Ertan. 2011. Optimization of parameters for α-amylase production under solid state fermentation by *Trichothecium roseum*. Romanian Biotechnological Letters 16(5): 6591–6600.

Bhatnagar, D., I. Joseph and R.P. Raj. 2010. Amylase and acid protease production by solid state fermentation using *Aspergillus niger* from mangrove swamp. Indian Journal of Fisheries 57(1): 45–51.

Bhatti, H. N., M.H. Rashid, R. Nawaz, M. Asgher, R. Perveen and A. Jabbar. 2007. Optimization of media for enhanced glucoamylase production in solid-state fermentation by *Fusarium solani*. Food Technology and Biotechnology 45(1): 51–56.

Castro, A.M., T.V. Andrea, L.R. Castilho and D.M.G. Freire. 2010. Use of mesophilic fungal amylases produced by solid-state fermentation in the cold hydrolysis of raw babassu cake starch. Applied Biochemistry and Biotechnology 162(6): 1612–1625.

Chamita, M.K., P. Sasidhar and S. Challa. 2010. Production of extracellular amylase from agricultural residues by a newly isolated *Aspergillus* species in solid state fermentation. African Journal of Biotechnology 9(32): 5162–5169.

Chen, J., Y.-Q. Zhang, C.-Q. Zhao, A.-N. Li, Q.-X. Zhou and D.-C. Li. 2007. Cloning of a gene encoding thermostable glucoamylase from *Chaetomium thermophilum* and its expression in *Pichia pastoris*. Journal of Applied Microbiology 103(6): 2277–2284.

Chen, L., Z.-M. Chi, Z. Chi and M. Li. 2010. Amylase production by *Saccharomycopsis fibuligera* A11 in solid-state fermentation for hydrolysis of cassava starch. Applied Biochemistry and Biotechnology 162(1): 252–263.

Diaz, A., C. Sieiro and T.G. Villa. 2003. Production and partial characterization of a β-amylase by *Xanthophyllomyces dendrorhous*. Letters in Applied Microbiology 36(4): 203–207.

González, C.F., J.I. Fariña and L.I.C. Figueroa. 2008. Optimized amylolytic enzymes production in *Saccharomycopsis fibuligera* DSM-70554: An approach to efficient cassava starch utilization. Enzyme and Microbial Technology 42: 272–277.

Gupta, R., P. Gigras, H. Mohapatra, V.K. Goswami and B. Chauhan. 2003. Microbial α-amylases: a biotechnological perspective. Process Biochemistry 38: 1599–1616.

Hema, A., U.B. Trivedi and K.C. Patel. 2006. Glucoamylase production by solid-state fermentation using rice flake manufacturing waste products as substrate. Bioresource Technology 97: 1161–1166.

Hölker, U., M. Höfer and J. Lenz. 2004. Biotechnological advantages of laboratory-scale solid-state fermentation with fungi. Appl. Microbiol. Biotechnol. 64: 175–186.

Kammoun, R., B. Naili and S. Bejar. 2008. Application of a statistical design to the optimization of parameters and culture medium for α-amylase production by *Aspergillus oryzae* CBS 819.72 grown on gruel (wheat grinding by-product). Bioresource Technology 99(13): 5602–5609.

Kumar, A. and J.S. Duhan. 2011. Production and characterization of amylase enzyme isolated from *Aspergillus niger* MTCC-104 employing solid state fermentation. International Journal of Pharma and Bio Sciences 2(3): B–258.

Kumar, S. and T. Satyanarayana. 2004. Statistical optimization of a thermostable and neutral glucoamylase production by a thermophilic mold *Thermomucor indicae-seudaticae* in solid-state fermentation. World Journal of Microbiology & Biotechnology 20(9): 895–902.

Kumar, S., P. Kumar and T. Satyanarayana. 2007. Production of raw starch-saccharifying thermostable and neutral glucoamylase by the thermophilic mold *Thermomucor indicae-seudaticae* in submerged fermentation. Appl. Biochem. Biotechnol. 142(3): 221–230.

Kunamneni, A., K.S. Kumar and S. Singh. 2005. Response surface methodological approach to optimize the nutritional parameters for enhanced production of α-amylase in solid state fermentation by *Thermomyces lanuginosus*. African Journal of Biotechnology 4(7): 708–716.

Li, S., W. Shen, X. Chen, G. Shi and Z. Wang. 2011a. Secretory expression of *Rhizopus oryzae* α-amylase in *Kluyveromyces lactis*. African Journal of Biotechnology 10(20): 4190–4196.

Li, S., S. Sing and Z. Wang. 2011b. Improved expression of *Rhizopus oryzae* α-amylase in the methylotrophic yeast *Pichia pastoris*. Protein Expression and Purification 79(1): 142–148.

Li, H., W. Sun, Y. Gao, Y. Wu, L. Huang, E.Z. Huang, A. Wang, X. Yin, Q. Wang, T. Xie and Z. Zeng. 2011c. Cloning, recombinant expression and characterization of a new glucoamylase gene from *Aureobasidium pullulans* NRRL 12974 and its potential application in raw potato starch degradation. African Journal of Biotechnology 10(45): 9122–9131.

Liu, Z., G. Zhang, Z. Long and S. Liu. 2005. Heterologous expression of amylase gene from *Saccharomycopsis fibuligera* in an industrial strain of *Saccharomyces cerevisiae*. Wuhan University Journal of Natural Sciences 10(6): 1041–1046.

Mohamed, L., M. Zakaria, A. Ali, W. Senhaji, O. Mohamed, E. Mohamed, B.E. El-Hassan and J. Mohamed. 2007. Optimization of growth and extracellular glucoamylase production by *Candida famata* isolate. African Journal of Biotechnology 6(22): 2590–2595.

Murthy, P. S., M.M. Naidu and P. Srinivas. 2009. Production of alpha-amylase under solid-state fermentation utilizing coffee waste. Journal of Chemical Technology and Biotechnology 84: 8, 1246–1249.

Negi, S. and R. Banerjee. 2010. Optimization of culture parameters to enhance production of amylase and protease from *Aspergillus awamori* in a single fermentation. African Journal of Biochemistry Research (AJBR). 4(3): 73–80.

Paciello, L., F. Romano, E. Alteriis, P. Parascandola and V. Romano. 2010. Glucoamylase by recombinant *Kluyveromyces lactis* cells: production and modelling of a fed batch bioreactor. Bioprocess and Biosystems Engineering 33(4): 525–532.

Pandey, A., P. Nigam, C.R. Soccol, V.T. Soccol, D. Singh and R. Mohan. 2000a. Advances in microbial amylases: review. Rev. Biotechnology Applied Biochemistry 31: 135–152.

Pandey, A., C.R. Soccol and D. Mitchell. 2000b. New developments in solid state fermentation: I Processes and products. Process Biochem. 35: 1153–1169.

Pavezzi, F.C., E. Gomes and R. DaSilva. 2008. Production and characterization of glucoamylase from fungus *Aspergillus awamori* expressed in yeast *Saccharomyces cerevisiae* using different carbon sources. Brazilian Journal of Microbiology 39(1): 108–114.

Pavezzi, F.C., A.A.J. Carneiro, D.A. Bocchini-Martins, H.F. Alves-Prado, H. Ferreira, P.M.M. Martins, E. Gomes and R. DaSilva. 2011. Influence of different substrates on the production of a mutant thermostable glucoamylase in submerged fermentation. Applied Biochemistry and Biotechnology 163(1): 14–24.

Peixoto-Nogueira, S.C., V.C. Sandrim, L.H.S. Guimarães, J.A. Jorge, H.F. Terenzi and M.L.T.M. Polizeli. 2008. Evidence of thermostable amylolytic activity from *Rhizopus microsporus* var. *rhizopodiformis* using wheat bran and corncob as alternative carbon source. Bioprocess and Biosystems Engineering 31(4): 329–334.

Perez Roses, R. and N.P. Guerra. 2009. Optimization of amylase production by *Aspergillus niger* in solid-state fermentation using sugarcane bagasse as solid support material. World Journal of Microbiology & Biotechnology 25(11): 1929–1939.

Ragunathan, R. and K. Swaminathan. 2005. Growth and amylase production by *Aspergillus oryzae* during solid state fermentation using banana waste as substrate. Journal of Environmental Biology 26(4): 653–656.

Rajoka, M.I. and A. Yasmeen. 2005. Induction, and production studies of a novel glucoamylase of *Aspergillus niger*. World Journal of Microbiology & Biotechnology 21(2): 179–187.

Ramesh, B., P.R.M. Reddy, G. Seenayya and G. Reddy. 2001. Effect of various flours on the production of thermostable β-amylase and pullulanase by *Clostridium thermosulfurogenes* SV2. Bioresource Technology 76(2): 169–171.

Randez-Gil, F. and P. Sanz. 1993. Expression of *Aspergillus oryzae* α-amylase gene in *Saccharomyces cerevisiae*. FEMS Microbiology Letters. 112(1): 119–123.

Reddy, P.R.M., B. Ramesh, S. Mrudula, G. Reddy and G. Seenayya. 2003. Production of thermostable β-amylase by *Clostridium thermosulfurogenes* SV2 in solid-state fermentation: optimization of nutrient levels using response surface methodology. Process Biochemistry 39(3): 267–277.

Sanghvi, G.V., R.D. Koyani and K.S. Rajput. 2011. Isolation, Optimization, and partial purification of amylase from *Chrysosporium asperatum* by submerged fermentation Journal of Microbiology and Biotechnology 21(5): 470–476.

Sasi, A., M. Kani1, A. Panneerselvam, G. Jegadeesh, K. Muthu and M.R. Kumar. 2010 Optimizing the conditions of α-amylase by an Esturian strain of *Aspergillus* spp. African Journal of Microbiology Research 4(8): 581–586.

Silva, T.M., R.F. Alarcon, A.R.L. Damasio, M. Michelin, A. Maller, D.C. Masui, H.F. Terenzi, J.A. Jorge and M.L.T.M. Polizeli. 2009. Use of cassava peel as carbon source for production of amylolytic enzymes by *Aspergillus niveus*. International Journal of Food Engineering 5(5): 1–12, Article 1.

Silva, T.M., M. Oliveira, A.F. Somera, J.A. Jorge, H.F. Terenzi, M.L.T.M. Polizeli and L.H.S. Guimaraes. 2011. Thermostable saccharogenic amylase produced under submerged fermentation by filamentous fungus *Penicillium purpurogenum*. Brazilian Journal of Microbiology 42(3): 1136–1140.

Spier, M.R., A.L. Woiciechowski, L.P.S. Vandenberghe and C.R. Soccol. 2006. Production and characterization of amylases by *Aspergillus niger* under solid state fermentation using agro industrials products. International Journal of Food Engineering 2(3): Article 6.

Su, X., X. Xiong, X. Tan, Q. Li, S. Luo and J. Yi. 2009. Optimization of raw-starch-hydrolysing amylase production by *Aspergillus niger* AF-1 in solid state fermentation. Journal of Hunan Agricultural University 35(2): 208–212.

Thorsen, T.S., A.H. Johnsen, K. Josefsen and B. Jensen. 2006. Identification and characterization of glucoamylase from the fungus *Thermomyces lanuginosus*. Biochimica et Biophysica Acta: Proteins and Proteomics 1764(4): 671–676.

Upadhyay, M.K. and A.K. Pandey. 2009. Production and optimization of α-amylase from *Aspergillus niger* MTCC 281 under solid state fermentation. Journal of Plant Science Research 25(1): 49–52.

van Ooyen, A.J.J., P. Dekker, M. Huang, M.M.A. Olsthoorn, D.I. Jacobs, P.A. Colussi and C.H. Taron. 2006. Heterologous protein production in the yeast *Kluyveromyces lactis*. FEMS Yeast Res. 6: 381–392.

Vijayaraghavan, P., C.S. Remya and S.G.P. Vincent. 2011. Production of α-amylase by *Rhizopus microsporus* using agricultural by-products in solid state fermentation. Research Journal of Microbiology 6(4): 366–375.

Wang, L., D. Ridgwaya, T. Gua and M. Moo-Youngb. 2005. Bioprocessing strategies to improve heterologous protein production in filamentous fungal fermentations. Biotechnology Advances 23: 115–129.

Wang, Q., X. Wang, X. Wang and H. Ma. 2008. Glucoamylase production from food waste by *Aspergillus niger* under submerged fermentation. Process Biochemistry 43(3): 280–286.

Wang, X., Q. Wang, H. Ma and W. Yin. 2009. Lactic acid fermentation of food waste using integrated glucoamylase production. Journal of Chemical Technology and Biotechnology 84(1): 139–143.

Wang, X., Q. Wang, Y. Liu and H. Ma. 2010. On-site production of crude glucoamylase for kitchen waste hydrolysis. Waste Management & Research 28(6): 539–544.

Ward, O.P. 2011. Production of recombinant proteins by filamentous fungi. Biotechnology Advances, doi:10.1016/j.biotechadv.2011.09.012.

Wu, B., G.-K. Hu, H. Feng, J.-M. Wu and Y.-Z. Zangh. 2007. Cloning and expression of an α-amylase gene from *Phanerochaete chrysosporium*. Current Microbiology 55(2): 105–113.

Xu, H., L. Sun, D. Zhao, B. Zhang, Y. Shi and Y. Wu. 2008. Production of α-amylase by *Aspergillus oryzae* As 3951 in solid state fermentation using spent brewing grains as substrate. Journal of the Science of Food and Agriculture 88(30): 529–535.

Yamada, R., Y. Bito, T. Adachi, T. Tanaka, C. Ogino, H. Fukuda and A. Kondo. 2009. Efficient production of ethanol from raw starch by a mated diploid *Saccharomyces cerevisiae* with integrated α-amylase and glucoamylase genes. Enzyme and Microbial Technology 44(5): 344–349.

Yang, L., X. Dai, J. Hou, C. Ma, C. Wang, Z. Wu and M. Li. 2008. Construction of a *Rhizopus arrhizus* glucoamylase gene suitable for expression in distinct host: introns spliced artificially by PCR. Molecular Biology Reports 35: 1, 9–15.

Yang, S., N. Jia, M. Li and J. Wang. 2011. Heterologous expression and efficient ethanol production of a *Rhizopus* glucoamylase gene in *Saccharomyces cerevisiae*. Molecular Biology Reports 38(1): 59–64.

Zambare, V. 2010. Solid state fermentation of *Aspergillus oryzae* for glucoamylase production on agro residues. International Journal of Life Sciences 4: 16–25.

Zeng, Q., C. Wei, J. Jin, C. Wu and B. Huang. 2011. Cloning of the gene encoding acid-stable α-amylase from *Aspergillus niger* and its expression in *Pichia pastoris*. African Journal of Food Science 5(12): 668–675.

Index

F

G

H

R

About the Editors

Dr. Maria de Lourdes T.M. Polizeli

Dr. Polizeli, PhD, is a Pharmaceutical Biochemist with a Masters and PhD in Biochemistry. As well as a postgraduate in Cellular Biology. She has 19 years of teaching and research experience as Professor of Cellular Biology of Biological Sciences Course in Department of Biology of Faculty of Philosophy, Science and Letters of Ribeirão Preto—São Paulo University, Brazil. Besides, she is the Professor responsible for disciplines, such as Signal Transduction in the Postgraduate Courses in Medicine Faculty of Ribeirão Preto and Comparative Biology in Faculty of Philosophy, Science and Letters of Ribeirão Preto, both at São Paulo University. In these Postgraduate Programs, she has guided many students in Scientific Initiation, Masters, Doctorates and Post-doctorates. She works with several enzymes produced by filamentous fungi, which degrade the cell wall of the plant, and for this, she has received funds from various agencies in Brazil, mainly the Foundation for Advancement of Science in the State of São Paulo (FAPESP) and the National Council for Scientific and Technological Development (CNPq) focusing on the application of bioethanol, animal feed and biobleaching cellulose pulp. She has published about a 100 research papers in Brazilian and foreign journals. She has also published various book chapters on biochemistry/cell biology aspects, and has published/edited two books, one of them dedicated to early childhood education in biology. Dr. Polizeli is member of Brazilian Society of Biochemistry and Molecular Biology and Brazilian Society for Microbiology. She has also received several researchers, students and professors from various national and international institutions and has had exchanges with international institutions from the United States, Spain, Portugal and many South American countries.

Dr. Mahendra Rai

Dr. Rai is Professor and Head of the Department of Biotechnology at SGB Amravati University in Maharashtra, India. He has published more than 220 research papers and 105 articles in Indian and foreign journals. He has also published/edited 25 scientific books. Dr Rai is member of several scientific societies, has been a national scholar for five years and received several prestigious awards, including the Father T.A. Mathias Award (1989) from the All India Association for Christian Higher Education and the Medini Award (1999) from the Ministry of Environment and Forest, Government of India. He also received a SERC Visiting Fellowship by the Department of Science and Technology (1996); the INSA visiting fellowship by Indian National Science Academy (1998); TWAS-UNESCO Associateship (2002), Italy; and fellowships from the Hungarian Scholarship Board, Hungary (2005 and 2008). He has been a visiting scientist at: Dipartimento Di Colture Arboree, University of Turin, Italy (1999); Laboratory of Bioenergetics, University of Geneva, Switzerland (2004); Department of Crop Protection, Debrecen University, Hungary (2005–06, 2008); Department of Chemical Biology, University of Campinas Under Indo-Brazil program (2009, 2010, 2011); and Nicolaus Copernicus University, Toruń, Poland (2012). He has 30 years of teaching and research experience. Presently, he is a visiting scientist at Department of Chemical Biology, Institute of Chemistry, UNICAMP, University of Campinas, Campinas.

Color Plate Section

Chapter 2

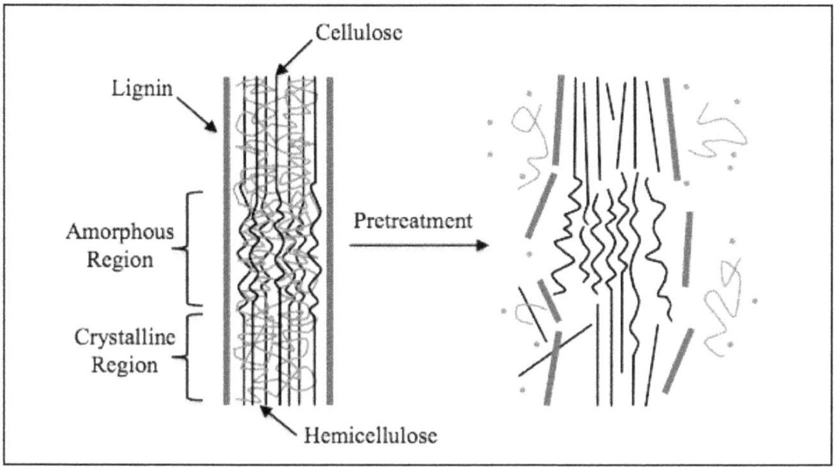

Figure 2.1 Scheme of pretreatment action on lignocellulosic residues (Adapted from Mosier et al. 2005).

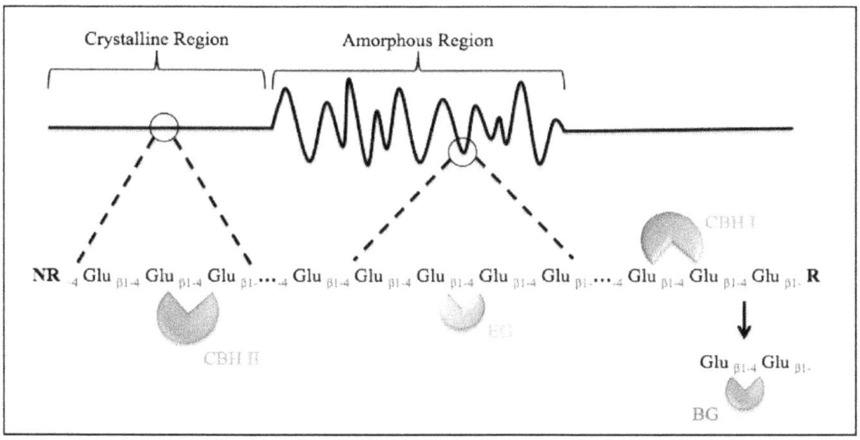

Figure 2.2 Schematic representation of cellulose structure and the action points of cellulolytic enzymes. CBH I: cellobiohydrolase I; CBH II: cellobiohydrolase II; EG: endoglucanase; BG: β-glucosidase; NR: non-reducing end; R: reducing end; Glu: D-glucose.

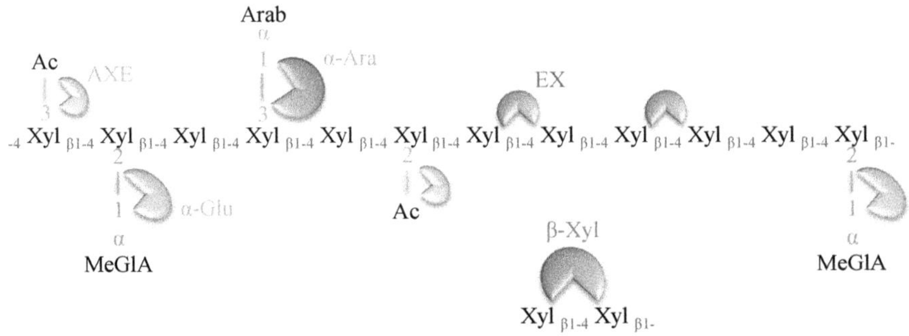

Figure 2.3 Schematic representation of xylan structure and the action points of xylanolytic enzymes. Ac: acetyl group; Arab: L-arabinofuranoside; MeG1A: 4-*O*-methyl-glucuronic acid; Xyl: D-xylose; AXE: acetyl xylan esterase; α-Glu: α-glucuronidase; α-Ara: α-arabinofuranosidase; EX: endoxylanases; β-Xyl: β-xylosidase. (Adapted from Biely 1985).

Chapter 9

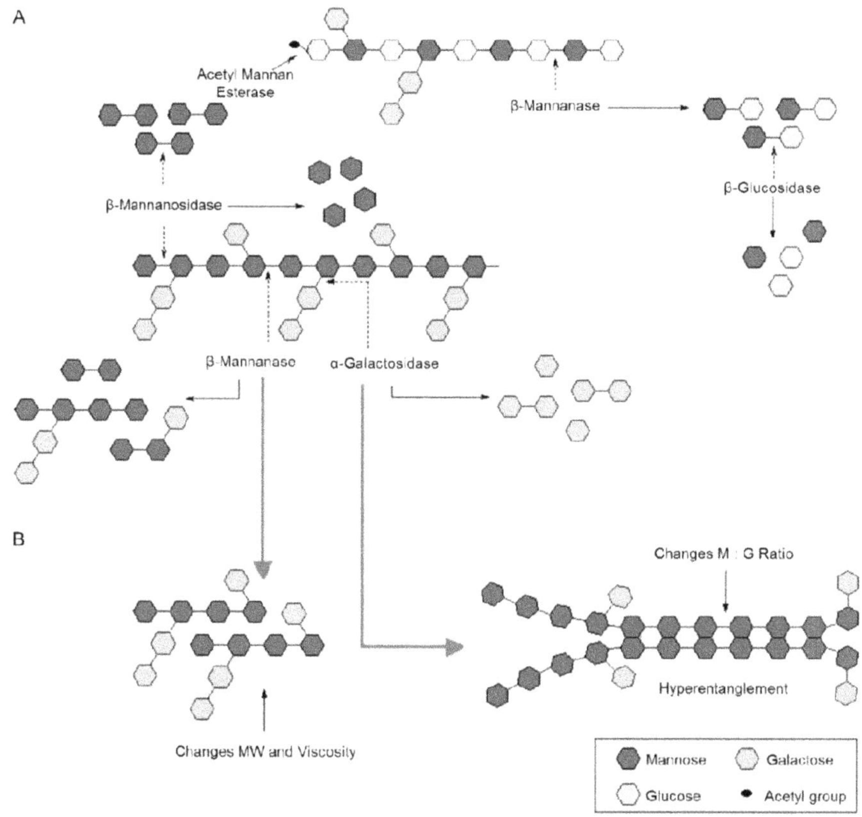

Figure 9.2 (A) Structure of different forms of mannan and enzymes required for their hydrolysis.The structure of mannan is hydrolyzed by β-mannanase, while α-galactosidase and acetyl mannan esterase release galactose and acetyl groups respectively. The products generated by the action of β-mannanase, mannose and glucomannose oligosaccharides, are then hydrolyzed by β-mannosidase and β-glucosidase to generate the monosaccharides mannose, glucose and galactose. (B) Enzymatic action of galactomannans. The β-1,4 linked polymannose chain replaced with α-1 ,6 galactose and the glycosidic links recognized by β-mannanase and α-galactosidase determining the hyperentanglement.

Chapter 10

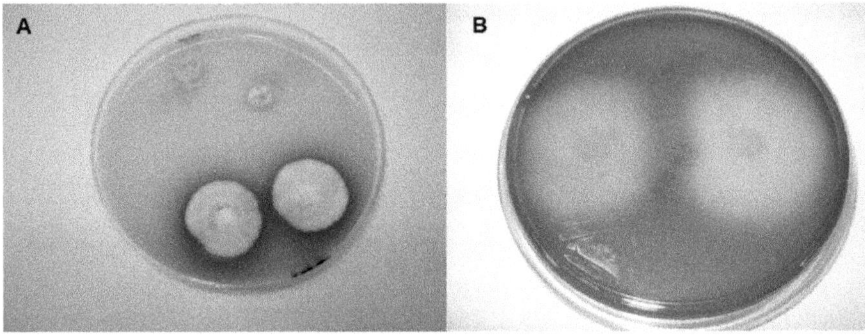

Figure 10.2 Evaluation of the ligninolytic activity on agar plates. In A: Test of oxidation of guaiacol. In B: Test of decolorization of the polymeric dye Poly R478.

Chapter 13

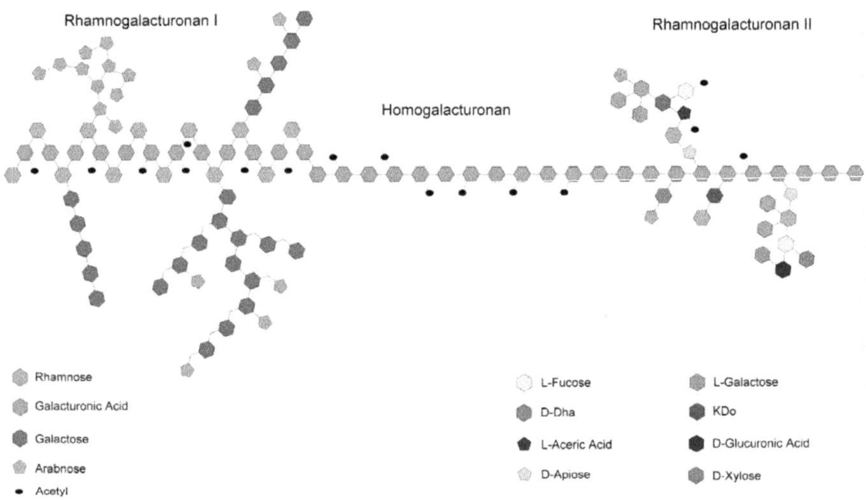

Figure 13.1 Representation of the structures of pectin. Dha: Deoxylyxoheptulopyranosylaric acid, KDo: Ketodeoxymannooctulopyranosylonic acid.

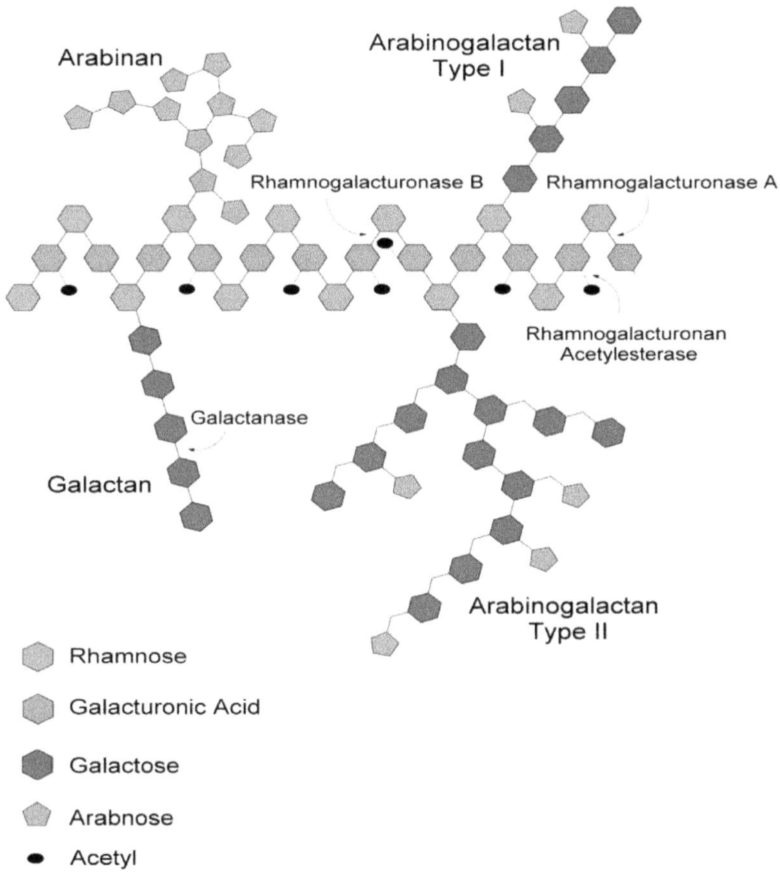

Figure 13.3 Enzymes and action sites on rhamnogalacturonan II structure.

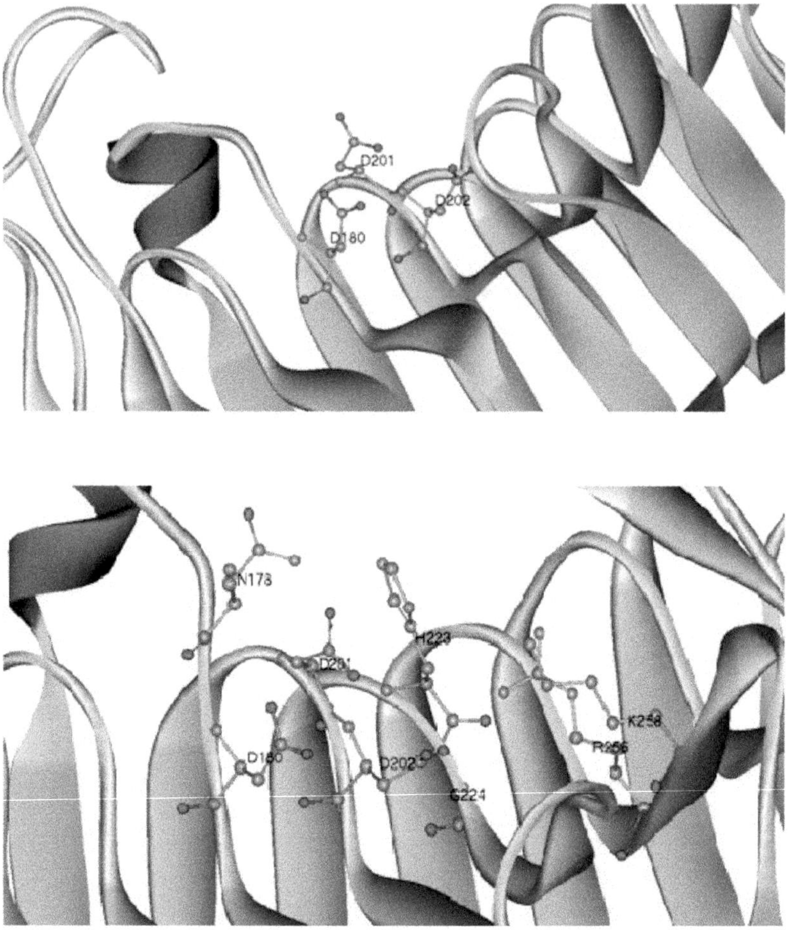

Figure 13.4 Ribbon representation of the conserved amino acid residues of the catalytic site of Polygalacturonase of *Aspergillus niger*: (A) representation of Asp-180, 201 and 202; (B) representation of the catalytic site residues.

www.ingramcontent.com/pod-product-compliance
Ingram Content Group UK Ltd.
Pitfield, Milton Keynes, MK11 3LW, UK
UKHW022213190926
475702UK00003B/328
9781466594548